HEAT TRANSFER TECHNOLOGIES AND PRACTICES

Edited by
G. Sam Samdani, Ph.D.
and
The Staff of Chemical Engineering

McGraw-Hill
New York San Francisco Washington, D.C. Auckland Bogota
Caracas Lisbon Longdon Madrid Mexico City Milan
Montreal New Delhi San Juan Singapore
Sydney Tokyo Toronto

1221 Avenue of the Americas, New York, New York 10020

Printed in the United States of America.

Library of Congress Cataloging-in-Publication Data

Heat transfer technologies and practices for effective energy management / edited by G. Sam Samdani and the staff of Chemical engineering.

p. cm.

Includes index.

ISBN 0-07-057769-2

1. Heat--Transmission. 2. Energy conservation I. Samdani. G. Sam. II. Chemical engineering (Albany, N.Y. : 1946)

TJ260.H415 1996

621'.402--dc20 96-14110

CIP

CONTENTS

Introduction

Heat transfer, an important unit operation in the chemical process industries (CPI), often allows significant cost savings and operational efficiency. The articles that follow have been selected for this volume from the recent issues of *Chemical Engineering* for their timeliness and potential utility to the reader. Indeed, they represent "best practices" for today's practicing engineers in the CPI.

The book is divided into six sections, each focusing on a particular theme. For example, Section I deals with the latest calculation methods used in the design and optimal operation of heat exchangers and associated systems. The equations provide quick estimates for various design parameters, which are useful for preliminary sizing and selection of heat-transfer equipment. Many of these calculation procedures can be easily incorporated into computer programs for repeated use. The articles provide guidelines as to their proper use and potential limitations. The underlying assumptions are also indicated.

Subsequent sections offer tips on design, operation and maintenance of heat-transfer equipment and processes. Safe and efficient operation often requires selection of the right heat-transfer fluids. The proper matching of the heat-transfer fluid with a given process is covered in Section IV.

The economics of heat transfer is an important topic. Therefore, Section V begins with an article on how to make every Btu count. Energy accounting is then covered in depth with a number of illustrative examples.

The book concludes with a section on the state of the art in effective energy management. Heat-recovery technologies and practices offer ample opportunities for cost savings and energy conservation in the CPI.

May the reader make the most out of the technologies and practices covered in these pages!

G. Sam Samdani, Ph.D.

Section I

Quick Calculation Methods for Heat Transfer

QUICK DESIGN AND EVALUATION: HEAT EXCHANGERS

Take the guesswork out of preliminary design of 13 different configurations of heat exchangers

Jeff Bowman
E.I. du Pont de Nemours Co.
Richard Turton
West Virginia University

Design of a heat exchanger is incomplete without calculation of its performance. Only when the calculated performance of the designed unit agrees with what is desired, can one decide upon the design. Thus, design and performance computations are really two sides of the same coin. In what follows, we present charts for quick design and performance evaluation of a number of shell-and-tube heat exchangers.

Quick estimates of heat-transfer-area requirements are often necessary during preliminary design of heat exchangers. To evaluate their performance, one needs ready estimates of the outlet temperatures as well. Charts for calculating the heat-transfer area are available, but they require trial-and-error solution for the performance evaluation. The charts presented here eliminate the need for a trial-and-error solution during evaluation of heat-exchanger performance, while retaining the direct approach to the calculation of the heat-transfer-area.

These charts are an extension of those presented in a previous article [*5*], and, as before, present the data in terms of dimensionless parameters derived from the following equations. Equation (1) represents a modified form of "Newton's law of cooling," which in effect defines U, the heat-transfer coefficient. Equations (2) and (3) are energy balances for the shell-side and tube-side fluids respectively.

$$Q = FUA\Delta T_{lm} \quad (1)$$

$$Q = \dot{M}\,C(T_1 - T_2) \quad (2)$$

$$Q = \dot{m}\,c(t_2 - t_1) \quad (3)$$

The efficiency factor, F, relates the true mean-temperature-difference driving force, ΔT_{mtd}, in the heat exchanger to the logarithmic temperature-difference, ΔT_{lm}, as follows:

$$F = \Delta T_{mtd}/\Delta T_{lm}$$

The remaining dimensionless parameters, namely R, the ratio of shell- to tube-side fluid temperature-differences, P, the thermal effectiveness, and

FIGURE 1. Design and performance chart for a 1-shell-pass and 3-tube-pass exchanger

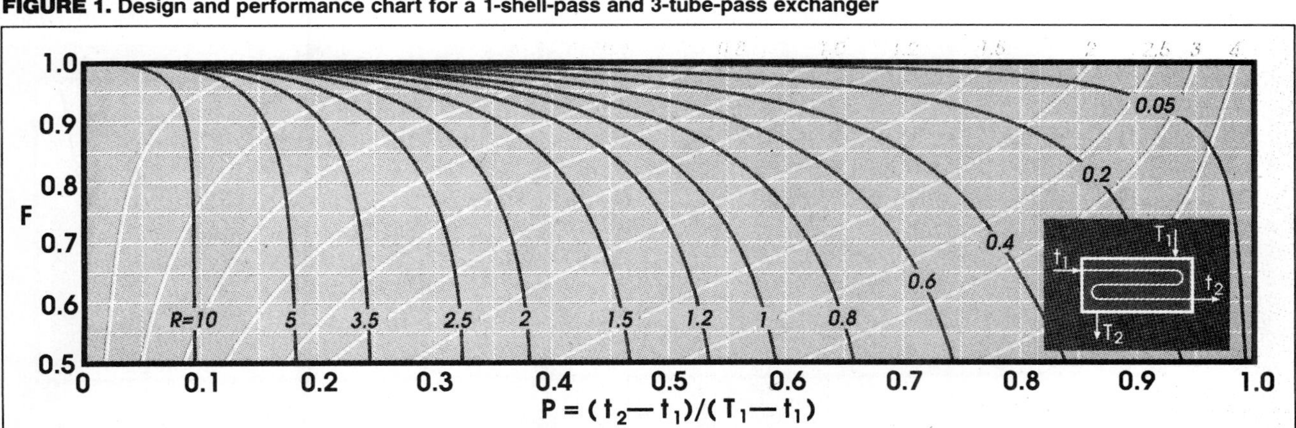

Originally published July 1990

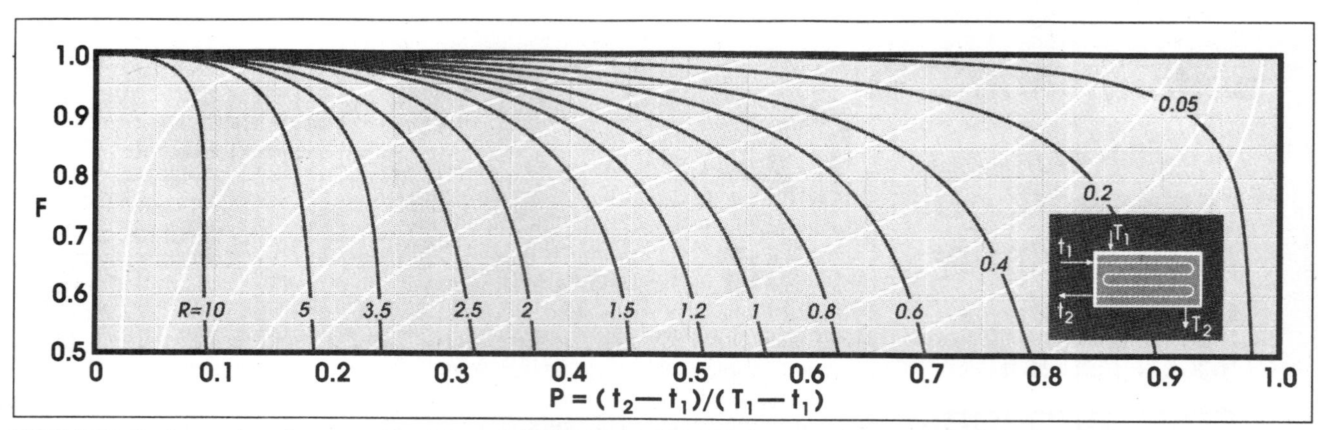

FIGURE 2. Design and performance chart for a 1-shell-pass and 4-tube-pass exchanger

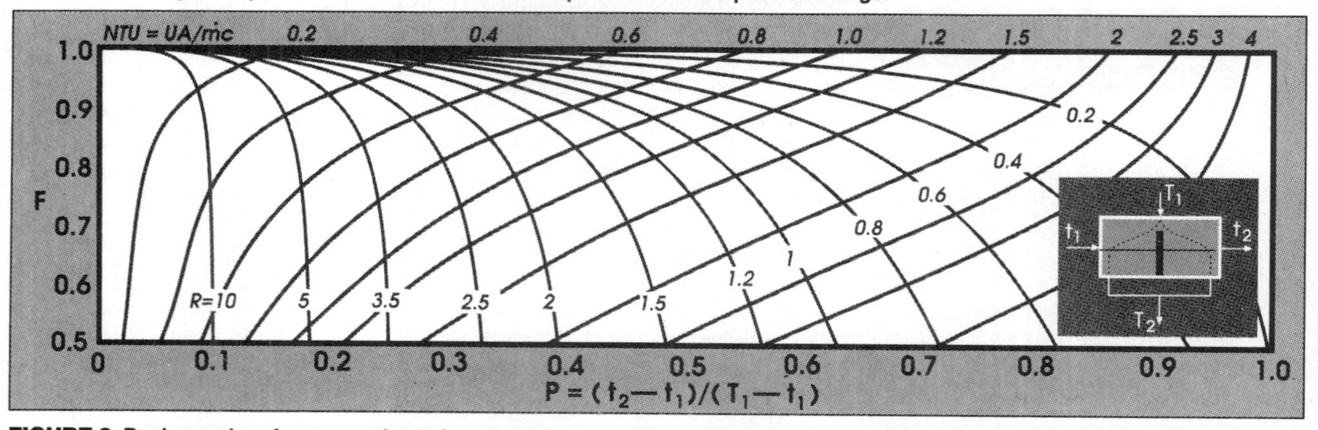

FIGURE 3. Design and performance charts for a 1-shell-pass and 1-tube-pass exchanger (above), and a 1-shell-pass and 2-tube-pass exchanger (below), with a vertical shell-side baffle

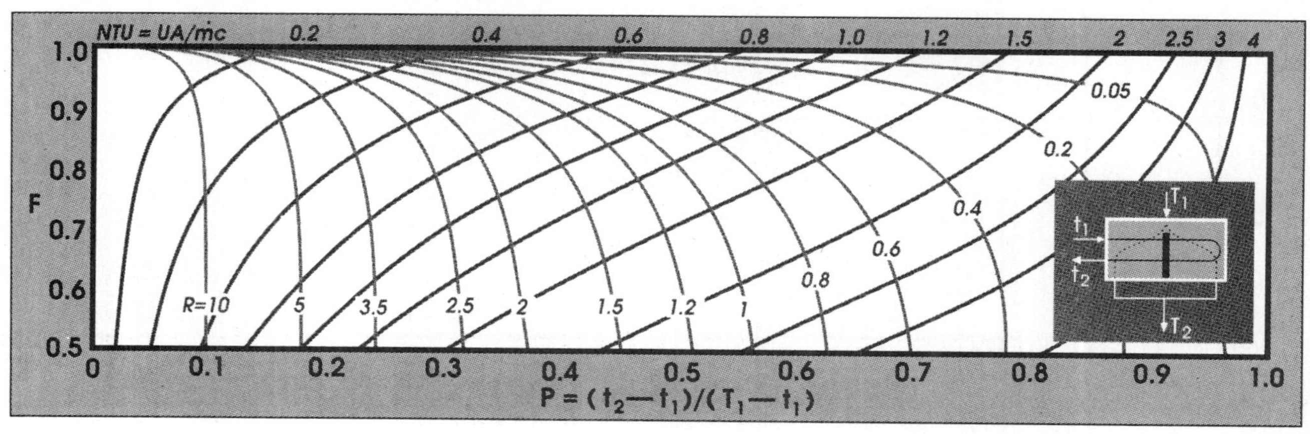

The charts presented here eliminate the need for a trial-and-error solution during evaluation of heat-exchanger performance

NTU, the number of transfer units are defined below:

$R = (T_1 - T_2)/(t_2 - t_1) = \dot{m}\,c/\dot{M}\,C$
$P = (t_2 - t_1)/(T_1 - t_1)$
$NTU = UA/\dot{m}\,c$

True mean-temperature-differences have been solved analytically for several heat-exchanger configurations [2 - 4]. In order to eliminate the trial-and-error calculations required during evaluation of heat-exchanger performance, we present these analytical solutions plotting the efficiency factor, F, as a function of P for constant values of both R and NTU. The "R curves" are similar to those presented by Bowman [1] for shell-and-tube and cross-flow heat-exchangers. Addition of the "NTU curves" to the present charts eliminates the need for a trial-and-error solution to the performance evaluation of a given heat-exchanger.

By specifying any two of the four parameters, the other two may be read directly from a chart. The design calculation, therefore, involves specifying R and P, reading off F and NTU, and then calculating the heat-exchanger surface area, A. The performance calculation, on the other hand, involves specifying R and NTU, reading off F and P, and calculating the outlet temperatures, T_2 and t_2.

Nomenclature

A	= Heat-exchanger surface area (m^2)
c	= Specific heat capacity, tube or cold side (kJ/kg-°C)
C	= Specific heat capacity, shell or hot side (kJ/kg-°C)
F	= Heat-exchanger efficiency (dimensionless)
$\dot{m}$	= Mass flowrate, tube or cold side (kg/h)
$\dot{M}$	= Mass flowrate, shell or hot side (kg/h)
NTU	= Number of transfer units (dimensionless)
P	= Thermal effectiveness (dimensionless)
Q	= Rate of heat transfer (kJ/h)
R	= Ratio of shell- to tube-side

Heat-exchanger configurations for which performance and design charts

FIGURE 4. Design and performance chart for a 1-shell-pass and 4-tube-pass exchanger with a vertical shell-side baffle

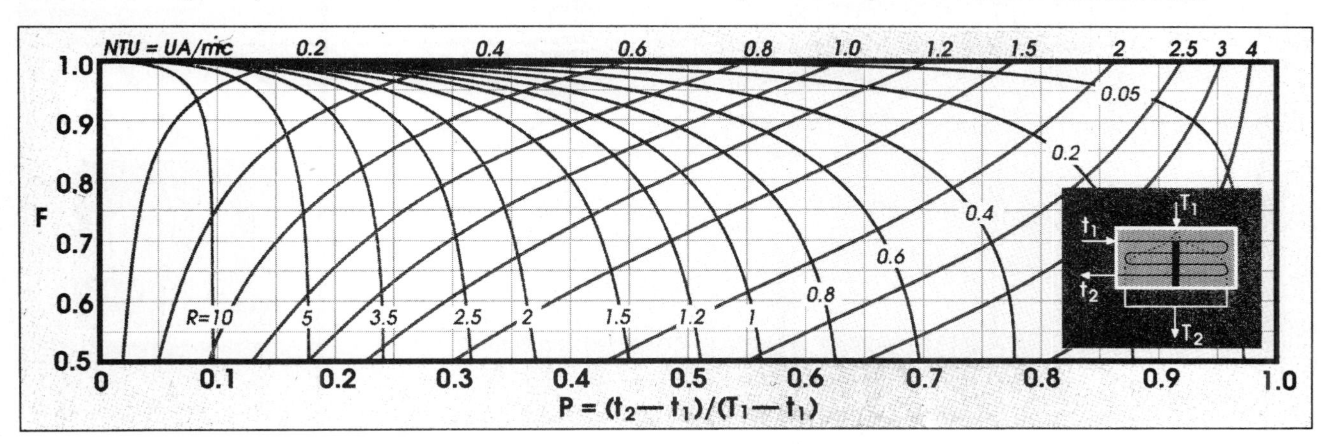

FIGURE 5. Design and performance chart for a 1-shell-pass and 2-tube-pass exchanger with a horizontal shell-side baffle

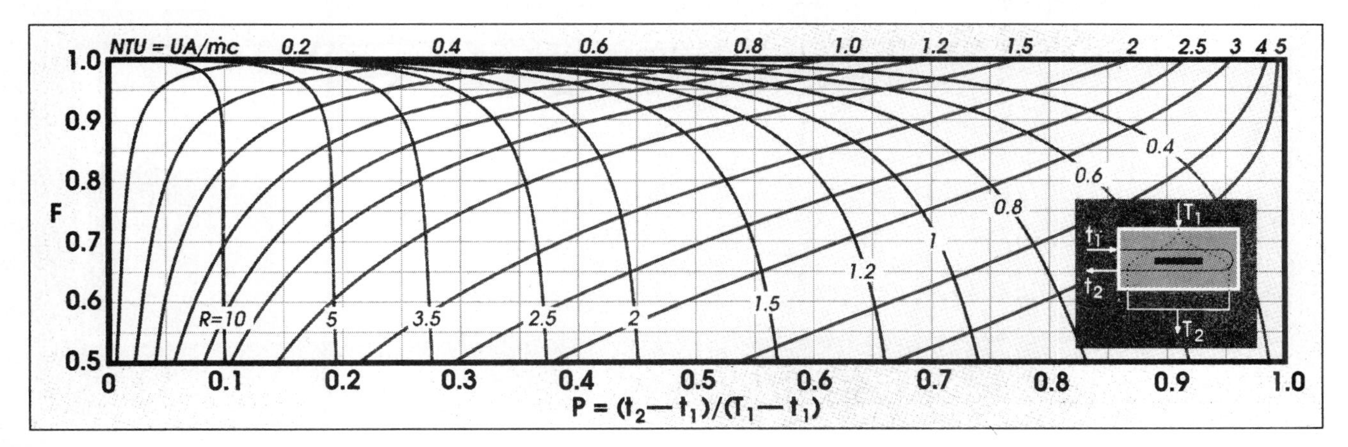

R = Ratio of shell- to tube-side fluid temperature-differences (dimensionless)
t = Temperature of tube-side fluid (°C)
T = Temperature of shell-side fluid(°C)
ΔT_{mtd} = True mean-temperature-difference (°C)
ΔT_{lm} = Logarithmic mean-temperature-difference for countercurrent flow (°C)
U = Overall heat-transfer coefficient (W/m²-°C)

Subscripts

1 = Inlet
2 = Outlet

have been compiled include: conventional 1-shell-pass-3-tube-pass and 1-shell-pass-4-tube-pass heat exchangers, four shell-and-tube heat exchangers with a divided flow-pattern on the shell side, and eight cross-flow heat exchangers with various tube-bank arrangements (see schematics on charts).

Applications of the 1-shell-pass-3-tube-pass and 1-shell-pass-4-tube-pass heat exchangers are simply those of conventional shell-and-tube heat exchangers: steam heating, heating of one process stream by cooling another, and condensation and cooling by a cooling-water utility. Where low shell-side pressure drops are necessary, a divided-flow heat-exchanger configuration is recommended. Shell-side pressure-drops through these exchangers are typically one-eighth of those through a conventional shell-and-tube heat exchangers.

The cross-flow heat exchangers, different from the conventional and divided-flow shell-and-tube heat exchangers, consist of various tube-bank arrangements over which another stream flows perpendicular to the tubes. Typical applications of such exchangers include air-cooling of overhead condensate streams and trim-product coolers.

The following two examples illustrate both the design and performance calculations that use the new design and performance charts. Additional examples illustrating use of the charts for outlet flowrate and temperature, and lowering of the overall heat-transfer coefficient because of fouling, are presented in the previous article [5].

EXAMPLE 1:
Heat-Exchanger Design

Given: $\dot{m}$ = 10,000 kg/h
c = 0.8 kJ/kg-°C
C = 1.0 kJ/kg-°C
t_1 = 250°C
t_2 = 150°C
T_1 = 25°C
T_2 = 50°C
U = 50 W/m²-°C(180 kJ/h-m²-°C)

FIGURE 6. Design and performance chart for a cross-flow exchanger with two single-pass tube rows

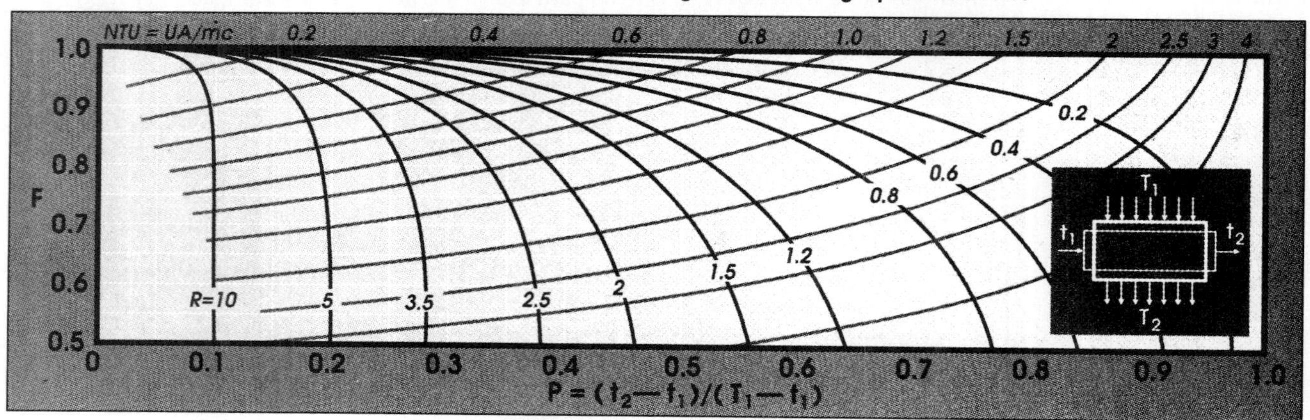

FIGURE 7. Design and performance chart for a cross-flow exchanger with three single-pass tube rows

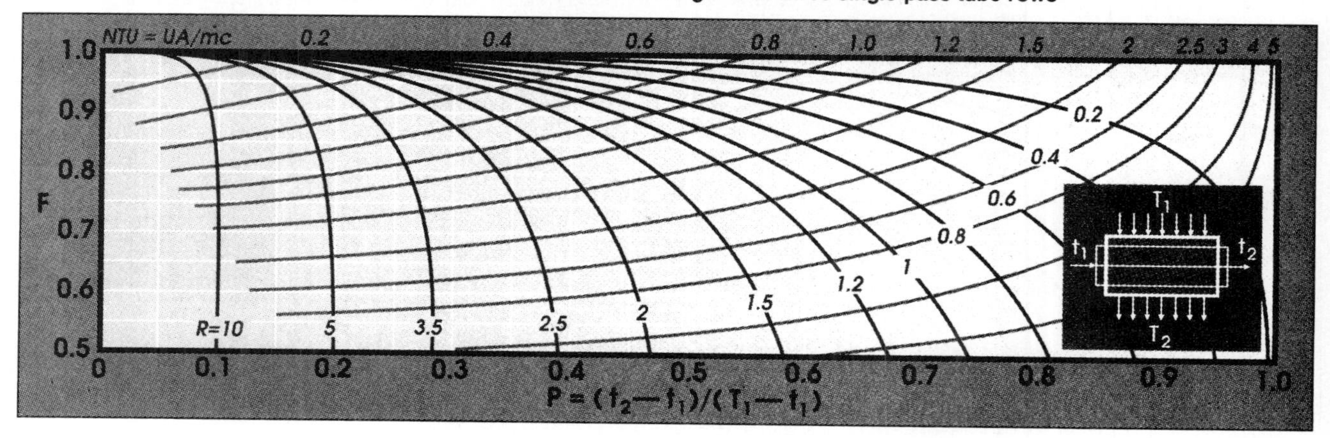

Required: Determine the surface area, A, and the shell-side flowrate required for a cross-flow heat-exchanger with four tube rows.

Solution: Calculate R and P

$$R = (T_1 - T_2)/(t_2 - t_1)$$
$$= (25 - 50)/(150 - 250)$$
$$= 0.25$$

$$P = (t_2 - t_1)/(T_1 - t_1)$$
$$= (150 - 250)/(25 - 250)$$
$$= 0.44$$

From the value of R, calculate the shell-side flowrate, $\dot{M}$.

$$\dot{M} = \dot{m}\, c/RC$$
$$= (10{,}000)(0.8)/(0.25)(1.0)$$
$$= 32{,}000 \text{ kg/h}$$

At the intersection of R and P on Figure 9, $NTU = 0.64$ and $F = 0.99$.

The value for F shows that this exchanger configuration has an efficiency close to pure countercurrent configuration ($F = 1.0$).

$$A = [NTU]\dot{m}\, c/U$$
$$= (0.64)(10{,}000)(0.8)/180$$
$$= 28 \text{ m}^2$$

EXAMPLE 2:
Heat Exchanger Performance

Given: $\dot{m} = 15{,}000$ kg/h
$c = 4.2$ kJ/kg-°C
$\dot{M} = 22{,}000$ kg/h
$C = 2.6$ kJ/kg-°C
$t_1 = 130$°C
$T_1 = 270$°C
$U = 70$ W/m²-°C (252 kJ/h-m²-°C)
$A = 200$ m²

Required: Find the exit temperature of the cold stream, t_2, and the exit temperature, T_2, for a 1-shell-pass-2-tube-pass divided flow shell-and-tube heat exchanger

Solution: Calculate R and NTU

$$R = \dot{m}\, c/\dot{M}\, C$$
$$= (15{,}000)(4.2)/(22{,}000)(2.6)$$
$$= 1.10$$

$$NTU = UA/\dot{m}c$$
$$= (252)(200)/(15{,}000)(4.2)$$
$$= 0.80$$

At the intersection of R and NTU on Figure 4, $P = 0.41$ and $F = 0.90$. The value of F corresponding to this exchanger configuration can be considered acceptable.

Now, calculate t_2 and T_2 from the values of P and R respectively.

$$t_2 = P(T_1 - t_1) + t_1$$
$$= 0.41(270 - 130) + 130$$
$$= 187°\text{C}$$

$$T_2 = T_1 - R(t_2 - t_1)$$
$$= 270 - 1.10(187 - 130)$$
$$= 207°\text{C}$$

■

References

1. Bowman, R. A., others, "Mean Temperature in Design," *Trans. A.S.M.E.*, pp. 284 - 293, May, 1940.
2. Fischer, F. K., "Mean Temperature Difference Correction in Multipass Exchangers," *Ind. Eng. Chem.*, Vol. 30, No. 4, pp. 377 - 383, Apr., 1938.

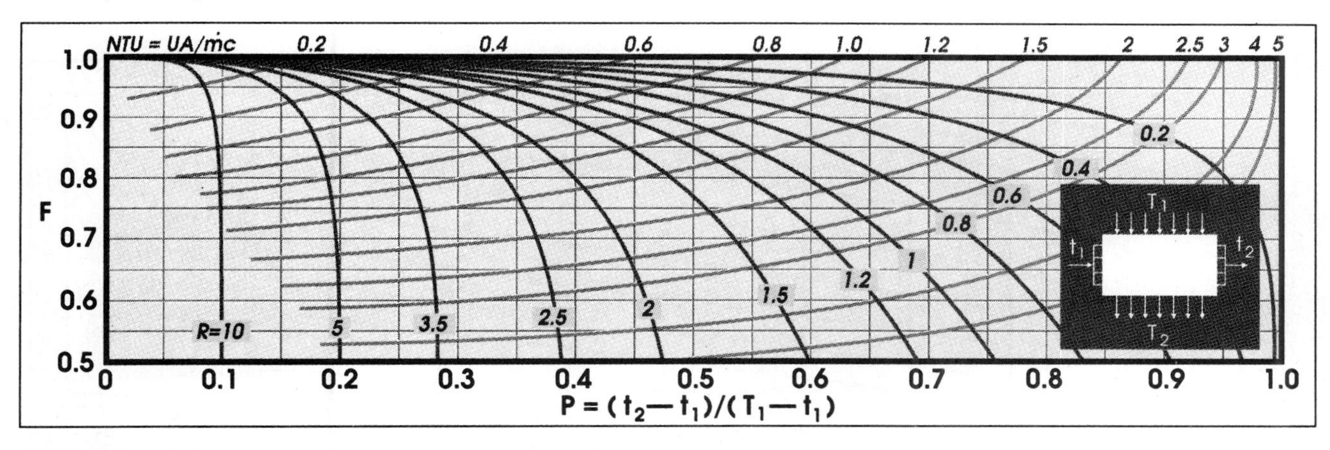

FIGURE 8. Design and performance chart for a cross-flow exchanger with four single-pass tube rows

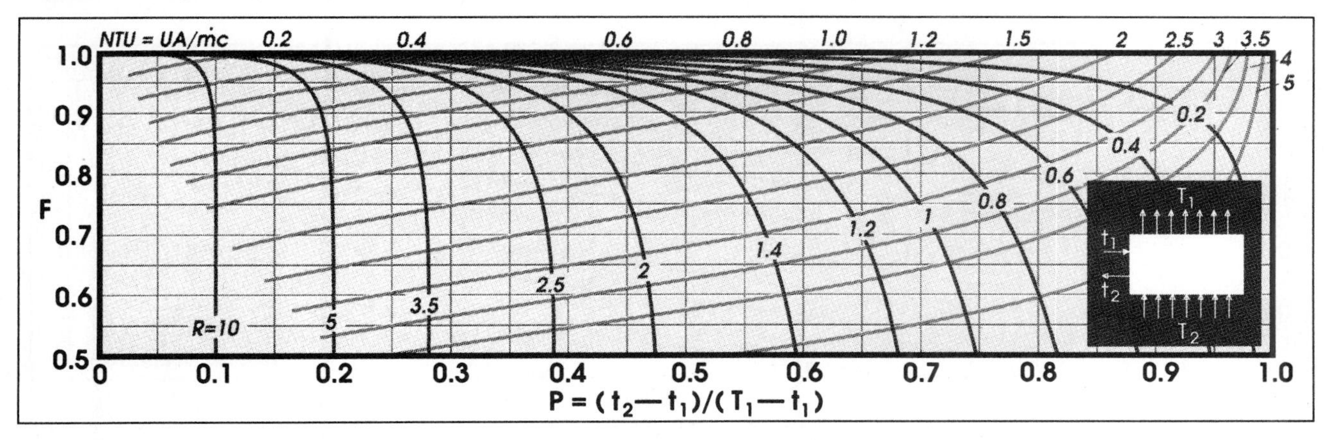

FIGURE 9. Design and performance chart for a cross-flow exchanger with a 2-tube pass

3. Jaw, L., "Temperature Relations in Shell-and-Tube Exchangers Having One-Pass Split-Flow Shells," *Trans. A.S.M.E.*, pp. 408 - 416, Aug., 1964.
4. Taborek, J., F and θ Charts for Cross-Flow Arrangements, in "Heat Exchanger Design Handbook," (ed. by Schlunder, E.), Hemisphere Pub. Corp., Washington, D.C., 1983.
5. Turton, R., and others, "Charts for the Performance and Design of Heat Exchangers," *Chem. Eng.*, pp. 81 - 88, Aug. 18, 1986.

The authors

Jeff Bowman received a B.S. in chemical engineering in 1989 from West Virginia University. He accepted a position in the Field Engineering Program with E.I. du Pont de Nemours Co., and began his first assignment as a process engineer at the company's Tyvek Div. in Richmond, Va., in June 1989. He is a member of AIChE.

Richard Turton received a B.S. in chemical engineering at Nottingham University (U.K.) and an M.S. and Ph.D. at Oregon State University. He spent several years in industry with M.W. Kellogg Co. and Fluor E&C Corp., and is currently an assistant professor of chemical engineering at West Virginia University, where he is carrying out research in the areas of fluidization, heat transfer and process design. He is a member of AIChE.

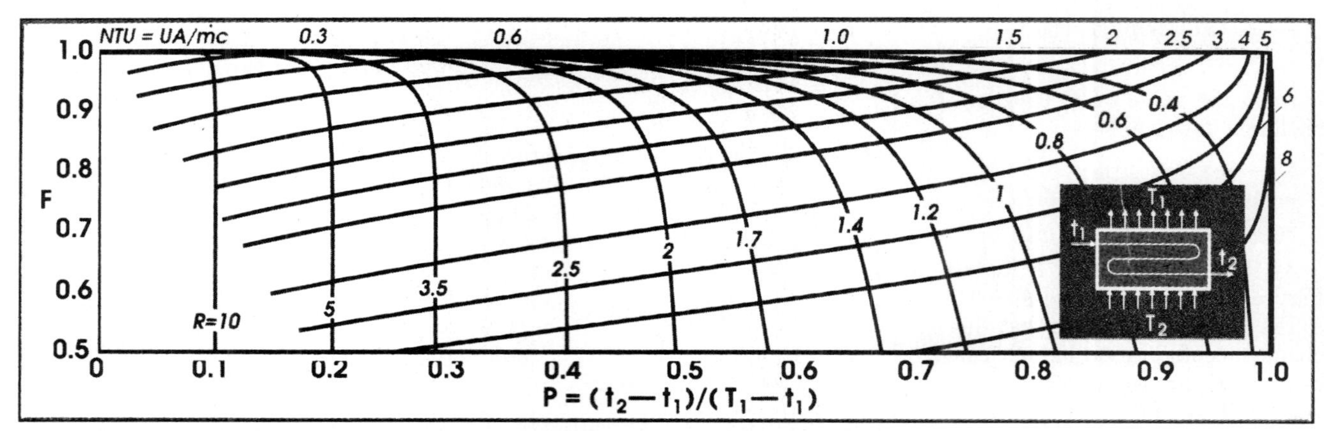

FIGURE 10. Design and performance chart for a cross-flow exchanger with a 3-tube pass

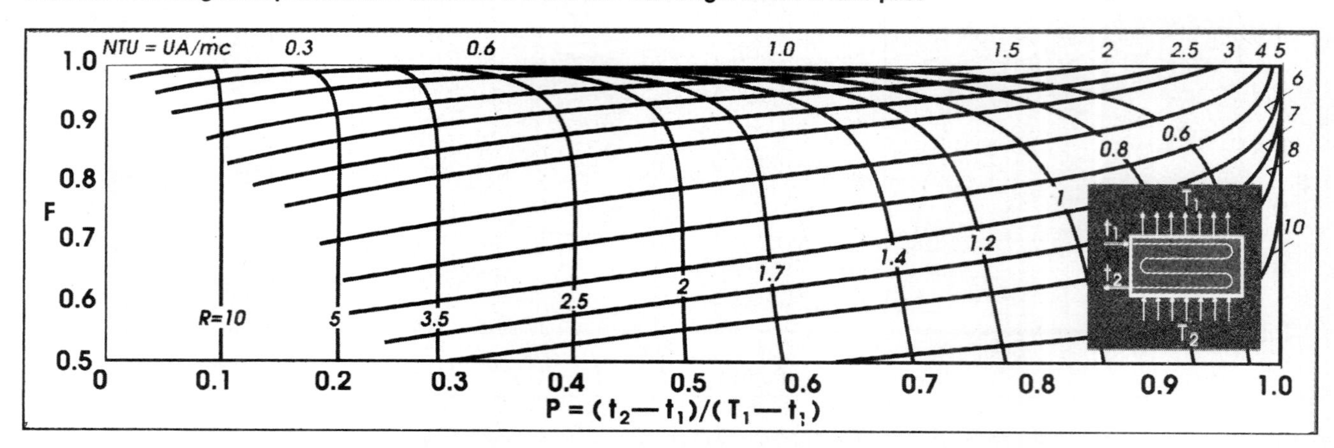

FIGURE 11. Design and performance chart for a cross-flow exchanger with a 4-tube pass

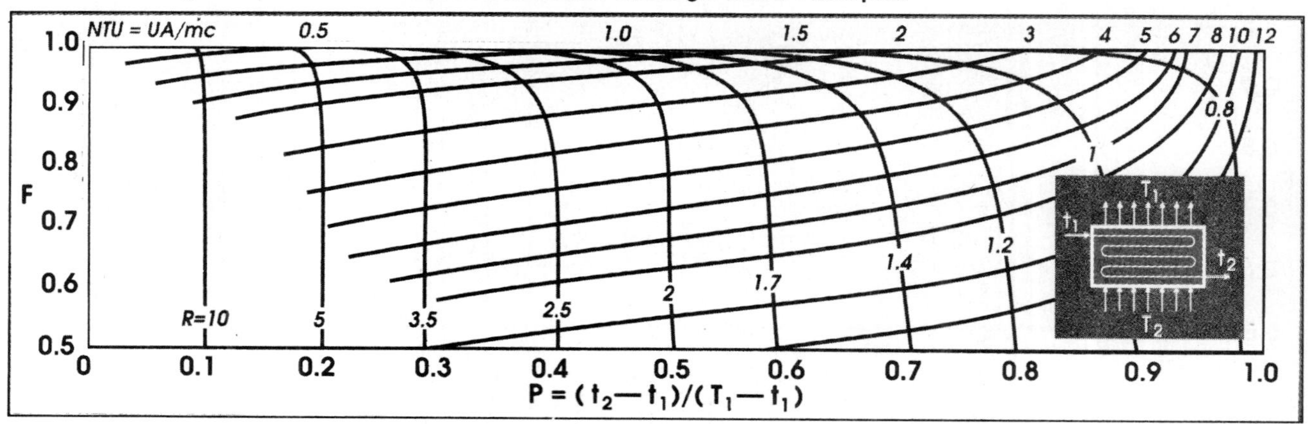

FIGURE 12. Design and performance chart for a cross-flow exchanger with a 5-tube pass

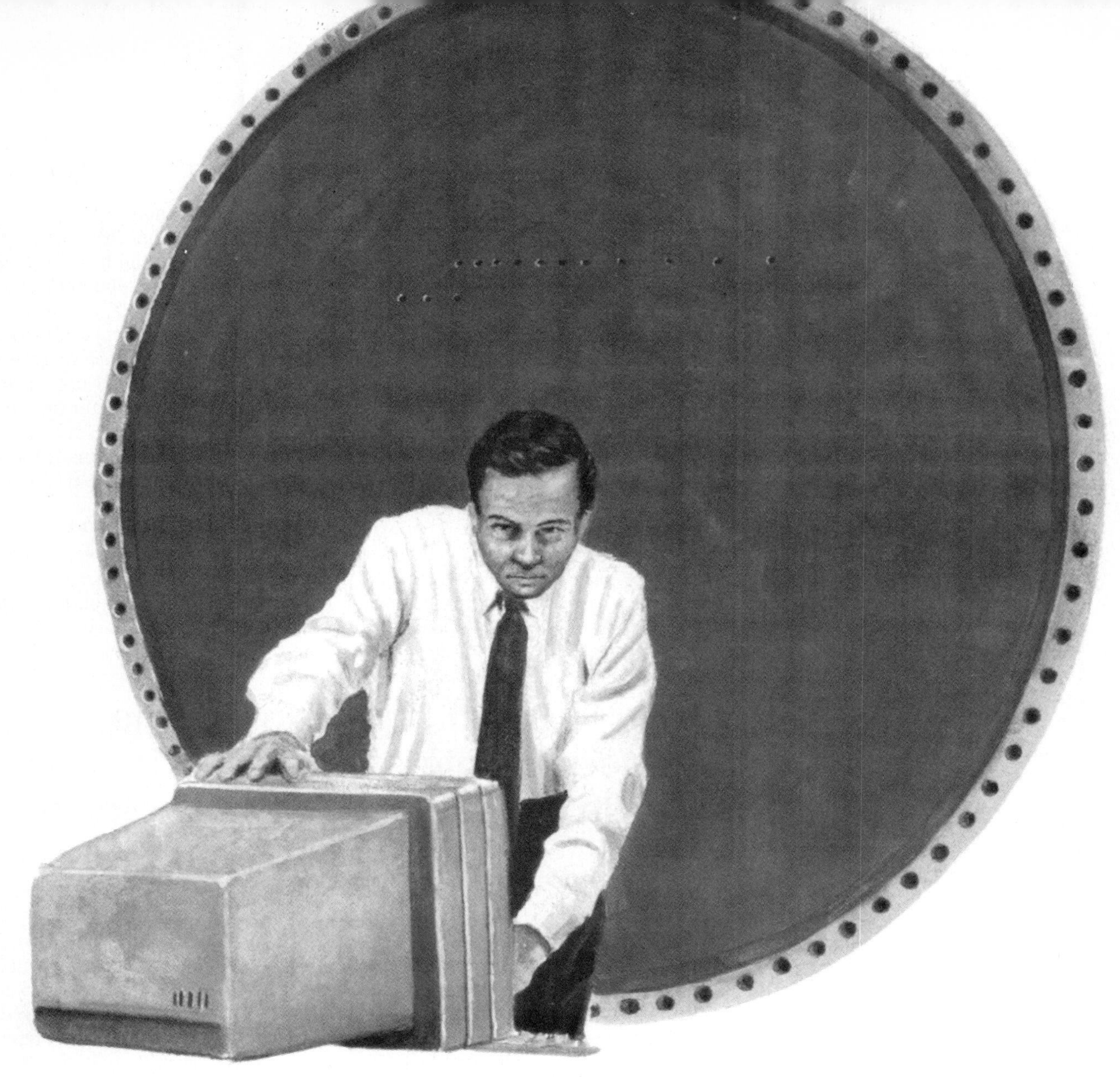

Part 1

EVALUATE HEAT-EXCHANGER PERFORMANCE

Chester A. Plants
Union Carbide Chemicals
and Plastics Technology Corp.

Originally published June 1992

Exact equations for the steady-state efficiencies do away with graphs and the unavoidable errors of interpolation

Analytical expressions for estimating the performance of a heat exchanger are very useful because they eliminate the tedious iterations otherwise required during the design and operation of such equipment. The explicit equations also simplify computer simulation* of individual units or heat-exchanger networks. The results are likely to be more accurate because one does not need to read the usual correction factors from charts or tables to use these equations, thereby eliminating the errors associated with the interpolation between parameter values.

The equations presented in this article cover nearly all the commonly used configurations of shell-and-tube heat exchangers (Sathes) operating at a steady state. A future article (Part 2) will describe the transient or time-dependent behavior of systems having multiple sources of heat transfer.

A shell-and-tube heat exchanger utilizes tubes arranged within a shell in such a way that one fluid, known as the tube fluid, flows within the tubes while the other fluid, known as the shell fluid, flows outside the tubes, but within the shell. Heat is transferred through the tube wall.

The shell fluid enters one of the shell nozzles, passes around the tubes, follows the path formed by shell baffles, and finally leaves through the other shell nozzle. The tube fluid enters one of the tube nozzles, and leaves through the other tube nozzle.

In many Sathes the tube fluid enters the tubes at one end, flows in only one direction through the tubes, and leaves via the exit nozzle. This arrangement is called single-pass, because the fluid passes from one end to the other only once. By adding partitions inside, the tube fluid can be made to have additional passes. Multipass arrangements are often used to obtain higher fluid velocities and longer flowpaths without increasing the length of the exchanger.

The baffles on the shell side are designed to create turbulence, and increase the shell fluid velocity. They also direct the shellside fluid in a path that promotes heat transfer by reducing the thickness of the stagnant fluid film developed near the tube walls. Typically, longitudinal baffles are used on the shell side to divide the flow into two or more parts, giving higher velocities for better heat transfer, or to provide an area for subcooling of the liquid, or cooling of noncondensible vapors as they leave the shell.

To prevent fluid bypassing, the longitudinal baffles must be effectively sealed at the shell wall. These baffles can also be designed to establish multipass flow of the shellside fluid.

The Tubular Exchanger Manufacturers Assn. (TEMA) has developed a standard for basic Sathe construction [14]. It covers essentially all the important features of Sathe fabrication. TEMA also has developed a naming convention for several flowpath configurations, such as TEMA E and TEMA G. A summary of the usual types of shells and attached channels can be obtained from TEMA (25 North Broadway, Tarrytown, NY 10591).

The heat transfer rate, or heat duty, depends on the basic type of unit considered, and thus on the particular flowpath geometry of each stream. Sathe design is based primarily on the overall heat-transfer coefficient defined by the rate equation: $Q = UA\,\Delta T_M$. The mean corrected temperature difference, ΔT_M, is the product of the log-mean temperature difference, ΔT_{LM}, and the correction factor, F; both are calculated from the heat-exchanger terminal temperatures.

The expression for F varies with the type of heat exchanger. In other words, F will vary depending on countercurrent, cocurrent, multipass, or split flows in the exchanger. The terminal temperatures are obtained from the heat balance, $w\,c\,(t_2 - t_1) = W\,C\,(T_1 - T_2)$. This formulation is sometimes called the F method or ΔT_M approach.

The ΔT_M method is suited for the design task, that is, calculation of the required area for the heat exchanger. The heat duty, Q, is obtained from the heat balance and ΔT_{LM} is evaluated from the terminal temperatures. The efficiency, $P = (t_2 - t_1)/\delta$, and the heat ratio, $R = w\,c/(W\,C)$, are used to obtain the correction factor, F, for the particular flow geometry. Then, ΔT_M is calculated from $F\,\Delta T_{LM}$, and finally the heat-transfer area, A, from $Q/(U\,\Delta T_M)$.

The efficiency method

The ΔT_M method is not suited for the rating of heat exchangers, that is, calculation of heat duty or outlet temperatures, given the unit's area. For this, the best approach is to use the so-called efficiency method. However, since the efficiency method is not well known, the ΔT_M is frequently employed, by trial and error with the outlet temperature, to satisfy the heat balance and rate equation.

Heat-exchanger performance may be described by the product of a maximum heat duty and thermodynamic efficiency:

$$Q = Q_M\,P \tag{1}$$

where Q_M $(= w\,c\,\delta)$ is the maximum heat duty based on the cold stream, δ $(= T_1 - t_1)$ is the maximum temperature difference, and P $(= (t_2 - t_1)/\delta)$ is the exchanger efficiency based on the cold stream. The expression for P represents the analytical solution for the following system of equations:

$$Q = w\,c\,(t_2 - t_1) = W\,C\,(T_1 - T_2) \tag{2}$$

$$Q = w\,c\,(t_2 - t_1) = U\,A\,\Delta T_M \tag{3}$$

The right side of Equation 1 above can be expressed as an explicit function of the flowrates, specific heats, and inlet temperatures of the two streams,

*For a glimpse of the widening options in the software tools for engineering calculations, turn to p. 173.

NOMENCLATURE

- A Total heat-transfer area
- A_c Isothermal condenser area
- A_s Subcooler area
- B Parameter in multipass efficiency equation, dimensionless $= S(K+1)/[(K-1)\times(1-R)]$
- C Specific heat of hot or shellside fluid
- c Specific heat of cold or tubeside fluid
- F Correction factor of ΔT_{LM}, dimensionless
- f_{BP} Fraction of total stream bypassing exchanger
- G Temperature approach or gap, dimensionless $= (T_2 - t_2)/\delta = 1 - P(1+R)$
- h Parameter in crossflow efficiency, dimensionless $= J/R$
- H Parameter in crossflow efficiency, dimensionless $= e^h$
- J Parameter in crossflow efficiency, dimensionless $= 1 - Z^{-R}$
- K Parameter in efficiency expression, dimensionless $= Z^{S/N_s}$
- L Parameter in G shell efficiency expression $= (1 + R_a P_c)/(1 - P_c)$ $min[a, b]$ The smaller value of a and b
- n Number of shells in series countercurrent flow
- N_s Number of shellside passes in series
- NTU Number of transfer units, dimensionless $= UA/(wc)$
- P Coldside temperature efficiency, dimensionless $= (t_2 - t_1)/\delta$
- P_a Efficiency of crossflow unit per pass
- P_c Efficiency of G shell countercurrent section
- P_E Isothermal condenser efficiency
- P_f Controlled heat-exchanger efficiency $= (t_D - t_1)/\delta$
- P_i Efficiency of shell i in series countercurrent flow
- P_o Efficiency of G shell cocurrent section
- P_s Subcooler efficiency
- Q Exchanger heat duty
- Q_M Maximum coldside heat duty $= wc\delta$
- Q_s Subcooling heat duty
- R Heat capacity ratio, dimensionless $wc/(WC)$
- R_a R value in two-pass G shell section $= 2 \times R$
- S Efficiency parameter, dimensionless $= (1 + R^2)^{1/2}$
- T_c Isothermal condensing temperature
- T_1 Inlet temperature for hot or shellside fluid
- T_2 Outlet temperature for hot or shellside fluid
- t_D Desired outlet temperature for process
- t_m Intermediate coolant temperature
- t_1 Inlet temperature for cold or tubeside fluid
- t_2 Outlet temperature for cold or tubeside fluid
- U Overall heat-transfer coefficient
- U_c Heat-transfer coefficient for the isothermal condenser section
- W Flowrate for hot or shellside fluid
- w Flowrate for cold or tubeside fluid
- X Parameter in series countercurrent efficiency
- Y_n Parameter in crossflow efficiency
- Z Parameter in efficiency expression, dimensionless $= e^{NTU}$
- λ Latent heat of condensation or boiling
- ΔT_{LM} Log-mean temperature difference
- ΔT_M Corrected, or actual, mean temperature difference $= F \times \Delta T_{LM} = \delta P/NTU$
- Δt_i Temperature change in cold fluid in shell i for series countercurrent flow
- δ Maximum temperature difference $= T_1 - t_1$
- δ_C Cold-end temperature approach, dimensionless $= (T_2 - t_1)/\delta = 1 - P$
- δ_H Hot-end temperature approach, dimensionless $= (T_1 - t_2)/\delta = 1 - PR$
- ϵ True thermodynamic efficiency $= PR/min[1, R]$

EFFICIENCY BY ANY OTHER NAME

Regarding the many literary works on this subject, a few comments cannot be overemphasized. The heat capacity ratio, R, and the number of transfer units, NTU, are independent, dimensionless variables resulting from simultaneous solution of the energy balances and the rate expression. They are not empirical factors, nor are they dependent on any characteristics other than their definition.

Secondly, the terms "efficiency," "thermodynamic efficiency," "performance parameter," "temperature efficiency," and "temperature effectiveness" are all synonyms, usually labeled P or ϵ, which are defined as the temperature change of one fluid divided by the difference between inlet temperatures. This efficiency parameter is the dependent variable that represents the solution to the heat transfer problem. □

and the exchanger area and overall heat-transfer coefficient, but not outlet temperature. The efficiency is a function of two dimensionless groups that are dependent on the stream inlet conditions and exchanger capacity. These groups are R, the heat capacity ratio of the two streams (defined as $wc/(WC)$), and NTU, the number of transfer units (defined as $UA/(wc)$).

This formulation is known as the efficiency or NTU method. The functional relationship between P, R, and NTU depends on the type of exchanger and flow geometry because the appropriate expression for the correction factor is always unique for each configuration.

Kern [7] suggests that there may be a lack of realism in an efficiency definition involving a temperature difference of zero. This is because an infinite heat-transfer area would be necessary to obtain Q_M, and the resulting ΔT_M, of course, would have to be zero. However, the presentation of heat-transfer performance on the efficiency basis can be justified for several reasons [6], whether one is designing or rating an exchanger:

- The parameters involved in the NTU method are analogous to those in the ΔT_M method. The heat ratio, R, and efficiency parameter, P, are used in both, and the correction factor, F, equals the ratio of a countercurrent NTU to the actual NTU (or equivalently, the ratio of the actual corrected efficiency to the corresponding countercurrent efficiency)
- The efficiency, P, is more simply defined, has thermodynamic significance, and aids in simplifying the mathematical description of a system of heat exchangers
- In the F-curve presentation, efficiency, P, does not stand alone as an independent variable. Instead, P appears directly in the abscissa and indirectly in the ordinate
- The ΔT_M method disguises the true notion of what is involved in exchanger performance, since the implication is that only a rate equation is required. The energy balance principles are hidden in the F factor
- The efficiency method reduces the mathematics involved to determine the capacity of a given unit operating in steady or transient states

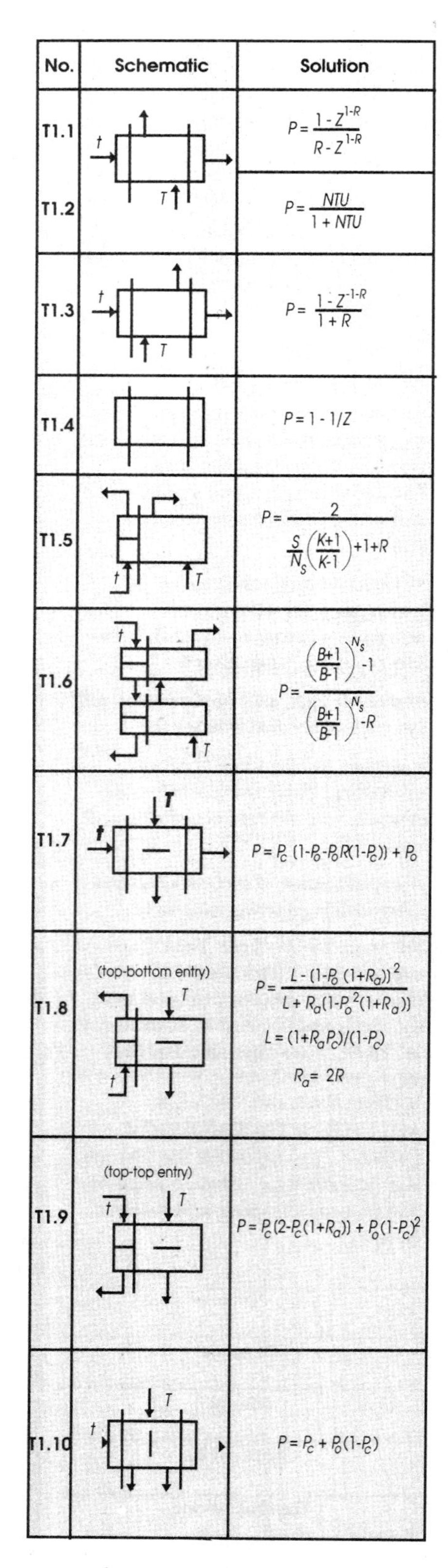

No.	Schematic	Solution
T1.1		$P = \frac{1 - Z^{1-R}}{R - Z^{1-R}}$
T1.2		$P = \frac{NTU}{1 + NTU}$
T1.3		$P = \frac{1 - Z^{-1-R}}{1 + R}$
T1.4		$P = 1 - 1/Z$
T1.5		$P = \frac{2}{\frac{S}{N_S}\left(\frac{K+1}{K-1}\right) + 1 + R}$
T1.6		$P = \frac{\left(\frac{B+1}{B-1}\right)^{N_s} - 1}{\left(\frac{B+1}{B-1}\right)^{N_s} - R}$
T1.7		$P = P_c\,(1 - P_o - P_o R(1 - P_c)) + P_o$
T1.8	(top-bottom entry)	$P = \frac{L - (1 - P_o\,(1 + R_a))^2}{L + R_a(1 - P_o^2(1 + R_a))}$ $L = (1 + R_a P_c)/(1 - P_c)$ $R_a = 2R$
T1.9	(top-top entry)	$P = P_c(2 - P_c(1 + R_a)) + P_o(1 - P_c)^2$
T1.10		$P = P_c + P_o(1 - P_c)$

TABLE 1A. The explicit equations for efficiencies of various TEMA shell-and-tube heat exchangers can be used for the direct calculation of a unit's performance

Directions of fluid flow

Ten Broeck [15] demonstrated in 1938 that the one-two (one shell-pass and two tube-passes) and two-four TEMA E [14] heat-exchanger equations can be solved analytically for the outlet temperature. Studies by Nagle [10], Underwood [16], and Bowman [2] indicate that the expression for a one-two TEMA E exchanger may be accurately employed for one shell-pass having any even number of tube passes.

Bowman [2] presented a general method for calculating F factors of TEMA E shells having N_s shell-passes and $2\,N_s$ tube passes from the F factor of one-two exchangers. Fischer [5] pointed out that ΔT_M is higher if the ratio of tubeside to shellside passes is an odd number, and if the tubeside fluid flows counter to the shellside fluid in over half the passes. Unfortunately, the greatest improvement comes at the lower values of F, where the predictions contain large errors.

Bates and Schindler [1] arranged equations to calculate the efficiency for a one-two split-flow exchanger (TEMA G shell). The tubeside fluid enters the bottom nozzle and the shellside fluid enters at the top. Singh and Holtz [12] derived the expressions for a one-four TEMA G exchanger. Murty [9] presented the mathematical analysis of a one-one and one-two TEMA G shell. This study, however, assumes the tube inlet nozzle is on top. Therefore, the one-two results are dependent on nozzle orientation.

The explicit equations for various TEMA shells are given in Table 1A. Each efficiency expression is the result of defining the correction factor according to the above-mentioned researchers' definition for each shell type, and simultaneously solving Equations 2 and 3. The TEMA X shell is covered in the crossflow section below. The geometries included in Table 1A are described in Table 1B.

When the heat-transfer area, overall heat-transfer coefficient as well as the inlet conditions of each fluid stream are known, Table 1A is used to calculate the efficiency, P, of the heat exchanger. With this, performance parameters for a heat exchanger can be directly calculated as follows:

Heat-transfer rate, $Q = Q_M P$

Cold-side outlet temperature, $t_2 = t_1 + P\delta$

Hot-side outlet temperatute, $T_2 = T_1 - R\,P\,\delta$

Mean temperature difference, $\Delta T_M = P\,\delta / NTU$

For $R \neq 1$, the correction factor for temperature difference, $F = (\ln[(1 - R\,P)/(1 - P)])/([1 - R] \times NTU)$

For $R = 1$, $F = P/[(1 - P) \times NTU]$

Cases in point

Since most F-factor expressions are quite complicated, many of the efficiency expressions are likewise. Results are tabulated to reassure the reader regarding the definitions of the many variables and their application. Assume that data are available per Table 1C. Then, the basic parameter values are:

$R = w\,c/(W\,C) = 50{,}000 \times 0.4/(140{,}000 \times 0.333) = 0.429$

$NTU = U\,A/(w\,c) = 150 \times 250/(50{,}000 \times 0.4) = 1.875$

$Z = e^{NTU} = 6.521$

The maximum temperature-difference is $\delta = T_1 - t_1 = 100 - 30 = 70°C$. The maximum heat duty is $Q_M = w\,c\,\delta = 50{,}000 \times 0.4 \times 1.8 \times 70 = 2{,}520{,}000$ Btu/h. For the example cases shown in Table 1D, results from substitution of required parameter values into the appropriate efficiency expression are summarized in Table 1E. Additional parameters, used only in specific examples, are:

Case 3. $S = (1 + R^2)^{1/2} = 1.088$
$K = Z^{S/N} = 7.69$

Case 4. $K = Z^{S/N_s} = (7.69)^{1/2} = 2.773$
$B = S(K + 1)/[(K - 1) \times (1 - R)] = 4.052$
$(B + 1)/(B - 1) = 1.655$

Case 5. Z (based on $A/2$) $= Z_2 = e^{NTU/2} = Z^{0.5} = 2.554$

Case 6. $R_a = 2 \times R = 0.857$
Z (based on $A/4$) $= Z_4 = e^{NTU/4} = Z_2^{0.5} = 1.6$

Crossflow geometry

Crossflow exchangers have the fluids flowing at right angles. As a result, the usual cocurrent or countercurrent equations do not apply. The fluid temperature becomes a two-dimensional variable because changes occur in a direction normal to its flow as well as in the direction of flow. Final outlet temperature is obtained by complete mixing of all infinitesimal stream sections that are subjected to different temperature ranges of the other fluid.

Recall that in most Sathes, the shell-side fluid is not in pure crossflow since the temperature gradient is parallel to the tubes. Many transverse baffles result in a negligible temperature change per baffle pass.

Mixing may take place within a crossflow exchanger. Performance, therefore, depends on whether either or both fluids are mixed or unmixed. In TEMA X shells (exhibiting pure crossflow) and air coolers, the fluids are more correctly characterized as unmixed, rather than completely mixed.

The one-pass crossflow bundle has been analyzed by Nusselt [11]. Mason [8] has developed a series solution that is conveniently expressed in closed form, which converges more rapidly than either series obtained by Nusselt. Smith [13] completes this case study, and extends it for two two-pass cases. Stevens, Fernandez, and Woolf [4] present the most complete crossflow survey to date. Solutions in the efficiency form cover one, two, and three passes with countercurrent and cocurrent crossflows.

Tables 2 through 6 contain solutions for cases not requiring iterative numerical techniques. An analytical solution can be obtained if at least one fluid is mixed in the bundle, or if both fluids are mixed between passes.

Case 8: Single-pass crossflow

This example can be applied to a single-pass air-cooler or TEMA X shell. Assume the tubeside fluid is "fluid A." The Y_n function of Table 2 may be reformulated into a more convenient expression for use in tabulating:

$$Y_n(NTU) = Y_{n-1} - NTU^n/(Z \times n!)$$

$$Y_n(R \times NTU) = Y_{n-1}(R \times NTU) - (R \times NTU)^n/(Z^R \times n!)$$

$$Y_0(NTU) = 1 - 1/Z \text{ and}$$

$$Y_0(R \times NTU) = 1 - Z^{-R}$$

Table 2A shows the desired summation. The crossflow efficiency per Equation T2.1 is given by:

$P = 0.5877/0.8036 = 0.731$

$Q = 1.84 \times 10^6$ Btu/h (same as Case 5)

It is worth noting that for this case (single pass and both fluids unmixed) the temperature factor is given by:

Eq. No.	Application
T1.1	Countercurrent flow, $R \neq 1$
T1.2	Countercurrent flow, $R = 1$
T1.3	Cocurrent flow
T1.4	Isothermal flow of one fluid
T1.5	Multipass TEMA E shell, or multipass TEMA E or F shells in series, $R = 1$
T1.6	Multipass TEMA E or F shells in series, $R \neq 1$
T1.7	One-tube-pass TEMA G shell
T1.8	Two-tube-pass TEMA G shell, top tube-inlet and bottom shell-inlet
T1.9	Two-tube-pass TEMA G shell, top tube- and shell-inlets
T1.10	One-tube-pass TEMA J shell

TABLE 1B. A wide variety of heat-exchanger flow geometries are covered in Table 1A. For further details on the configurations, refer to the TEMA standards [14]

TABLE 1C. The heat exchangers used as illustrative examples in this article have a given set of steady-state parameters

The parameters	Tube side	Shell side
Flowrate, lb/h	50,000	140,000
Specific heat, Btu/lb-°F	0.4	0.333
Inlet temperature, °C	30.0	100.0
Overall heat-transfer coefficient, Btu/h-ft²-°F	150	
Exchanger total surface, ft²	250	

Case No.	Flow pattern
1	Countercurrent
2	Isothermal on the shell side
3	Multipass shellside, single-pass shellside TEMA E shell
4	Case 3, except two shells in series-counterflow
5	Single-pass on the tube side, and split flow on the shell side
6a	Two tube-passes, split flow on the shell side, top inlet on the tube side and bottom inlet on the shell side
6b	Two tube-passes, split flow on the shell side, top tube-side- and shellside-inlets
7	Sinle-pass on the tube side, divided flow on the shell side

TABLE 1D. The example cases have the specified flow patterns governed by the heat-exchanger geometries

Case No.	Equation No.	P	$Q = Q_M P$ 10⁶ Btu/h	t_2 °C	T_2 °C
1	T1.1	0.771	1.94	83.9	76.9
2	T1.4	0.847		89.3	100.0
3	T1.5	0.704	1.77	79.3	78.9
4	T1.6	0.753	1.9	82.7	77.4
5	T1.1	$0.554 = P_C$			
	T1.3	$0.517 = P_O$			
	T1.7	0.729	1.84	81.1	78.1
6a	T1.1	$0.501 = P_C$			
	T1.3	$0.313 = P_O$			
	T1.8	0.754	1.9		
6b	T1.1	$0.326 = P_C$			
	T1.3	$0.444 = P_O$			
	T1.9	0.656	1.65	75.9	80.3
7	T1.1	$0.501 = P_C$	(Same as 6a)		
	T1.3	$0.444 = P_O$	(Same as 6b)		
	T1.10	0.723	1.82	80.6	78.3

TABLE 1E. The results from substitution of the required parameter values can be summarized for the example cases indicated in Table 1D

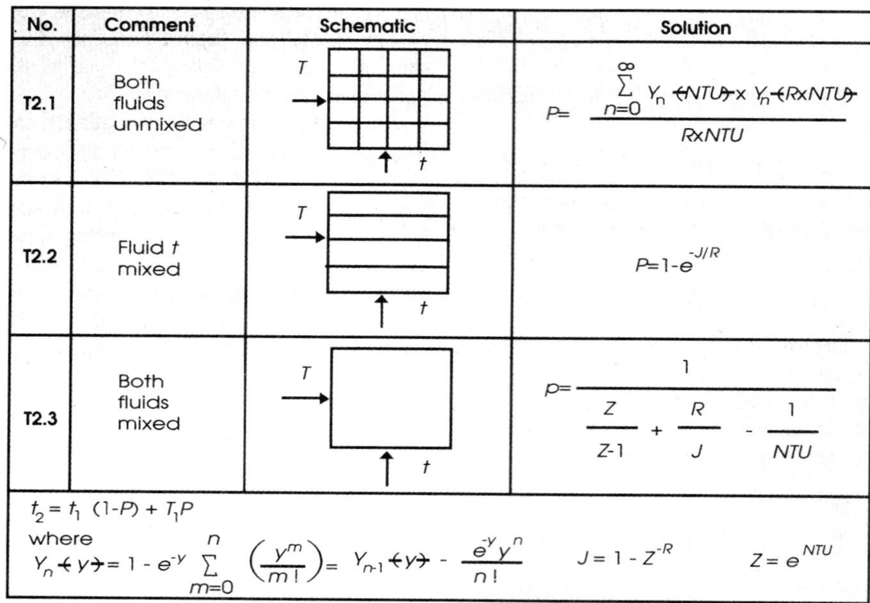

No.	Comment	Schematic	Solution
T2.1	Both fluids unmixed		$P = \dfrac{\sum_{n=0}^{\infty} Y_n(NTU) \times Y_n(R \times NTU)}{R \times NTU}$
T2.2	Fluid t mixed		$P = 1 - e^{-J/R}$
T2.3	Both fluids mixed		$P = \dfrac{1}{\dfrac{Z}{Z-1} + \dfrac{R}{J} - \dfrac{1}{NTU}}$

$t_2 = t_1(1-P) + T_1 P$

where

$Y_n(y) = 1 - e^{-y} \sum_{m=0}^{n} \left(\frac{y^m}{m!}\right) = Y_{n-1}(y) - \frac{e^{-y} y^n}{n!}$ $\quad J = 1 - Z^{-R}$ $\quad Z = e^{NTU}$

TABLE 2. Efficiency equations for single-pass crossflow exchangers

n	$Y_n(NTU)$	$Y_n(R \times NTU)$	$Y_n(NTU) \times Y_n(R \times NTU)$	$\Sigma\, Y_n(NTU) \times Y_n(R \times NTU)$
0	0.84665	0.55228	0.46759	0.46759
1	0.55912	0.19250	0.10763	0.57522
2	0.24639	0.04794	0.01181	0.58703
3	0.07792	0.00922	0.00072	0.58775
4	-0.00105	0.00144	-0.0000015	0.58774

TABLE 2A. The Y_n function of Table 2 is tabulated for Example Case 8

No.	Solution
T3.1	$P = \dfrac{2P_a - P_a^2(1+R)}{1 - P_a^2 R}$
T3.2	$P = 1 - \dfrac{1}{\dfrac{J}{2} \quad (1 - \dfrac{J}{2})H^2}$
T3.3	$P = 1 - \dfrac{1}{H^{-2} - JhH}$

TABLE 3. Efficiency equations for two-pass, counter-crossflow exchangers

No.	Solution
T5.2	$P = (1 - \frac{J}{2})(1 - H^{-2})$

TABLE 5. Efficiency expressions for two-pass, co-crossflow exchangers

$F = \ln[1 - R\,P)/(1-P)]/ [(1-R)\,NTU)] = 0.87$

For a two-pass side-by-side air cooler, the results are nearly the same. For a two-pass over-and-under air cooler (the usual geometry) or a TEMA E shell with two baffle spaces, using Equation T2.1 yields an approximation of F that is 9% conservative, while the above relation for F vs. P implies only a 3.5% difference between estimated and actual efficiencies. Therefore, use of Equation T2.1 for almost any air-cooler geometry will give engineering accuracy for heat-duty estimates.

Generalizing on the examples

Kays and London [6] take great care to distinguish between the stream that has the highest and lowest heat capacity. As a result, their efficiency equation represents the true thermodynamic efficiency and Q_M, based on the minimum heat capacity, is the true maximum heat duty for an infinite area. However, the user is required to keep track of which side has the minimum heat capacity. The outlet temperature or heat duty is calculated for this minimum heat capacity stream.

The results in Table 1A are identical to those of Kays and London [6] where *wc* is used for the minimal heat capacity, irrespective of whether it is or not. Also, the traditional nomenclature usually implies that lower-case symbols represent cold-fluid conditions. However, the solutions of Table 1A are correct if all the upper- and lower-case designations are switched. Confusion and ambiguity may be eliminated by observing the following rules:

- The results of Table 1A are valid, under the stated assumptions, regardless of which fluid has the minimum heat capacity
- If all occurrences of hot- and cold-stream conditions are interchanged, Table 1A is valid for the hot side rather than the usual cold side
- When one fluid is isothermal, heat capacity terms refer to the stream undergoing temeprature change. The isothermal results, of course, pertain to pure condensation as well as pure evaporation. The heat-duty and outlet-temperature equations are for the nonisothermal fluid in the exchanger

The primary implication of these

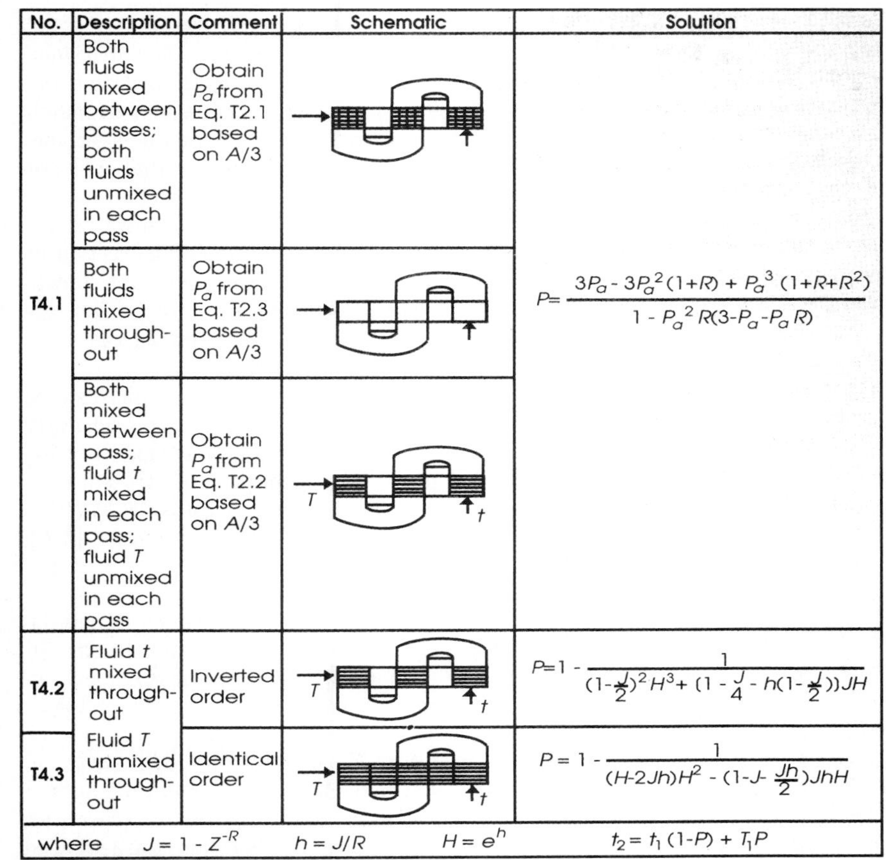

No.	Description	Comment	Schematic	Solution
T4.1	Both fluids mixed between passes; both fluids unmixed in each pass	Obtain P_a from Eq. T2.1 based on A/3		$P = \frac{3P_a - 3P_a^2(1+R) + P_a^3(1+R+R^2)}{1 - P_a^2 R(3-P_a-P_a R)}$
	Both fluids mixed throughout	Obtain P_a from Eq. T2.3 based on A/3		
	Both mixed between pass; fluid t mixed in each pass; fluid T unmixed in each pass	Obtain P_a from Eq. T2.2 based on A/3		
T4.2	Fluid t mixed throughout	Inverted order		$P = 1 - \frac{1}{(1-\frac{J}{2})^2 H^3 + (1 - \frac{J}{4} - h(1-\frac{J}{2}))JH}$
T4.3	Fluid T unmixed throughout	Identical order		$P = 1 - \frac{1}{(H-2Jh)H^2 - (1-J-\frac{Jh}{2})JhH}$
where $J = 1 - Z^{-R}$		$h = J/R$	$H = e^h$	$t_2 = t_1(1-P) + T_1 P$

TABLE 4. Efficiency expressions for three-pass, counter-crossflow exchangers

rules is that the efficiency expressions may be based on, or referred to, either of the single-phase stream as long as the heat-capacity term that appears in the numerator of the expression for R, the denominator of the expressions for *NTU* and the heat duty are for the same stream:

$R = WC/(wc)$ and $NTU = UA/(WC)$, yielding $Q = WC(\delta P)$, where $P = (T_1 - T_2)/\delta$; or $R = wc/(WC)$ and $NTU = UA/(wc)$, yielding $Q = wc(\delta P)$, where $P = (t_2 - t_1)/\delta$

This result is useful in select applications, and aids in comparing the content of numerous publications. For example, the traditional hot- and cold-side nomenclature is inappropriate for divided or split-flow exchangers.

Scope of the method

In what follows, the use of the efficiency expression will be illustrated for exchanger design. In addition, the method is shown to be suitable for applications involving the log-mean temperature correction factor, series counterflow stages with unequal areas and coefficients, heat exchanger networks, stream bypasses, and condensers with subcooling.

Working with three-pass counter-crossflow data, Stevens, Fernandez and Woolf [4] have concluded that the efficiency relation for true countercurrent flow may be used with very little error for four or more counter-crossflow passes. The true parallel-flow equation should give accurate results for four or more co-crossflow passes. These observations are justification for using the ΔT_{LM} formula in Sathe applications with four or more baffles.

The fairly simple relations for "both fluids mixed between passes and unmixed in passes" may be used with excellent accuracy to predict the performance of any counter-crossflow exchanger. This result is helpful in analyzing air coolers or multipass TEMA X shell and tube exchangers. Equation T2.1 is the analytical expression for performance of a single-pass TEMA X shell. The multipass TEMA X shell or multipass air cooler with over-and-under passes requires a numerical solution technique. However, little error is introduced by using the above recommendation instead.

The correction factor, *F*, plays the role of exchanger efficiency arbitrarily defined, relative to countercurrent flow. Bowman, Mueller, and Nagle [3] first adopted this method of comparison for a collection of multipass and crossflow heat-exchanger geometries. They coordinated the results published prior to 1940, representing nearly all the *F* curves frequently used today.

The most noticeable feature of the popular F curves is the fact that they all have a very steep gradient at lower values. A relatively small change in conditions will cause a large change in *F*, and therefore *Q*. Operation in the range of correction factors less than 0.8 would appear questionable.

The *F* factor should be checked when predicting performance, as well as when designing or obtaining data. Wales [17] defined a dimensionless temperature-gap parameter ($G = [T_2 - t_2]/\delta$), and showed that the *F* factor decreases moderately with decreasing, positive G values, but F falls dramatically at the "threshold" condition ($G = 0$, implying equal outlet temperatures), or with slightly negative gaps (small temperature crosses).

The gap parameter can be simply expressed as $G = 1 - P(1 + R)$ for nonisothermal flow, and $G = 1 - P$ for isothermal flow of one fluid. When the efficiency equals $1/(1 + R)$, the outlet temperatures are equal. Therefore, reliable multipass predictions should not be expected for industrial applications when the exchanger efficiency exceeds this value.

Heat exchanger networks

The previous equations do not apply to a system of heat exchangers with different individual coefficients or areas. The following solution is for such a system in overall counterflow when inlet conditions to the first and last shell are known, as well as each shell's heat-transfer coefficient, area and flow geometry. In this case, t_1 represents the cold-fluid temperature entering shell number 1, and T_1 is the hot-fluid temperature entering shell number n. For n shells in overall counterflow with different efficiencies

$P = X_n/(1 + X_n)$

where $X_i = [P_i + X_{i-1}(1 - R\,P_i)]/(1 - P_i)$, $X_0 = 0$, and P_i = efficiency of shell i. After the system heat-duty is obtained, the following formula may be used to calculate the cold-fluid temperature for the stream exiting shell number i:

$(\Delta t)_i = (T_2 - t_1)_i/[(1/P) - R]_i$

For a cocurrent series arrangement, the gap parameter may be used to express overall heat-transfer efficiency of a set of components with different areas and coefficients:

$$P = \sum_{i=1}^{n} P_i \left(\prod_{j=1}^{i-1} G_j\right)$$

$$= \sum_{i=1}^{n} P_i \prod_{j=1}^{i-1} [1 - P_j(1 + R)]$$

Case 9: 'Two body' system

An additional single-pass TEMA E shell with 300-ft² surface for heat transfer, and overall coefficient of 120 Btu/h-ft²-°F is placed downstream of Case-1 exchanger. These units are to be in series-counterflow. The efficiency of this new item, per Equation T1.1 is:

$Z = e^{300\times 120/(50{,}000\times 0.4)} = 6.05$, and

$P_2 = 0.759$

For two shells, in series, with differing efficiencies:

$X_0 = 0$

$X_1 = (P_1 + 0)/(1 - P_1) = 0.771/(1 - 0.771) = 3.37$

$X_2 = [P_2 + X_1(1 - R \times P_2)]/(1 - P_2) = 12.6$

The overall efficiency is:

$P = X_2/(1 + X_2) = 0.926$

$Q = 2.33\times 10^6$ Btu/h

$t_2 = 94.8$°C

$T_2 = 72.2$°C

To obtain the intermediate temperatures between shells, begin with the first shell, and calculate each temperature using the following expression:

$(\Delta t)_j = [(T_2 - t_1)/((1/P) - R)]_j$

$(\Delta t)_1 = (72.2 - 30)/((1/0.77) - 0.429)$
$= 48.6$°C

$Q_1 = 1.75\times 10^6$ Btu/h

$t_2 = 78.6$°C

$T_2 = 93.0$°C

Heat-exchanger network design depends on the minimum temperature difference approach. This difference is defined as the smaller of the hot-end approach $(T_1 - t_2)$ and the cold-end approach $(T_2 - t_1)$. Each can be expressed in dimensionless form, and referred to the maximum temperature-difference, δ, as functions of efficiency:

$\delta_H = (T_1 - t_2)/\delta = [T_1 - (t_1 - t_1) - t_2]/\delta = (\delta - \Delta t)/\delta = 1 - P$

$\delta_C = (T_2 - t_1)/\delta = [T_2 - (T_1 - T_1) - t_1]/\delta = (\delta - \Delta T)/\delta = 1 - PR$

Whether the minimum approach is on the hot or cold end depends on which stream has the minimum heat capacity:

$R > 1$ implies $PR > P$ implies $PR - 1 > P - 1$ implies $1 - PR < 1 - P$ implies $\delta_C < \delta_H$

Likewise, "R less than unity" implies δ_H is less than δ_C. Most heat-integration schemes define efficiency relative to the "true" thermodynamic maximum heat-duty:

$\epsilon = w\,c\,P\,\delta/[(min[WC, wc]) \times \delta]$
$= P/(min[(1/R), 1]) = PR/(min[1, R])$

Then, the minimum approach can always be expressed with the same relation:

$R > 1$ implies $\epsilon = PR$ implies $\delta_{min} = \delta_C = 1 - PR = 1 - \epsilon$

$R < 1$ implies $\epsilon = P$ implies $\delta_{min} = \delta_H = 1 - P = 1 - \epsilon$

The explicit equations simplify computer simulation of heat-exchanger networks as well as individual units

Heat-exchanger network optimization actually entails determination of system efficiency that satisfies a desired objective function. The efficiency equations may be used to define the flow arrangement necessary to match the network solution, or the equations can indicate geometries that will not achieve the desired recovery.

Frequently, process control is accomplished by bypassing a fraction of one stream. When constant downstream temperature t_D is desired, the heat-exchange efficiency is fixed:

$P_f = (t_D - t_1)/(T_1 - t_1)$

The bypassing fraction f_{BP} can be approximated for a known exchanger efficiency by using the heat-duty expression as a function of efficiency:

$Q = Q_M\,P = (1 - f_{BP})\,w\,c \times [\delta\,P]$

But Q is controlled at $w\,c \times [\delta\,P]$, therefore

$f_{BP} = 1 - (P_f/P)$

The exchanger efficiency, P, can be obtained from a flowrate estimate. Iteration may be needed if greater accuracy is desired.

Case 10: Controlled bypass

Let us consider the amount of bypass necessary to control the tubeside outlet temperature at 70°C in Case 1. The fixed efficiency is:

$P_f = (t_D - t_1)/(T_1 - t_1) =$
$= (70 - 30)/(100 - 30) = 0.571$

Assuming that the efficiency in Case 1 is correct, the fraction of cold stream bypassed is given by

$f_{BP} = 1 - (P_f/P) = 1 - 0.571/0.771$
$= 0.259$

The exchanger parameters, considering bypassing, are:

$R = 0.429 \times (1 - f_{BP})$

$NTU = 1.875/(1 - f_{BP})$

Table 7 shows the series of iterations performed until the error between assumed and calculated efficiencies is less than 2%. Cold fluid exiting the exchanger at 94°C mixes with bypassed material to yield a final temperature of $94\,(1 - 0.375) + 30 \times 0.375 = 70$°C.

Bypassing alters the exchanger efficiency a great deal. Considerations need to be made regarding minimum allowable velocity, or maximum allowable process temperature anytime flow control is utilized.

No.	Description	Comment	Schematic	Solution
T6.1	Both fluids mixed between passes; both fluids unmixed in each pass	Obtain P_a from Eq. T2.1 based on $A/3$		
	Both fluids mixed throughout	Obtain P_a from Eq. T2.3 based on $A/3$		$P = 3P_a - 3P_a^2(1+R) + P_a^3(1+R)^2$
	Both mixed between pass; fluid t mixed in each pass; fluid T unmixed in each pass	Obtain P_a from Eq. T2.2 based on $A/3$		
T6.2	Fluid t mixed throughout	Inverted order		$P = 1 - (1 - \frac{J}{2})^2 H^{-3} - (\frac{1}{2} + (\frac{1}{2} + h)(1 - \frac{J}{2}))JH^{-1}$
T6.3	Fluid T unmixed throughout	Identical order		$P = 1 - H^{-3} - 2JhH^{-2} - (1 - J + \frac{Jh}{2})JhH^{-1}$
where $J = 1 - Z^{-R}$	$h = J/R$		$H = e^h$	$t_2 = t_1(1-P) + T_1 P$

TABLE 6. Efficiency equations for three-pass, co-crossflow exchangers

f_{BP}	R	NTU	P
0.259	0.318	2.53	0.871
0.345	0.281	2.86	0.905
0.369	0.271	2.97	0.914
0.375			

TABLE 7. For Example Case 10, one needs to carry out iterations until the error between assumed and calculated error is less than 2%

The parameters	Tube side	Shell side
Flowrate, lb/h	8,000	300,000
Specific heat, Btu/lb-°F	0.4	0.333
Inlet temperature, °C	130.0	30.0
Latent heat, Btu/lb	300.0	
Exchanger total surface, ft²	250	
Overall heat-transfer coefficient, Btu/h-ft²-°F	150.0 (during condensation)	
	10.7 (during subcooling)	

TABLE 8. To determine the extent of subcooling in Example Case 11, use the following data

Guessed $t_m - t_i$	t_m	P_E	A_c	A_s	P_s	Calculated t_m	Error
40.00	70.00	0.2224	167.6	82.4	0.0077	30.77	51.016
0.77	30.77	0.1345	96.2	153.8	0.0128	31.28	-0.399
1.28	31.28	0.1352	96.7	153.3	0.0128	31.28	0.002

TABLE 9. The power of the efficiency method is evident from the following data, showing nearly zero error after only three trials for Example Case 11

An isothermal condenser with significant liquid subcooling can be evaluated by a trial-and-error combination of two-efficiency equations. The countercurrent condenser and subcooler is considered to be a condensate cooler in series with a total condenser. For an assumed intermediate coolant temperature, t_m, between the two zones, the necessary condenser efficiency, P_E, is easily calculated from

$$Q/Q_M = P_E = W\lambda / [w\,c\,(T_c - t_m)]$$

From the expression T1.4, for isothermal efficiency, the needed condenser area is

$$A_c = -w\,c \times \ln(1 - P_E)/U_c$$

The exchanger area remaining for subcooling is

$$A_s = A - A_c$$

The efficiency, P_s, for a countercurrent subcooler may now be calculated, to obtain an intermediate temperature

$$t_m = t_1(1 - P_s) + T_c P_s$$

Iteration with these four equations is fast, easy, and produces results that are much more accurate than reading the *P-NTU* curves.

Case 11: Tubeside condenser

A TEMA E exchanger serves as a single-component condenser, and a subcooler if sufficient area exists. The expected degree of subcooling can be calculated from the data given in Table 8. First, check that enough condensing area exists. Assuming no subcooling, one gets

$$P_E = 1 - e^{-150 \times 250/(300{,}000 \times 0.333)}$$
$$= 0.3127$$

$$Q = 300{,}000 \times 0.333 \times 1.8 \times 0.3127\,(130 - 30)$$
$$= 5.63 \times 10^6 \text{ Btu/h}$$

W_c = 18,762 lb/h (which indicates too much condensation)

$$t_2 = 30 + 0.3127\,(130 - 30) = 61.3°C$$

Therefore, subcooling occurs. Consequently, the actual cooling-medium outlet temperature will be lower (since the heat-transfer capacity, and therefore the medium temperature rise, will be less than the condensing section). Assume that the medium temperature rise is 40°C in the subcooling section (an unusually high guess, but intended to highlight the trial-and-error convergence speed):

$$t_m = 30 + 40 = 70°C$$

The condensing load is $8{,}000 \times 300 = 2.4 \times 10^6$ Btu/h. For the estimated intermediate temperature, the required efficiency in the condensing zone is directly obtained as follows

$$P_E = Q/Q_M = W_c \lambda/[w\,c\,(T_c - t_m)] = 2.4 \times 10^6/[0.3 \times 10^6 \times 0.333 \times 1.8 \times (130 - 70)] = 0.2224$$

The necessary condensing area is

$$A_c = -w\,c \times [\ln(1 - P_E)]/U_c = -0.3 \times 10^6 \times 0.333 \times [\ln(1 - 0.22)]/150 = 168 \text{ ft}^2$$

This leaves 82 ft² in the subcooling zone. The efficiency can be calculated as follows

$$R = 300{,}000 \times 0.33/(8{,}000 \times 0.4) = 31.25$$

$$K = e^{[10.7 \times 82\,(1 - 31.25)/(300{,}000 \times 0.333)]} = 0.7658$$

$$P_s = (1 - 0.7658)/(31.25 - 0.7658) = 0.0077$$

The calculated intermediate temperature and error is

$$t_m = 30 + 0.0077\,(130 - 30) = 30.77°\text{C}$$

$$\text{Error} = (t_m - t_1)_{guess}/(t_m - t_1)_{calc} - 1 = (70 - 30)/(30.77 - 30) - 1 = 51$$

Table 9 shows nearly zero error after only three trials. The heat duties and outlet conditions are

$$Q_s = 300{,}000 \times 0.333 \times 1.8 \times (31.28 - 30) = 230{,}349 \text{ Btu/h}$$

$$Q = 2.4 \times 10^6 + 230{,}349 = 2{,}630{,}349 \text{ Btu/h}$$

$$t_2 = 30 + 2{,}630{,}349/(300{,}000 \times 0.333 \times 1.8) = 44.6°\text{C}$$

$$T_2 = 130 - 230{,}349/(8{,}000 \times 0.4 \times 1.8) = 90.0°\text{C}$$

From all the example cases cited above, it is clear that the efficiency formulation is usually superior to the ΔT_M method. Simulations and optimizations nearly always deal with given coefficients, flowrates, and physical properties. The efficiency expressions allow areas, inlet temperatures, or outlet temperatures to be calculated to satisfy desired constraints. In addition to a single unit, predictions are readily possible for systems of exchangers of differing component capacities connected in various configurations.

Edited by Gulam Samdani

Multipass arrangements provide higher fluid velocities and longer flowpaths without increasing a unit's length

References

1. Bates, H. T., and Schindler, D. L., True Temperature Difference in a One-Two Divided Flow Heat Exchanger, Third Natl. Heat Transfer Conf., ASME-AIChE, Preprint 121, 1959.
2. Bowman, R. A., Mean Temperature Difference Correction in Multipass Exchangers, *Ind. Engr. Chem.*, 28, pp. 541 – 544, 1936.
3. Bowman, R. A., others, Mean Temperature Difference in Design, Trans. ASME, 62, pp. 283 – 294, 1940.
4. Fernandez, J., others, Mean Temperature Difference in One-, Two-, and Three-Pass Crossflow Heat Exchangers, Trans. ASME, pp. 287 – 297, 1957.
5. Fischer, F. K., Mean Temperature Correction in Multipass Exchangers, *Ind. Engr. Chem.*, 30, pp. 377 – 383, 1938.
6. Kays, W. M., and London, A. L., "Compact Heat Exchangers," McGraw-Hill, 1964.
7. Kern, D. Q., "Process Heat Transfer," McGraw-Hill, 1950.
8. Mason, J. L., Heat Transfer in Crossflow, Proc. Appl. Mechanics, 2nd ed., U.S. Natl. Congress, pp. 801 – 803, 1954.
9. Murty, K. N., Heat Transfer Characteristics of One- and Two-Tube Pass Split Flow Heat Exchangers, *Heat Transfer Engineering*, 4, Nos. 3 – 4, pp. 26 – 34,1983.
10. Nagle, W. M., Mean Temperature Difference in Multipass Heat Exchangers, *Ind. Engr. Chem.*, 25, 1933.

11a. Nusselt, W., Der Warmeubergang im Kreuzstrom, *Zeitschrist des Vereines deutscher Ingenieur*, 55, pp. 604 – 609, 1944.

11b. Nusselt, W., Eine neue Formel fur den Warmedurchang im Kruezstrom, *Technische Mechanik und Thermodynamik*, 1, 1930.

12. Singh, K. P., and Holtz, M. J., Generalization of the Split Flow Heat Exchanger Geometry for Enhanced Heat Transfer, AIChE Symp. Series, 75, No. 189, pp. 219 – 226, 1979.
13. Smith, D. M., Mean Temperature Difference in Crossflow, *Engineering*, 138, pp. 471 – 481, 606 – 607, 1934.
14. Standards of Tubular Exchanger Manufacturers Assn. Inc. (25 North Broadway, Tarrytown, NY 10591), Fourth ed., 1959.
15. Ten Broeck, H., Multipass Exchanger Calculations, *Ind. Engr. Chem.*, 30, pp. 1041 – 1042,1938.

16a. Underwood, A. J. V., The Calculation of the Mean Temperature Difference in Multipass Heat Exchangers, *Inst. Petro. Tech. Journal*, 20, pp. 145 – 158, 1934.

16b. Underwood, A. J. V., Graphical Computation of Logarithmic-Mean Temperature Difference, *Industrial Chemistry*, 9, pp. 167 – 170, 1933.

17. Wales, R. E., Mean Temperature Difference in Heat Exchangers, *Chem. Eng.*, February 23, pp. 77 – 81, 1981. ■

The author

Chester A. Plants is a staff engineer at the Heat Transfer and Fluid Dynamics Group of Union Carbide Chemicals and Plastics Technology Corp. (P.O. Box 8361, South Charleston, WV 25303; tel. (304) 747-3710). For the past four years, he has been involved in design and troubleshooting of such fluid dynamics equipment as gas-liquid separators, and safety-relief systems; hydraulic surge analysis; compressible and two-phase noncritical and critical flows. His previous sixteen years were devoted to the design and troubleshooting of shell-tube and plate-frame heat exchangers; computer modeling of radiation, and the mechanisms of free and forced convection. His process engineering assignments include plant debottlenecking and capital projects, technology development, economic feasibility studies, and design and analysis of plant safety systems. He holds B.S. degrees in physics and mathematics from West Virginia State College, and an M.S. in chemical engineering from West Virginia College of Graduate Studies.

Part 2

EVALUATE HEAT-EXCHANGER PERFORMANCE

Chester A. Plants*
Union Carbide Chemicals
and Plastics Technology Corp.

*For author information, see Part 1 (*CE*, June, p. 110).

Originally published July 1992

Gauge the dynamics of a unit during batch processing and many other transient operating conditions

Heat exchangers operate in a transient or unsteady fashion during startup, shutdown and upset conditions. Transient modes also prevail during batch processing, and in operations with deliberate changes in flowrates or inlet temperatures.

The procedure — dubbed the efficiency method — presented in Part 1 of this article (starting on page 9) for calculating heat duties at steady state is easily extended to determine the transient or time-dependent behaviors of heat exchangers. It requires reading of no charts or tables to obtain the values of the associated parameters, thereby avoiding the inevitable errors of interpolation. The method is also very convenient for simultaneously calculating the dynamics of systems involving more than one heat-exchange area.

Such "multiple source" systems are of great practical utility in the chemical process industries. For example, engineers are likely to encounter combinations of four unrelated heat sources: ambient heat input, heat generation from internal source (that is, heat of reaction), an internal coil, and an external exchanger (jacket).

Derivations based on the efficiency method are independent of the type of heat exchanger. Consequently, a clearer insight into the significance of system parameters along with their application to any surface geometry is possible. For example, consider the temperature of a uniformly agitated batch that is circulated through an external heat exchanger. The rate of heat addition or removal equals the heat exchanger's duty:

$$M\,C\,dt/d\theta = Q \qquad (1)$$

$$Q = w\,c\,(t_2 - t) = W\,C\,(T_M - T_2) = U\,A\,F\,\Delta T_{LM}$$

The last three equalities above represent the same mathematical system as that described by Equations 2 and 3 in Part 1 of this article, and they are the basis for Tables 1 through 6 [2]. Therefore, these tabulated results can be used to our advantage by expressing Q in terms of the independent variables, and the batch temperature, t. The unsteady heat balance (Equation 1), making use of Equation 1 (from Part 1 of this article [2]), becomes

$$M\,C\,dt/d\theta = (W\,C)_B\,(T_M - t)\,P \qquad (2)$$

The right side of Equation 2 is a function of batch temperature t in the explicit temperature-difference term only. Integration yields

$$t = T_M - (T_M - t_0)\,e^{-P\theta/\tau} \qquad (3)$$

where T_M is the inlet temperature from the utility (for steam or cooling water, depending on whether the application involves heating or cooling), t_0 is the original batch temperature, P is the exchanger efficiency based on the heat capacity for the batch, θ is the time to achieve batch temperature t, and τ $(= M/W)$ is batch "turnover" time. Note that this result is true, under the stated assumptions, for heating or cooling, regardless of exchanger type.

Various flow geometries can be represented by using the appropriate efficiency expression presented in Part 1 [2]. The derivations and the results should apply to a range of broadly applicable heat-transfer problems. Their scope includes the eight individual cases of external exchangers described by Kern [1], N-2N shell bodies, as well as crossflow exchanger systems composed of the elements described in Tables 1 through 6 (Part 1).

It can be shown that the various types of accessories connected to a batch process have a significant impact on the time constant of the batch heat-exchanger system. For example, the internal exchanger (tank coil or jacket) with a nonisothermal medium can be described by Equation 2. In this case, the heat capacity of interest must be that of the utility fluid.

Because batch temperature is assumed constant with respect to any position on the exchanger surface, the efficiency expression is described by the equation for the isothermal case of Table 1A (Equation T1.4 in Part 1). The integrated result can be expressed by Equation 3, except that the time constant must be redefined for an internal exchanger system as $\tau = (M\,C)_B/(w\,c)_m$. This extension of the efficiency method draws from the heating and cooling analysis of Kern [1].

Example 1. An 80,000-lb agitated batch of shellside material must be cooled from 100°C to 50°C using the fluid with the following characteristics:

Flowrate = 140,000 lb/h
Specific heat = 0.4 Btu/lb-f
Overall heat-transfer coefficient = 150 Btu/h-ft²-°F
Exchanger surface area = 250 ft²

The point in question is this: How significant are the differences in efficiencies of the exchangers on the required cooling time?

Because the heat-exchangers considered here must be installed externally, Equation 3 can be solved to yield

$$P\,\theta = -\tau \times \ln[(50 - 30)/(100 - 30)]$$
where $\tau = M/W = 80{,}000/140{,}000 = 0.57$ h

It follows that $P\,\theta = 0.72$ h, or $\theta = 0.72/P$ h. Clearly, the choice of exchanger will have a significant impact on the time requirement, which is inversely proportional to the chosen unit's efficiency.

Handling complex geometries

When the inlet stream conditions and surface capacity of complex geometries (that is, area, overall heat-transfer coefficient, and configuration) of a heat-exchanger are known, the heat balance, $w\,c\,(t_2 - t_1) = W\,C\,(T_1 - T_2)$, and the rate equation, $Q = U\,A\,F\,\Delta T_{LM}$ can be solved simultaneously yielding an expression for heat-transfer rate as a function of known data:

NOMENCLATURE

Symbol	Definition
A	Total heat-transfer area
C	Batch specific heat, or specific heat of hot fluid
c	Specific heat of cold fluid
E	Activation energy (for reactive systems)
F	Correction factor of ΔT_{LM}, dimensionless
h_i	Batch film heat transfer coefficient
M	Mass of batch
$(MC)_B$	Batch heat capacity
NTU	Number of transfer units, dimensionless, $= UA/(WC)$
NTU_j	$(UA)_j/(WC)_j$
NTU_{js}	$(UA)_{js}/(WC)_f$
P	Heat transfer efficiency, dimensionless
P_f	Overall efficiency of external exchanger X and internal exchanger JS in series, $= P_x + P_{js} - P_x P_{js}$
P_{IF}	Overall efficiency of utility fluid circulation system, $= 1/[(1/P_f) + (1/P_{ux}) - 1]$
P_j	Efficiency of batch internal coil or jacket
P_{js}	Efficiency of batch jacket JS
P_{ux}	Efficiency of external system utility heat exchanger
P_x	Efficiency of batch external recirculation heat exchanger
Q	Heat duty
Q_G	Internal heat generation rate
Q_s	Maximum available heat duty from utility
R	Heat capacity ratio, dimensionless $wc/(WC)$
T_a	Constant ambient temperature
T_j	Jacket utility inlet temperature
T_M	Batch process design utility inlet temperature
T_m	Batch process utility inlet temperature as a function of time
T_{m0}	Initial batch process utility inlet temperature
T_w	Inside temperature of the batch wall
T_1	Inlet temperature for hot fluid
T_2	Outlet temperature for hot fluid
T_{js1}	Inlet temperature for batch internal exchanger JS
T_{x1}	Inlet temperature to batch external exchanger X
t	Batch temperature
t_k	Known batch temperature at a known time θ_k
t_m	Batch temperature at the end of control period
t_0	Batch temperature at $\theta = 0$
t_1	Inlet temperature for cold fluid
t_2	Outlet temperature for cold fluid
U	Overall heat-transfer coefficient
$(UA)_a$	$U \times A$ product for ambient heat input
$(UA)_j$	$U \times A$ product for independently supplied internal coil J
$(UA)_{js}$	$U \times A$ product for internal coil or jacket JS in-series circulation system
$(WC)_B$	Batch recirculation heat capacity
$(wc)_m$	Utility heat capacity in unsteady-state expressions
$(WC)_f$	Heat capacity of intermediate fluid circulating through jacket or external exchanger
$(WC)_j$	Heat capacity of jacket
w	Flowrate for cold fluid
Z	Parameter in efficiency expression, dimensionless, $= e^{NTU}$
Z_j	Z parameter for batch jacket or coil J
Z_{js}	Z parameter for batch series jacket or coil
τ_a	Ambient time constant, $= MC/(UA)_a$
τ_{aj}	Overall time constant for ambient and jacket J exchangers, $= 1/[(1/\tau_a) + (1/\tau_j)]$
τ_f	Time constant for jacket in series with external exchanger X, $= MC/(WC)_f$
τ_j	Jacket time constant, $= MC/[(U \times A)_j \times P_j]$
τ_p	Overall process time constant for unsteady-state systems, $= 1/[(1/\tau_{aj}) + (1/\tau_{ux})]$
τ_{ux}	Time constant for overall medium fluid, $= \tau_f/P_{IF}$
θ	Batch heating or cooling time
θ_k	Known time corresponding to known batch temperature
θ_m	Time interval corresponding to the limit of batch utility
Δ_c	Specified, constant heating or cooling rate for a batch
Δ_F	Heating rate parameter for a batch with multiple heat sources, $= (T_a/\tau_a) + \Delta_G + (T_j/\tau_j) + (T_m/\tau_{ux})$
Δ_{FS}	Heating rate parameter for a batch with multiple heat sources, $= (T_a/\tau_a) + \Delta_G + (T_j/\tau_j) + \Delta_s$
Δ_G	Batch heating rate due to internal heat generation, $= Q_G/(MC)$
Δ_m	Rate of change of utility temperature
Δ_o	Initial rate of change of batch temperature
Δ_s	Batch heating rate due to fixed utility capability, $= Q_s/(MC)$
ΔT_{LM}	Log-mean temperature difference
δ_i	Specified, constant film temperature difference for the batch, $= t - T_w$
δ_j	Specified, constant jacket-to-batch temperature difference, $= T_{js1} - t$

$Q = WC(T_1 - t_1)P$ or
$Q = wc(T_1 - t_1)P$

As indicated earlier (Part 1 [2]), the explicit rate equation can be referred to the hot or cold fluid. That is, per the efficiency method, the dimensionless inlet conditions can be defined either as $R = WC/(wc)$ and $NTU = UA/(WC)$ to use $Q = WC(T_1 - t_1)P$; or $R = wc/WC$ and $NTU = UA/(wc)$ to be able to calculate the heat duty via $Q = wc(T_1 - t_1)P$.

The previous article (Part 1) contains an extensive tabulation of P — a function of R and NTU — for nearly all types of shell-and-tube heat exchangers. It also details the overall efficiency for combinations of commonly used countercurrent and cocurrent units that have unequal UA values.

The Figure (p. 107) shows the following multiple, unrelated sources: ambient heat input, internal heat generation rate (that is, heat of reaction), an internal coil with efficiency P_j, and an external exchanger having efficiency P_x with provision for utility fluid circulation though a second internal exchanger (jacket) with efficiency P_{js}. Subscripts j and js are advisedly chosen to highlight two notions: P_j requires no modification if the internal coil is replaced by a jacket supplied with medium $(WC)_j$ at temperature T_j; P_{js} signifies an internal surface (jacket or coil) in series with exchanger X. Then, energy accumulation for the batch system is expressed by:

$$MC\,dt/d\theta = (UA)_a \times (T_a - t) + Q_G e^{E/t} + (WC)_j \times (T_j - t) P_j + (WC)_f \times [T_{x1} - t) P_x + (T_{js1} - t) P_{js}] \quad (4)$$

Using the explicit expression for exchanger performance allows one to write the differential equation as a sum of at least four parts linear in temperature (except, of course, for the term representing heat generation via chemical reaction or some other means). Let us define the time constant for the coil as $\tau_j = MC/[(WC)_j P_j]$, the batch temperature rise due to internal heat generation as $\Delta_G = Q_G/(MC)$, the ambient time constant as $\tau_a = MC/(UA)_a$, the overall series jacket (external exchanger) efficiency as $P_f = P_x + P_{js} - P_x P_{js}$, and the time constant for the jacket in series with exchanger X as $\tau_f = MC/(WC)_f$. Then, the batch temperature change can be expressed as

$$dt/d\theta = (T_a - t)/\tau_a + \Delta_G e^{E/t} + (T_j - t)/\tau_j + (T_{x1} - t) P_f/\tau_f$$

Additional generalization includes two alternatives for the inlet temperature, T_{x1}, of the external exchanger:

- Constant temperature T_M of the utility supply (shown as heavy flow lines in the Figure, with no flow for the dotted lines)
- Outlet temperature of a closed-loop, circulating fluid from a utility exchanger having efficiency P_{ux} (shown by the dotted lines; no flow for the heavy lines)

The alternatives can be presented in a general form by equating the heat duty in exchangers X plus JS to that of unit UX. The relationship between these exchangers, defining $1/P_{IF}$ as $1/P_f + 1/P_{ux} - 1$, is expressed by $P_f(T_{x1} - t) = P_{IF}(T_M - t)$. For the open-loop, once-through utility arrangement, P_{ux} is unity, since T_{x1} equals T_M for this alternative. Moreover, it turns out that design for steady-state control at batch temperature t requires an overall fluid efficiency of

$$P_{IF} = [(T_a - t)/\tau_a + \Delta_G e^{E/t} + (T_j - t)/\tau_j] \tau_f/(t - T_M)$$

Note that efficiency requirements for integral components (with efficiency P_f), or the closed-loop utility exchanger (having efficiency P_{ux}), may be determined from this steady state P_{IF}.

Define the overall time constant for the utility fluid, τ_{ux}, as τ_f/P_{IF}, the inverse of the time constant for the ambient and coil subsystems as $1/\tau_{aj} = 1/\tau_a + 1/\tau_j$, the system time constant as $1/\tau_p = 1/\tau_{aj} + 1/\tau_{ux}$ and a fixed rate (applicable when the heat generation term is constant, that is $E = 0$) of temperature change as $\Delta_F = T_a/\tau_a + \Delta_G + T_j/\tau_j + T_M/\tau_{ux}$. These definitions yield the following simplification of Equation 4 for energy accumulation:

$$dt/d\theta = \Delta_F - t/\tau_p \quad (5)$$

If the coil utility fluid is isothermal (via condensation or vaporization), then $P_j = 1$. In addition, $\tau_j = MC/(UA)_j$, which is analogous to the terms used for the time constant of the ambient heat-transfer subsystem.

The solution to describe various surface-area geometries and heat sources is given in terms of two constants of a system:

$$t = \Delta_F \tau_p - (\Delta_F \tau_p - t_k) e^{(\theta_k - \theta)/\tau_p} \quad (6)$$

where t_k is a known batch temperature at a known time θ_k (usually $t_k = t_0$ at $\theta_k = 0$). If only one of the heat sources is present, Equation 6 can be simplified appropriately by omitting the other terms in the defining expressions for τ_p and Δ_F.

exchanger X (possibly by using preheated or tempered water, or steam) of the Figure below.

The time interval for heating rate control is $\theta_m = (T_M - T_{m0})/\Delta_m$, where T_M is the design or target temperature of the utility fluid for the controlled stream, T_{m0} is its initial value, and Δ_m is the rate of change of utility temperature by the control system. The rate of temperature change and the initial utility temperature are given by

$$\Delta_m = \tau_{ux}\Delta_c/\tau_p$$

STEVE HAIMOWITZ

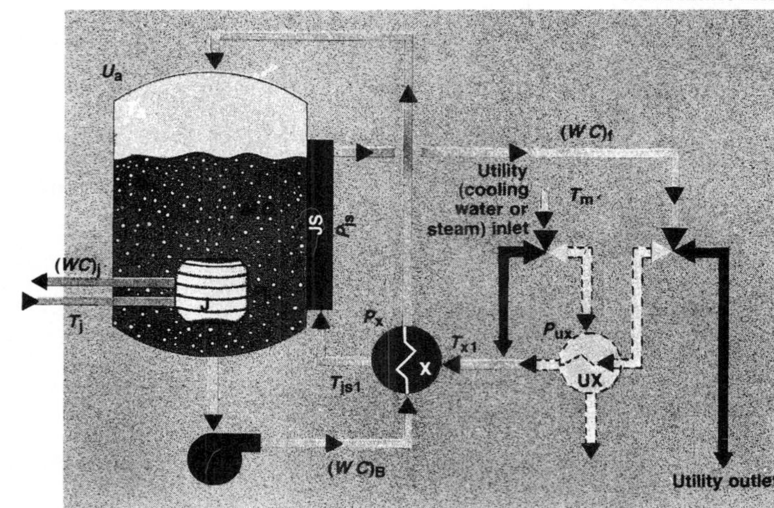

FIGURE. The various heat sources or sinks are connected in series and parallel to a batch reactor or process vessel to illustrate the possible combinations of heat-exchanger configurations

Dealing with three imperatives

In what follows, three operating requirements will be imposed separately to the general-purpose results presented thus far:

- Constant rate of change for batch temperature
- Constant value for $(t - T_w)$, which represents the batch film temperature difference
- An upper limit on the heat-transfer capacity of the utility supply (be it steam or cooling water)

Often, especially during startup, one can manipulate a parameter or two of the utility stream via a control system to achieve the specified heating rate ($dt/d\theta = \Delta_c$). Assume that the control system varies the utility temperature, T_{x1}, of the utility fluid into external

$$T_{m0} = (t_0/\tau_p + \Delta_c - \Delta_G - T_a/\tau_a - T_j/\tau_j) \tau_{ux}$$

During the control period ($\theta \lesseqgtr \theta_m$, where θ_m is the time required to achieve the target value T_M), the batch and utility temperatures are described by

$$t = t_0 + \Delta_c \theta \quad (7)$$
$$T_m = T_{m0} + \Delta_m \theta$$

After the design utility temperature is reached, the batch temperature change rate will be less than Δ_c (the critical rate). One can then apply Equation 6 with $\theta_k = \theta_m$ and t_k evaluated from Equation 7 at $\theta = \theta_m$.

Certain specialized equipment, such as glass-lined vessels, must be operated within prescribed limits of heat flux. Consider a glass-lined batch unit that is required to maintain a fixed temperature difference ($\delta_j = T_{js1} - t$) between the jacket JS and the batch. Usually, the specification of maximum temperature difference is the so-called film tem-

TIME DEPENDENCE OF TEMPERATURE

$$t = \Delta_F \tau_p - (\Delta_F \tau_p - t_0)\, e^{-\theta/\tau_p}$$

Condition	Δ_F	$1/\tau_p$
Closed-loop fluid circulates with external exchanger (X) and jacket (JS) in series, and a second medium through internal coil (J)	$T_j/\tau_j + T_a/\tau_a + \Delta_G + T_m/\tau_{ux}$	$1/\tau_{aj} + 1/\tau_{ux}$

When the batch fluid circulates through external exchanger (X) and jacket (JS) with second utility fluid through coil (J); that is, a once-through system without utility exchanger, the above result simplifies to

$$P_{IF} = P_f \text{ and } \tau_{ux} = \tau_f$$

When there is batch circulation through external exchanger (X) with second utility fluid through coil (J); that is, no jacket (JS) in series with external exchanger (X), the simplification is

$$P_f = P_x$$

When nonisothermal fluid passes through either internal coil (J) or jacket (JS), one can define

$$P_j = 1 - 1/Z_j;\ \tau_j = MC/(WC)_j;\ Z_j = e^{NTU_j}$$
$$P_{js} = 1 - 1/Z_{js};\ Z_{js} = e^{NTU_{js}}$$

When isothermal fluid passes through either internal coil (J) or jacket (JS), define

$$P_j = 1;\ \tau_j = MC/(UA)_j$$
$$P_{js} = 1 \text{ (implies fluid "f" is isothermal)}$$

TABLE 1. The general solution to the heat-balance equation incorporates a number of important special cases

THE TRANSIENT SOLUTIONS AT A GLANCE

Efficiency requirement for batch temperature at steady state	$P_{IF} = [(T_a - t)/\tau_a + \Delta_G e^{E/t} + (T_j - t)/\tau_j]\ \tau_f/(t - T_M)$
Given the fixed rate of change of batch temperature, $\Delta_c = dt/d\theta$	
Batch temperature	$t = t_0 + \Delta_c \theta$
Rate of change of utility temperature	$\Delta_m = \tau_{ux} \Delta_c/\tau_p$
Initial medium temp.	$T_{m0} = (t_0/\tau_p + \Delta_c - \Delta_G - T_a/\tau_a - T_j/\tau_j)\ \tau_{ux}$
Utility temperature	$T_m = T_{m0} + \Delta_m \theta$
Control time period	$\theta_m = (T_M - T_{m0})/\Delta_m$
Given the fixed jacket-batch temperature difference, $\delta_j = T_{js1} - t$	
Relationship between utility and batch temp.	$T_m - t = P_f\delta_j/[(1 - P_x) \times P_{IF}] = T_M - t_m$
Rate of change of initial batch temp.	$\Delta_o = \Delta_G + T_a/\tau_a + T_j/\tau_j + T_{m0}/\tau_{ux} - t_0/\tau_p$
Batch temperature	$t = t_0 + \Delta_o\tau_{aj}(1 - e^{-\theta/\tau_{aj}})$ or, $t = t_0 + \Delta_o\theta$ (when ambient and jacket energies are negligible)
Control time period	$\theta_m = -\tau_{aj}\ln[1 - (t_m - t_0)/(\Delta_o \tau_{aj})]$ or, $\theta_m = (t_m - t_0)/\Delta_o$ (when ambient and jacket energies are negligible)
Given the maximum heat-duty capacity, Q_s, for the utility (cooling water or steam)	
	$\Delta_s = Q_s/(MC)$
	$\Delta_{FS} = \Delta_G + T_a/\tau_a + T_j/\tau_j + \Delta_s$
Batch temperature	$t = \Delta_{FS}\tau_{aj} - (\Delta_{FS}\tau_{aj} - t_0)\, e^{-\theta/\tau_{aj}}$ or, $t = t_0 + \Delta_{FS}\theta$ (when ambient and jacket energies are negligible)
	$t_m = T_M - \tau_{ux}\Delta_s$
Control time period	$\theta_m = \tau_{aj}\ln[(\Delta_{FS}\tau_{aj} - t_0)/(\Delta_{FS}\tau_{aj} - t_m)]$ or, $\theta_m = (t_m - t_0)/\Delta_{FS}$ (when ambient and jacket energies are negligible)

TABLE 2. Use of the appropriate equations to various batch operating modes allows quick estimation of parameters

perature difference $(T_w - t)$ inside a glass-lined vessel. The fixed difference δ_j may be determined from the inside (h_i) and overall heat-transfer coefficients (U_{js}) and the maximum inside film temperature difference δ_i as follows

$$\delta_j = h_i\, \delta_i / U_{js}$$

Starting with the batch temperature rate Equation 5, the utility temperature during the control period will be

$$T_m = t + P_f\delta_j/[(1 - P_x) \times P_{IF}] \quad (8)$$

The initial rate of batch temperature change, Δ_0, (given by $dt/d\theta$ at $\theta = 0$) is determined from the specified initial conditions. The result is then used to calculate batch temperature during the control period from the following solution:

$$t = t_0 + \Delta_0\, \tau_{aj}\,(1 - e^{-\theta/\tau_{aj}})$$

It must be noted that when ambient and jacket energy contributions are negligible, the above result simplifies substantially to the following form:

$$t = t_0 + \Delta_0\, \theta$$

The control period turns out to be

$$\theta_m = -\tau_{aj} \ln[1 - (t_m - t_0)/(\Delta_0 \times \tau_{aj})]$$

When the contributions of ambient and jacket energy are negligible, the above expression also reduces to

$$\theta_m = (t_m - t_0)/\Delta_0$$

The batch temperature at the end of the control period t_m is obtained from Equation 8 using the design or target value of the utility temperature, T_M.

If utility is a limiting factor

Quite often, during the early stages of a batch production, options may be limited by the maximum amount of utility fluid (that is, steam or cooling water) available. For example, the temperature of the utility fluid entering the exchanger system of the Figure (p. 107) may be constrained because one can allow only a fixed, maximum heat duty Q_s.

Such situations are encountered in practice during isothermal heat transfer via condensation of steam or boiling of a refrigerant. Under some circumstances, nonisothermal tempered water systems can also fall into this category.

When the above situations apply, the

fixed heat term Δ_F in the energy accumulation equation requires modification as follows:

$$\Delta_{FS} = T_a/\tau_a + \Delta_G + T_j/\tau_j + \Delta_s$$

where $\Delta_s = Q_s/(M\,C)$

Batch temperature is then described by

$$t = \Delta_{FS}\tau_{aj} - (\Delta_{FS}\,\tau_{aj} - t_0)\,e^{-\theta/\tau_{aj}}$$

However, if ambient and jacket energy contributions are negligible, the above expression simplifies to

$$t = t_0 + \Delta_{FS}\,\theta$$

In either case, when the load capacity of the utility stream exceeds that of the system (in other words, utility availability is no longer a limiting factor) the batch temperature is given by

$$t_m = T_M - \tau_{ux}\Delta_s$$

The corresponding time period is

$$\theta_m = \tau_{aj}\ln[\Delta_{FS}\,\tau_{aj} - t_0)/(\Delta_{FS}\tau_{aj} - t_m)]$$

or $\theta_m = (t_m - t_0)/\Delta_{FS}$ when ambient and jacket energy contributions are negligible.

Note that the way τ_{ux} appears in the previous analysis implies that fluid "f" is a nonisothermal utility fluid, since it is defined relative to the heat capacity of that stream. However, the above equations are applicable to closed-loop, intermediate fluid circulation with an isothermal fluid entering the utility exchanger UX.

For the once-through flow scheme ($P_{IF} = P_f$) with an isothermal fluid, either the external exchanger X or the series jacket JS will probably be omitted in practice. Then, P_{IF} will be either the efficiency of the external unit based on the batch heat capacity ($P_f = P_x$), or P_{IF} will be unity with $\tau_{ux} = \tau_f$ defined as $M\,C/(U\,A)_{js}$.

The general transient solutions along with the common simplifications that frequently occur are summarized in Table 1. Table 2 lists the characteristics of various operating modes. The primary focus for the reader, however, should be the ways of using the heat balance in the efficiency form to meet numerous ends.

Example 2. An 80,000-lb agitated batch is currently cooled from 100°C to 50°C using 20,000 lb/h of – 10°C brine in the coil of an insulated vessel. Specific heat of the batch and brine are 0.333 Btu/lb-°F and 0.5 Btu/lb-°F, respectively. The current time requirement is excessive (8 h), and future demands necessitate cooling twice the current volume in two hours or less.

The new vessel will be uninsulated to take advantage of cooling by an environment assumed at 20°C. The area available for ambient cooling and its heat-transfer coefficient are estimated to be 500 ft² and 2 Btu/h-ft²-°F, respectively. An external exchanger circulating system with the medium at 30°C is available. The system is 77.1% efficient, based on a 1.14-h batch turnover time. The issue at hand is this: Can this system satisfy the new operating conditions?

The time constant for the coil based on current data is

$$\tau_j = -8/\ln([50 - (-10)]/[100 - (-10)]) = 13.2\ \text{h}$$

and for the new volume, the ambient time constant is

$$\tau_a = 2 \times 80{,}000 \times 0.333/(2 \times 500) = 53.3\ \text{h}$$

The existing coil and proposed ambient cooling combination performance is

$$\tau_p = \tau_{aj} = 1/(1/\tau_j + 1/\tau_a) = 1/(1/13.2 + 1/53.3) = 10.6\ \text{h}$$

$$\Delta_F = -10/13.2 + 20/53.3 = -0.38°\text{C/h}$$

$$\Delta_F \times \tau_p = -4°\text{C}$$

$$\theta = -\tau_p \times \ln[(50 - (-4))/(100 - (-4))] = 6.9\ \text{h}$$ is the required cooling time

With the external system, performance characteristics are:

$\tau_{ux} = \tau_f =$ batch turnover time, and $P_{IF} = P_f = P_x$

$$\tau_p = 1/(0.771/1.14 + 1/10.6) = 1/(1/1.5 + 1/10.6) = 1.3\ \text{h}$$

$$\Delta_F = -0.38° + 30/1.5 = 19.6°\text{C/h}$$

$$\Delta_F \times \tau_p = 25.5°\text{C}$$

$$\theta = -\tau_p \times \ln[(50 - 25.5)/(100 - 25.5)] = 1.5\ \text{h}$$ is the required cooling time ■

Edited by Gulam Samdani

References

1. Kern, D. Q., "Process Heat Transfer," McGraw-Hill, 1950.
2. Plants, C. A., Evaluate Heat-Exchanger Performance, Part 1, *Chem. Eng.*, pp. 100 – 110, June, 1992.

HOW TO DESIGN BAYONET HEAT-EXCHANGERS

A. Hernandez-Guerrero, and
A. Macias-Machin
Oregon State University

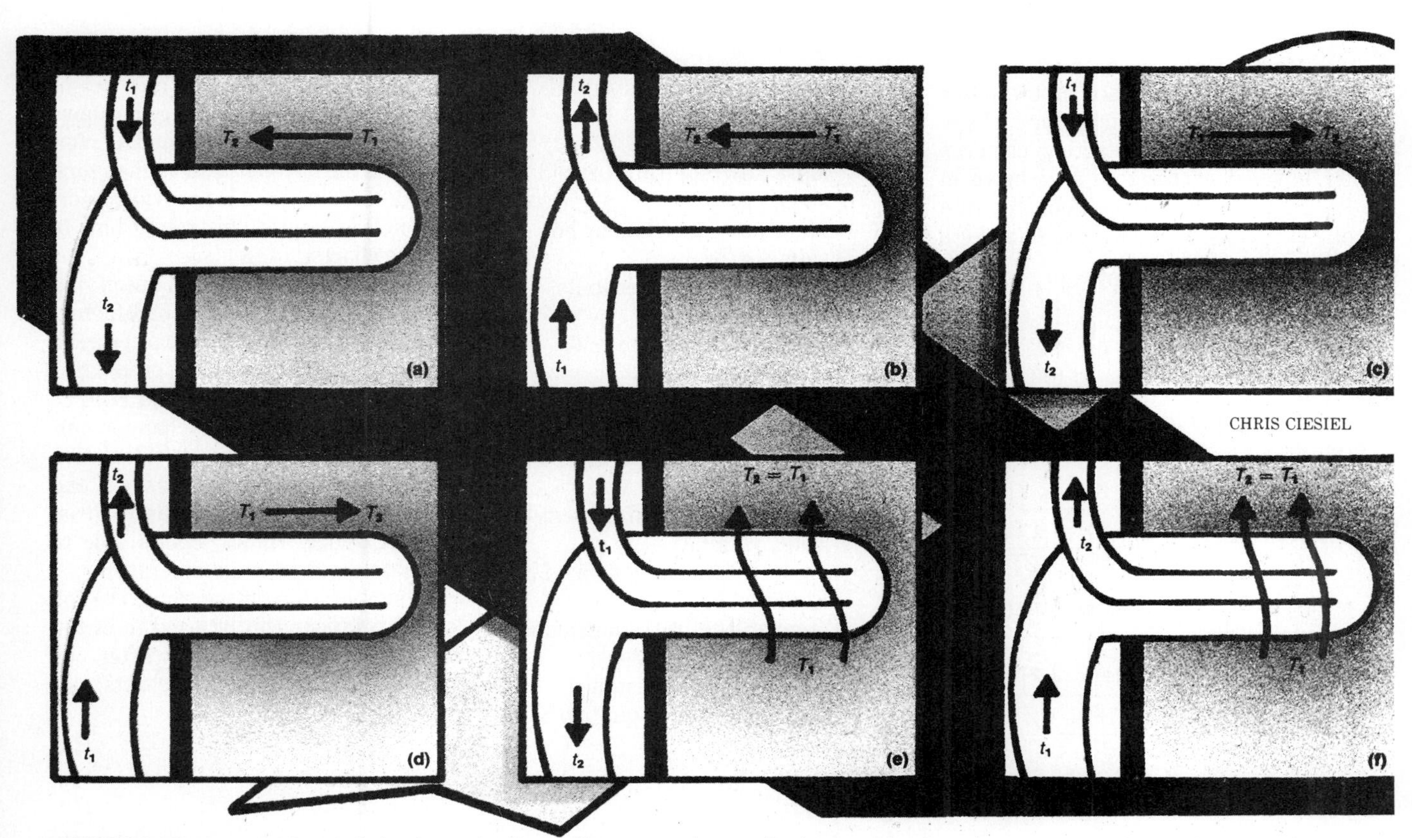

FIGURE 1. There are at least six basic contacting patterns or schemes for bayonet heat-exchangers

Originally published April 1991

Charts make for quick and easy calculations

Bayonet heat-exchangers are particularly suited for condensation of vapors under moderate and very-high vacuum. They are also used in operations where there are large temperature differentials between the hot and cold fluids. Graphs for determining only the heat-transfer area for some special cases of bayonet heat-exchangers have been published [1, 2].*

But no published data exist for the practical situations where boiling or condensation occurs, nor for crossflow or for very high flowrate of fluid outside the bayonet. Moreover, for the reverse problem of determining the temperature of flow streams leaving the bayonet exchangers of known area, no charts are available.

Made up of a pair of concentric tubes, the outer of which has one end sealed, the bayonet is also known as the field tube. Both the inner and outer tubes extend from separate stationary tubesheets either into shells, or directly into vessels. But the surface of the outer tube is the actual heat-transfer surface. The various ideal contacting patterns for bayonet exchangers are shown in Figure 1. For these situations, the rate of heat transfer from fluid to fluid is given by:

$$q = U_o A_o \Delta T \tag{1}$$

where the expression for ΔT, for the different cases, is given as follows [5]:

- Cases (a) and (d) representing cocurrent flow

$$\Delta T = \frac{(t_2 - t_1)[(X+1)^2 + 4Z]^{1/2}}{\ln \dfrac{Y + [(X+1)^2 + 4Z]^{1/2}}{Y - [(X+1)^2 + 4Z]^{1/2}}} \tag{2}$$

- Cases (b) and (c) representing countercurrent flow

$$\Delta T = \frac{(t_2 - t_1)[(X-1)^2 + 4Z]^{1/2}}{\ln \dfrac{Y + [(X-1)^2 + 4Z]^{1/2}}{Y - [(X-1)^2 + 4Z]^{1/2}}} \tag{3}$$

- Cases (e) and (f) representing boiling, condensation, crossflow, or very high flowrate

$$\Delta T = \frac{(t_2 - t_1)(1 + 4Z)^{1/2}}{\ln \dfrac{Y + (1 + 4Z)^{1/2}}{Y - (1 + 4Z)^{1/2}}} \tag{4}$$

In Equations 2 to 4, whenever

$$Y < [(X \pm 1)^2 + 4Z]^{1/2} \tag{5}$$

there is no real solution to ΔT, which means that it is impossible to achieve the desired outlet temperature with the heat exchanger.

The dimensionless parameters X, Y and Z are defined by:

$$X = \frac{T_1 - T_2}{t_2 - t_1} = \frac{\dot{m}\, c}{\dot{M}\, C} \tag{6}$$

$$Y = \frac{(T_1 + T_2) - (t_2 + t_1)}{t_2 - t_1} \tag{7}$$

$$Z = \frac{U_i A_i}{U_o A_o} \tag{8}$$

A heat balance around the two flowing streams gives:

$$q = \dot{m}\, c\,(t_2 - t_1) = \dot{M}\, C\,(T_1 - T_2) \tag{9}$$

The number of transfer units can then be obtained by combining Equation 9 with Equation 1,

$$\frac{1}{NTU} = \frac{\dot{m}\, c}{U_o A_o} = \frac{\Delta T}{t_2 - t_1} \tag{10}$$

Using the charts

Let us now introduce the new set of charts that will aid in solving the above equations for the design and performance of the bayonet heat-exchangers. In what follows, we also provide guidelines and examples illustrating how to use the charts.

Setting a value of Z, and specifying any two of the three other parameters X, Y, NTU, the unknown can be read from the chart. For a design problem, if we know X and Y, we can find $1/NTU$, and then evaluate the exchanger surface area. The performance calculation involves specifying $1/NTU$ and X, and then from the chart one reads Y. From these X and Y values, one is able to calculate the terminal temperatures.

To simplify the calculations, the following relations can be used to evaluate the outlet temperatures after one has read X and Y from the charts:

$$t_2 = t_1 + \frac{2(T_1 - t_1)}{1 + X + Y} \tag{11}$$

$$T_2 = T_1 - X(t_2 - t_1) \tag{12}$$

However, for cases (e) and (f), by setting $X = 0$, one obtains the outlet

Nomenclature

Variables

A	Tube surface area, m^2
c	Specific heat capacity (tube- or cold-side fluid), kJ/kg-°C
C	Specific heat capacity (shell- or hot-side fluid), kJ/kg-°C
D	Tube diameter, m
q	Heat transfer rate, W
$\dot{m}$	Mass flowrate (tube or cold side), kg/h
$\dot{M}$	Mass flowrate (shell or hot side), kg/h
NTU	Number of transfer units, dimensionless
t	Temperature of tube-side fluid, °C
T	Temperature of shell-side fluid, °C
ΔT	Proper mean temperature difference, °C
U	Overall heat-transfer coefficient, W/m^2-°C
X	Temperature ratio, defined in Equation 6
Y	Temperature ratio, defined in Equation 7
Z	Heat-transfer resistance ratio, defined in Equation 8

Subscripts

1	Inlet
2	Outlet
i	Inner tube
o	Outer tube

*Charts for the design and performance of different kinds of shell-and-tube and compact exchangers are readily available in the literature [3, 4].

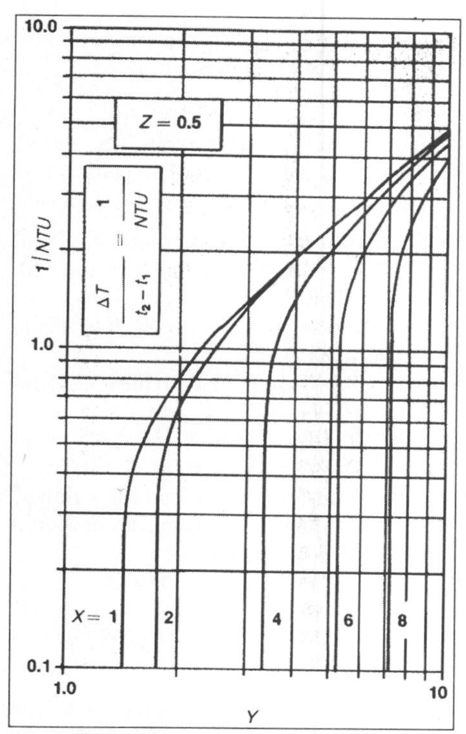

FIGURE 2a. The performance curves correspond to countercurrent exchangers with $Z = 0.5$. The configurations belong to schemes (b) and (c) in Fig. 1

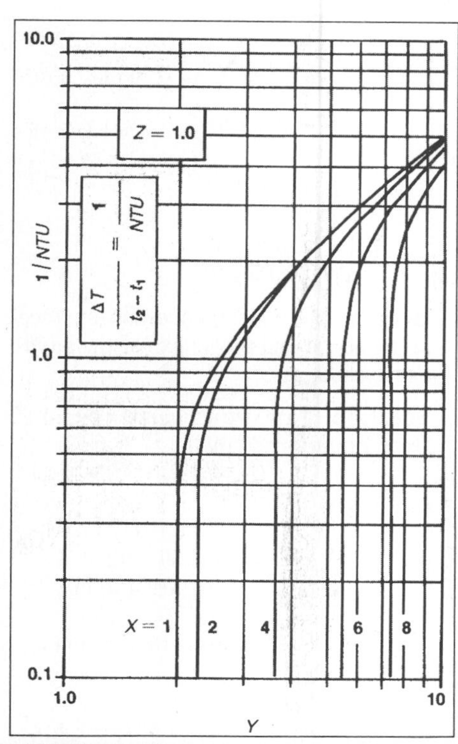

FIGURE 2b. The performance curves correspond to countercurrent exchangers with $Z = 1.0$. The configurations belong to schemes (b) and (c) in Fig. 1

temperature t_2 directly from Equation 11.

The curves in Figures 2a – 2d and Figure 4 apply to schemes (b), (c), (e) and (f) from Figure 1. For the selected values of Z, they represent the performance curves for the indicated configurations of bayonet heat-exchangers.

Figure 3 compares the behavior of cocurrent exchangers represented by cases (b) and (c), for $Z = 1$, and two different values of X, namely 1 and 2. From both sets of curves, it is clear that the countercurrent exchanger is always more efficient than its cocurrent counterpart. In other words, the countercurrent exchanger provides a larger mean temperature driving force between the inside and outside fluids, thus requiring a smaller heat exchange area.

This finding is quite general for all values of the parameters. Consequently, one need not consider the charts for the cocurrent schemes (a) and (d). The following examples illustrate the use of these charts for the practically important cases.

Example 1. Finding the needed size of an exchanger: schemes (b) and (c)

Given: $T_1 = 232°C$
$T_2 = 176°C$
$t_1 = 154°C$
$t_2 = 140°C$
$D_i = 2.54$ cm
$D_o = 5.08$ cm
$U_i = 114$ W/m²-°C $= 410$ kJ/h-m²-°C
$U_o = 57$ W/m²-°C $= 205$ kJ/h-m²-°C
$\dot{M} = 1{,}000$ kg/h
$c = 2.427$ kJ/kg-°C
$C = 2.594$ kJ/kg-°C

Required: To determine the surface area A, and the tube-side flowrate

Solution: First determine the value of the parameters X, Y and Z:

From Equation 6, $X = (T_1 - T_2)/(t_2 - t_1) = (232 - 176)/(154 - 140) = 4$

From Equation 7, $Y = [(232 + 176) - (154 + 140)]/(154 - 140) = 8.14$

From Equation 8, $Z = (U_i\,A_i)/(U_o\,A_o) = (114/57)(1/2) = 1$

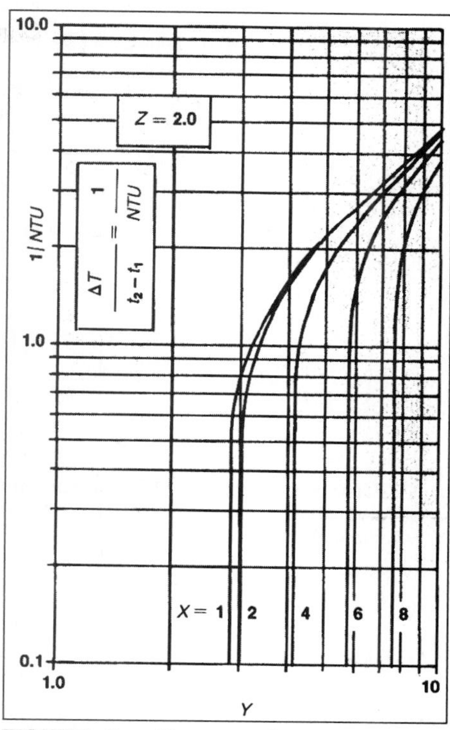

FIGURE 2c. These performance curves correspond to countercurrent exchangers with $Z = 2.0$. The configurations belong to schemes (b) and (c) in Fig. 1

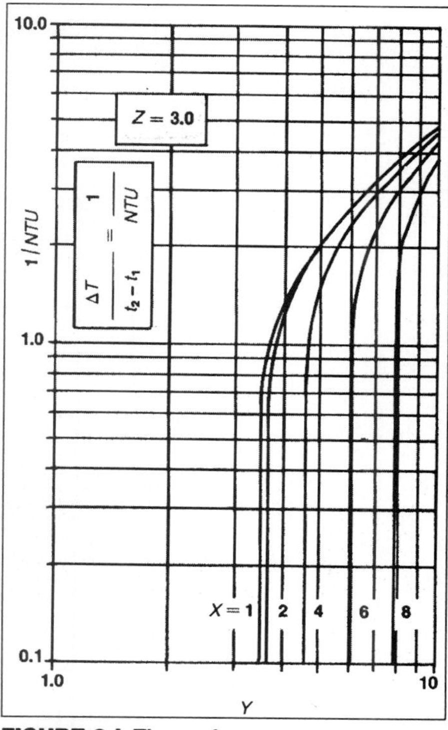

FIGURE 2d. The performance curves represent countercurrent exchangers with $Z = 3.0$. The configurations belong to schemes (b) and (c) in Fig. 1

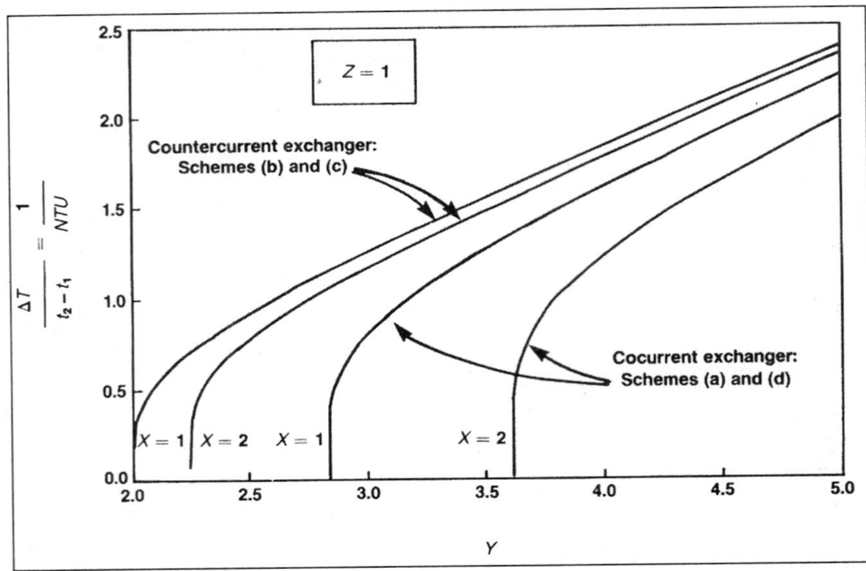

FIGURE 3. Comparison of the performance curves of cocurrent exchangers (schemes (a) and (d)) with those of countercurrent systems (schemes (b) and (c)) indicates higher efficiency of the latter

From Equation 9, $\dot{m} = (X \dot{M}\ C)/c = (4 \times 1{,}000 \times 2.594)/2.427 = 4{,}275$ kg/h

Now we use Figure 2b with the parameters X and Y to get

$\Delta T/(t_2 - t_1) = 1/NTU = 3.79$

Then, using Equation 10, the surface area required is

$A_o = (NTU \dot{m}\ C)/U_o = (0.263 \times 4275 \times 2.427)/205 = 13.31\ m^2$

For a 19-tube hexagonal array, the length of the tube will be

$L = A_o/(2\,\pi\,R_o\,n) = 13.31/(2\,\pi \times 2.54 \times 10^{-2} \times 19) = 4.4$ m

For the cocurrent arrangement with the above feed conditions, Figure 3 gives $\Delta T/(t_2 - t_1) = 3.38$, while in this example we found it to be 3.79. Thus the cocurrent arrangement requires 12% more area that the countercurrent system.

Example 2. Evaluating the performance of a given exchanger: schemes (b) and (c)

Given: $T_1 = 200°C$

$t_1 = 23°C$

$D_i = 2.54$ cm

$D_o = 5.08$ cm

$U_i = 193\ W/m^2\text{-}°C = 694\ kJ/h\text{-}m^2\text{-}°C$

$U_o = 96.5\ W/m^2\text{-}°C = 347\ kJ/h\text{-}m^2\text{-}°C$

$\dot{m} = 2{,}470$ kg/h

$\dot{M} = 1{,}000$ kg/h

$c = 2.05$ kJ/kg-°C

$C = 2.53$ kJ/kg-°C

$A_o = 9.19\ m^2$

Required: To determine the outlet temperatures of the cold stream t_2, and the exit temperature T_2 of the shell side

Solution: First we calculate X, Z and $1/NTU$, from Equations 6, 8 and 10, respectively

$X = (\dot{m}\ c)/(\dot{M}\ C) = (2470 \times 2.05)/(1{,}000 \times 2.53) = 2$

$Z = (U_i\,A_i)/(U_o\,A_o) = (694 \times 2.54)/(347 \times 5.08) = 1$

$1/NTU = (\dot{m}\ c)/(U_o\,A_o) = (2470 \times 2.05)/(347 \times 9.19) = 1.587$

From Figure 2b, we find the value of Y to be about 3.68. Then, from Equations 11 and 12, we get

$t_2 = t_1 + 2\,(T_1 - t_1)/(Y + X + 1) = 23 + 2\,(200 - 23)/(3.68 + 2 + 1) = 76°C$

$T_2 = T_1 - X\,(t_2 - t_1) = 200 - 2\,(76 - 23) = 94°C$

Example 3. Heat-exchanger size for condensation: schemes (e) and (f)

Given: $T_1 = T_2 = 118°C$

$t_1 = 30°C$

$t_2 = 60°C$

$D_i = 3$ cm

$D_o = 6$ cm

$U_i = 908\ W/m^2\text{-}°C = 3{,}268\ kJ/h\text{-}m^2\text{-}°C$

$U_o = 454\ W/m^2\text{-}°C = 1{,}634\ kJ/h\text{-}m^2\text{-}°C$

$\dot{m} = 2{,}111$ kg/h

$\dot{M} = 400$ kg/h

$c = 4.184$ kJ/kg-°C

$C = 2.343$ kJ/kg-°C

Required: To determine the surface area needed and the length of the bayonet exchanger

Solution: First we calculate Y and Z from Equations 7 and 8:

$Z = (U_i\,A_i)/(U_o\,A_o) = (3{,}268 \times 3)/(1{,}634 \times 6) = 1$

$Y = [2\,T_1 - (t_1 + t_2)]/(t_2 - t_1) = [(2 \times 118) - (30 + 60)]/(60 - 30) = 4.86$

We use values in Figure 4 to get

$\Delta T/(t_2 - t_1) = 1/NTU = 2.25$

Thus, from Equations 10, the surface area required is

$A_o = (NTU \times \dot{m}\ C)/U_o = (0.444 \times 2{,}111 \times 4.184)/1{,}634 = 2.4\ m^2$

Whence we can determine the total length needed in the exchanger:

$L = A_o/(2\,\pi\,R_o)$

$= 2.4/(2\,\pi\,3 \times 10^{-2}) = 12.73$ m

For the seven-tube array, length of each tube will be 1.81 m

CARLA MAGAZINO

FIGURE 4. Chart shows performance curves of bayonet exchangers belonging to schemes (e) and (f)

Example 4. Heat-exchanger performance for condensation:
schemes (e) and (f)

Given: $T_1 = T_2 = 55°C$
$t_1 = 380°C$
$D_i = 3.5$ cm
$D_o = 5$ cm
$U_i = 3{,}240\ W/m^2\text{-}°C = 11{,}665\ kJ/h\text{-}m^2\text{-}°C$
$U_o = 1{,}135\ W/m^2\text{-}°C = 4{,}086\ kJ/h\text{-}m^2\text{-}°C$
$\dot{m} = 5{,}048$ kg/h
$\dot{M} = 500$ kg/h
$c = 4.184$ kJ/kg-°C

Required: To determine the outlet temperature of the cold stream t_2

Solution: First we calculate Z and $1/NTU$ from Equations 8 and 10

$$Z = (U_i A_i)/(U_o A_o) = (3{,}240 \times 3.5)/(1{,}135 \times 5) = 2$$

$$1/NTU = (\dot{m}\ c)/(U_o A_o) = (5{,}048 \times 4.184)/(4{,}086 \times 4.84) = 1.06$$

With these values, we make use of Figure 4 to obtain $Y = 3.36$. Substituting this value of Y in Equation 11, with $X = 0$ gives

$$t_2 = t_1 + 2\,(T_1 - t_1)/(Y + 1) = 18 + 2\,(55 - 18)/(3.36 + 1) = 36°C$$

Edited by Gulam Samdani

Acknowledgments

The authors would like to thank professor Octave Levenspiel for his expert assistance and helpful criticism.

References

1. Hurd, N. L., Mean Temperature Difference in the Field or Bayonet Tube, *Ind. Engr. Chem.*, vol. 38, no. 12, pp. 1266 – 1271, 1946.
2. Kern, D. Q., "Process Heat Transfer," McGraw-Hill, New York, 1950.
3. Turton, R., others, Charts for the Performance and Design of Heat Exchangers, *Chem. Eng.*, pp. 81 – 88, Aug., 1986.
4. Bowman, J., and Turton, R., Quick Design and Evaluation: Heat Exchangers, *Chem. Eng.*, pp. 92 – 99, July, 1990.
5. Levenspiel, O., "Engineering Flow and Heat Exchange," Plenum Press, New York, 1984.

The authors

Abel Hernandez-Guerrero is a Ph.D. candidate at the Department of Mechanical Engineering, Oregon State University, Rogers Hall, Corvallis, OR 97331-6001; telephone: (503) 737-2832. His research focuses on a study of self-propagating high-temperature synthesis (SHS) of powders. At the end of his studies, he will will return to his alma mater in Mexico as an assistant professor of mechanical engineering. His general areas of interest are heat transfer and combustion. He received his B.S. in mechanical engineering from the Universidad de Guanajuato (Mexico), and a Master's degree also in mechanical engineering from Oregon State University.

Agustin Macias-Machin is an assistant professor at the Polytechnic University of Canary Islands (Spain). He has just completed a two-year visiting faculty appointment at the Oregon State University. His research interest lies exclusively in solar energy storage in phase-change materials, because the Canary Islands are energy-poor except for sunlight, which they get more than 300 days a year. ■

Alfa-Laval Thermal Inc.

In general, plate heat exchangers demonstrate less fouling than shell-and-tube units. Here, a fouled unit is being cleaned in an agitated liquid-nitrogen bath

SIZING PLATE HEAT EXCHANGERS

The right fouling factor is the key to specifying plate heat exchanger areas correctly

Jeff Kerner, P.E., Alberts & Associates, Inc.

Since their commercial debut in the 1930s, plate heat exchangers have found widespread use in the chemical process industries (CPI). Today, more than two dozen firms market this space-saving and highly efficient type of heat exchanger.

One reason for the popularity of plate heat exchangers is that their overall heat-transfer coefficient (U) is superior to that of shell-and-tube heat exchangers [*1,2,3,4*]. In clean water-to-water service, for example, a shell-and-tube heat exchanger has a U value of 350 Btu/ft^2-h-°F, much lower than the 1,000 of a plate design at the same pressure drop.

However, the plate heat exchanger's much higher U values also mean that fouling factors have a much greater effect on calculations of exchanger surface area. For example, a fouling resistance of 0.001 ft^2-h-°F/Btu will increase the surface area of a shell-and-tube unit by about 35%, but will increase that of a plate heat exchanger by 100%.

A good deal of literature has been published cautioning against specifying excess area for heat exchangers because this may be the cause of fouling beyond what is considered in the design [*5*]. It is, therefore, important not only to understand the concept of fouling factors for plate heat exchangers, but also to know how the factors are categorized by different suppliers.

To calculate the effects of fouling on surface area, manufacturers of plate heat exchangers employ many different terms:

- Percent oversurfacing
- Excess area
- Safety factor
- Safety margin
- Cleanliness factor
- Fouling factor

As a result, users are often confused as to the differences between the various terms and, more importantly, the right way to employ them in carrying out design calculations for specifying the units. To distinguish between the terms, and to understand the interelationship between them, it helps to list the basic equation:

$$Q = UA\Delta t_{lm} \quad \textbf{(1)}$$

In Equation (1),

Q = Heat transferred, Btu/h

U = Overall heat-transfer coefficient, Btu/ft^2-h-°F

A = Heat transfer area, ft^2

Δt_{lm} = Log mean temperature difference, dimensionless

For a clean, new heat exchanger,

$$Q = U_c A_c \Delta t_{lm} \quad \textbf{(2)}$$

where the subscript c refers to a "clean" condition

In order to allow the heat exchanger to transfer the required heat after it is in service and has become dirty, a new, lower U value and, correspondingly, higher heat-transfer area have to fit the same equation:

$$Q = U_s A_s \Delta t_{lm} \quad \textbf{(3)}$$

where the subscript s refers to "service" or "dirty" conditions.

Relating the two equations,

$$U_c A_c \Delta t_{lm} = U_s A_s \Delta t_{lm} \quad \textbf{(4)}$$

$$\text{and, } A_s/A_c = U_c/U_s \quad \textbf{(5)}$$

For the heat exchanger designed with fouling in mind, the term A_s/A_c, or the ratio of the "service" (dirty) area to the design or "clean" area is always greater than 1. Converting this to a

Originally published November 1993

percentage and subtracting 100 gives the term commonly called by most manufacturers as "percent oversurfacing."

It is correct to use this interchangeably with "percent excess area," and is an indication of how much excess area over the clean area is being provided in the heat exchanger design. It may also be called the "safety factor" or "safety margin."

Most manufacturers use values for these terms ranging from 5 to 15%, depending upon the type of fluids being handled. Plant engineers can specify other values, depending on the degree of fouling in an application.

"Cleanliness factor" can be defined as U_s/U_c, the ratio of the values of the overall heat-transfer coefficient in "service" to that of the overall heat-transfer coefficient in "clean" conditions. This ratio is less than 1 (or equal to it). It is a term commonly used in the power industry for water-to-water applications, where the factors usually range between 0.75 and 0.90. The ratio is commonly expressed as a percentage — in the power industry, the range is 75 to 90%.

Not the same

At first glance, a specifying engineer may think that a 90% cleanliness factor is equivalent to a 10% oversurfacing or a 10% safety factor (i.e., 100% minus 90% = 10%). However, because of the reciprocal relationship between the two definitions, there is a difference in the resulting area calculation.

Thus, for a 90% cleanliness factor,

$C_f = U_s/U_c$

$= [Q/(A_s\ \Delta t_{lm})]/[Q/(A_c\ \Delta t_{lm})]$

$= 0.90$

Simplifying,

$A_c/A_s = 0.90$

$A_s = A_c/0.90$

$A_s/A_c = 1/C_f$

$= 1.11$

On the other hand, using a 10% oversurfacing or safety factor,

$A_s/A_c = 1.10$

Thus, a 90% cleanliness factor represents a 11% excess area, not 10% as might have been assumed. While this represents only a 1% difference in area, the difference becomes greater as the cleanliness factors become lower. For example, at a cleanliness factor of 85%, the excess area is 17%, not 15%, and at a cleanliness factor of 75%, the excess area becomes 33%, not 25%.

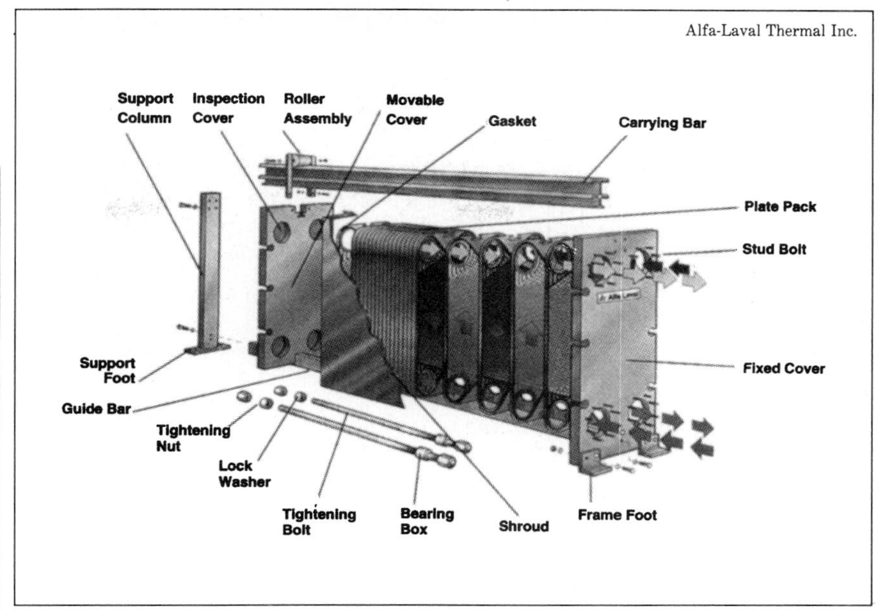

In specifying plate heat exchangers, fouling is a key consideration since a small increase in the fouling factor can lead to a large increase in exchanger area

Moreover, the cleanliness factor defined is obviously a function of the clean U value itself. This can be shown as follows:

$$U_s = [(1/h_1)+(1/h_2)+(x/k)+ R_f)^{-1} \quad \textbf{(6)}$$

In Equation (6),

h_1 and h_2 = Individual film coefficients for the two fluids

x = Thickness of wall separating the fluids

k = Thermal conductivity of wall

R_f = Total fouling resistance, both fluids

Since the "clean" U value, U_c, is defined as:

$U_c = [(1/h_1) + (1/h_2) + (x/k)]^{-1}$

It can easily be shown that:

$(1/U_s) - (1/U_c) = R_f$

Solving for the ratio U_s/U_c yields:

$$C_f = U_s/U_c = [1/(R_f U_c + 1)] \quad \textbf{(7)}$$

Obviously, U_c changes with the channel velocity of the fluid, and with different manufacturers' plate-heat-exchanger designs. As a result, the cleanliness factor for a given application has relevance only if U_c is constant from case to case, as can be seen in Equation 7. If the engineer evaluates different cases using a constant cleanliness factor, then Equation 7 implies that U_c and the fouling factor are inversely related.

Calling the cleanliness factor a constant, C_f, in Equation 7:

$C_f = 1/(R_f U_c + 1)$

$(C_f\ R_f\ U_c) + C_f = 1$

$C_f\ R_f\ U_c = 1 - C_f$

$R_f\ U_c = (1 - C_f)/C_f$

Since C_f has been assumed to be constant for this calculation,

$R_f\ U_c$ = constant

By this inverse relationship between R_f, the total fouling factor, and U_c, the overall "clean" heat-transfer coefficient, it is obvious that as the "clean" U value increases, the fouling factor must decrease. However, there is no reason to expect any relationship at all between the manufacturer's "clean" U value and the actual fouling of the heat exchanger.

No relationship

For example, if the manufacturer increases the "clean" U value by using a more efficient plate, there is no reason to expect a decrease in the fouling tendency of the fluids. Therefore, one must caution against using the "cleanliness factor" as a means of specifying additional surface area to allow for fouling on a plate heat exchanger. Instead, in most cases, "percent oversurfacing," "excess area," "safety factor," or "safety margin" should be used.

The last term, "fouling factor," is an acceptable one provided that the fouling factors used are those for plate heat exchangers and not for shell-and-tube units (the latter are commonly known as Tubular Exchanger Manufacturers Assn., or TEMA, factors). It has been repeatedly stated [*2,3*] that fouling factors for plate heat exchangers are much lower than those for shell-and-tube equipment. For instance, in water applications, the plate heat exchanger values are, typically, 25–30%

of those of shell-and-tube units.

The lower values are primarily because the wall shear stresses for flow between narrowly spaced parallel plates, such as found in a plate heat exchanger, are several times greater than the wall shear stresses in a tube, even at equal flowrates in the channels. Since it has been shown that wall shear stress contributes the most toward the removal of fouling deposits [*6*], it is easy to understand why plate heat exchangers exhibit lower values of fouling factors than tubular heat exchangers.

In the absence of other data, the fouling factors published by Heat Transfer Research, Inc. [*7*] are useful for water-to-water applications. Since fouling factors for fluids other than water are not publicly available, users should rely on values selected by experienced plate-heat-exchanger vendors.■

Edited by Jayadev Chowdhury

References

1. Marriott, Jan, Where and How to Use Plate Heat Exchangers, *Chem. Eng.*, Apr. 5, 1971, pp. 127–134.
2. Cross, P. H., Preventing Fouling in Plate Heat Exchangers, *Chem. Eng.*, Jan. 1, 1979, pp.87–90.
3. Cooper, A., Suitor, J. ., Usher, J.D. Cooling Water Fouling in Plate Heat Exchangers, *Heat Tran. Eng.*, Vol. 1, No. 3, January–March 1980, pp. 50–55.
4. Burley, J. R., Don't Overlook Compact Heat Exchangers, *Chem. Eng.*, August 1991, pp. 90–96.
5. Knudsen, J., Fouling of Heat exchangers: Are We Solving the Problem?, *Chem. Eng. Progr.*, February 1984, pp. 63–69.
6. Novak, L., Fouling in Plate Heat Exchangers and its Reduction by Proper Design, *Heat Exchangers: Theory and Practice,* Taborek, Hewitt and Afghan, eds. Hemisphere Publishing Corp., 1983, New York.
7. *Design Manual D7.lPHE,* July 1984, Heat Transfer Research Institute, Inc., College Station, Tex.

The author

Jeff Kerner is vice president and partner in Alberts & Associates, Inc. (1042 Norvelt Drive, Philadelphia, PA 19115. Tel: 215-673-8290; Fax: 215-632-9248), an authorized representative of Alfa-Laval Thermal Inc. A chemical engineering graduate of The Cooper Union (New York City), he received a master's from the University of Michigan. Previously, he worked at Du Pont and Rohm & Haas Co. Jeff has been active in the American Institute of Chemical Engineers, having served as chairman of the Philadelphia subsection, and on AIChE's national committee on employment guidelines. He is a professional engineer registered in Pennsylvania, Delaware and New Jersey.

Robert H. Hedrick, P.E.
Chemical Engineer*

CALCULATE HEAT TRANSFER FOR TUBES IN THE TRANSITION REGION

This correlation is valid for Reynolds numbers between 2,000 and 10,000

Until now, there has been no good correlation for determining heat transfer in tubes for fluids in the transition region between laminar and turbulent flow. The following equations can be used to calculate the inside film coefficient (based on the outside tube diameter), h_{io}:

$$B_i = (-3.08 + 3.075X + 0.32567X^2 - 0.02185X^3)(10d_i/L)^{[1-(X/10)^{0.256}]}$$

$$h_{io} = (16.1/d_o)[B_i k(c\mu/k)^{1/3}(\mu/\mu_w)^{0.14}]$$

where: $X = N_{Re}/1{,}000$
$_{Re}$ = Reynolds number
d_i = inside tube dia., in.
d_o = outside tube dia., in.
k = thermal conductivity, Btu/h-ft-°F
c = specific heat, Btu/lb-°F
μ = viscosity, cP
μ_w = viscosity at the wall, cP;
L = tube length, ft

This relationship produces smooth curves for all values of L/d between 2 and 50 over the Reynolds number range of 2,000 to 10,000. Using this correlation along with the Sieder-Tate correlations for laminar and turbulent flow, one can calculate the inside film coefficient for forced-circulation convection over a wide range of Reynolds numbers. ■

* 2706 Lakeland Trail, Birmingham, AL 35243

Inside film coefficients for water at 200°F, in an 18 BWG tube with an outside diameter of 2 in.

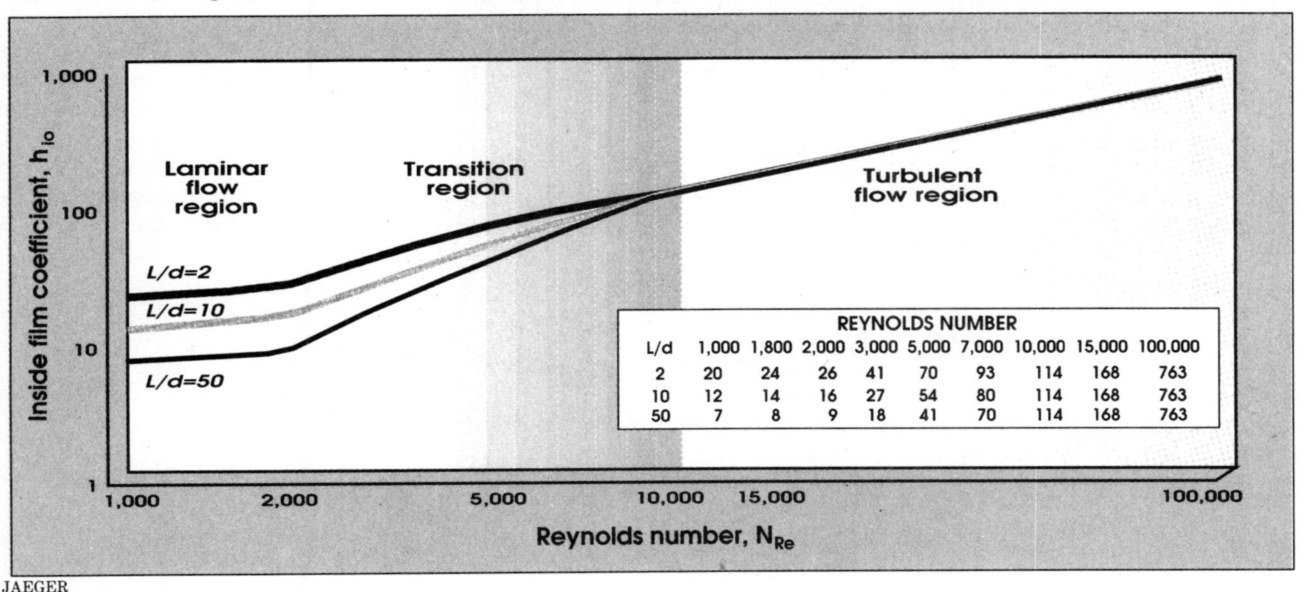

	REYNOLDS NUMBER								
L/d	1,000	1,800	2,000	3,000	5,000	7,000	10,000	15,000	100,000
2	20	24	26	41	70	93	114	168	763
10	12	14	16	27	54	80	114	168	763
50	7	8	9	18	41	70	114	168	763

JAEGER

Originally published June 1990

JIM STONE

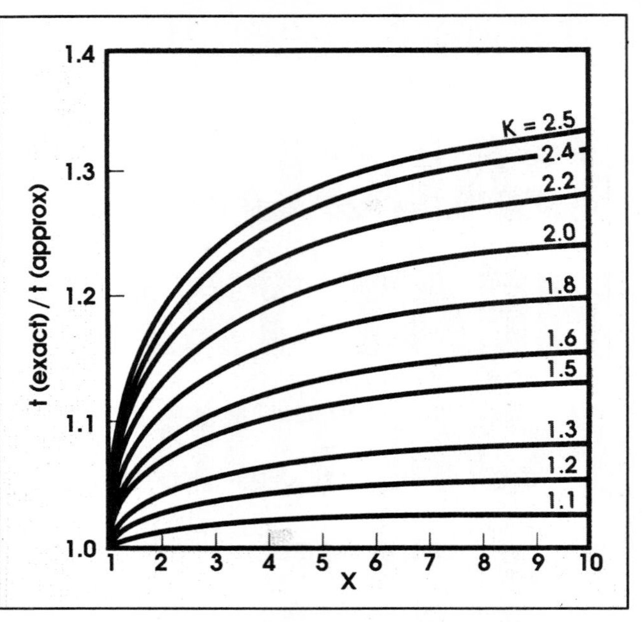

How to deal with heat-transfer-medium temperatures that are not constant

PREDICT HEATING AND COOLING TIMES ACCURATELY

Ismail Tosun
Ilhan Aksahin
Middle East Technical Univ.

The heating and cooling of liquids in jacketed reactors is a common CPI operation. When the temperature of the heat-transfer medium is constant—as it is for steam—a simple calculation determines the time needed to reach the desired temperature.

However, when the heat-transfer medium does not have a constant temperature, the calculation changes. Engineers will often take a "shortcut" and use the wrong equation to save steps. In some cases this is acceptable, but in others, using the wrong equation can result in errors of 10% or more.

This article describes the equation used in each case, and helps determine the proper one to use. It also contains a graph that measures the accuracy of the calculation.

When the temperature of the heat-transfer medium is constant, the time is calculated by the following equation:

$$t = \frac{MC_p}{UA} \ln \left[\frac{T_s - T_1}{T_s - T_2} \right] \quad (1)$$

where t = time

T_s = heat-transfer medium temperature

T_1= initial reactor-fluid temperature

T_2= final reactor-fluid temperature

M= mass of fluid

C_P= heat capacity of the fluid

U= overall heat-transfer coefficient

A= heat-transfer area

When the heat-transfer medium's temperature is not constant, the medium's flowrate, heat capacity and inlet and outlet temperatures must be considered. In this case:

$$t = \frac{MC_p}{UA} \left[\frac{K}{K-1} \right] [\ln K] \ln \left[\frac{T_{in} - T_1}{T_{in} - T_2} \right] \quad (2)$$

where T_{in}= inlet temperature of the heat-transfer medium, and

$$K = \exp \left[\frac{UA}{wc} \right] = \frac{T_{in} - T_2}{T_{out} - T_2} \quad (3)$$

where w= mass flowrate of the heat-transfer medium

c= heat capacity of the heat-transfer medium

T_{out}= outlet temperature of the heat-transfer medium

Because Equations 2 and 3 are more complicated than Equation 1, it is tempting to use Equation 1 when the heat-transfer medium's temperature is not constant. Many engineers simply use the average temperature as T_S, making Equation 1:

$$t = \frac{MC_p}{UA} \ln \left[\frac{\left[\frac{T_{in} + T_{out}}{2} \right] - T_1}{\left[\frac{T_{in} + T_{out}}{2} \right] - T_2} \right] \quad (4)$$

Previous literature stated that this equation can be used only when the difference between the inlet and outlet temperatures of the heat-transfer medium is less than 10% of the log-mean temperature difference between the average temperature of the heat-transfer medium and the temperature of the liquid in the reactor [1]. Expressing this condition mathematically:

$$\ln \left[\frac{\left[\frac{T_{in} + T_{out}}{2} \right] - T_1}{\left[\frac{T_{in} + T_{out}}{2} \right] - T_2} \right] \leq 0.10 \left[\frac{T_2 - T_1}{T_{in} - T_{out}} \right] \quad (5)$$

However, this condition does not correctly predict the cases in which Equation 4 can be used. The above graph shows the danger of using the wrong equation. If Equations 2 and 4 are considered as exact and approximate, respectively, dividing Equation 2 by Equation 4 gives the following ratio:

$$\frac{t\,(\text{exact})}{t\,(\text{approx.})} = \left[\frac{K}{K-1} \right] \frac{\ln K \ln X}{\ln \left[\frac{2X + (1/K) - 1}{1 + (1/K)} \right]} \quad (6)$$

where X is defined as

$$X = \frac{T_{in} - T_1}{T_{in} - T_2} \quad (7)$$

The graph shows this ratio as a function of X and K. The error introduced by choosing the wrong equation is

Originally published November 1993

[t(exact) – t(approx.)]/t(exact).

Example

50,000 lb of liquid in a jacketed reactor are heated from 68 to 131°F. The heat-transfer medium enters the jacket at 194°F and leaves it at 172°F. Determine the time taken to raise the temperature. The liquid's heat capacity is 0.4 Btu/(lb)(°F), the overall heat-transfer coefficient is 150 Btu/(h)(ft²)(°F) and the heat-transfer area is 100 ft².

First, calculate the left and right terms of Equation 5:

$$\ln\left[\frac{\left[\frac{T_{in}+T_{out}}{2}\right]-T_1}{\left[\frac{T_{in}+T_{out}}{2}\right]-T_2}\right] = \ln\left[\frac{\left[\frac{194+172}{2}\right]-68}{\left[\frac{194+172}{2}\right]-131}\right] = 0.79$$

$$0.10\left[\frac{T_2-T_1}{T_{in}-T_{out}}\right] = 0.10\left[\frac{131-68}{194-172}\right] = 0.29$$

Since 0.79>0.29, Equation 4 should not be used. If it is used, the heating time is incorrectly calculated as:

NOMENCLATURE

A	Heat-transfer area
C_p	Heat capacity of the liquid in the tank
c	Heat capacity of the heat-transfer medium
K	Parameter defined by Equation 3
M	Mass of the liquid to be heated
T_{in}	Inlet temperature of the heating medium
T_{out}	Outlet temperature of the heating medium
T_s	Isothermal heating-medium tem perature
T_1	Initial temperature of the liquid to be heated
T_2	Final temperature of the liquid to be heated
t	Time
U	Overall heat-transfer coefficient
w	Mass flowrate of the heating medium
X	Parameter defined by Equation 7

$$t_{(approx.)} = \frac{(50,000)(0.4)}{(150)(100)} \times \ln\left[\frac{\left[\frac{194+172}{2}\right]-68}{\left[\frac{194+172}{2}\right]-131}\right] = 1.05\text{ h}$$

The values of X and K are:

$$X = \frac{T_{in}-T_1}{T_{in}-T_2} = \frac{194-68}{194-131} = \frac{126}{63} = 2$$

$$K = \frac{T_{in}-T_2}{T_{out}-T_2} = \frac{194-131}{172-131} = \frac{63}{41} = 1.54$$

From the graph, the value of t(exact)/t(approx.) is 1.08. The error in the heating-time approximation is 7.4%. The exact heating time is calculated from the approximation by the following route:

$$t_{(exact)} = (1.08)\,(1.05) = 1.13\text{ h}$$

This result is verified by Equation 2.■

References

1. Bondy, F., and Lippa, S., Heat transfer in agitated vessels, Chem. Eng., 90, 7, pp. 62-71, Apr. 4, 1983.

Ismail Tosun is a professor of chemical engineering at the Middle East Technical Univ., and Ilhan Aksahin is a Ph. D. candidate there. They can be reached at the Chemical Engineering department, Ankara 06531, Turkey, telephone (90)4-233-71-00, fax (90)312-210-1264

HOW TO PREDICT BATCH-REACTOR HEATING AND COOLING

John McEwan, PE
Paducah, Ky.

Calculating temperatures for jacketed batch reactors

Starting up a jacketed batch reactor requires control of the heat-up and cool-down rates. To do this, the jacket heat-transfer-fluid temperatures have to be determined and set. This can be done by trial-and-error experiments, but it is often quicker and more straightforward to simply make a trial heat-up and then plug the results into time-dependent heat-transfer equations. Here's how to do this for steam or hot-water jacketed reactors.

The time required for isothermally heating or cooling the reactor can be calculated by the equation

$$(1) \quad \ln\left(\frac{T_1 - t_1}{T_1 - t_2}\right) = \frac{UA\phi^{(1)}}{MCp} \quad [1]$$

Rearranging to solve for the heat-up time gives

$$(2) \quad \phi = \ln\left(\frac{T_1 - t_1}{T_1 - t_2}\right)\frac{MCp}{UA}$$

Eq. (2) can also be used to calculate the heat-up time for nonisothermal heating (such as hot-water jacketing), provided that the difference between the outlet and inlet jacket temperatures is not greater than 10% of the difference between the batch and average jacket water temperature [2].

The heat transfer area, reaction mass and heat capacity of the vessel contents are generally known. The overall heat transfer coefficient, however, is a function of five resistances and can be difficult to estimate.

These resistances include the inside-the-jacket film heat-transfer coefficient (HTC) and fouling factor, the inside-the-reactor HTC and fouling factor, and the reactor-wall resistance.

Assuming U, M, Cp and A are constant, Eq. (2) can be rewritten as follows:

$$(3) \quad \phi = \ln\left(\frac{T_1 - t_1}{T_1 - t_2}\right) \times \frac{1}{K}$$

where K = UA/MCp

Rearranging Eq. (3) to solve for the jacket temperature as a function of time gives

$$(4) \quad T_1 = \frac{t_1 - t_2 e^{K\phi}}{1 - e^{K\phi}}$$

By taking a series of readings during a trial heat-up, K can be calculated. The heat-up and cool-down times for varying jacket temperatures can then be predicted.

Nomenclature

A = heat transfer area
Cp = heat capacity of reactor contents
K = constant = UA/MCp
M = mass of the reactor contents
T = average jacket temperature
t_1 = initial temperature of reactor contents
t_2 = final temperature of reactor contents
U = overall heat transfer coefficient
Φ = time

Example: The overall cycle time for a batch reaction is to be 8 h. The cycle time will include 2 h for heat-up and 3 h for cool-down. The batch will be heated from 20°C to the reaction temperature of 60°C, then cooled to 40°C. What are the jacket temperatures required for heat-up and cool-down?

Using a hot water jacket temperature of 75°C, it was found that it took 20 min to heat the batch from 20°C to 30°C.

Then, following Eq. (3),

$$\ln\left(\frac{75° - 20°}{75° - 30°}\right) = K \times 20 \text{ min},$$

$$K = \frac{\ln\left(\frac{55°}{45°}\right)}{20 \text{ min}}$$

$$K = \frac{0.01}{\text{min}}$$

From Eq. (4),

$$T_1 = \frac{20 - 60e^{(0.01/\text{min} \times 120 \text{ min})}}{1 - e^{(0.01/\text{min} \times 120 \text{ min})}}$$

T_1 = 78°C for heat-up

$$T_1 = \frac{60 - 30e^{(0.01/\text{min} \times 180 \text{ min})}}{1 - e^{(0.01/\text{min} \times 180 \text{ min})}}$$

T_1 = 24°C for cool-down

References

1. Kern, D. Q., *Process Heat Transfer,* (New York: McGraw-Hill Book Co., 1950), p.628-637.
2. Bondy, F. and Lippa, S., "Heat transfer in agitated vessels," *Chem. Eng.,* April 4, 1983, p. 62.

Originally published May 1989

CALCULATE THE HEAT LOSS FROM PIPES

A. R. Konak*
Southern Alberta Institute of Technology

Heat lost from piped fluids is energy wasted. Even a fluid-temperature drop of 1°C or less from the pipe's inlet to its outlet is a sign of heat loss. This raises energy costs because heat must usually be added back to the fluid later.

If the pipe-wall temperature is not unduly high, convection can be assumed to be the main cause of heat loss. A simple method quantifies this heat loss. When applied successively to bare and insulated pipes, it shows the heat savings provided by insulation.

DENISE EARDMAN

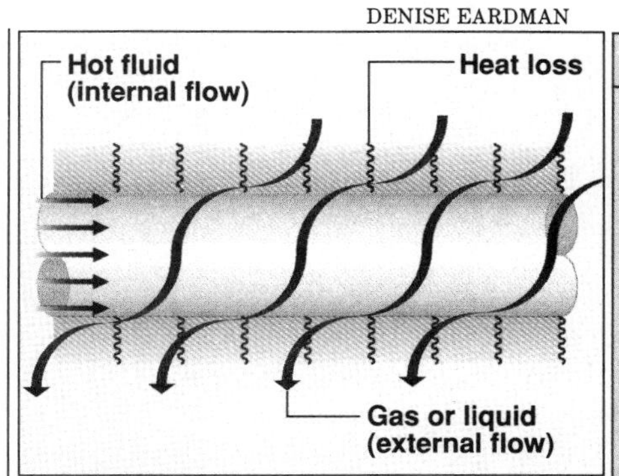

The temperature and velocity of the fluid outside the pipe (air in this case) are all that determine the heat loss from pipes with nearly constant fluid temperatures

A four-step process determines the dissipation per unit length

In general, heat loss is determined by the properties of the fluids inside and outside of the pipe (figure above). However, when the temperature drop within the pipe (between its upstream and downstream ends) is small, and when the pipe is of metal or other material with high heat conductivity, the heat loss can be assumed to depend very strongly on the external conditions, and the engineer need focus only on the external flow. In this situation, the temperature of the pipe's outer wall is assumed to be the same as the bulk temperature of the fluid within the pipe.

The equation used to calculate heat loss from pipes is called the Hilpert correlation:

$$Nu = \frac{h\,D}{k} = C(Re)^m\,(Pr)^{0.33} \qquad (1)$$

where Nu is the Nusselt number, a ratio of the fluid's total heat transfer to its conductive heat transfer

h = the external fluid's heat-transfer coefficient at the outer pipe wall

D = pipe diameter

k = thermal conductivity of the external fluid

Re is the Reynolds number for the flow outside the pipe; this is calculated by the same formula as that for flow inside a pipe: $Re = D\,v\,\rho\,/\,\mu$ where v is the velocity of the external fluid, ρ is its density and μ its viscosity

Pr is the Prandtl number, a ratio of momentum transfer to heat transfer. It equals $(C_P)\,(\mu)/k$ where C_P is the specific heat of the external fluid at constant pressure. Pr is generally between 0.65 and 0.75 for air and other gases

C and m are constants based upon the value of Re. Values for C and m are given in the table on p. 168

The Hilpert correlation is a specific

NOMENCLATURE

A	Heat-transfer area of the pipe
C	Constant in the Hilpert correlation
C_p	Specific heat of the fluid
D	Diameter of the pipe
h	Heat-transfer coefficient of the film
k	Thermal conductivity of the fluid
L	Pipe length
m	Constant in the Hilpert correlation
n	Constant in the Nusselt-number correlation
Nu	Nusselt number
Pr	Prandtl number
q	Heat loss
Re	Reynolds number
T_f	Film temperature
T_w	Pipe-wall temperature
T_∞	External-fluid temperature
ΔT	Temperature difference between fluid and air
v	Fluid velocity
μ	Fluid viscosity
ρ	Fluid density

*This is first of two articles on heat-loss calculations. Part 2 will discuss heat losses from pipelines in which fluid temperatures change significantly from the inlet to the outlet. (pp. 39–40)

Originally published September 1993

version of the well-known Nusselt-number correlation that is applicable to either external or internal forced convection: $Nu = C\,(Re)^m\,(Pr)^n$, where n is a constant

The external fluid forms a boundary layer, or "film," at the pipe wall. Assume that all the resistance to heat transfer is located in this layer. The film temperature is simply the average of the pipe-wall and external-fluid temperatures:

$$T_f = (T_w + T_\infty)/2 \qquad (2)$$

where T_f = average film temperature

T_w = pipe-wall temperature (temperature of the piped liquid)

T_∞ = temperature of the external fluid away from the pipe wall

The properties of the external fluid at this temperature can be taken from literature [1, 2]. This allows the Reynolds number to be calculated and Equation 1 to be solved for h, the film's heat-transfer coefficient. Then, insert h into the standard heat transfer equation below, solving for q:

CONSTANTS FOR EQUATION 1, THE HILPERT CORRELATION		
Re	*C*	*m*
0.4–4	0.989	0.330
4–40	0.911	0.385
40–4,000	0.683	0.466
4,000–40,000	0.193	0.618
40,000–400,000	0.0266	0.805

$$q = h\,A\,\Delta T \qquad (3)$$

where q = heat loss

A = area of the pipe through which heat loss occurs

ΔT = temperature difference between the external fluid and the pipe wall, $T_w - T_\infty$

Noting that the pipe area A is equal to $\pi\,D\,L$, where L is the pipe length, determine heat loss per unit length of pipe from the equation:

$$\frac{q}{L} = h\,\pi\,D\,\Delta T \qquad (4)$$

Example 1. An 0.5-m-dia. pipe in a refinery carries oil at 50°C. Wind blows across the pipe at 30 km/h (8.33 m/s) and –25°C. The pipe is not insulated. Find the heat loss per meter of pipe.

For uninsulated pipe, the wall temperature is nearly the same as the liquid temperature, 50°C. The film temperature is the average of the pipe-wall and air temperatures, 12.5°C. The physical constants for air are taken at this temperature, allowing the Reynolds number to be calculated:

$\mu = 1.80\times10^{-5}$ kg/(m)(s)
$\rho = 1.236$ kg/m^3
$k = 0.0248$ W/(m)(°C)
$Pr = 0.71$

$$Re = \frac{D\,v\,\rho}{\mu} = \frac{(0.5)(8.33)(1.236)}{1.80\times10^{-5}} = 2.86\times10^5$$

From Table 1, $C = 0.0266$ and $m = 0.805$. Solve Equation 1 for h:

$$h = \left(\frac{k}{D}\right) C\,(Re)^m\,(Pr)^{0.33}$$

$$= \left(\frac{0.0248}{0.5}\right)(0.0266)\,(2.86\times10^5)^{0.805}\times(0.71)^{0.33}$$

$$= (0.0496)\,(0.0266)\,(2.47\times10^4)\times(8.93\times10^{-1}) = 29.10\,\frac{\text{W}}{(\text{m}^2)(°\text{C})}$$

The heat loss per meter of pipe is calculated by inserting h, D and ΔT into Equation 4:

$$\frac{q}{L} = h\,\pi\,D\,\Delta T = (29.10)(3.14)(0.5)(75) = 3{,}426.53 \text{ W/m}$$

Even a minimal temperature drop indicates heat loss

Example 2. The pipe in Example 1 is wrapped with 2.5-cm-thick insulation. The temperature of the insulation surface adjacent to the air is 10°C. Find the heat loss from this pipe, given the same fluid and air conditions.

The film temperature is now –7.5°C, so the new constants are:

$\mu = 1.65\times10^{-5}$ kg/(m)(s)
$\rho = 1.329$ kg/m^3
$k = 0.023$ W/(m)(°C)
$Pr = 0.72$
$Re = 3.69\times10^5$

From Table 1, C and m remain the same and $h = 30.22$ W/(m^2)(°C). Substituting the insulated pipe's diameter, 0.55 m, into Equation 4:

$$\frac{q}{L} = h\,\pi\,D\,\Delta T = (30.22)\,(3.14)\,(0.55)\times(35) = 1{,}826.65 \text{ W/m}$$

Insulating the pipe has reduced the heat loss by 46.7%. ■

References

1. Incropera, F. P. and DeWitt, D. P., Fundamentals of Heat and Mass Transfer, 3rd ed., Wiley, New York.

2. Holman, J. P., Heat Transfer, 7th ed., McGraw-Hill, New York.

The author

A. Riza Konak is an instructor in chemical engineering technology at the Southern Alberta Institute of Technology, 1301 16th Ave. N. W., Calgary, AB, T2M 0L4 Canada, phone (403)284-8126. He holds B. Sc. and Ph. D. degrees in chemical engineering from the Univ. of Birmingham (England). Prior to teaching, Konak spent nearly 13 years with Imperial Oil Ltd. (Calgary), as head of various research sections. He has authored or co-authored five patents and 25 papers.

A. R. Konak*
Southern Alberta Institute of Technology

CALCULATE THE HEAT LOSS FROM PIPES

Combine the film coefficients for the internal and external fluids

Hot fluids cool as they move through pipelines. Calculating the extent of cooling can determine the energy needed to restore the heat to the fluid.

In Part 1, heat losses were calculated in pipes where the internal-fluid temperature does not change appreciably from the inlet to the outlet. In these pipes, the heat loss is determined only by the velocity and properties of the fluid surrounding the pipe.

When the internal-fluid temperature changes significantly, its properties must be included in the heat-loss calculations. This is because a film of internal fluid at the inner pipe wall resists heat transfer just as a film of external fluid resists heat transfer at the outer wall.

The heat-transfer coefficients of these films are combined to give an overall heat-transfer coefficient. This value and the temperature difference between internal and external fluids are plugged into an energy-balance equation to determine the fluid temperature at any distance from the inlet.

The Reynolds number of the piped fluid, a function of its mass flowrate and viscosity, is calculated first. The Reynolds number determines whether the internal flow is laminar or turbulent, and the heat-transfer coefficient of the film:

$$Re = 4M/\pi D\mu \quad (1)$$

where Re = Reynolds number
M = mass flowrate
D = pipe diameter
μ = fluid viscosity

If Re is greater than 4,000, there is turbulent flow in the pipe. Convection is the primary method of heat loss.

The Reynolds number is used to calculate the Nusselt number, Nu, a ratio of the fluid's total heat transfer to its conductive heat transfer. The Dittus-Boelter equation (a variations of the Nusselt-number equation given in Part 1) calculates Nu:

$$Nu = 0.023\,(Re)^{0.8}\,(Pr)^{n} \quad (2)$$

where Pr = Prandtl number, which 0.65–0.75 for air and other gases and 10–1,000 for liquids
n = a constant. If the fluid is losing heat, $n = 0.3$ and if the fluid is gaining heat, $n = 0.4$.

From there, calculate the heat-transfer coefficient of the film as in Part 1.† If Re is less than 2,000, there is laminar flow inside the pipe, and heat loss is mostly conductive.‡ In this case, the Nusselt number is a constant:

$$Nu = h_i D/k = 3.66 \quad (3)$$

where h_i = internal-film coefficient
k = thermal conductivity of the internal fluid

Now that h_e and h_i have been calculated, they are added together to give U, the overall heat-transfer coefficient. Because we are measuring the system's resistance to heat loss, we add the reciprocals of the films' heat-transfer coefficients, neglecting the heat of resistance of pipe metal and the difference between the internal and external pipe areas:

$$\frac{1}{U} = \frac{1}{h_i} + \frac{1}{h_e} \quad (4)$$

For buried pipelines, the overall heat-transfer coefficient ranges between 0.5 and 5 W/(m²)(°C), depending upon the type of soil, its moisture content and the depth at which the pipeline is buried [1].

Last, calculate the temperature difference between the internal and external fluids. When the internal-fluid temperature does not remain constant, the difference is not constant over the length of the pipe. Instead, the log-mean temperature difference is calculated:

$$(T - T_\infty)_{LM} = \frac{(T_1 - T_\infty) - (T_2 - T_\infty)}{\ln\left[(T_1 - T_\infty)/(T_2 - T_\infty)\right]} \quad (5)$$

where T = internal-fluid temperature at any point
T_1 = initial temperature of the internal fluid
T_2 = final temperature of the internal fluid
T_∞ = external-fluid temperature

This expression is plugged into the heat-loss equation, which reflects the movement of heat from the internal fluid to the external fluid:

$$q = (M)(C_P)(T_1 - T_2) = U\,\pi\,D\,L\,(T - T_\infty)_{LM} \quad (6)$$

where C_P = specific heat of the internal fluid
L = pipe length

Combining Equations 5 and 6 gives:

$$\frac{T_2 - T_\infty}{T_1 - T_\infty} = \exp - \left[\frac{U\,\pi\,D\,L}{MC_P}\right] \quad (7)$$

To calculate the distance at which the liquid temperature has reached some value T_2 below the initial value T_1, insert a value for T_2 and solve Equation 7 for L. To calculate the internal-fluid temperature at a distance L from the inlet, solve Equation 7 for T_2.

Although these equations address changes in internal-fluid temperatures, fluid properties are taken at the internal fluid's average bulk temperature, $(T_1 + T_2)/2$. A more-rigorous but time-

*Part 1 appears on pp. 37–38.

† In this article, we will refer to the h calculated in Part 1 as h_e because we are calculating the heat-transfer coefficients for two films, internal and external.

‡ When Re falls between 2,000 and 4,000, the flow is assumed to be a combination of laminar and turbulent. The Nusselt number can be calculated using either an average of the two methods or by a method chosen from the user's experience.

Originally published October 1993

consuming approach is to divide the pipeline into segments with nearly constant temperatures, calculate and combine the heat losses for each segment.

Insulation increases the length over which heat losses occur. The overall heat-transfer coefficient of an insulated pipeline can be as much as one order of magnitude less than that of an uninsulated one.

Example

Crude oil at 70°C flows through a 0.5-m-dia. pipeline at a volumetric flowrate (Q) of 15,000 m^3/d. Calculate the distance down the line at which the oil temperature is 30°C. The air temperature is −25°C, as in Part 1, example 1, and h_e is 29.10 W/(m^2)(°C). The liquid properties are taken at 50°C:

$\rho = 872$ kg/m^3
$\mu = 0.220$ kg/(m)(s)
$k = 0.143$ W/(m)(°C)
$C_P = 1{,}991$ J/(kg)(°C)

NOMENCLATURE

- C_P Specific heat of the fluid
- D Diameter of the pipeline
- h_i Heat-transfer coefficient of the internal film
- h_e Heat-transfer coefficient of the external film
- k Thermal conductivity of the fluid
- L Pipe length
- M Mass flowrate of the fluid
- Nu Nusselt number
- Q Volumetric flowrate of the fluid
- q Heat loss
- Re Reynolds number
- T_1 Initial temperature of the liquid
- T_2 Final temperature of the liquid
- T_∞ External-fluid temperature
- U Overall heat-transfer coefficient
- μ Fluid viscosity
- ρ Fluid density

Calculate the Reynolds number of the internal fluid to determine the nature of the flow. First, multiply the volumetric flowrate and the fluid density to get the mass flowrate:

$M = (Q)(\rho) = (15{,}000)(872)/(86{,}400 \text{ s/d}) = 151.39$ kg/s

Calculate the Reynolds number:

$Re = 4M/\pi D\mu = (4)(151.39)/(3.14)(0.5)(0.220) = 605.56/0.35 = 1730$

Re is below 2,000, so the flow is laminar and the Nusselt number is a constant. Solve Equation 3 for h_i:

$$h_i = \frac{(3.66)(0.143)}{0.5} = 1.05 \frac{W}{(m^2)(°C)}$$

Now solve Equation 4 to determine the overall heat-transfer coefficient. Using the given value of h_e:

$$\frac{1}{U} = \frac{1}{1.05} + \frac{1}{29.10} = 0.986$$

$$U = 1.014 \frac{W}{(m^2)(°C)}$$

Finally, Equation 7 can be solved:

$$\frac{30 + 25}{70 + 25} = \exp - \left[\frac{(1.014)(\pi)(0.5)(L)}{(151.39)(1{,}991)}\right]$$

$L = 1.04 \times 10^5$ m or 104 km. ■

References

1. Smith, W., "Guidelines set out for pumping heavy crudes," *Oil Gas J.*, May 28, 1979

CORRECTION FACTORS FOR DOUBLE-PIPE HEAT EXCHANGERS

Charts provide a speedy way to determine temperature differences

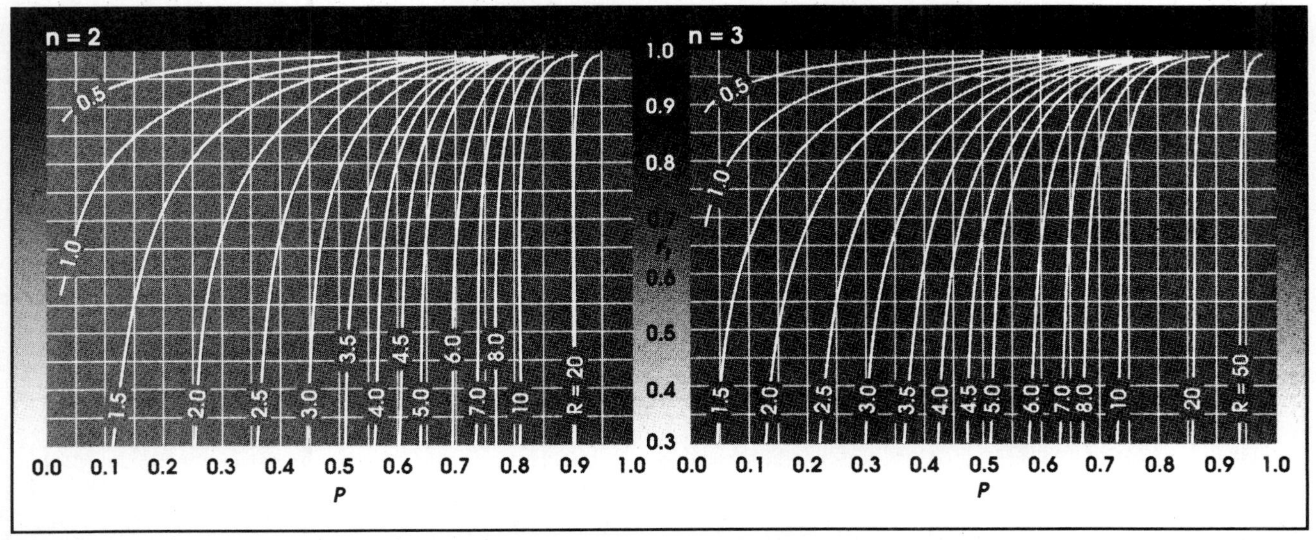

Figures 1 & 2 Temperature-related parameters, P and R, are used to determine the correction factor

JIM STONE

Alejandro Anaya Durand and **Juan Manuel Maldonado**
Instituto Mexicano del Petróleo*

Double-pipe series-parallel arrangements are commonly used for services where small heat duties are exchanged and noticeable differences are found in the mass flowrates of fluids. The true temperature difference, Δt, for the heat-transfer equation, $Q = UA\Delta t$, can be obtained from the logarithmic mean temperature difference, *LMTD*, multiplied by a correction factor, F_t:

$$\Delta t = LMTD \times F_t \qquad (1)$$

where

$$LMTD = \frac{(T_1 - t_2)(T_2 - t_1)}{\ln\left[\frac{(T_1 - t_2)}{(T_2 - t_1)}\right]}$$

There are several charts present here for determining F_t, based on the temperature-related parameters P and R. These parameters apply to a heat exchanger with one hot-stream pipe and two to six parallel cold streams, and vice versa.

Defining parameters

The parameters P and R in these charts for the case of one hot stream and n parallel cold streams are defined as:

$$P = \frac{(T_2 - t_1)}{(T_1 - t_1)} \qquad R = \frac{n(t_2 - t_1)}{(T_1 - T_2)}$$

For the case of one cold stream and n parallel hot streams, the parameters are

$$P = \frac{(T_1 - t_2)}{(T_1 - t_1)} \qquad R = \frac{n(T_1 - T_2)}{(t_2 - t_1)}$$

Then, correlations for F_t were derived from equations of a similar correction factor γ, defined by Kern (*1*):

*Eje Central Lazaro Cardenas No. 152, Apartado Postal 14-805, 07730 Mexico, D.F.

Originally published April 1991

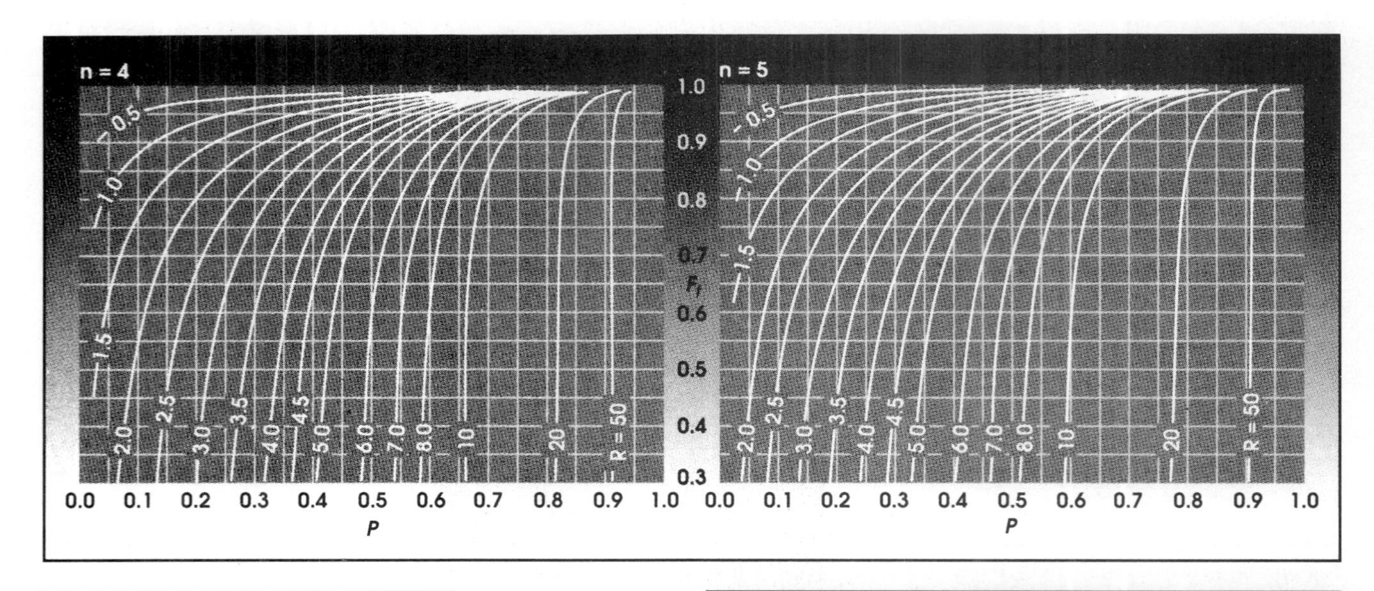

Nomenclature

A	=	heat transfer surface
$LMTD$	=	logarithmic mean temperature difference
F_t	=	$LMTD$ correction factor
P, R	=	dimensional parameters
n	=	number of parallel streams
U	=	overall heat transfer coefficient
T_1, T_2	=	inlet and outlet hot stream temperatures
t_1, t_2	=	inlet and outlet cold stream temperatures
Δt	=	true temperature difference
γ	=	Kern's correction factor

Correction factors for obtaining true temperature difference can be picked right off the charts

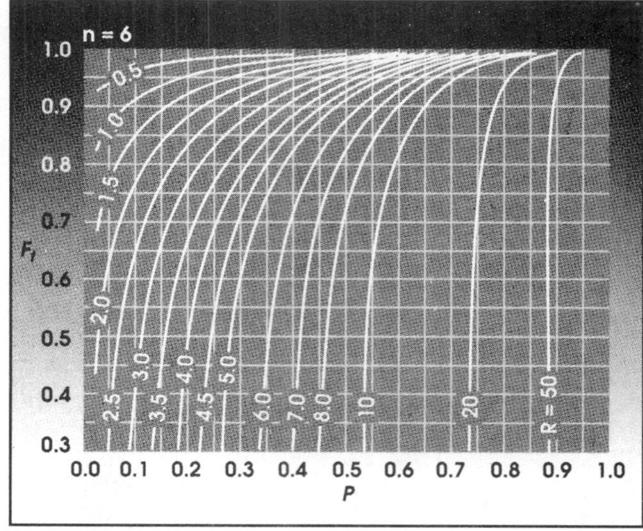

$$\Delta t = \gamma \times (T_1 - t_1) \quad (2)$$

From combining Equations 1 and 2, one obtains:

$$F_t = \gamma \times (T_1 - t_1)/LMTD$$

Although the application of factor γ is simpler, factor F_t is more commonly used to take into account any deviations that might occur with respect to countercurrent flow.

Example 1

A double-pipe heat exchanger cools gasoline, which flows in a series arrangement through the unit, from $T_1 = 212°F$ to $T_2 = 100.4°F$. The gasoline is being cooled by water flowing in five parallel cold streams ranging from $t_1 = 89.6°F$ to $t_2 = 114.8°F$. The problem: Determine the true temperature difference.

$$P = \frac{100.4 - 89.6}{212 - 89.6} = 0.088$$

$$R = \frac{5(114.8 - 89.6)}{212 - 100.4} = 1.129$$

from the chart for $n = 5$, $F_t = 0.87$. Then $LMTD = 39.32°F$, and

$$\Delta t = (39.32)(0.87) = 34.2°F$$

Example 2

A double-pipe exchanger bank operates with the hot fluid in series from $T_1 = 300°F$ to $T_2 = 200°F$, and the cold fluid in six parallel streams from $t_1 = 190°F$ to $t_2 = 220°F$. What is the true temperature difference (problem 6.2 from reference [1])?

$$P = \frac{200 - 190}{300 - 190} = 0.091$$

$$R = \frac{6(220 - 190)}{300 - 200} = 1.8$$

from the chart for $n = 6$, $F_t = 0.79$. Then $LMTD = 33.6°F$, and

$$\Delta t = (33.6)(0.79) = 26.6°F$$

References

1. Kern, D.G, *Process Heat Transfer, International Student Edition* (McGraw-Hill Co., New York, 1950), pp. 116–120. ■

Quick Estimating for THERMAL CONDUCTIVITY

S. R. S. Sastri and K. K. Rao,
Regional Research Laboratory, Bhubaneswar

Accurate values for thermal conductivity — an important engineering property used in heat transfer calculations of liquids — are not as readily available as those for other physical properties. Therefore, it often becomes necessary to use estimated data. A new estimating method combines ease of use with an accuracy that is generally better than existing procedures.

Reid [1,2] and Birkett [3] reviewed many of the available methods for estimating thermal conductivity. They concluded that all of the estimation techniques are empirical, and generally need input data on other physical properties such as heat capacity, density, heat of vaporization and the like. Very often, these physical properties also need to be estimated. While no method was found suitable for the region above normal boiling point, the methods of Latini and coworkers [4–9] and Sato and Riedel [1,3] were recommended for temperatures below the normal boiling point.

However, even these methods have many limitations. For example, the Latini method is not applicable to many liquids containing nitrogen, sulfur or aldehyde groups, and the Sato-Riedel technique gives poor results for the low-molecular-weight hydrocarbons, the branched hydrocarbons, and polar liquids [1].

These limitations indicate the potential value of the new method, based on chemical structure and normal boiling point. The proposed correlation is applicable to a wider range of liquids and covers practically the entire saturated-liquid region except near the critical temperature. At temperatures below the normal boiling point, it is comparable to the recommended methods. At higher temperatures, average and maximum errors are less than 10% and 35%, respectively.

How to select terms

The thermal conductivity of organic liquids, except those containing multiple hydroxy groups, generally decreases as temperature rises. The parameter T_{ref}/T has been used to correlate thermal conductivity at different temperatures [2,10]. In the present study, the normal boiling point, T_B, has been selected as T_{ref}, since it is near the middle of the liquid state and it has been successfully used as a starting point for correlation of many physical properties of liquids, such as the heat of vaporization [11], viscosity [12] surface tension [13] and thermal conductivity itself [1].

In developing the new estimating method, the relationship between thermal conductivity and T_B/T has been established using the data of Miller [14] for 37 organic liquids, covering almost the entire saturated liquid region. From these data, it is found that

$$k = k_B(T_B/T)^{0.5} \text{ for } T \leq T_B \quad (1)$$

and

$$k = k_B(T_B/T)^{1.15} \text{ for } T \geq T_B \quad (2)$$

Further, when thermal conductivity data for members of any homologous series are plotted against their respective T_B/T values, a single line can be drawn to represent the whole range. This linear relationship indicates that all members of the homologous series have the same thermal conductivity, k_B, at their respective normal boiling points.

Combining this observation with the temperature relationships established above, the thermal conductivities of liquids belonging to different groups at their respective normal boiling points have been calculated using available data [14–20]. The contributions of different structural groups to k_B have also been determined from these data, and are given in Table 1. During this study, it was noted that the lower members (1–3 carbon atoms in the molecule) of each series require some correction; these are given in Table 2. In practice, the correction factors are dealt with first, and then the square root value that corresponds to k can be calculated from the sum of the data.

Testing the correlation

From the k_B values estimated from Tables 1 and 2 and the temperature relationships given by Equations 1 and 2, thermal conductivity data are estimated and compared with experimental values for 23 organic liquids, at 1–3 temperature points. The highest relative overestimate is for propane (16%) at 323 K, and the lowest underestimate is for benzene at 293 K (–16%). Overall, the values compare favorably with the alternative estimating techniques.

The new method's reliability over a temperature range was tested in another way. A comparison was made of calculated results with those compiled by Miller et al. [14] over a wide range of temperature conditions, covering almost the entire saturated liquid region.

Analysis of the data show that more than 80% of the errors are less than 15%

NOMENCLATURE

k	Thermal conductivity of liquid at temperature T, W/mK
k_B	k at normal boiling point
k_{cal}	k calculated at T
k_{exp}	k determined experimentally
T	Liquid temperature, K
T_B	Liquid's normal boiling point
T_c	Liquid's critical temperature
T_F	Liquid's freezing point
T_R	Liquid's reduced temperature, T/T_c, dimensionless

Originally published August 1993

and the average error remains below 10% except at T_R of 0.95. This is considered satisfactory, especially as the variations between different sets of experimental data are reported to be under around 5–10% [1-3].

Note that even though the temperature range indicated in the compilation extends to about 0.95 T_C for some liquids, Miller et al. suggest that actual use of the data should be limited to T_R of 0.9 only. Thus the use of any thermal conductivity data beyond T_R of 0.9 should be done with caution.

An advantage the present method has over the others lies in its applicability to liquids with more complex structures, such as those having more than one fundamental group and those containing nitrogen and hydroxyl groups. Typical data for such compounds show that even in such cases the estimated data do not deviate from the general trend for hydrocarbons. Two examples for calculating thermal conductivity by the proposed method are as follows.

Example 1

Calculate the thermal conductivity of propionic acid, CH_3CH_2COOH, at 285 K. The normal boiling point is 414 K [2]. The experimental k value is 0.173 W/mK [1]. From Tables 1 and 2:

$k_B = 0.0545 + 0.0630 + 0.0250$
$= 0.1425$ W/mK

Since $T_B/T > 1$, $k = 0.1425\ (414/285)^{0.5} = 0.172$ W/mK. The error is (0.172 − 0.173)/0.173 × 100, or −0.6%. The corresponding errors for Latini et al and Sato and Riedel are −8.9% and −3.4% respectively [1].

Example 2

Calculate k for chlorobenzene, C_6H_6Cl, at 233 K. The normal boiling point is 404.9 K [2]. The experimental value of k is 0.141 W/mK [1]. Then

$k_B = 5(0.019) + (-0.0375) + 0.0545$
$= 0.112$ W/mK

Since $T_B/T > 1$, $k = 0.112(404.9/233)^{0.5} = 0.148$ W/mK. The error is (0.148 − 0.141)/0.141 × 100, or 5%. ■

Edited by Nicholas Basta

Acknowledgement:
The authors wish to acknowledge the Director, Regional Research Laboratory, Bhubaneswar (India) for his kind permission to publish this paper.

Table 1: GROUP CONTRIBUTIONS TO THERMAL CONDUCTIVITY

HYDROCARBONS	Value, W/mK
Nonaromatic	
$-CH_3$	0.0545
$-CH_2-$	0.0000
$-\overset{\vert}{\underset{\vert}{C}}H-$	-0.0585
$-\overset{\vert}{\underset{\vert}{C}}-$	-0.1275
$=CH_2$	0.0630
$=CH-$	0.0020
$=\underset{\vert}{C}-$	-0.0630
Alicyclic ring (e.g., cyclopropane)	0.1130
Aromatic	
$=C<^{H}$	0.0190
$=C<$	-0.0375
$=C<$ (fused, such as in a naphthalene ring)	-0.0135
HALOGENS	
$-F$	0.0545
$-Cl$	0.0545
$-Br$	0.0440
$-I$	0.0250
OXY GROUPS	
$-O-$	0.0085
$O<_{ring}$	0.0335
$-OH$	0.0730
$-CO-$	0.0230
$-CHO$	0.0670
$-COOH$	0.0630
$-COO-$	0.0125
NITROGEN GROUPS	
$-NH_2$	0.1090
$-NH-$	0.0250
$-\underset{\vert}{N}-$	-0.0650
OTHER GROUPS	
$-S-$	0.0020
$-NO_2$	0.0730
$-H$ (e.g., methane, formate)	0.0670
ring (e.g., ethylene oxide)	0.1130

Table 2: CORRECTIONS FOR k_B FOR LOWER ALIPHATICS

COMPOUND	CORRECTION, W/mK
Hydrocarbons containing	
2 carbon atoms	0.0670
3 carbon atoms	0.0250
One methyl (CH_3) group attached to nonhydrocarbons (e.g., methyl chloride)	0.0670
Two methyl groups attached to nonhydrocarbons (e.g., acetone)	0.0250
One methylene ($-CH_2-$) group attached to nonhydrocarbon (e.g., methylene chloride)	0.0250
One ethyl group (CH_3CH_2-) attached *only* to oxygen-containing groups (e.g., ethanol, nitroethane)	0.0250

References

1. Reid, R. C., et al., *The Properties of Gases and Liquids, 4th ed.*, (New York: McGraw-Hill, 1987).
2. Reid, R. C., et al., *The Properties of Gases and Liquids, 3rd. ed.*, (New York: McGraw-Hill, 1977).
3. Birkett, J. D., "Thermal Conductivity," in Lyman, W. J., et al., *Handbook of Chemical Property Estimation Methods*, (New York: McGraw-Hill, 1982).
4. Baroncini, D., Di Filippo, P., and Latini, G., "Comparison between predicted and experimental thermal conductivity values ...", presented at the workshop on thermal conductivity measurement, IMEKO, Budapest, March 14–16, 1983.
5. Baroncini, D., Di Filippo, P., and Latini, G., *Intern. J. Refrig.*, 6 (1):60, 1983.
6. Baroncini, D., Di Filippo, P., and Latini, G., *Intern. J. Thermophys.*, 1 (2):159, 1980.
7. Baroncini, D., Di Filippo, P., and Latini, G., *Intern. J. Thermophys.*, 2 (1):21, 1981.
8. Baroncini, D., Di Filippo, P., and Latini, G., *Thermal Conductivity 1981 Proceedings* (New York: Plenum Publishing, 1983), p. 285.
9. Baroncini, D., Latini, G., and Pierpaoli, G., *Intern. J. Thermophys.*, 5 (4):387, 1984.
10. Narasimhan, K. S., et al., *Chem. Eng.*, 82 (8):83, 1975
11. Sastri, S. R. S., Ramana Rao, M. V., et al., *Brit. Chem. Eng.*, 14 (7):957, 1969.
12. Sastri, S. R. S., and Rao, K. K., *Chem. Eng. J.*, 50 (1):9-25, 1992.
13. Ramana Rao, M. V., et al., *Hydrocarbon Processing*, 47 (1):151, 1968.
14. Miller, J. W., Jr., et al., *Chem. Eng.*, 83 (23):133, 1976.
15. Riedel, L., *Chem. Ing. Tech.*, 21, 349, 1949; 23, (59), 321, 465, 1951.
16. Sadiadis, B. C., and Coates, J., *AIChE J.*, 1, 275, 1955.
17. Sadiadis, B. C., and Coates, J., *AIChE J.*, 3, 121, 1957.
18. Dean, J. A., ed., *Lange's Handbook of Chemistry*, (New York: McGraw-Hill, 1972).
19. Weast, R. C., ed., *CRC Handbook of Chemistry and Physics, 60th ed.*, (Boca Raton, FL: CRC Press, 1979).
20. Gray, D. E., ed., *Am. Inst. Physics Handbook*, (New York: McGraw-Hill, 1963).
21. DiGullo, R. M., et al., *J. Chem. Eng. Data*, 37, 242, 1992.

The authors

S. R. S. Sastri is an assistant director at the Regional Research Laboratory (RRL), Council of Scientific and Industrial Research, Bhubaneswar 751013, India. Before joining the present organization, in 1968, he worked at the National Chemical Laboratory (Pune). His research interests include process design and development, mineral properties, and the properties of liquids. He has published more than 50 technical papers on these subjects. He is a life member of the Indian Institute of Chemical Engineers, the Indian Institute of Metals, and the Indian Institute of Mineral Engineers. He received his B.S.Ch.E. from the Andhra University, Waltair, in 1962.

K. K. Rao is an assistant director and Leader of the Computer Applications Group at RRL. Before joining RRL in 1978, he was at the Indian Institute of Technology (Madras). He has been involved in technology forecasting, planning and physical-properties research, having published more than 20 technical papers. He is a member of the Indian Institute of Chemical Engineers, the Indian Institute of Metals, the Computer Soc. of India, and the Thermal Analysis Soc. of India. He earned a B.S.Ch.E. at the Annamalai University (Annamalai Nagar) in 1972, and a master's from Mysore University in 1974.

HOW FIN CONFIGURATION AFFECTS HEAT TRANSFER

Don't assume more surface area equals more energy transfer

V. Ganapathy,
ABCO Industries, Inc.*

Finned tubes are used extensively in heat-recovery boilers, fired heaters and related heat-transfer equipment. Several variables are involved in their selection, including size, cost and allowable gas pressure drop. One aspect that many designers overlook, however, is the effect of fin configuration and fin density on surface-area requirements. They often fall into the trap of selecting a particular design simply because it has more surface area. While this could be true with bare-tube heat-transfer equipment for comparable gas velocities, it is not always so with finned tubes.

The better approach is to examine the product of overall heat-transfer coefficient (U) and total surface area (A) and *not* the surface area alone. This is because the basic equation for energy transfer [*1, 2, 3*] takes the form:

$$Q = UA(\Delta T) \tag{1}$$

It will be shown later that a large fin density or height can result in lower U value, and the product UA may be comparable with another design with a lower fin density or height, even though the surface areas could be significantly different.

Also, finned tubes are effective when the ratio of external to internal heat transfer coefficients (h_g to h_i) is low. For example, in boiler tubes, the (internal) tube-side coefficient can be in the range of 2,000 to 2,500 Btu/ft²h°F, while the (external) gas-side coefficient can be in the range of 10–20 Btu/ft²h°F. Finned tubes are justified in such cases, provided fouling or slagging is

Effect of fin configuration on heat transfer, pressure drop

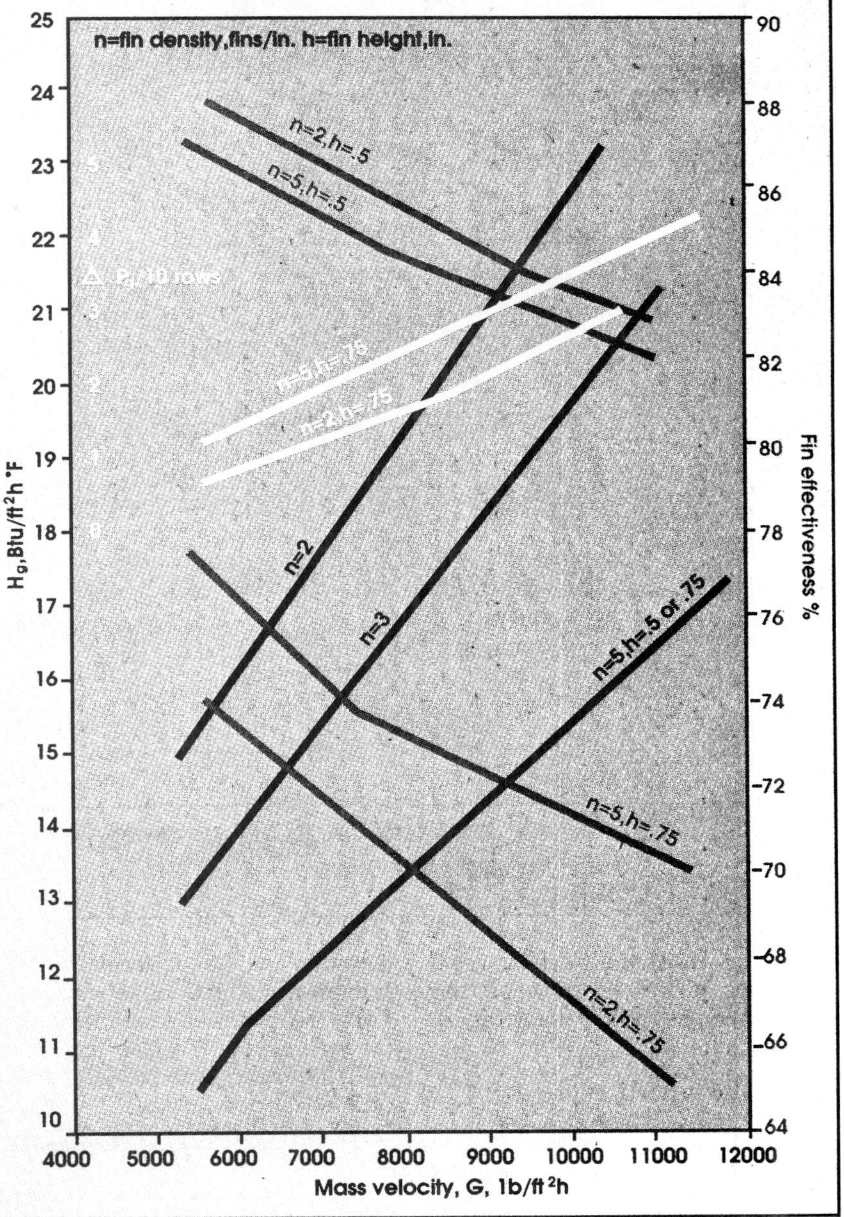

n=fin density,fins/in. h=fin height,in.

*P.O. Box 268, Abilene, TX 79604

Originally published March 1990

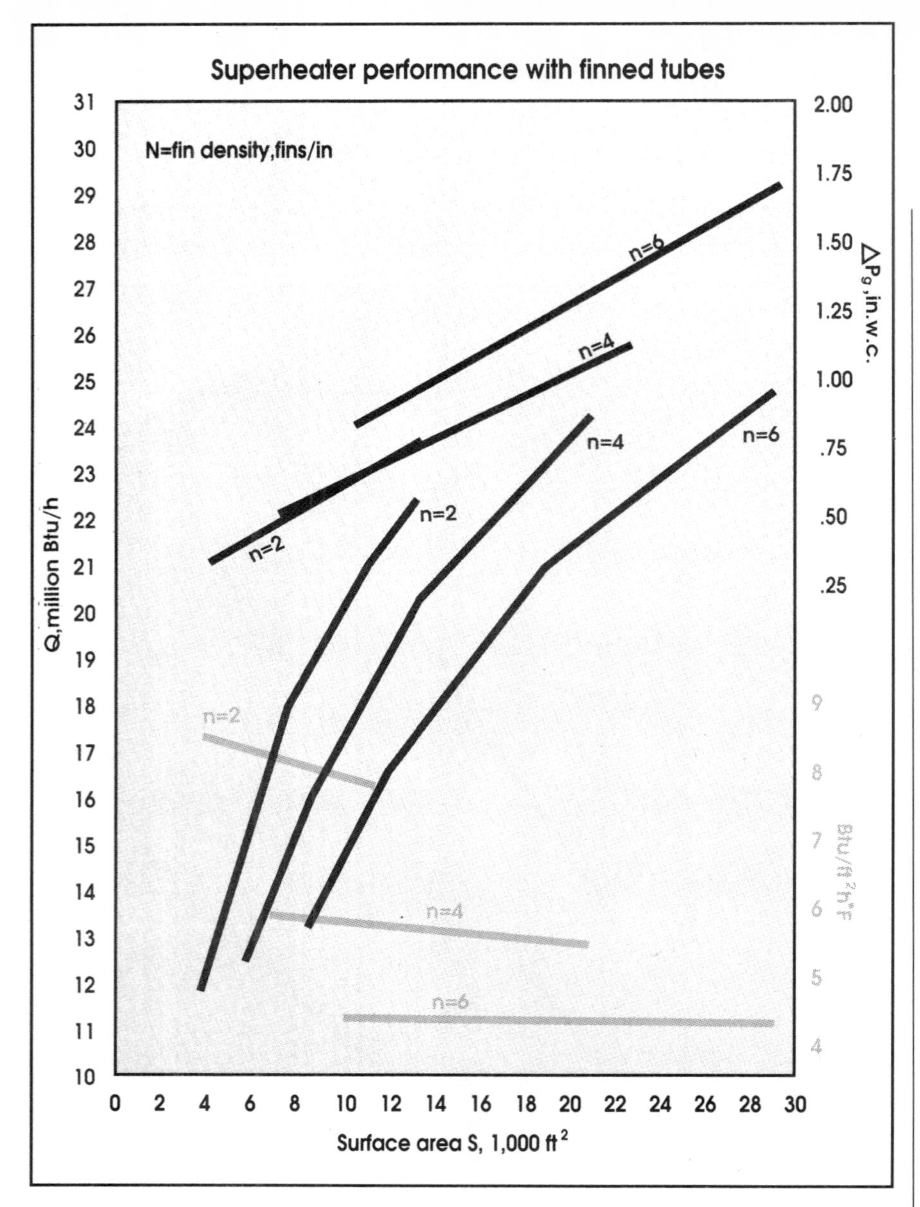

not a concern. As the tube-side coefficient drops, so does the impact of using fins. This can be seen by analyzing the expression for U [1]:

$$1/U = 1/h_i(A_T/A_i) + ff_i + ff_o + (A_T d/A_w 24K_m) \times (\ln d/d_i) + 1/\eta h_g \quad (2)$$

The gas-side coefficient, h_g, may be obtained from correlations available in manufacturers literature. Figure 1 shows the results of calculations using ESCOA data [4]. It can be seen that as the fin density increases, h_g decreases, resulting in a lower U. Also, the gas pressure drop increases.

Simplifying and rewriting Equation 2 for U_i (the overall heat-transfer coefficient based on tube inner diameter), results in:

$$1/U_i = 1/h_i + (A_i/A_T)/\eta h_g \quad (3)$$

One can compute U_i for different fin densities and for different h_i values, and draw the following conclusions:

- As the tube-side coefficient drops, the ratio of the U_i values (at 5 fins/in. vs. 2) decreases. With an h_i of 2,000, the ratio is 1.74; at h_i = 20, it is 1.11. Therefore, as h_i decreases, the benefit of adding more external surface is not attractive. At h_i = 20, one can have 2.325 times the surface area but only 1.11 times the increase in U_i
- The higher the fin density or height, the greater the wall temperature of the tubes. A simple calculation will show this. Let the gas temperature be 900°F and the fluid temperature 600°F. With a fin density, n, of 2, U_i = 39000171eat flux q_i = (900-600) × 39.28 = 11,784 Btu/ft²h. The drop across the tube-side film (neglecting metal resistance and fouling factors) is 11,784/100 = 118°F. The tube-wall temperature is then 600°F + 118°F = 718°F. With n = 5, a similar calculation shows that the wall temperature is 600°F + 161°F = 761°F. Similar results can be shown with variations in height
- The ratio of gas pressure drop (after adjusting for U_i values and the consequent reduction in number of rows) shows that as h_i drops, the ratio between the 5- and 2-fin/in. cases increases. It is 1.3 for h_i = 100 and 1.02 for h_i = 2,000

These calculations lead to one simple fact: A large ratio of external to internal surface area results in lower thermal performance, higher tube-wall temperatures, and larger gas pressure drop. Also, it is not prudent to use large external-surface-area configurations (say, 5 or 6 fins/in.) in superheaters or air heaters (air heaters have low h_i values). These configurations may be justified in the cases of economizers or evaporators.

Case in point

The following example shows the effects of fin configuration on the performance of a steam superheater. Process conditions are:

gas flow = 550,000 lb/h
gas inlet T = 1,000°F

Nomenclature

U = heat transfer coefficient, Btu/ft²h°F
h_i = tube-side (internal) U
h_g = gas-side (external) U
A_T = total heat-exchange surface area, ft²
A_i = tube surface area, internal, ft²
A_w = average tube surface area, ft²/ft
d = tube outside diameter, in.; d_i, inside diameter, in.
ff_i, ff_o = inside and outside fouling factors, ft²h°F/Btu
K_m = metal thermal conductivity, Btu/ft(h)°F
η = fin effectiveness
T = temperature, °F
Q = duty, Btu/h
q_i = tube-side heat flux, Btu/ft²h

gas composition (vol%) = 3.5% CO_2; 10.2% H_2O; 73.6% N_2; and 12.7% O_2

steam flow = 100,000 lb/h at 640 psig, saturated, at inlet

2-in. outer dia., 0.120-in. wall thickness tubes with a square pitch of 4.0 in., 27 tubes per row, 27 streams (carrying steam), and a length of 25 ft.

Vary the fin density from 2 to 6 fins/in.; use a fin height of 0.75 in. and a thickness of 0.05 in.

The gas flow, velocity and composition all affect the value of h_g, which in turn affects the value of U. (See Reference 5 for a method for calculating h_g.) The results of the calculations are shown in Figure 2.

As the fin density increases from 2 to 6 fins/in., U decreases significantly. For the 2-fin/in. case, it is around 8 Btu/ft²h°F, while for the 4-fin/in. case, it is 4.3. The gas pressure drop also increases with fin density. For gas turbine boilers, the increase in gas pressure drop would reduce electrical power output.

In the case of an 18-million-Btu/h superheater, it can be seen that the surface area with a 2-fin/in. design is 7,400 ft², while it is 14,000 ft² for the 6-fins/in. case. The gas pressure drop increases from 0.5 in. w.c. for 2 fins/in., to 0.95 in. w.c. for 6 fins/in.

Additional studies could be performed by varying the fin height. However, the basic fact to be kept in mind is that more surface area with finned tubes does not *automatically* imply more energy transfer. On the contrary, it could mean problems by way of increased gas pressure drop, or higher tube or fin-tip temperatures.

Are there reasons to use higher fin surface area even when it is not efficient? Yes — it could result in fewer rows, thereby saving some manufacturing costs in some cases. ■

References

1. Ganapathy, V., *Applied Heat Transfer*, (Pennwell Books: Tulsa, Okla., 1982) pp 454–476.
2. *ibid.*, "Charts help evaluate finned tube alternatives," *Oil and Gas J.*, Dec. 3, 1979, pp. 74–77.
3. *ibid.*, "Charts simplify spiral finned-tube calculations," *Chem. Eng.*, April 25, 1977, pp. 117–121.
4. *ESCOA Fin Tube Manual*, pub. by ESCOA Corp., Tulsa, Okla.
5. Ganapathy, V., *Applied Heat Transfer*, *op. cit.*, p. 228–245.

P. Ramakrishna, Tata Tea Ltd.

HOW OFTEN TO CLEAN AN EVAPORATOR

Calculations are based on the easily measured rates of vapor condensation

STEVE HAIMOWITZ

Widely used for concentrating a solution made up of a nonvolatile solute and a volatile solvent, evaporators find applications in a variety of jobs, such as the processing of sugar, dye, fruit juice, pulp and paper, and caustic soda. During operation of the evaporator, some solutions deposit scale on the heating surfaces of the boiling tubes, resulting in a drop in heat-transfer efficiency. To revamp performance to the design level, the evaporators must be shut down and the tubes cleaned.

A single-effect evaporator can be wasteful of available thermal energy if the vapor from the boiling liquid is condensed and discarded. A multiple-effect unit, on the other hand, uses the heating value of the vapor by feeding it into the steam chest of a second evaporator. The vapor from the second unit is then reused in a third evaporator, and so on. A number of these units or "effects" put together in series allow repeated use of the vapor between the steam supply and the condenser, increasing the evaporation per unit mass of steam.

Both single- and multiple-effect evaporators require periodic cleaning to remove the scale from the heat-transfer surfaces, and thus maximize their thermal efficiency and capacity. In particular, when the amassed scale is hard and insoluble, the cleaning may be difficult and expensive. Furthermore,

Originally published July 1991

frequent changes in the process liquor and its characteristics may require appropriate modification in the cleaning cycles. As a result, the engineer needs some quick method for determining the optimum cleaning schedule under changing production and processing requirements.

The mathematical formulas developed below for easy calculation of the optimum cleaning cycles of multiple-effect evaporators are based on the assumption that the amount of scale buildup is proportional to the total amount of heat transferred to the surface since the start of the operation [5]. Elaborate graphical procedures for such computation are available elsewhere [1 – 4]. Also presumed here is that the heat economy — the ratio of heat utilized to the heat supplied — is identical for each effect. The special-case computation for a single-effect unit readily follows as long as the underlying assumptions hold true.

Under constant operating conditions, one can define the overall heat transfer coefficient for each effect as follows

$$U_i = 1/[(1/U_{0,i}) + a_i\, Q_i] \qquad (1)$$

Assuming identical surface area for heat transfer in each effect; in other words, $A_i = A$, the amount of heat transferred in time t, for the ith effect, is given by

$$dQ_i/dt = U_i\, A\, \Delta T_i \qquad (2)$$

Eliminating Q_i from Equations 1 and 2, one gets

$$(1/U^2_i) - (1/U^2_{0,i}) = 2\, a_i\, A\, \Delta T_i\, t \qquad (3)$$

Considering W, instead of U, for each effect, it follows

$$(1/W^2_i) - (1/W^2_{0,i}) = (2\, a_i\, \lambda^2_i/[A\, \Delta T_i])\, t \qquad (4)$$

It is clear from Equations 3 and 4 that a plot of $1/U^2_i$ or $1/W^2_i$ against t will be a straight line.

Combining Equations 1 and 2 gives

$$dQ_i/dt = A\, \Delta T_i/[(1/U_{0,i}) + a_i\, Q_i] \qquad (5)$$

Rearranging and integrating, one obtains

$$t = Q_i/[A\, \Delta T_i\, U_{0,i}] + a_i\, Q^2_i/[2\, A\, \Delta T_i] \qquad (6)$$

The assumption of identical heat

NOMENCLATURE

Symbols

A	Heat transfer area, m²
a	Constant
b	Heat economy
C	Intercept of the straight line representing Equation 4
M	Slope of the straight line for Equation 4
Q	Heat transferred in time t, kJ
q	Rate of heat transfer, kJ/s
ΔT	Temperature difference, K
U	Overall heat-transfer coefficient at any time t, kW/m²K
W	Condensate rate at any time t, kg/s
t	Time, s
λ	Latent heat of vaporization, kJ/kg

Subscripts

avg	Averaged over operating, as well as cleaning, time
c	Cleaning
i	Individual effect in n-effect evaporator
n	Number of effects in multi-effect evaporator
total	Total of the quantity in time t, summed over all the effects
0	At the start of operation
1, 2, 3	At points 1, 2, 3, . . . after start

Superscript

*	At optimum

economy for each effect leads to

$$Q_n = b\, Q_{n-1} = b^2\, Q_{n-2} = \ldots = b^{n-1} Q_1$$

Thus, for any effect i,

$$Q_n = b^{n-i} Q_i \qquad (7)$$

By summing individual effects, one gets the total amount of heat transferred

$$Q_{total} = \sum_{n=1}^{n} b^{\,n-i} Q_i = [Q_i/b^{i-1}] \sum_{n=1}^{n} b^{n-1} = [Q_i/b^{i-1}] \times [1 - b^n]/[1 - b] \qquad (8)$$

Therefore, the average rate of heat transfer is given by

$$q_{avg} = Q_{total}/[t + t_c] = (1/b^{i-1}) \times ([1 - b^n]/[1 - b]) \times Q_i/([Q_i/(A\, \Delta T_i\, U_{0,i})] + [a_i\, Q^2_i/(2\, A\, \Delta T_i)] + t_c) \qquad (9)$$

Knowing the cleaning time, t_c, which is normally constant, q_{avg} is plotted against t to get t^* at which q_{avg} is maximum. Thus, t^* denotes the "high time" to initiate the cleaning cycle, because after this point in time, there will be a continuing decrease in q_{avg} as a result of scale buildup. Setting $dq_{avg}/dQ_i = 0$, and checking d^2q_{avg}/dQ^2_i at Q^*_i, it can be shown that

$$Q^*_i = [2\, A\, \Delta T_i\, t_c/a_i]^{1/2} \qquad (10)$$

Substituting Q^*_c in Equation 6, and rearranging, one gets

$$t^*/t_c = 1 + 2\,[1/(U^2_{0,i} \times 2\, a_i\, A\, \Delta T_i \times t_c)]^{1/2} = 1 + 2\,[(1/W^2_{0,i}) \times (A\, \Delta T_i/[2\, \lambda^2_i\, a_i]) \times (1/t_c)]^{1/2} = 1 + 2\,[(C_i/M_i) \times (1/t_c)]^{1/2} \qquad (11)$$

where M and C are the slope and intercept of the straight line representing Equation 4, and can be calculated easily from measuring condensate rates as follows:

$$M_i = [(1/W^2_{2,i}) - (1/W^2_{1,i})]/(t_2 - t_1) = [\Delta(1/W^2_i)]/\Delta t \qquad (12)$$

$$C_i = (1/W^2_{0,i}) = (1/W^2_{2,i}) - M_i\, t_2 = (1/W^2_{1,i}) - M_i\, t_1 \qquad (13)$$

Keep in mind that the proposed method is applicable only when the amount of scale formed is proportional to Q_i in each effect of the multiple-effect evaporator. This condition is valid as long as the measured condensate rates abide by the following rule derived from Equation 4:

$$(1/W^2_{2,i}) - (1/W^2_{1,i}) = (1/W^2_{3,i}) - (1/W^2_{2,i}) \qquad (14)$$

Also, the heat economy should be equal in all the effects so that the optimum cleaning schedule of any one of the effects is the optimum for the multiple-effect evaporator. This is possible only when each effect has the same value of C_i/M_i. This condition can be checked easily from the measurement of condensate rates from each of the effects, using Equations 12 and 13.

The primary advantage of this proposed "quick method" of estimating the optimum cleaning cycles is that it requires easily measurable conden-

DATA FROM A TRIPLE-EFFECT EVAPORATOR USED IN THE ILLUSTRATIVE EXAMPLE

t min	Δt	W kg/min	$1/W$	$1/W^2$	$\Delta(1/W^2)$	M	C	C/M
First effect:								
50	—	1.0540	0.9489	0.90	—	—	—	—
100	50	0.8333	1.2000	1.44	0.54	0.0108	0.36	33.33
150	50	0.7106	1.4073	1.98	0.54	0.0108	0.36	33.33
Second effect:								
50	—	1.0050	0.9950	0.990	—	—	—	—
100	50	0.7945	1.2586	1.584	0.594	0.01188	0.396	33.33
150	50	0.6776	1.4758	2.178	0.594	0.01188	0.396	33.33
Third effect:								
50	—	0.9622	1.0392	1.080	—	—	—	—
100	50	0.7607	1.3145	1.728	0.648	0.01296	0.432	33.33
150	50	0.6489	1.5414	2.376	0.648	0.01296	0.432	33.33

sate rates from the various effects, provided the heat economy is equal in each effect. Another plus for this technique is that the condition for its applicability can be tested from the same condensate rates. It can be inferred from Equation 11 that the minimum time of operation for a multiple-effect evaporator is equal to the time required for its cleaning.

Illustrative example

The condensates collected over 50-min intervals after the start of a triple-effect evaporator concentrating a scaling liquor are 1.0540, 0.8333, and 0.7106 kg in 1.0 min in the first effect. The condensate rates are 1.0050, 0.7945, and 0.6776 kg/min in the second, and 0.9622, 0.7607, and 0.6489 kg/min in the third effect, respectively. If the time required to remove the scale from the evaporator is 100 min, estimate the optimum time of operation to get the maximum overall-heat-transfer.

First, check for applicability. From the Table of data, it can be verified that in each effect, i,

$$(1/W^2_{2,i}) - (1/W^2_{1,i}) = (1/W^2_{3,i}) - (1/W^2_{2,i})$$

Similarly, it can be shown that the value of C/M is the same in all three effects, establishing that the method is indeed applicable.

The quick calculation. The optimum operating time can be computed as follows

$$C/M = 33.33$$
$$C/(M \times t_c) = 0.3333$$
$$[C/(M \times t_c)]^{1/2} = 0.5773$$
$$t^*/t_c = 2 \times 0.5773 + 1 = 2.1546$$
$$t^* = 2.1546 \times 100.0 = 215.46 \text{ min} = 3.59 \text{ h}$$

Edited by Gulam Samdani

References

1. Badger, W. L., and Othmer, M. F., Studies in evaporator design – VIII; optimum cycles for liquids which form scale, *Trans. Am. Inst. Chem. Eng.*, 19, pp. 3 – 11, 1925.
2. Epstein, N., Optimum evaporator cycles with scale formation, Canadian J. Chem. Eng., 52, pp. 659 – 661, 1979.
3. Epstein, N., On minimum fouling of heat transfer surfaces, in "Heat Exchanger Source Book," (ed. by Palen, J. W.), Hemisphere Publishing Co., pp. 699 – 705, 1986.
4. McCabe, W. L., Economic side of evaporator scale formation, Chem. Met. Eng., 33, p. 86, 1926.
5. McCabe, W. L., and Robinson, C. S., Evaporator scale formation, Ind. Eng. Chem., 16, p. 478, 1924.

The author

P. Ramakrishna is a senior manager at Tata Tea Ltd., Munnar-685612, Kerala, India. He is in charge of new projects in the processing of tea and other agricultural products, such as coffee, spices and medicinals. Prior to joining Tata Tea in 1990, he worked at Food Technological Research Institute, Mysore, India, for 15 years as a scientist in process engineering and plant design. He specializes in process scaleup, and design of food-industry equipment. He holds a master's degree in chemical engineering, and is a life member of the Indian Institute of Chemical Engineers and the Assn. of Food Scientists and Technologists, India. ■

Section II

Selection and Design of Heat-Transfer Equipment

CUSTOMIZING PAYS OFF IN STEAM GENERATORS

Energy and environmental concerns dictate special boiler designs

V. Ganapathy
ABCO Industries

Packaged steam generators are the workhorses of chemical process plants, power plants and cogeneration systems. They are available as oil- or gas-fired models, and are used to generate either high-pressure superheated steam (400 to 1,200 psig, at 500 to 900°F) or saturated steam at low pressures (100 to 300 psig).

The maximum steam capacity for boilers assembled completely in shops is limited to about 250,000 lb/h because of restrictions in shipping size. With some field work, however, steam generators up to 400,000 lb/h capacity can be built.

In today's emission- and efficiency-conscious environment, steam generators have to be custom designed. Gone are the days when a boiler supplier – or for that matter an end user – could look up a model number from a list of standard sizes and select one for a particular need.

Customized designs are preferable because equipment buyers have to compromise to some degree on a boiler's specifications, such as emissions, efficiency and operating costs, if they select an off-the-shelf system. Custom designs are also needed because environmental regulations vary from area to area. For example, a boiler designed for operation in Louisiana may not be suitable for a site in California, which has stricter environmental regulations, even though all the other steam parameters, such as flow, pressure and temperature, may be identical.

Basic parameters

Thus, before selecting a system, it is desirable to know the features of oil- and gas-fired steam generators, and the important variables that influence their selection, design and performance (Table). It is imperative that all of these data are supplied to the boiler supplier so that the engineers may come up with the right design.

Duty. The duty (or the energy absorbed by steam) of a boiler can be obtained if the flow, pressure and temperature of the steam and water, and the amount of blowdown, are given. The data on blowdown is often not provided to vendors, as sufficient information is not available in the early stages of a project. This can lead to errors in the boiler's duty of as much as 2 to 5%. However, boiler designers can estimate the blowdown quantity if the feedwater analysis is provided.

Steam temperature. Specifiers of generators should ask the steam consumer if steam temperature should be controlled or whether it can "float." Often, small variations in steam temperature can be tolerated, which eliminates the need for the added expense of a steam-temperature control system. Also, if the degree of superheat is small, say 50°F, a control system is not recommended, since the end user only wants to be assured that superheated steam and not saturated steam will be available. Further, it is difficult to control the temperature when the degree of superheat is so small.

In case the superheated steam is to be controlled, the user should state the load range over which it should be controlled. Typically, this is between 60 and 100%, but one can encounter cases where the load range is 10–100%.

The steam pressure drop through the superheater varies as the square of the flow. Therefore, at 10% load, the pressure drop will be 1/100th smaller.

In order to maintain a reasonable flow distribution through the tubes if, say, a 2- to 5-psi drop is chosen, then at 100% load, the pressure drop will be 200–500 psi. This is extremely high considering the pressure levels of the systems (600–1,200 psig).

Further, gas flow distribution across

Originally published January 1995

the tubes will be non-uniform, and it will be difficult to predict heat transfer at such small load, because of the low Reynold's number. Thus, in general, generator performance cannot be predicted with good accuracy below 50% load.

Steam temperatures are often controlled using a desuperheater or attemperator. The latter device controls temperature by injecting demineralized water between the two stages of a superheater. If demineralized water is not available and softened water is used, solids added to the steam can deposit inside superheater tubes or in steam and gas turbines during the expansion process, causing problems. In such situations, other methods of steam temperature control, such as bypassing of the superheater, should be considered.

Steam purity. If the steam generated is to be used in a steam or gas turbine, then the steam purity should be high, on the order of 0.02 to 0.2 ppm of total dissolved solids. The recommendations of the gas- and steam-turbine supplier should be considered.

Cooling, heating or process applications employing saturated steam usually tolerate up to 1 ppm of total dissolved solids. Also, the steam's purity dictates the type of drum internals to be used. High-pressure superheated steam (above 600 psig) requires a combination of cyclones and V-bank Chevron internals, while Chevrons or demister pads may be adequate for low-pressure saturated steam boilers.

Emissions vs. boiler design

The single most important factor influencing the design of steam generators today is the emission of pollutants, mainly nitrogen oxides and carbon monoxide. Each country, state or county has a regulation that specifies the maximum amount of such pollutants that can be be tolerated. Several methods are available for controlling these pollutants:

- Fluegas recirculation
- Excess air manipulation
- Air- and fuel-staging during combustion
- Steam injection into the burner
- Selective catalytic reduction (SCR)
- Selective non-catalytic reduction

The first four methods generally decrease NOx levels by 30 to 50%, while the latter two can cut it by 80 to 90%. However, the latter methods are expensive, and are used when very low levels of NOx and CO, on the order of 6 to 10 ppm, are desired. For more on these techniques, see Reference 1.

In practice, burner manufacturers model the combustion process for a particular fuel, and suggest the amount of excess air and fluegas recirculation that has to be used to meet desired the levels of CO and NOx. For example, the strategy for a fuel that has a higher hydrogen content, with its higher flame temperature, will be different than that for a low-Btu fuel.

Fuel oils that have fuel-bound nitrogen have to be dealt with differently than gases, since fluegas recirculation has less influence on fuel-bound NOx. Therefore, excess air and fluegas recirculation, which affect the quantity of fluegas through the boiler, may vary from site to site, even though the steam parameters may be identical.

Excess air levels of 15% and fluegas recirculation rates of 25% are not uncommon in packaged steam generators. The mass flow through these units is much higher (by about 1.37 times) than boilers designed decades ago, when 5% excess air was typical. The larger fluegas quantity affects the thermal performance, and increases the gas pressure drop through the convection section, superheater, economizer and duct work, which, in turn, impacts fan size and operating costs.

A 1-in. increase in gas pressure drop through the boiler in a 100,000-lb/h unit adds 5–7 kW in fan power consumption. This translates to nearly $2,000 per year. A 40°F-change in the exit gas temperature from a boiler results in about a 1% change in efficiency. At $3/million Btu, this is equivalent to $25,000 per year.

Staging combustion air can also reduce NOx. However, this process changes the flame shape and increases its length, requiring the furnace to be designed to avoid flame impingment.

SCRs operate efficiently over a particular gas temperature range, usually 600 to 750°F. This requires that some means of gas temperature control be used to achieve the temperature profile at varying loads. Internal or external gas-bypass systems, such as those shown in Figure 1, are used for this purpose.

As the boiler duty decreases, the gas temperature entering the SCR decreases [2,3]. Therefore, more fluegas

BASIC PARAMETERS FOR DESIGN OF STEAM GENERATORS

1. STEAM
- Pressure
- Temperature
- Feedwater temperature
- Blowdown
- Quantity of superheated steam and saturated steam: Net steam from the boiler
- Control of steam temperature (If temperature is to be controlled, determine the load range)
- Steam purity
- Feedwater, make-up water analysis

2. FUEL
- Fuels to be used (gas or oil)
- Complete fuel analysis (including such constituents as nitrogen, sulfur and ash)
- Pressure at which each fuel is available

3. EMISSIONS REQUIREMENTS
- Levels of carbon monoxide, volatile organic compunds, nitrogen oxides for each fuel (in lb/million Btu or ppm)

4. SITE CONDITIONS
- Ambient temperature range
- Elevation
- Wind velocity, earthquake considerations
- Electrical power availability (voltage and frequency)
- Space and tranportation limitations

5. COSTS
- Fuel cost
- Electricity cost
- Times of operation (% load)

6. PURPOSE
- Standby boiler
- Steady-state operation

is bypassed around the convection section to increase the gas temperature at the SCR. If it is not possible to locate the SCR at the exit of the convection section, the economizer has to be split up into two sections and the SCR placed in between.

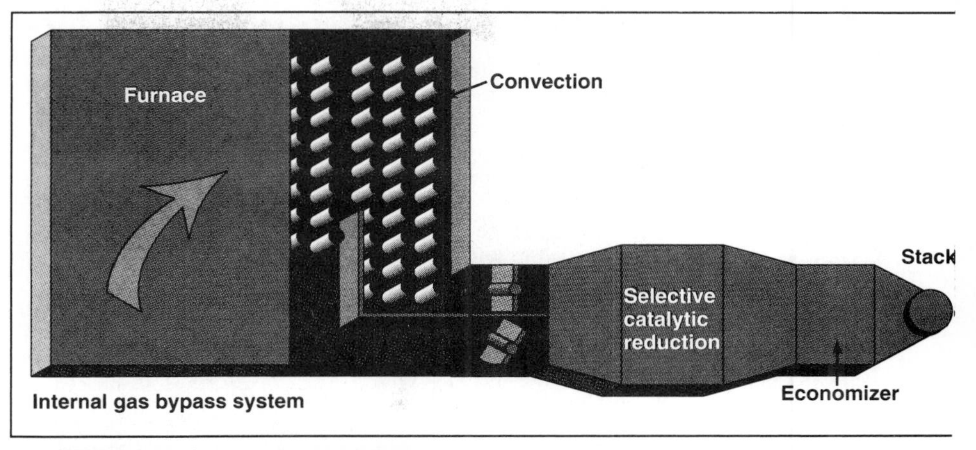

FIGURE 1. Internal or external gas-bypass systems are used to control the temperature to selective catalytic reduction systems

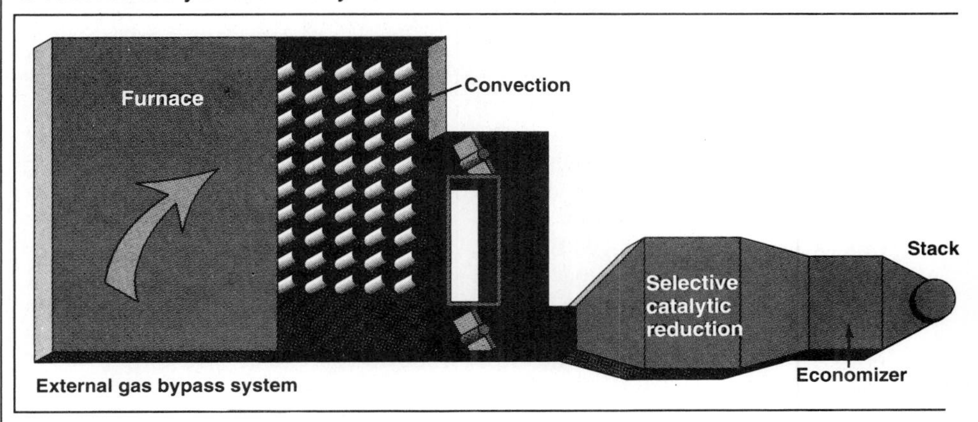

Furnace design

A boiler's furnace is its heart. Besides providing heat at its rated duty, a furnace should ensure that combustion is complete before the gases leave the furnace (i.e., the O_2 levels are below a given threshold level). The desired emission levels should be attained not only by proper combustion, but also by making sure that combustion products don't leak or bypass into the convection section across the partition wall.

The furnace envelope should be designed so that that the flame does not impinge on the side walls, roof or floor at any load. This is of concern when different fuels are fired, or when multiple burners are used. Also, the heat release rate – the net heat input divided by the effective, projected radiant surface – should be such that the maximum heat flux in the furnace does not cause departure from nucleate boiling (DNB) conditions.

In conventional furnaces, the side walls are water cooled, while the front and rear walls are lined with refractories. A completely water-cooled furnace that has membrane walls offers several advantages since it lowers heat-release rates (for the same furnace volume) and heat flux.

Since nearly 80–90% of NOx is formed in the first 10–15% of the flame's length, conventional refractories reradiate energy back to the flame, thus offering no help in reducing the local flame temperature, which directly impacts NOx formation. Water-cooled designs, on the other hand, cool the flame better and, thus, provide an environment for reducing NOx levels.

Water-cooled furnaces also eliminate maintenance problems associated with refractories, and startup rates need not be constrained by refractory failure concerns. This is particularly important in standby boilers in cogeneration plants, which have to be started up fast to meet steam demands.

The skin-cased partition wall design in a conventional furnace can leak combustion products from the furnace to the convection section, thus raising CO values. In such cases, CO levels can be 200–300 ppm higher at the boiler exit than at the furnace exit. In contrast, a membrane-wall design of the partition wall (Figure 2) offers a leak-proof partition that channels the combustion products through the furnace and on to the convection section.

In addition to ensuring that flame impingment does not occur, the furnace envelope should be chosen to give the required superheated steam temperature over the desired load range. A smaller furnace results in higher furnace-exit gas temperatures, thus requiring a smaller superheater. However, the higher temperature of the tube wall requires the tubes to be made of more-expensive metallurgical alloy grades.

The amount of energy transferred in a furnace depends on the fuel, the amount of excess air and fluegas recirculation. The furnace's exit-gas temperature is higher with natural gas fuel, while the emissivity of a fuel oil flame is much higher than that of gas. Therefore, fuel oil transfers more energy at the same duty and, as a result, the furnace exit-gas temperature is lower.

Thus, if a boiler is to guarantee superheated steam temperature with both fuels, it should be sized for fuel oil firing. With gas firing, the steam temperature will be higher because of the higher gas temperature entering it.

Typically, membrane-wall construction consists of 2-in. (outer dia.) carbon steel tubes at a pitch of 4 in. However, when steam pressure exceeds 900–1,000 psig, the tube wall and fin tip temperature increase beyond the limits of carbon steel, due to the higher saturation temperature; in these situations, a lower fin spacing is required to minimize the temperatures.

Heat-release rates, usually expressed in Btu/ft^{2}h of effective projected radiant surface (EPRS) on a lower heating value basis, give an idea of the furnace size. Typically, the values range from 100,000 to 200,000 Btu/ft^{2}h. A more important parameter is the heat flux – the energy absorbed by the furnace divided by the EPRS – which is responsible for departure from nucleate boiling when it reaches a critical value.

Generally, the higher the heat release rate, the higher the heat flux but not in the same proportion. There are conditions when the heat-release rates can be very high, in the range of 150,000–200,000 Btu/ft^{2}h, while the heat flux is very low, in the range of 30,000–50,000 Btu/ft^{2}h. This situation arises when the flame temperature is low and, hence, the radiant energy transfer is low.

Critical heat flux is a function of steam pressure, quality or circulation ratio, tube size, water chemistry and roughness of tubes [2]. The higher the steam pressure or the higher the steam quality, the lower should be the heat flux. Typical heat-flux values range from 60,000 to 120,000 Btu/ft^{2}h, depending on the boiler capacity and furnace dimensions.

The volumetric heat-release rate, expressed in Btu/ft^{3}h, reflects the residence time for combustion of the fuel in the furnace. It is more significant for fuels that are difficult to burn, such as solids-based fuels.

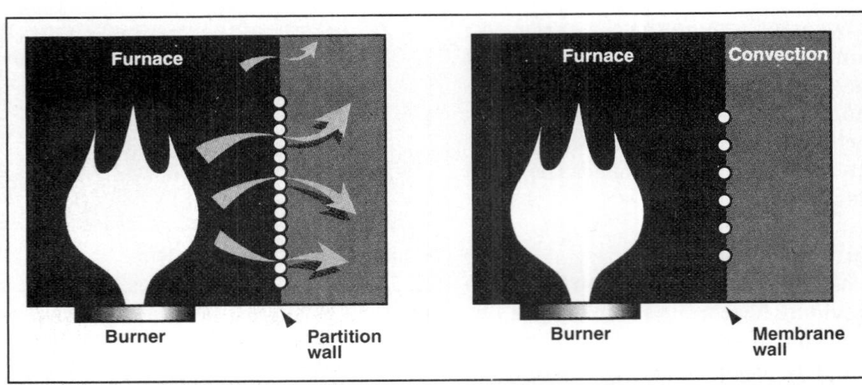

FIGURE 2. Combustion products leak to the convection section in a tangent-tube partition-wall furnace (left), but not in a membrane-wall unit

Why custom designs?

As can be seen, conventional boiler designs have several performance limitations. For one, emission levels will be much higher. Further, if proper excess air and fluegas recirculation quantities are used to meet the required emissions level, the exit-gas temperature will be higher, resulting in lower efficiency.

Also, the pressure drop of the fluegases will be higher, resulting in higher power consumption and operating costs for fans. It is unlikely that the "standard" fans in conventional designs will be able to handle the larger flows and pressure drops.

Customized designs that include the following modifications offer better performance:

1. The furnace is engineered based on burner design, excess air, gas recirculation, and flame shape. Completely water-cooled furnaces help minimize emission problems
2. The convection section is designed to handle the addition of fluegases via increases in tube height and size, such as by using more tubes per row, and changing the tube size and pitch to minimize gas pressure drop and improve duty
3. In the convection section, tubes with extended surfaces may be used to minimize gas pressure drop and improve performance if the products of combustion are clean, such as with gas and distillate fuels. This concept is not new, and is widely used with gas-turbine-exhaust from heat-recovery steam generators (HRSGs). The use of extended surfaces offers several advantages, including compactness, low weight and small pressure drops for the same steam generator duty [2,3]
4. Superheaters can be located in an appropriate zone based on the degree of superheat. Standard designs typically locate the superheater in a radiant zone, irrespective of the steam temperature desired. However, the tube wall's temperature and the life of the superheater can be improved if the superheater is designed on a case-to-case basis

For example, if the steam temperature is not high, say, around 50 to 100°F of superheat, it is not necessary to locate the unit in the radiant zone. The superheater can be buried deep in the convection section or it can be located between the evaporator and economizer, resulting in cooler tube-wall temperatures. Customized designs locate the superheater in an optimum temperature zone based on such aspects as tube wall temperature and performance over the load range

5. Depending on layout requirements, the economizer can be horizontal or vertical and the boiler's height, width and length can be modified to match the availability of space. In contrast, it may be difficult to locate an off-the-shelf design with its standard dimensions within a given plot plan
6. The size of the forced draft fan is based on the excess air and fluegas recirculation used. Usually the fluegases are induced into the suction of the fan, which, naturally, impacts the suction temperature and density. This, in turn, affects the fluegas volume. One also has to consider the effect of site elevation and ambient temperature variations on the gas pressure drop through the boiler. Thus, fan size may vary from site to site depending on the above factors, even though the steam duty may be the same

Superheater and economizer

The gas temperature entering the convection section typically ranges from 2,000 to 2,350°F at 100% load, depending on the fuel used, furnace size, and excess air and fluegas-recirculation quantities. Gas temperature decreases as the boiler duty decreases.

If a superheater is used, it is preferable to locate it behind a screen section, to shield it from the radiant energy of the furnace. The surface area requirements of the superheater, which increase if the unit is located in a cooler zone, have to be balanced against the material requirements. Typically, low-alloy to stainless-steel tubes are used, depending on the steam temperature, which, in turn, impacts the tube wall temperature.

In semi-radiant or convective type of superheaters, the steam temperature decreases with load. Hence, if the steam temperature is required to be maintained from, say, 50 to 100% load, the superheater is designed to provide the steam temperature at 50% load (with fuel-oil firing, and if both natural gas and fuel oil are to be used). At higher loads, the steam temperature will be higher, which can then be controlled using an attemperator.

Interstage attemperation is preferred over downstream attemperation for two reasons. The maximum steam temperature and, therefore, the tube wall temperature will be lower. If attemperation is performed downstream of the superheater, it is likely that water particles can get carried away into the piping system and downstream equipment. Superheaters should be designed to be drained. Generally, smaller tube sizes result in lower tube-wall temperatures, and should be considered when the steam temperature is high, say, above 800°F.

When the degree of superheat is small (around 50°F), it is preferable to locate the superheater deep down in the convection section where the gas temperature is around 1,000°F. Another possible location for a low-temperature superheater is between the evaporator and economizer.

The evaporator tubes are usually of carbon steel. Electric resistance welded (ERW) tubes are used in low pressure units running at pressures up to 900 psig. Seamless steel tubes are preferred for higher pressures. The tube length and spacing can be varied to miminmize the gas pressure drop and at the same time to transfer the desired duty. Extended surfaces may be used when a larger duty is to be transferred with a low gas-pressure drop. This concept, used when the gas stream is clean, is applied in gas turbine HRSGs.

The tubes are usually rolled and expanded into the drum in low pressure units (<900 psig). At higher pressure, tubes are welded to steam and mud drums. Also, when frequent startups and shutdowns are required, welded-tube construction is preferred.

The economizer, if used, is usually of extended surface design. Depending on layout considerations, it could be horizontal or vertical. Steaming in the economizer is not a concern in packaged steam generators, as in gas turbine HRSGs [2,3]. The main reason is that the ratio of gas and steam quantities does not change significantly with load in packaged boilers. As a result, the approach point – the temperature difference between saturated steam in the evaporator and the entering water – increases at lower loads, when the gas flow and gas temperature to the economizer are lower than the values at full-load conditions.

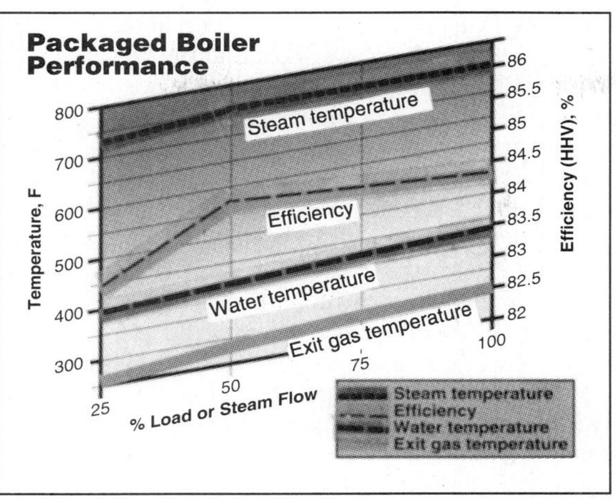

FIGURE 3. The efficiency of steam generators varies with load, due to variations in radiation and fluegas losses

When fuel oils with sulfur content are fired, one has to be concerned about dewpoint corrosion. However, if the feed water temperature is close to the dew point, then condensation of acid vapors is unlikely.

Air heaters are generally not recommended for packaged gas- or oil-fired boilers, since they are more expensive, subject to higher gas and air pressure drops, and lead to higher combustion temperatures. This, in turn, results in higher NOx. Air heaters may be necessary in special situations, such as when low-Btu fuels or a difficult-to-fire fuel, such as a solid, are used.

Performance

Unless the packaged boiler is used as a standby unit that operates for less than 1,000–2,000 h/yr, it is desirable to design it for high efficiency and low pressure drop. Efficiency is specified on both higher and lower heating value basis (hhv and lhv, respectively), and the relationship between them is:

$$(E_{hhv})\,(hhv) = (E_{lhv})\,(lhv)$$

Thus, if a steam generator firing natural gas has an efficiency of 93% on a low heating value basis, and the lower and higher heating values are 21,000 and 23,000 Btu/lb, the efficiency on a higher heating value basis is:

93 (21,000/23,000) = 84.9%.

The efficiency of steam generators varies with load, as illustrated in Figure 3. This is due to the fact that radiation loss as a percentage increases as load decreases, while the fluegas loss decreases at lower loads because of the lower exit-gas temperature. At very low loads, the radiation losses are more significant and, hence, the inverted U- curve.

Steaming in the economizer is not a problem in generators. If the economizer is designed with an approach temperature of, say, 40°F at full load, steaming will be much higher at lower loads due to the reduction in gas flow and its temperature at lower loads. In gas turbine HRSGs, on the other hand, the gas flow is nearly the same at all loads, which is responsible for the steaming problem at low loads. ■

Edited by Jayadev Chowdhury

References

1. Makansi, J., "Clean air act amendments-The engineering response," Power, June 1991.
2. Ganapathy, V. "Steam plant calculations manual," Revised and expanded edition, September 1993, Marcel Dekker, New York
3. Ganapathy, V. "Waste heat boiler deskbook," Fairmont Press, Atlanta, 1991

The author

V. Ganapathy is the Heat Transfer Specialist at ABCO Industries, Inc. (2675 E. Hwy 80, P.O. Box 268, Abilene, TX 79604-0268; Tel: 915-677-2011; Fax: 915-677-1420). He is responsible for process and thermal engineering, performance calculations and software development related to packaged boilers and waste-heat boilers. He has authored more than 200 articles on boilers, heat recovery, steam generation and related topics. He holds a B. Tech degree in Mechanical Engineering from the Indian Institute of Technology, Madras, and a Master's degree in Boiler Technology from Madras University. He is the author of the books "Steam plant calculations manual," "Waste heat boiler deskbook for steam plant engineers" and the software "HRSGS program for simulation of design and off-design performance of HRSGS," copies of which are available from him. He has also written several chapters for the Encyclopedia of Chemical Processing and Design.

Giovanni S. Crisi
Consulting Engineer

Engineers are generally well schooled in the process aspects related to selecting, specifying and obtaining heat exchangers, and much literature on the subject is available.* Apart from this technical knowledge, however, an extensive amount of useful practical lore has accumulated on how to acquire exchangers effectively and without mishap.

The discussion here relates mainly to heat-exchange devices, such as shell-and-tube, double-pipe and plate exchangers, that regularly accommodate process fluids on both the hot and cold sides. However, many of the same considerations apply to other heat-transfer devices used in the CPI, including air coolers, heating or cooling coils for vessels, and barometric condensers.

Acquiring a heat exchanger generally consists of following these steps:

- Establish the process conditions
- Make the thermal balance
- Select the type of exchanger
- Size the exchanger
- Choose the materials of construction
- Determine key construction details
- Write the Data Sheet and the General Specification
- Prepare the vendor list
- Send the inquiries
- Make the bid analysis
- Check vendor drawings
- Follow manufacture and testing

The first two steps are mainly a matter of straightforward process engineering and stoichiometry, and are not commented upon here. The third step, selecting the type of exchanger, can however be a bit more involved.

Making the choice

Selecting the type of exchanger entails process considerations, capacity, ease of maintenance, suitability for the site, cost, and perhaps other factors. Some can be resolved fairly readily.

For others, however, the engineer may wish to get more information by talking with suppliers. This is particularly true when he or she has no experience in choosing exchangers for the type of service at hand, and when the available literature is inadequate.

The first step in contacting suppliers is to assess their previous experience in equal or similar services. Begin by asking for references among the supplier's past customers having similar plants.

Check out these references. Do not take it for granted that they are valid just because they have been included in a list. Occasionally, the author has consulted referenced users who were *not* satisfied with the product, or even who had never in fact bought it.

Whenever possible, visit the actual customer site and talk directly with the personnel who operate and maintain the equipment. When a visit is not possible, make at least a phone call.

Also keep in mind the possibility of a quite different kind of pitfall during the selection step: your own subjective preferences. For example, the plate heat exchangers, invented in Sweden in the 30's, were long ignored by U.S. engineers — they preferred the shell-and-tube type because they were more accustomed to it.

Size the exchanger

The procedure for determining the required size depends on the equipment type. Plate and double-pipe exchangers, for instance, are generally protect-

*For example, see these *Chemical Engineering* articles: "Don't Overlook Compact Heat Exchangers," August 1991, pp. 90-96; "How to Design Bayonet Heat Exchangers," April 1991, pp. 122-128; "Quick Design and Evaluation: Heat Exchangers." July 1990, pp. 92-96; "The Right Metal for Heat Exchanger Tubes," January 1990, pp. 120-124; "Speed Up Heat Exchanger Design," Mar. 14, 1988, pp. 143-146; "Charts for the Performance and Design of Heat Exchangers," Aug. 18, 1986, pp. 81-88; "Perspective on Shell-and-tube Heat Exchangers," July 25, 1983. pp. 46-80.

Many of these articles appear in this volume.

Originally published February 1992

ed by patents, so their sizing (indeed, their overall process and mechanical design) is carried out by the suppliers or the patent or license owners, based on the process conditions.

Shell-and-tubes exchangers generally are not covered by similar protection. However, the CPI engineer attempting to size one is likely to run into another complication, concerning the overall heat-transfer coefficient.*

In principle, the coefficient can be calculated. But in practice it commonly is not, because the calculated results can differ greatly from those encountered in actual practice.

Data on heat-transfer coefficients are available in tables and books. Unfortunately, these data (often consisting of maxima and minima, sometimes far apart) are only broad indicators, maybe suitable for comparisons but not precise enough for design.

One reliable basis for obtaining a heat-transfer coefficient is previous experience in an identical or very similar application. Another reliable source consists of specialized institutions, such as the U.S.'s HTRI (Heat Transfer Research Institute), that supply their

Problem-free acquisition of an exchanger goes far beyond the normal heat balances and specifications

members with data on heat-transfer coefficients, fouling factor and others parameters. Membership in such an institution requires paying a fee, and may not be worthwhile if the user-company needs only relatively few exchangers during a year.

In such a case, it may be desirable to appoint a consultant. He or she might not only determine the coefficient but also carry out the process and mechanical design of the exchanger. A similar alternative is to assign the overall task to a reliable supplier having process-design and test capabilities.

If the engineer does have a reliable heat-transfer coefficient at hand, the sizing process can be carried out directly. Help in this is available from numerous excellent publications (see, for instance, some of the articles referenced earlier). The mechanial design can be carried out the same way. In addition, there are computer programs available that carry out such tasks in a matter of minutes.

Materials of construction

Take care when using the corrosion curves and tables available in the literature. Usually such data are obtained in laboratory tests either with pure liquids or at concentrations different from those to be encountered by the heat exchanger itself. Apart from the concentration question, the *type* of diluent (or contaminant) present can also affect the corrosivity.

Moreover, corrosion data are usually gathered under static conditions. Conditions inside the heat exchanger are instead dynamic, which can highly increase the rate of attack. The possibili-

*This coefficient, generally symbolized by the letter U, crops up in the standard heat-transfer equation

$$A = Q/U\Delta T$$

where A is the required exchanger area, Q the required heat flowrate, U the overall heat-transfer coefficient and ΔT the logarithmic mean temperature difference across the exchanger. U embodies the ease (or lack of ease) with which heat flows through the exchanger surface and through the fluid films on the two sides of that surface, and it includes a factor that takes into account fluid fouling.

DENISE EARDMAN

ty of physical abrasion must also be kept in mind.

Tables and curves should thus be used only for approximate information. As with heat-transfer coefficients, more-reliable information comes from CPI operating experience in similar service, data supplied by the aforementioned research institutes, or input from specializec consultants.

In shell-and-tube exchangers, the tubes are the most subject to corrosion. Accordingly, input from tube manufacturers can be advantageous, especially when the user is considering special, corrosion-resisting materials such as titanium, stainless steel and proprietary alloys. Do not, of course, neglect to check the references.

Important structural details

For shell-and-tube exchangers, almost all the details of construction are set during mechanical design, usually in accordance with TEMA (Tubular Exchanger Mfrs. Assn.) codes. However, it is also necessary to address some items not covered by the codes, such as:

- Cradles for supporting the exchanger. It is necessary to specify both the cradle type and the shape
- Lifting lugs, which must be specified in accordance with the weight and dimensions of the exchanger and the amount of overhead space available
- Holes in the base plate for the anchor bolts, which must be slotted to allow the exchanger to expand
- Supports or clips for insulation, if applicable
- The name plate, usually stainless steel, containing information necessary to identify the equipment

Some of these items also apply to double-pipe exchangers. Accordingly, they must be included in the specifications given to the double-pipe-exchanger manufacturer who sizes the device and carries out the process and mechanical design, as mentioned earlier.

Data Sheet, general specs

With all the aforementioned information on hand, the engineer prepares a Data Sheet. It is also necessary to prepare a General Specification, unless one is already at hand.

The General Specification sets out the general conditions of design, manufacture, inspection and tests that will be required of the supplier in order for its proposal to be technically acceptable. Many companies that frequently buy exchangers prepare one General Specification for each major type (e.g., shell-and-tube, double-pipe, plate).

The Data Sheet, by contrast, includes the specific requirements and details that pertain to a particular purchase. These cover such points as the process conditions, the thermal balance, the exchanger TEMA classification, the main mechanical features, and the materials of construction.

When filling in a Data Sheet, the engineer should leave blank the items with respect to which he or she does not have a specific requirement. This leaves different suppliers free to comply with those items in the ways that they individually feel most suitable for the bid at hand. The resulting differences can be remarkable, particularly in cases (such as double-pipe or plate exchangers) where the design is largely left up to the supplier.

A typical Data Sheet for use with plate heat exchangers appears on these pages. For shell-and-tube exchangers, the TEMA Code, Section 3, has a standard Data Sheet that can be supplemented with additional items considered necessary by a given user.

Do not forget to specify the tests (witnessed or not by the purchaser) that the supplier must perform, as well as the kinds of inspection the purchaser will carry out, either personally or via an inspection firm. This allows suppliers to take into account the corresponding costs when preparing their offers. It also forestalls possible claims based on ignorance of these items when the price was calculated.

All exchangers without exception should be hydrostatically tested after manufacture. Depending on the appli-

PLATE TYPE HEAT EXCHANGER
DATA SHEET #1

CUSTOMER:
TAG NUMBER:
NAME:
DATE:

CONSTRUCTION
CODE:
INSPECTION BY:
DESIGN PRESSURE:
DESIGN TEMPERATURE:
TEST PRESSURE:

FRAME TYPE:
No. in parallel:
No. in series:
Material:
Surface finish:
Drawing no.:
Dimensions:
Length:
Width:
Height:

PLATE TYPE:
No. /frame:
No. of passes
Warm:
Cold:
Material:

GASKET MATERIAL:

CONNECTIONS:
In warm:
Out warm:
In cold:
Out cold:

Data sheet, filled out when specifying the exchanger, covers structural details . . .

cable codes, X-ray inspections of the welding may also be required.

For plate heat exchangers, the purchaser may also demand performance tests under operating conditions after startup. This is especially appropriate if the supplier has no experience in manufacturing exchangers for the type of service at hand.

Inspection provisions vary according to the complexity of the manufacturing method and the nature of the service in which the exchanger will perform. For simple equipment in non-demanding service, it may be enough to require inspection after fabrication, with non-witnessed tests. In these cases, the manufacturing is not followed by the purchaser or the inspection firm. At delivery, the supplier hands over a certificate of the tests results.

The next alternative in terms of increasing severity is inspection after fabrication, *with* witnessed tests. In this case the tests are made in the presence of the customer's inspector.

More severe is partial inspection during fabrication. In-process product leaving each key manufacturing step is subject to inspection, which might consist of visual examination, dimension control, radiography of welds, control of tube expansion, and other items. The tests will be witnessed, and the shipment of the equipment must be approved by the client's inspector.

Most severe of all is intensive manufacturing inspection. Output from each successive manufacturing step is scrutinized, and the supplier cannot commence a new step until the previous one is approved. This is applicable to exchangers for, e.g., nuclear plants.

The vendor list

It is sometimes believed that the engineer has no useful role to play while the list of prospective vendors for a particular purchase is being prepared. That belief is inaccurate.

True, a Procurement or Purchasing Department (particularly in a large firm) generally maintains a vendor list for each type of equipment, the list carrying all vendors that at one time or another have been registered with the firm according to its procedures. Unfortunately, there is a tendency for such lists to give the most weight to the business and commercial practices of the vendor while attaching less importance to its technical abilities.

This tendency can be avoided, of course, by bringing engineers into the list-maintenance process. If such ongoing engineer participation is, unfortunately, not company practice, engineers involved in a given project should insist on a chance to review the list of bidders who have been contacted.

In either case, here are some points to keep in mind when assessing vendors for inclusion on the list. First, make sure that the equipment falls within the vendor's normal production line. For instance, it is unwise to buy big shell-and-tube exchangers from a supplier that has made only small oil coolers for lubrication systems.

Second, confirm the supplier's actual experience concerning the desired equipment. Vendors' catalogs are not always are reliable for this purpose, because some firms will include equipment that they have in fact never manufactured. When in doubt, ask the manufacturer for a reference list and check it out along the lines already described.

When a manufacturer is being qualified for the first time to supply a given kind of exchanger, the steps to be taken are different for shell-and-tube exchangers from those for plate and double pipe exchangers. In the shell-and-tube case, it will be enough to visit the

PLATE TYPE HEAT EXCHANGER
DATA SHEET #2

CUSTOMER:
TAG NUMBER:
NAME:
DATE:

DESIGN DATA:

PRODUCT:
Warm:
Cold:
HEAT EXCHANGED:
FLOW **Warm:**
Cold:
TEMP. **In Warm:**
Out Warm:
In cold:
Out cold:
DENSITY **Warm:**
Cold:
VISCOSITY **Warm:**
Cold:
SPECIFIC HEAT **Warm:**
Cold:
THERMAL COND. **Warm:**
Cold:
OPER. PRESS **Warm:**
Cold:
PRESSURE DROP:
LMTD:
EXCHANGE SURFACE:
FOULING FACTOR
TYPE EXCHANGER:
WEIGHT OPERATION:
EMPTY:
HYDRO TEST:

. . . and process-design information. Details not explicitly required should stay blank

manufacturing plant to evaluate its capabilities. Among other things, check the competence of the welders. And learn how the quality control system operates — it is very important, for instance, that the quality-control team (which assesses the work of the production department) not be subordinate to the production department.

As for plate and double pipe exchangers, where the knowhow is more proprietary, supplement the plant visit by checking the operation of exchangers supplied by the same manufacturer for similar services. It is important to verify at least one case, following the procedures already explained.

In short, a well-made vendor list consists of companies that have experience in similar plants, a technical reputation in the marketplace and industrial capacity compatible with the scope of supply, and that can offer references as well as customer service and availability of parts. Preparation of such a list requires care and discretion.

The engineer plays a valuable role throughout the exchanger-purchase process

In the mail

Once the vendor list for the project at hand is ready, each firm on it receives a invitation to bid. The main contents of this invitation include the Data Sheet and General Specification, commercial documents (not discussed in this article), a list of the paperwork that the bidder must submit with the bid, and a list of the documents that the winning bidder will have to submit after being awarded the sale.

Items on the first of those two lists should normally include a complete and detailed description of the offered equipment, catalogs, brochures, and the Data Sheet filled out with respect to the items that had been left blank by the purchaser.

As for the list to be subsequently supplied by the winning bidder for customer approval, it should consist of: general arrangement drawings; shop drawings (if applicable); cross-sectional drawings; a parts list; a load diagram including weights (of the exchanger when empty, in normal operation, and full of water) and anchor-bolt location; information about allowable loads in the nozzles; erection, operation and maintenance instructions, in case of plate and double-pipe exchangers; welder qualifications and welding procedures. Subsequently, of course, material certificates are furnished at inspection-times during fabrication.

Analyze the bids

After the proposals have been received, the first step is to check whether they include all the documents requested in the inquiry. Then the technical analysis can start. For that purpose, it is useful to draw up a bid-analysis comparison chart, whose left-hand side lists the requirements set forth in the Data Sheet and the General Specification and whose separate columns indicate how each bidding vendor offers to meet those requirements.

It is usual, after filling out the space corresponding to a certain item, to add a code valuation expressing whether this entry (i.e., this response by the supplier in question) is technically acceptable, or acceptable with reservations, or not acceptable. Occasionally the response will instead be a "valid alternative," i.e., different from what had been specified but nevertheless acceptable.

In general, the chart can be filled out by even a relatively inexperienced engineer. Of course, its subsequent analysis requires experience.

Once the technical requirements have been satisfied, the final step in selecting the winning bidder is to analyze the bids in accordance with commercial considerations. In some cases, it may be desirable to first "equalize" the proposals; i.e., to put all of the suppliers on the same level from a technical standpoint. This is done by giving bidders a chance to correct the unacceptable items, revise the acceptable-with-reservations ones to make them totally acceptable, and explain any ambiguous items.

The final steps

After the purchase order is issued but before manufacture gets under way, the winning supplier will send for customer approval the drawings and documents mentioned earlier. The customer must analyze these in detail, to confirm their adherence to the conditions set forth in the purchase order and its attached documents.

On the basis of this analysis, the documents can be rated in one of three classes:

- Approved. Customer has no comments to make, and fabrication can start immediately
- Approved with comments. Customer has objections of only minor importance, and fabrication can start provided that the supplier complies with the objections
- Not approved. Customer has serious objections and the supplier must satisfy them and send new drawings for approval. Fabrication cannot begin before this approval

Once the manufacturing process gets under way, it is accompanied and followed by inspection as summarized earlier in this article. And when the exchanger is delivered, the supplier usually furnishes a "Data Book." This contains not only documents prepared before fabrication but also operation and maintenance instructions. ■

Edited by Nicholas P. Chopey

The author

Giovanni S. Crisi, R. Mateus Grou 365, Apt. 151, 05415 Sao Paulo, Brazil, Tel. 55-11-853-8419, Fax. 55-11-240-1099, has spent nearly 30 years in Brazil and Argentina on the design and construction of large process plants, with particular emphasis during the past eight years on projects that involve environmental protection. He has written more than 20 magazine articles, and delivered numerous papers at international conferences. He is a member of IBP, the Brazilian petroleum institute, for whom he coordinates an annual course on erection of process plants, and has been vice president of the Sao Paulo section of ABEQ, the Brazilian association of chemical engineering. His degree in chemical engineering is from Universidad Nacional del Litoral (Argentina).

HEAT EXCHANGER DATABASES ACCELERATE PROCESS DESIGN & COSTING

Quickly evaluate heat-exchanger area, utilities consumption and project economics

Stephen G. Hall and
Steve W. Morgan
The M. W. Kellogg Co.

STEVEN HAIMOWITZ

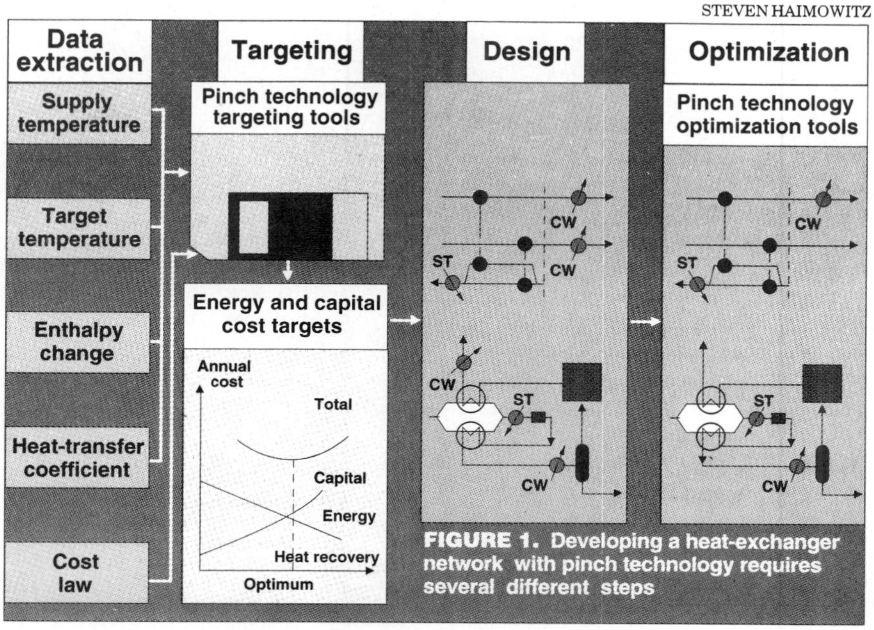

FIGURE 1. Developing a heat-exchanger network with pinch technology requires several different steps

During the last 15 years, the various analytical and representational tools of pinch technology have repeatedly shown their value in helping engineers gain a comprehensive understanding of the heat flows in a process. It is this understanding that then leads to the generation of efficient, heat-integrated process designs [*1,2*].

An important activity undertaken during a pinch study of a process is the evaluation of optimal performance targets for utilities consumption, heat exchange area and project economics. The data required to perform such an evaluation include heat-transfer coefficients of process streams and heat exchanger costs. The cost data are employed in cost equations – otherwise known as cost laws – to relate the heat-transfer area in a heat exchanger to its purchased cost.

One way to determine such data is by working with heat exchanger designers and cost estimators. However, such interaction frequently takes place at a stage when a project's overall schedule requires the process configuration to be set as quickly as possible.

Pinch studies help determine a process configuration. But there is an obvious benefit to be gained by speeding up the data collection, such as via a database of historical heat-exchanger information.

Having such information at one's fingertips minimizes interaction with heat-exchanger designers and cost estimators at the preliminary design stage, and so accelerates the design of practical heat-exchanger networks. Such design can often be done with acceptable accuracy, and little or no effect on project schedule.

A heat-exchanger database that speeds up heat integration studies using pinch technology has been developed by The M. W. Kellogg Co. (Houston, Tex.). The database and database program have already proven their viability on several recent projects.

Process targets

In designing an energy-efficient process, the extent of heat recovery is a variable that can significantly affect a project's profitability. Higher heat recoveries can reduce energy costs, but these gains come at the expense of higher capital investment for larger or additional heat exchangers, or both. Figure 1 outlines the steps for developing a heat exchanger network.

Basically, pinch technology employs a concept called "targeting" to determine the most economic level of heat recovery prior to detailed plant design [*3*]. Here, engineers first determine the process and utility stream

Originally published July 1994

temperatures, enthalpy changes, heat-transfer coefficients, and heat-exchanger cost laws.

These data enable estimates of the energy and capital costs associated with a range of heat recoveries to be carried out. In turn, these cost values identify the extent of heat recovery corresponding to a minimum total cost. Thus, once the optimum level of heat recovery has been targeted, the information can be used to design the heat exchanger network [4].

Whither the data?

At present, there are several ways for engineers to obtain heat-transfer coefficients (HTCs) and heat exchanger costs for use in pinch studies. One way is to use published information from books or journals. Another is by interacting with in-house experts, such as heat exchanger designers and cost estimators.

Publicly available literature gives a range of HTCs, with values independent of type of process and exchanger flow side (shell or tube). Heat exchanger costs are generally more specific, and are easy to find for shell-and-tube exchangers. However, cost laws for other exchanger designs, such as plate-and-frame or spiral-plate, are more difficult to come by, and these HTC and cost data will most likely be less accurate.

Heat-transfer coefficients and exchanger costs obtained from in-house experts are generally more accurate than publicly available data. However, obtaining in-house information may require a significant amount of time, and could result in delays to the project.

Delays may also occur as the project progresses and additional design information becomes available. Such data may lead heat-exchanger designers or cost estimators to revise their initial estimates for HTCs and exchanger cost laws, and engineers to change their heat-exchanger network designs. All these changes require additional commitments of time.

However, these design activities and changes generally occur at a stage in the project's overall schedule when it is desirable to fix the process configuration as quickly as possible. A database can quickly provide more-accurate heat exchanger data, minimize interaction with in-house experts at the preliminary design stage, and speed up the project schedule.

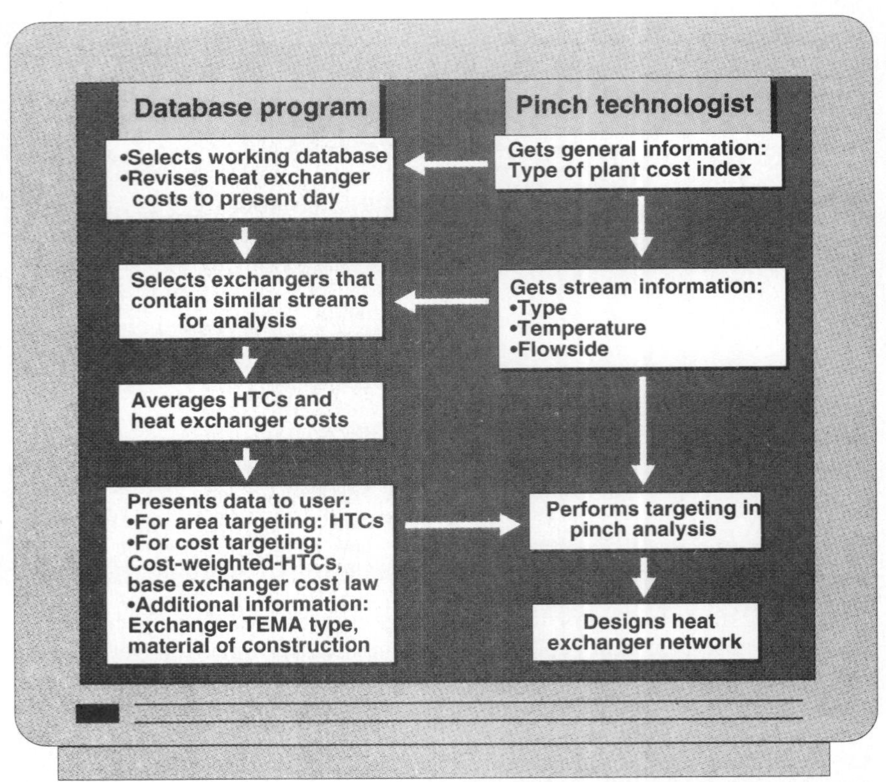

FIGURE 3. The sequence of events needed to design a heat exchanger network with the database

FIGURE 2. An example of the information that is stored on the database for one heat-exchange application

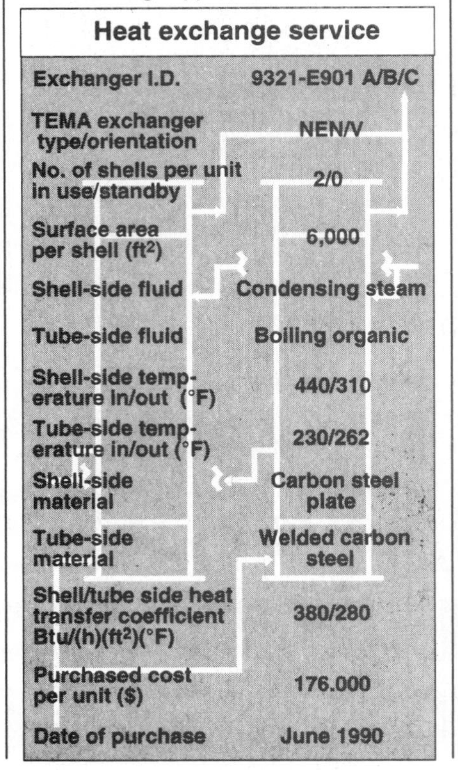

Heat exchange service	
Exchanger I.D.	9321-E901 A/B/C
TEMA exchanger type/orientation	NEN/V
No. of shells per unit in use/standby	2/0
Surface area per shell (ft²)	6,000
Shell-side fluid	Condensing steam
Tube-side fluid	Boiling organic
Shell-side temperature in/out (°F)	440/310
Tube-side temperature in/out (°F)	230/262
Shell-side material	Carbon steel plate
Tube-side material	Welded carbon steel
Shell/tube side heat transfer coefficient Btu/(h)(ft²)(°F)	380/280
Purchased cost per unit ($)	176.000
Date of purchase	June 1990

The database approach is not intended to lead to a shutdown in communication with other engineering groups. Rather, it allows preliminary information to be obtained quickly, so as to expedite the targeting phase of pinch analysis and, finally, the design of the heat-exchanger network itself.

The heat-exchanger database

Historical information for hundreds of recently built heat exchangers from all over the chemical process industries can easily be stored in a standard PC-based database system. The stored data can be used not only in design, but also in other applications, such as maintenance and troubleshooting.

In designing a database, it is important to keep in mind all possible uses for the data, since this will minimize future modifications to the database's structure. The database variables relevant for a particular application should be chosen after carefully considering many factors. In heat-exchanger design, these include the information available to the engineers in the early stages of design, the required accuracy of the heat-transfer coefficients and exchanger costs, and the requirements for traceability of data.

Each database record represents a single heat exchanger, and contains the exchanger's item number, service, design type, associated fluids with their inlet and outlet temperatures and pressures, materials of construction, surface area, purchase price and date of purchase. For shell-and-tube heat exchangers, the design type indicates the corresponding classification from TEMA (Tubular Exchanger Manufacturers Assn.). If the exchanger is not a shell-and-tube, the design type is a code indicating either an air cooler, spiral exchanger or plate-and-frame unit.

A typical heat exchanger record is presented in Figure 2. Single values for the HTC of each stream in the exchanger are also stored.

Each HTC contains an allowance for heat-transfer resistance due to fluid film, fouling, exchanger wall resistance and, if required, an allowance for design margin. The various resistances to heat transfer are usually calculated with a high degree of accuracy during the detailed design of a heat exchanger, using computer programs such as those from Heat Transfer Research, Inc. (HTRI). The overall heat-transfer coefficient for a heat exchanger can be evaluated solely from the two coefficients stored on the database.

Filtering the data

Engineers can determine the heat-transfer coefficient and exchanger costs for a particular application by viewing the database records of heat exchangers in similar processes. However, the sheer volume of stored information can often lead to confusion and may in itself lead to a significant amount of analysis.

Therefore, it is useful to have a database program to filter and analyze the relevant data. A database program can also process the stored data and present it in a pertinent form.

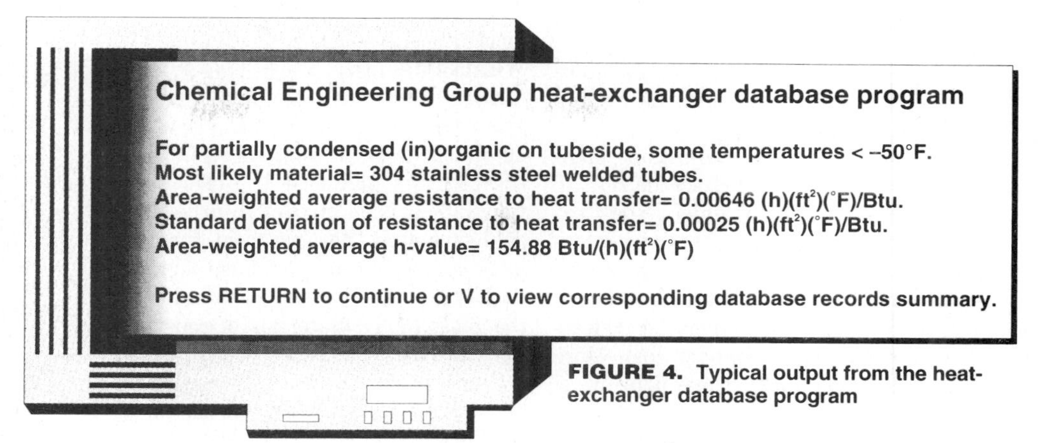

FIGURE 4. Typical output from the heat-exchanger database program

Figure 3 outlines how a typical database program interfaces with a user to generate stream HTCs and heat exchanger costs. First, the database asks the user to characterize the type of process being considered, and the appropriate equipment index within the "Chemical Engineering Plant Cost Index" that is used for scaling the heat-exchanger costs (p. 174).

The type of process can range widely – it might be an ethylene plant, an ammonia facility or a petroleum refinery. By specifying the type of plant, the program knows which database to consider in the analysis.

Next, the user is asked for a description of the particular stream being considered. For example, this may be a condensing organic, a liquid organic, condensing steam, or cooling water. The temperature range of the fluid is also input, together with available information about the stream's flow side (for example, either shell or tube side for a shell-and-tube heat exchanger).

The database program selects the heat exchanger records containing streams with the characteristics described in the user's input. The heat-transfer coefficients of the selected records are then averaged, taking into account the surface area of the various exchangers through which a stream of the specified type passes (area weighting). Averaging the HTCs on an area-weighted basis should, when considered in the context of the whole heat-exchange system, give the highest accuracy for heat-transfer area targets.

HTCs for the same type of stream can sometimes vary significantly between heat exchangers. To give a user some idea of the variability of the stored data, the program can generate a standard deviation of thermal resistance to heat transfer. Then, if the standard deviation is large, the user can view the heat exchangers used to calculate the average HTC, and can delete any unusual heat exchanger records that may not correspond to the case being studied.

The database program can also extract other information. One example is the most likely material of construction, based on the maximum number of times that material occurs in the records under consideration.

Thus, after inputting very preliminary stream information into the database program, the user will be given a single HTC for that stream based on representative historical data. The HTCs for all the streams can then be used to determine heat-transfer area targets for a whole process-heat exchange system. A typical output from a database program is shown in Figure 4.

Targeting networks

A limitation of the generally accepted method for cost targeting of heat exchanger networks in pinch technology is that a single cost law must be applied across all network exchangers, irrespective of design type or materials of construction, or both. One way to overcome this is to "cost weight" the stream heat-transfer coefficients [*6*].

For example, consider a heat exchanger network of several carbon-steel heat exchangers and a single stainless-steel exchanger. In the traditional method of network cost targeting, the user might cost the exchangers assuming that they are all made of carbon steel. However, such an approach would not correctly reflect the additional cost due to the exchanger made of stainless steel.

To overcome this limitation, engineers can artificially reduce the HTC of the stream in the stainless steel exchanger. Thus, in the cost target calculations, this increases the heat exchanger's associated surface area in such a way as to correctly reflect the actual increase in exchanger cost. More details about such cost-weighting of HTCs, including a discussion on the expected accuracy of such a method, are described in Reference [*6*].

To evaluate a cost target for a heat exchanger network requires a heat-exchanger cost law and cost-weighted stream HTCs. The database program then determines a base-case cost law for network cost targeting that should be valid for all exchangers made in the most common construction materials.

To find the cost-weighted HTC for a particular stream, the program scans the exchanger database to determine the most likely heat-exchanger material that is associated with that stream. The program then finds the cost law for all exchangers made of such a material, and then uses this cost law to calculate a cost-weighted HTC. Using this coefficient will maximize accuracy when targeting the capital cost of a heat exchanger network made up of a variety of materials of construction.

What of design?

Just as in targeting, the database program can generate information with an emphasis on preliminary design. Here, the program first asks the user questions about each stream in a heat exchange match. The program then selects corresponding database records, and after analyzing this data indicates the most likely materials of construction, stream HTCs and cost law.

Since the cost law for each exchanger is evaluated separately, there is no need to cost-weight HTCs. This enables the cost of each exchanger to be predicted with acceptable accuracy.

Such design is in no way a replacement for analysis by exchanger designers or cost estimators. Instead, it is a guide for engineers to screen uneconomic or "not done before" designs. More-accurate cost laws also allow capital and energy trade-offs in a heat exchanger network to be optimized. ■

Edited by Jayadev Chowdhury

References

1. Linnhoff, B. and Vredeveld, D.R. "Pinch Technology Has Come of Age," *Chemical Engineering Progress*, July 1984, pp. 33–40.
2. Korner, H. *Chemie-Ingenieur-Technik*, 60, pp. 511–518, 1988.
3. Linnhoff, B. and Ahmad, S. "Cost Optimum Heat Exchanger Networks – 1. Minimum Energy and Capital Using Simple Models for Capital Cost," *Computers and Chemical Engineering*, 14, 7, pp. 729–750, 1990.
4. Linnhoff, B. and Hindmarsh, E. "The Pinch Design Method of Heat Exchanger Networks," *Chemical Engineering Science*, 38, 5, p. 745, 1983.
5. Linnhoff, B. and Ahmad, S. "Cost Optimum Heat Exchanger Networks – 2. Targets and Design for Detailed Capital Cost Models", *Computers and Chemical Engineering*, 14, 7, pp. 751–767, 1990.
6. Hall, S. G., Ahmad, S. and R. Smith. "Capital Cost Targets for Heat Exchanger Networks Comprising Mixed Materials of Construction, Pressure Ratings and Exchanger Types," *Computers and Chemical Engineering*, 14, 3, pp. 319–335, 1990.

The authors

Stephen Hall is a senior engineer in the chemical engineering technology group at The M.W. Kellogg Co. (601 Jefferson Ave., P.O. Box 4557, Houston, TX 77210–4557; Tel: 713-753-3348). Since joining Kellogg in 1990, he has been involved in the application of advanced process-design technologies, such as pinch analysis, dynamic simulation, and distillation-column analysis. He has a B.S. in chemical engineering from the University of Bradford (U.K.) and M.S. and Ph.D. degrees from the University of Manchester Institute of Science and Technology (U.K.). He is a registered P.E. in Texas, and is a member of AIChE.

Stephen Morgan is chief technology engineer in the advanced process-control group at The M.W. Kellogg Co. He has more than 20 years of experience in process design technology, the last 19 with Kellogg. Currently, he is responsible for applications involving model-referenced advanced process control, plantwide optimization, operator training simulators and CIM. The author of several papers, Morgan holds a patent for a process for steam cracking of feed-gas saturation. He holds a B.S. in chemical engineering from North Carolina State University. He is a member of AIChE, and a member of the administration committee of AIChE's Process Data Exchange Institute (PDXI).

Arthur H. Tuthill,
Tuthill Associates Inc.

Attention to this checklist of selection factors will materially reduce heat-exchanger failures

When designing a heat exchanger, an engineer first calculates the surface area needed to carry the heat load. Next, he or she develops the design to meet the standards of the Tubular Exchanger Manufacturers Assn. (TEMA) for heat exchangers, or other codes, and the company's standards. He or she then makes comparative cost estimates, factoring in knowledge from experience, and selects the best tubing metal for the service.

Most unexpected failures of heat exchangers can be traced to a factor that had not been fully taken into account when the tube material was selected. In Tables 1, 2 and 3, these factors are arranged according to, respectively, water quality, the character of operation and maintenance, and exchanger design. Tube materials considered are copper alloys, stainless steels (Types 304 and 316), and high alloys (6% molybdenum, superferritics* and titanium).

The impact of each factor is noted without it necessarily being related to that of other factors. In the tables, the impact of each factor is rated one of three ways: green means the tubing alloy or alloy group has given good performance under the stated conditions; yellow designates that the tubing may give good performance, but may require a closer study of the conditions at the site and relationships with other factors; and red signifies that the material has not performed well under the stipulated conditions, and special precautions are required to achieve good performance.

Water quality

Water quality encompasses: cleanliness, and the content of chloride, dissolved oxygen and sulfides, and residual chlorine and manganese. It also includes pH, temperature and scaling tendency (Table 1).

Water cleanliness — Design engineers tend to assume that cooling water will be clean. This occurs only if the right screens and filters have been installed and operators have made sure that they work properly. Debris (such as sticks and stones) and sediment (such as sand and mud) that are passed through or around the screens and filters have been responsible for many tube failures.

Long term, copper-alloy and stainless-steel tubes perform excellently in clean water (i.e., free of sediment, debris and fouling organisms). Too often, however, sediment and debris find their way into exchanger tubes. Corrosion under sediment is common with tubes of these two materials. A high

*A new family of stainless steels high in chromium content (25-30%).

THE RIGHT METAL FOR HEAT EXCHANGER TUBES

Originally published January 1990

concentration of sand can abrade the protective film on copper-alloy tubes.† Such service, therefore, requires a 70-30 CuNi-2Fe-2Mn alloy in the copper-alloy series, or stainless steel. A lodgment of sticks, stones and shell fragments creates downstream turbulence, which can cause pinholes in copper-alloy tubes.

The obvious cure for these problems is to screen the water better. As a safeguard, the newer 85-15 CuNi with 0.5 Cr, C72200 alloy and stainless-steel tubes do well in withstanding the effects of lodgment turbulence. Copper-alloy tubes effectively keep organisms from becoming attached, and are preferred when environmental restrictions on the use of biocides would require frequent manual cleaning of stainless-steel and higher-alloy tubes.

Chlorides — These provide a convenient framework for differentiating the stainless-steel alloys. Type 304 stainless steel resists crevice corrosion at chloride-ion concentrations of less than about 200 ppm, and Type 316 does so at levels up to about 1,000 ppm. (The chloride content of U.S. fresh water is typically less than 50 ppm, for which Type 304 stainless steel is therefore normally adequate.)

The 4½% molybdenum alloys suffer crevice attack at chloride-ion concentrations from 2,000 to 3,000 ppm. Both 4½% molybdenum-alloy and duplex (another new family of stainless steels) tubes have been subjected to sea water (2,000 ppm chloride content) and have undergone substantial crevice corrosion beneath fouling. On the other hand, the 6% molybdenum and superferritic alloys, and titanium have proved resistant to crevice and beneath-sediment corrosion in salt water.

Dissolved oxygen and sulfides — Copper-alloy and stainless-steel tubes perform best in water having enough oxygen (about 3-4 ppm) to keep fish alive. These tubes also do well in deaerated water, such as that used in water-flood oil wells. Copper-alloy tubes do not stand up well in severely polluted water in which dissolved oxygen has been consumed in the decay

†All alloys form a protective film in corrosive liquids; on stainless steel, it is chromium oxide; on copper alloy, it is cuprous hydroxychloride.

TABLE 1A -- Poor water quality can cut off the service life of exchanger tubes

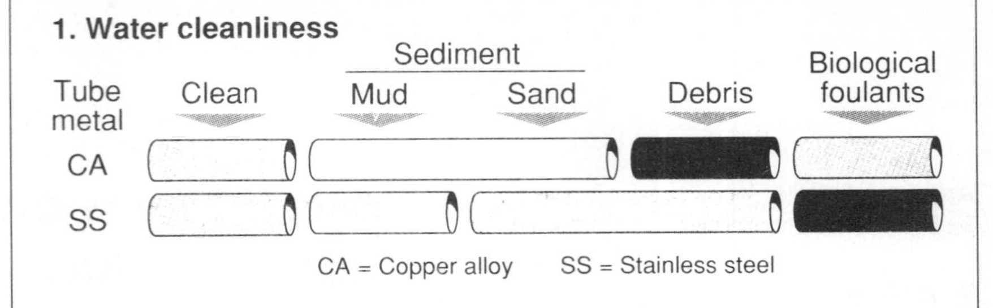

TABLE 1B -- Poor water quality can cut off the service life of exchanger tubes

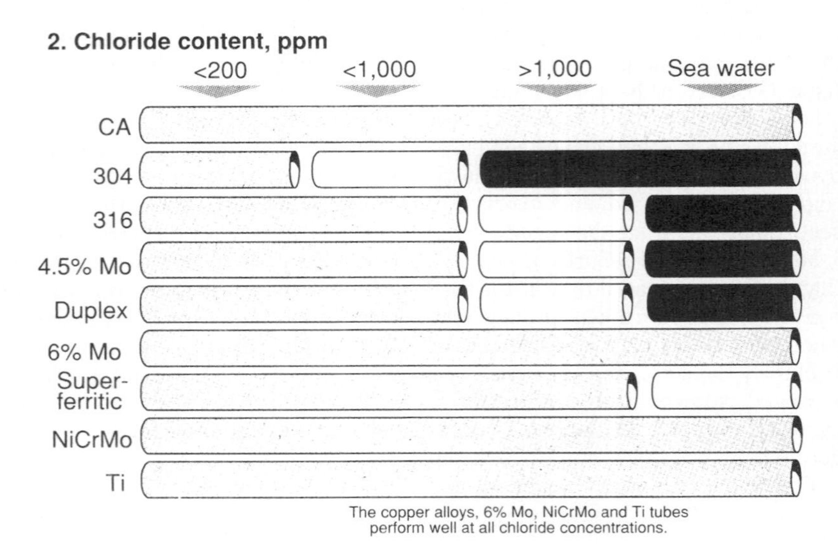

3. Dissolved oxygen and sulfides
>3-4 ppm O_2 | Dearated | Sulfide polluted
CA
SS

4. Residual chlorine
<2 ppm | 2-10 ppm | >10 ppm
CA
SS

5. Manganese content
Present | Absent
CA
SS

6. Water pH
6-8 | <5 | >9
CA
SS

7. Temperature, °F
32-45 | 60 | 120-140
CA
SS

CA = Copper alloy SS = Stainless steel

process and sulfides are present. Tubes of higher-alloyed stainless steel or titanium have served successfully in such water.

Residual chlorine — Both copper-alloy and stainless-steel tubes have performed well in water containing up to 2 ppm residual chlorine, and have failed in heavily chlorinated water. Although the usual objective is to keep residual chlorine at about 0.5 ppm at the inlet tubesheet, this level is sometimes exceeded and the residue is normally higher at the point of injection.

Acidity — In aerated water of pH less than about 5, a protective film does not easily form on copper-alloy tubes, so they corrode and thin rapidly. In deaerated water of low pH, copper-alloy tubes resist corrosion well. For high-pH water, copper-nickel or stainless-steel tubes are preferred to admiralty- (71% copper, 28% brass, 1% tin) or aluminum-brass tubes, which tend to corrode under highly alkaline conditions. Stainless-steel tubes have performed well at a pH less than 5 and greater than 9.

Temperature — A protective film readily forms on copper alloys in warm water (in about ten minutes at 60°F), but forms very slowly in cold water. It develops almost instantaneously on stainless steel in both warm and cold water.

Manganese — Type 304 stainless steel tubes have failed in fresh water having an appreciable manganese content. However, copper-alloy and higher-alloyed tubes have fared well in such water.

Leaving an exchanger full, or even only wet, invites corrosion

Scaling tendency — Copper-alloy and stainless-steel tubes perform well in both hard (scaling) and soft (nonscaling) water. The Langelier saturation index* is frequently used to distinguish scale-forming from corrosive (to carbon steel) water.

Operation and maintenance

Among these factors are type of operation (regular vs. intermittent) and the frequency of cleaning (Table 2).

TABLE 2-- Operation and maintenance practices that extend tube life

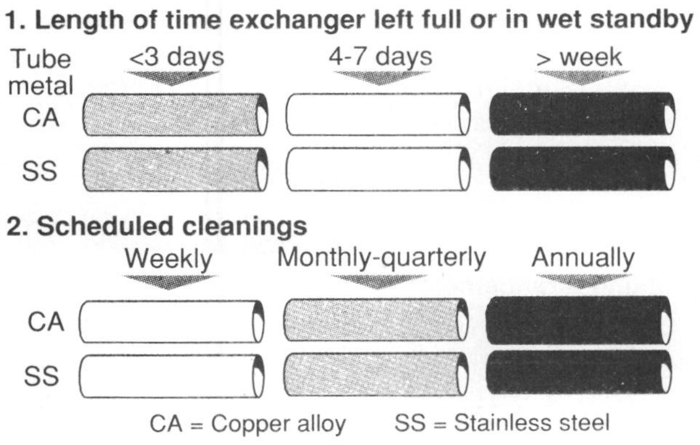

1. Length of time exchanger left full or in wet standby

Tube metal	<3 days	4-7 days	> week
CA			
SS			

2. Scheduled cleanings

	Weekly	Monthly-quarterly	Annually
CA			
SS			

CA = Copper alloy SS = Stainless steel

Character of operation — Lengthy startups have been responsible for many failures of copper-alloy and stainless-steel tubes. These occurred because water had been left in, or it had only been partially drained from, tubes for a long time. Such failures have also been caused by extended outages from normal operations.

Leaving an exchanger full, or even only wet, invites corrosion. The water will become foul, creating an environment in which bacteria thrives, promoting microbiologically induced corrosion. Corrosion will also take place under sediment. If an exchanger is to be left full for more than 2 or 3 days, water should be pumped through it once a day to displace the stagnant water. If an exchanger will be down for at least a week, it should be drained and blown dry.

Cleaning schedule — Heat exchangers should be periodically flushed out, opened and brush cleaned, to remove sediment and debris and to restore heat transfer capability. Exchangers handling water high in biological foulants or sediment should be cleaned weekly. Micro-organisms will corrode stainless steel if they and other deposits are not removed.

Monthly or even quarterly mechanical cleanings of copper-alloy and stainless-steel tubes are adequate with most waters. The optimum interval can be critical because the protective film on copper alloy tubes can be damaged by cleaning with metallic brushes and some types of abrasive blasting. Mechanical cleaning is sometimes put off for a year, and even longer, particularly if restoring an exchanger's heat-transfer rate is not critical to plant performance. Be cautioned, however, that corrosion beneath sediment frequently occurs when sediment removal is delayed by more than three months.

Exchanger design

The principal design factors that influence tube performance are water velocity, tube diameter, shape (i.e., once-through or U-bend), orientation, venting, tubesheet material, channel (waterbox) material and channel inlet arrangement (Table 3).

Fluid velocity — At velocities of less than 3 ft/s, sediment deposit, debris buildup and biological fouling in and on tubes can be excessive, resulting in the need for frequent mechanical cleaning, which can cause copper-alloy and stainless-tube tubes to fail prematurely.

Copper alloys can be conveniently differentiated according to their water-velocity tolerance. Approximate maximum design velocities for tubes of copper-base alloys and stainless

*This index indicates the tendency of a water solution to precipitate or dissolve calcium carbonate. It is calculated from total dissolved solids, calcium concentration, total alkalinity, pH and solution temperature.

steel in salt water are listed in Table 3. Maximum velocity is usually arrived at as a compromise between the cost of pumping and the advantage in heat transfer. The design velocity usually falls in the range of 6-8 ft/s, and may reach 12-15 ft/s when extra cooling demand is placed on an exchanger, such as in summer.

Copper-nickel tubes stand up reasonably well to variations in velocity, although some erosion and corrosion may occur at the inlet. The C72200 alloy resists inlet-end erosion and corrosion, as well as corrosion downstream of lodgments. The 2Fe-2Mn modification of the C71500 alloy resists inlet erosion and corrosion excellently. Copper-nickel tubes, after their protective film has aged, withstand considerable velocity excursions without significant inlet erosion. Stainless-steel tubes perform best at high velocity and are useful up to velocities that induce cavitation. Copper-alloy tubes tolerate high velocities, 10 ft/s being a common design velocity for copper in air-conditioning coolers and condensers.

Tube diameter — Tubes of large diameter are preferred for heat exchangers because any solids that pass through screens will also flow through the tubes. By one rule of thumb, tube diameters should be at least twice the diameters of the screen openings. Tubes should not be less than ½ inch if the water to the exchanger is not filtered.

Once-through or U-bend — Because U-tube bundles are difficult to clean, the water must be very clean (e.g., boiler feedwater quality), or at least well-filtered. Both stainless-steel and copper-alloy tubes are liable to corrode beneath sediment. U-tubes are particularly prone to such corrosion if sediment and debris are not removed from their bends. Once-through and 2- to 4-pass bundles are easily cleaned by flushing, water lancing, or brushing.

Orientation — Heat exchangers, particularly condensers, are normally placed horizontally, with water flowing through the tubes. Both stainless-steel and copper-alloy tubes perform well in such exchangers. In those unusual circumstances that require that

TABLE 3 -- Exchanger design features that influence tube performance

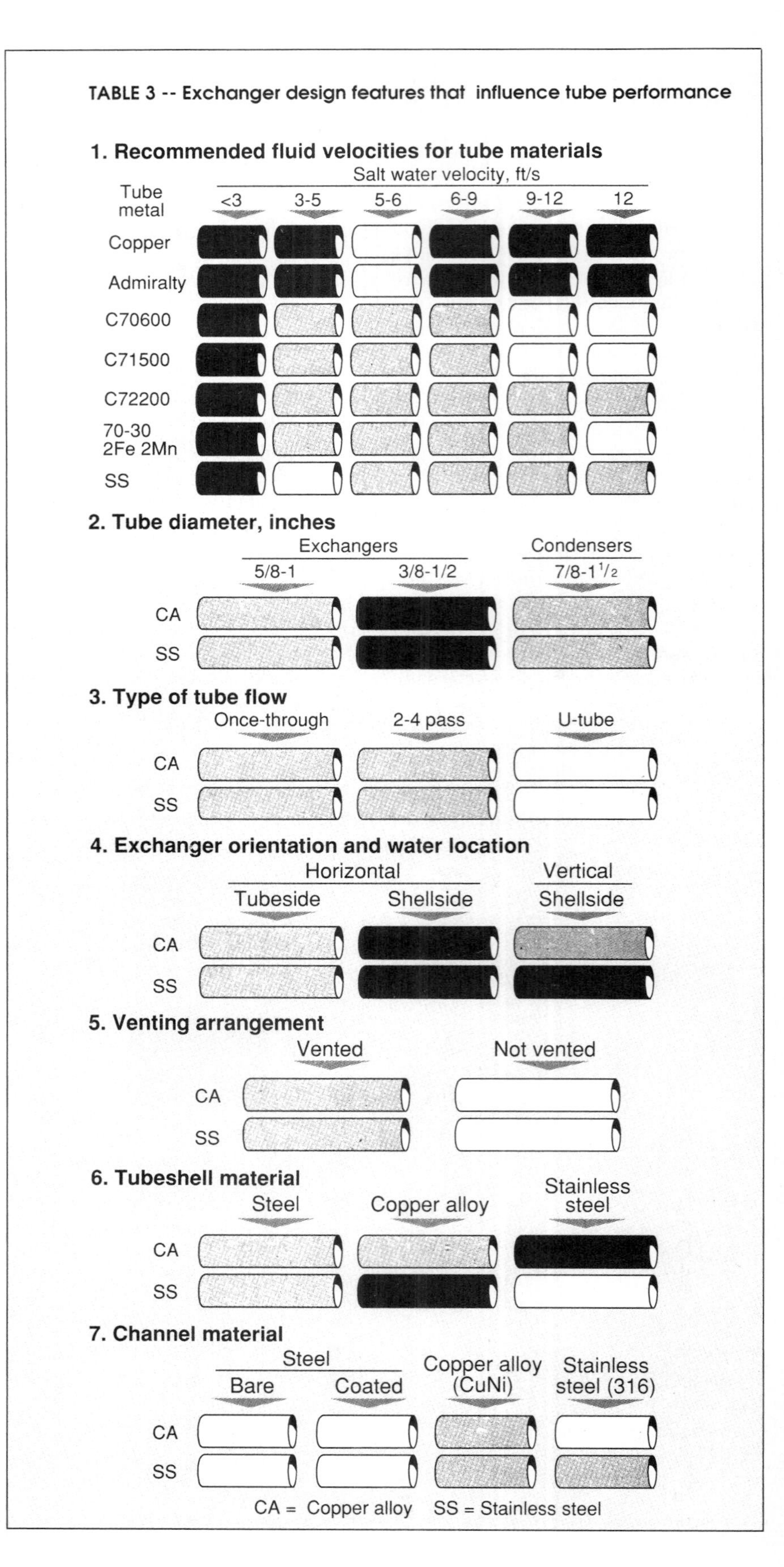

the cooling-water flow on the shellside of a horizontal exchanger, sediment and debris cannot be kept from building up around the support plates and lower tubes. This results in corrosion beneath the sediment. The buildup can be eliminated, or at least restrained, by upstream filters, but these will not prevent fouling organisms from thriving in low-flow areas. When the water, even clean water, must be on the shellside, the tubes should be of high alloy.

A condenser is sometimes oriented vertically, with the cooling water on the shellside, to improve heat transfer. This allows noncondensable gases to collect under the top tubesheet, letting the temperature of the tube walls in the gas pocket get almost as hot as the incoming gas being condensed, evaporating the water and depositing scale on the hot tube surfaces. Sediment also tends to build up in the bottom of a vertical condenser.

In vertical installations, both Type 304 and Type 316 stainless-steel tubes are prone to stress corrosion and cracking just under the top tubesheet. Remedies include: venting gas through the top tubesheet; changing to copper-alloy or copper alloy-stainless steel bimetallic tubing or to high-alloy tubing, or orienting the exchanger horizontally. Although the TEMA standards recommend that exchangers be installed level, exchangers should really be sloped slightly so that they will drain completely when shut down, avoiding corrosion where tubes sag between support plates and retain water even after being drained.

Venting — Exchangers are normally fitted with vent cocks so they can be purged to clear gas pockets. Condensers, particularly when chlorine is used as a biocide, tend to suffer corrosion when gases are not vented.

Tubesheet material — Tubesheets of carbon steel, common copper alloy, Muntz metal (a brass composed of 58-61% copper, up to 1% lead, with the remainder zinc), admiralty brass and aluminum bronze are anodic to copper-alloy tubes. The galvanic protection that these tubesheet materials afford copper-alloy tubes does not eliminate all inlet-end corrosion, but often does help to keep it at a tolerable level.

Stainless-steel tubesheets are rarely used with copper-alloy tubes, because stainless steel is cathodic to copper alloy. Installing stainless-steel or titanium tubes in copper-alloy tubesheets has accelerated the corrosion of the tubesheets. Stainless steel and titanium polarize so readily that the entire inside surface of the tubes (not just the part adjacent to the tubesheet) must be considered as the cathode to the copper-alloy tubesheet anode.

The normal cure for this anode-cathode problem is to protect the copper-alloy tubesheet with an impressed-current cathodic protection unit. The unit's potential must be controlled to avoid hydrogen embrittlement of ferritic stainless steel, or hydriding of titanium. An alternative is to resort to tubes of 6%-molybdenum, which resist hydrogen embrittlement without the need to carefully control an applied potential.

Tubesheets for austenitic stainless steel should be of identical composition. Tubesheets of high-alloy austenitic stainless steel are used with tubes of ferritic stainless steel, because tubesheets of matching composition are not available. Solid or clad titanium tubesheets are preferred with titanium tubes.

Channel material — Corrosion products from bare-steel and cast-iron channels (exchanger inlet chambers) have occasionally caused the corrosion failure of copper-alloy and stainless-steel tubes. Coatings applied to steel and cast-iron channels for the purpose of reducing corrosion sometimes deteriorate and spall, leading to tube corrosion and failure. Coated channels are also liable to deep local corrosion at pinholes in coatings brought on by an adverse galvanic couple between the steel channel and the alloy tube and tubesheet.

Copper-nickel and aluminum-bronze (solid or weld-overlaid surface) channels are preferred with copper-alloy tubes, and may also be used with high-alloy tubes. The adverse galvanic effect is diminished in such installations, because the area of the channel is larger than that of the tubesheet, and the channel and tubes are not contiguous. Channels of Type 316 stainless steel (solid or overlaid) are preferred with stainless-steel tubes.

Channel inlet arrangement — Uneven flow, restricted water passages and poor inlet-piping entry arrangements have caused numerous failures of copper-alloy tubes, as well as a few failures of stainless-steel tubes. Basically, the entry and flow pattern in both large and small channels should distribute the water uniformly to all tubes with as little swirling as possible.

If a review of the foregoing factors indicates that an exchanger's tubes must be of a metal that is even more corrosion resistant than a copper alloy or stainless steel, an engineer frequently resorts to a more highly alloyed metal, such a 6%-molybdenum stainless steel, a superferritic or titanium. These materials provide outstanding resistance to corrosion in crevices and beneath sediment. ■

TEMA standards recommend that exchangers be installed level, but they should really be sloped

The author

Arthur H. Tuthill, a materials and corrosion consultant (2903 Wakefield Dr., Blackburg, VA 24060; tel.: 703-953-2626), has had extensive experience in the fabrication, use and performance of metals, particularly involving heat exchangers, piping and pumps, in a wide range of industries. Holder of an M.S. in metallurgical engineering from Carnegie-Mellon University and a B.S. in chemical engineering from the University of Virginia, he is a licensed engineer (metallurgy). Among the many articles that he has had published is a five-part series in Chemical Engineering: "Installed Cost of Corrosion-Resistant Piping" (Mar. 3 and 31, Apr. 28, May 26 and June 23, 1986).

HEAT TRANSFER TOPICS

Errors are frequently made during specification of shell-and-tube heat exchangers. Many of the common mistakes can be avoided

Frank L. Rubin
Practical Heat Transfer Consultants

There is an immense array of literature on heat-exchanger design and specification. Even so, numerous aspects that can have significant consequences are covered rarely or not at all. Awareness of and attention to these aspects can lead to better design and performance.*

Heat load — Which one?

The most common problem, which is never discussed in the literature, is the confusion as to which quantity of heat to use for designing the exchanger. There are three heat loads, and they almost always differ.

First, there is the specified heat load, which typically appears on the heat-exchanger specification sheet on the line for "heat exchanged." Second is the calculated heat released by the hot fluid, and third is the calculated heat absorbed by the cold fluid.

Even in the best of circumstances, these quantities will rarely be identical. However, if any one of them differs by more than 10% from another, thermal design of the heat exchanger should not proceed until the discrepancy is resolved. At any rate, the design should be based upon the specified "heat exchanged" (and upon the fluid temperatures of the specification sheet).

LMTD imprecision

Charts showing correction factors for log mean temperature differences (LMTDs), such as Figure 1, can be hard to read. For instance, consider Example 7.6 from p. 162 of Kern's "Process Heat Transfer" [2], in which a potassium phosphate solution is to be cooled from 150 to 90°F using well water rising from 68 to 90°F.

The temperature change of the hot fluid is 60 degrees, that of the cold fluid is 22 degrees, and the extreme temperature range for the system is 150 – 68, i.e., 82 degrees. The ratio of cold-fluid change to extreme temperature change, 22/82 or 0.268, is the abscissa for Figure 1. (This value is designated as S in Kern, but instead as P in the Tubular Exchanger Mfrs. Assn. [TEMA] standards [3].) And the ratio of fluid-temperature changes, 60/22 or 2.73, is the parameter R in Figure 1.

The LMTD correction factor is read from Figure 1 as the ordinate that corresponds to R and P. In the Kern example, this factor has been read as 0.81. In my opinion, however, users of Figure 1 might read any value between 0.78 and 0.82. A precise reading cannot be readily made.

Often, however, a simple manipulation of the variables can lessen this difficulty. Many heat exchangers are stream-symmetric, which means that the "T" and the "t" of the correction-factor diagrams (as in Figure 1) can be applied to either the shell- or tubeside fluids. In such cases, one can replace R by $1/R$ and P by the product PR when using the diagram.

Thus, in our example, R of 2.73 is replaced by 0.366 and P is replaced by 0.732. With these values, the diagram can be read more easily, and a correction factor of 0.815 appears applicable.

*These and other topics are discussed more fully in the soon-to-be-published "Heat Exchangers, Economics and Management;" for details, see Reference [1].

Originally published August 1992

ROBERT NEUBECKER

The TEMA correction factor diagram itself indicates whether a given exchanger is stream-symmetric.

Design-dimension challenges

Numerous ambiguities and other problems can crop up with regards to the dimensions for the heat exchanger.

Information not available in tabulations. One such complication is that the dimensions for minimum-wall tubes are not tabulated. "Characteristics of Tubing" in the TEMA standards and in most other publications are applicable to seamless, average-wall tubes, which have thicknesses of BWG (Birmingham Wire Gauge) dimensions. Minimum-wall tubes are thicker, and therefore both their weight and their moment of inertia are greater. As a rule of thumb, the smallest thickness of a seamless, minimum-wall steel tube approaches the tabulated thickness of the BWG tube with the next thicker wall.

A similar difficulty is that "Characteristics of Steel Tubing" in the "Stan-

OFTEN OVERLOOKED

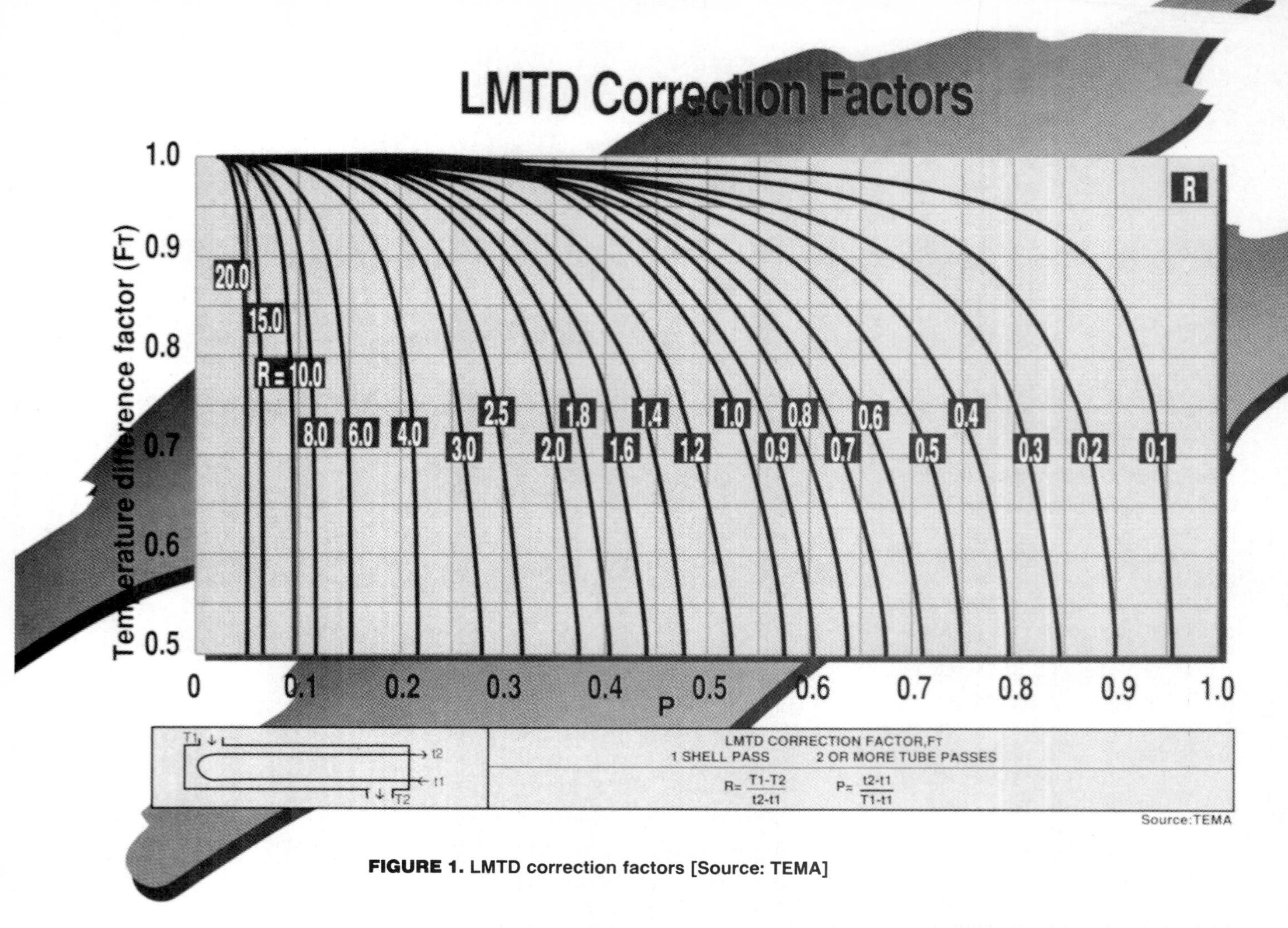

FIGURE 1. LMTD correction factors [Source: TEMA]

dard for Power Plant Heat Exchangers" [4] apply to welded, average-wall tubes. These tubes are available in thickness ranging from 0.050 to 0.150 in., in increments of 0.005 in. The dimensions for seamless tubing are different. Furthermore, the rolled strip used to produce welded tubing is not available in thicknesses corresponding to those of the Birmingham Wire Gauge.

Baffles and their tube holes. A source of difficulty is the inability to drill baffle holes in precisely identical size. Test results for pressure drop and heat transfer generally ignore this fact-of-life, instead going on the assumption that the baffle-hole diameters coincide exactly with the drill diameter.

As it happens, the TEMA standards

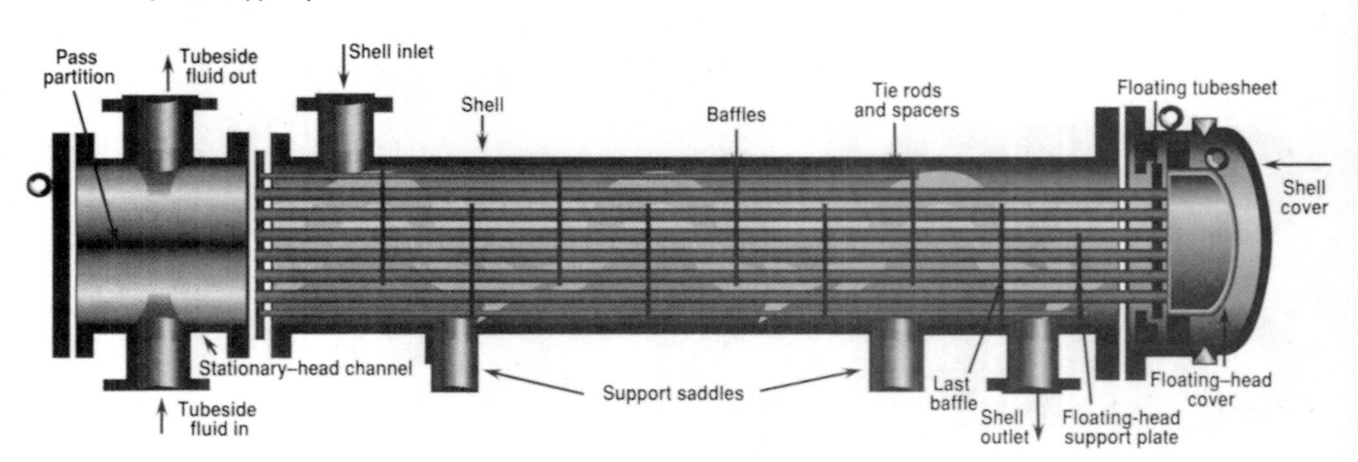

FIGURE 2. In this Type AES internal floating-head heat exchanger, the shell outlet nozzle is located between the last baffle and the floating-head support plate

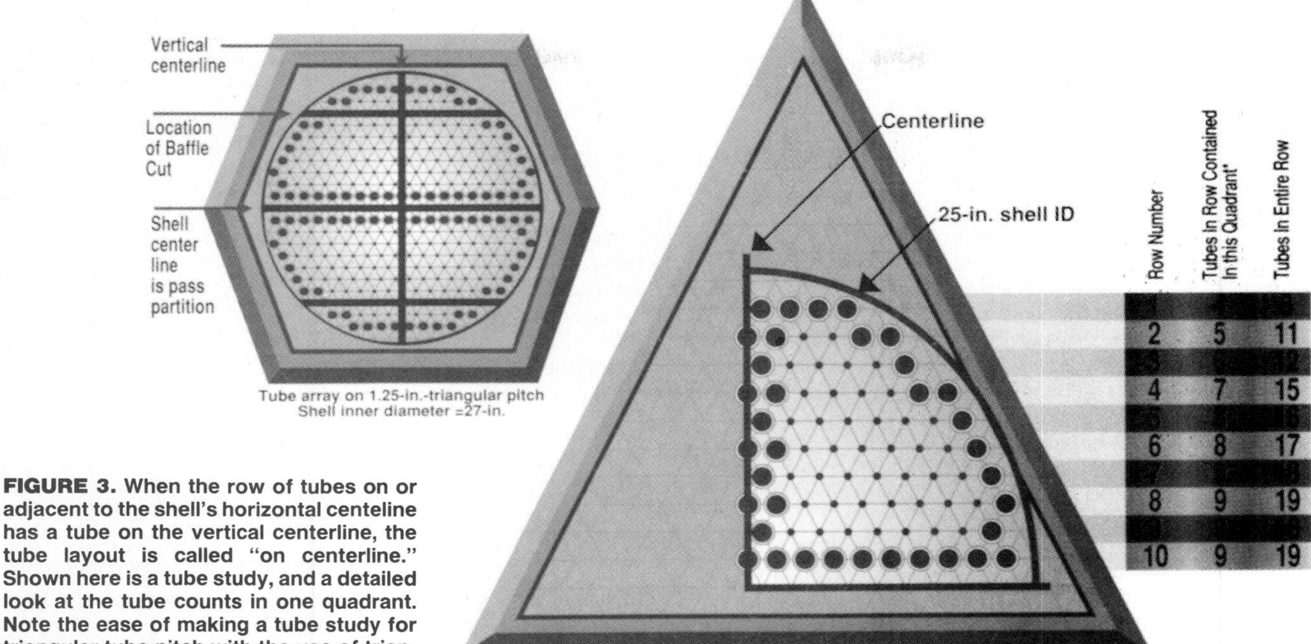

FIGURE 3. When the row of tubes on or adjacent to the shell's horizontal centeline has a tube on the vertical centerline, the tube layout is called "on centerline." Shown here is a tube study, and a detailed look at the tube counts in one quadrant. Note the ease of making a tube study for triangular tube pitch with the use of triangular graph paper

do permit 0.010 in. as oversize tolerance on holes in baffles, with 4% of the holes allowed to be 0.011 to 0.015 in. oversize. In my opinion, however, the probable maximum average hole diameter is only 0.007 in. greater than the drill diameter when no more than 4% of the tube holes are larger than drill diameter plus 0.010 inch. Thus, it is impossible to have an average hole size that equals drill diameter plus 0.010 inch and still comply with the TEMA requirements.

Another point should be kept in mind regarding baffles. Bypassing of the baffles depends upon inside shell diameter and the baffle diameter. These design dimensions should be employed in all calculations, unless the fabrication firm has demonstrated that its baffle holes have a lesser over-tolerance.

Be aware that the radial distance between baffle and rolled shells, as well as that between baffles and pipe shells, has zero over-tolerance. Both ends of heat exchanger shells are welded at the "design ID [inside diameter]" dimension to flanges or tubesheets. The diameter of the shell cylinder cannot change significantly from the design ID.

Design of exchangers by computer. Proper heat-exchanger design with the assistance of a computer program requires a knowledgeable, experienced thermal engineer. Also needed is close cooperation from the mechanical designer of the exchanger.

Be alert to a variety of problems that can arise involving exchanger dimensions

Ideally, a computer program would input the customer's thermal and mechanical requirements and then produce both the thermal and mechanical design. However, even if a such a program could be prepared, the multiplicity of dimensional options allowed by the ASME Code for Unfired Pressure Vessels [5] and other standards would prevent the program from being put to use.

For instance, one manufacturer's computer program for the design of hub flanges develops five alternate solutions having different hub lengths, thicknesses and tapers. The mechanical-design engineer must select the version that is most suitable for the particular installation. Similarly, the ASME Code allows twelve options for reinforcing of cylinders into which many nozzles are attached, and the mechanical designer must choose among these options.

The engineer with the background adequate to select the best hub flange and nozzle reinforcement is in fact unlikely to be using the proposed design program in the first place. Such an engineer's knowledge and experience will result in promotion to greater responsibilities, either by the current employer or by a new one.

As it happens, a complication that involves hub flanges and nozzle reinforcement also prevents computer thermal-design programs for shell-and-tube heat exchangers from maximizing heat transfer in one computer run. The heat transfer depends in part on the location of terminal baffles, but this itself is dependent upon subsequent mechanical design for the hub-flange length and for the type of nozzle reinforcing.

Another case in which a mechanical-design uncertainty hampers thermal-design computer programs (or, for

A STEP-BY-STEP STROLL THROUGH TABLE 1

(1) Consider Row 4 as an example
(2) The row is 8.453 in. above the centerline
(3) The distance from the outermost tube to the shell on the row centerline is 0.58 in., and is the "bundle bypass distance"
(4) There are 14 tubes in Row 4
(5) There are 272 tubes in the exchanger, since the three upper rows (and the symmetrical rows of the lower pass) are omitted to provide shell inlet and outlet areas. This arrangement is suitable both with and without an impingement plate at the inlet nozzle
(6) When there is an impingement plate, there may be enough shell exit area to permit one extra row below the centerline. The tube count thus rises to 283
(7) An 18-in. nozzle is the largest that can be installed when there is no impingement plate
(8) A 12-in. nozzle is the largest that can be installed when there is an impingement plate. The square impingement plate has 15-in. sides
(9) Velocity with no impingement plate is 4.59 ft/s in 18-in. nozzle
(10) Velocity with next smaller nozzle is 4.90 ft/s in 16-in. nozzle
(11) Velocity with impingement plate and biggest nozzle plus extra row in lower pass: 5.79 ft/s in 12-in. nozzle. Tube count is 283
(12) Velocity with impingement plate and smaller nozzle plus extra row in lower pass: 7.58 ft/s in 10-in. nozzle
(13) Velocity with impingement plate and biggest nozzle with symmeterical rows of tubes: 5.79 ft/s in 12-in. nozzle. Tube count is 272
(14) Velocity with impingement plate and smaller nozzle with symmetrical rows: 7.84 ft/s in 10-in. nozzle.

TABLE 1. Two-pass tube-count table

Shell ID	Row no. (1)	Vert. dist. (2)	Edge dist. (3)
25 in.	1	11.700	3.90
	2	10.618	1.72
	3	9.535	1.33
	4	8.453	0.58
	5	7.370	0.85
	6	6.288	0.93
	7	5.205	0.86
	8	4.123	0.68
	9	3.040	0.37
Clear	10	1.958	1.22
5.94	11	0.875	0.72

that matter, human thermal designers) arises with Type AES internal-floating-head exchangers, one of which appears in Figure 2. This exchanger differs from the typical heat exchanger, in which the shell outlet nozzle is located between the tubesheet and the last baffle.

In the Type AES exchanger, the outlet nozzle instead lies between the floating-head support plate and the last baffle. The location of the support plate is not known during the thermal design because it depends on the subsequent mechanical design. As the distance between the support plate and the adjacent baffle is thus not known, the thermal designer almost always ignores the pressure drop developed in the flow between them. Frequently, however, this is in fact larger than the pressure loss between each of the uniformly spaced baffles.

Ignoring this pressure drop affects the accuracy of the thermal calculation. The relationships between heat transfer and pressure drop can be determined from the factors of the Delaware Method, which is discussed later in this article.

Before leaving dimensional problems, it is well to digress and comment

Thermal design is complicated by uncertainties not resolved before the mechanical design

briefly on pressure drop in another context. The sharp rise in energy cost in the latter years of this century must be kept in mind as regards not only the heat-transfer efficiency of the exchanger but also the energy cost of pumping fluid through it.

When half of the "allowable pressure drop" is sufficient to meet the thermal design requirements for a given exchanger, is it desirable to double the pressure drop if the overall heat transfer rate increases by, say, 0.1%? By 0.5%? By 4%? Since the most common change in plant requirements is to increase the flow quantities, the engineer should exercise caution before increasing the number of baffles, because of enormous pressure drops at increased flows. Near the end of this article, we examine a situation where doubling the shell-side pressure drop provides very minimal increase in heat transfer. In many applications, the small benefit would not be economically desirable.

In fact, the concept of "allowable pressure drop" might well be discarded throughout much of process-industries manufacture. Bidders who are selected during heat-exchanger inquiries should be required not only to comply with specifications but also to offer the lowest *total* for (a) the delivered cost of the heat exchanger plus (b) the present value of the electricity required for the pressure losses during the entire projected plant life.

Tube counts

The Downingtown Iron Works probably published the first tube-count tables [6] for several types of shell-and-tube heat exchangers. Crucial dimensions for Outer Tube Limit (OTL) and minimum spacing of tubes across the pass partition, apparently available in only one paper in the literature [7], make up the basis for Figure 3. The "on-centerline" tube study in that figure has a tube on the vertical centerline of the row adjacent to the horizontal pass partition. In the "off-centerline" layout shown in Figure 4, the tubes of that row straddle the vertical centerline.

The information in these two figures assumes usage of a maximum number of tubes. So do confidential tube-count drawings and tables for internal floating head construction prepared by TEMA. On the other hand, the tube-count tables of "Perry's

Tubes	Total tube count		——Nozzle——		——Velocities (ft/s)——					
in row (4)	NoP (5)	IMP (6)	ZNB (7)	ZPB (8)	VNB (9)	VNS (10)	VPB+ (11)	VPS+ (12)	VPB (13)	VPS (14)
1	312	312	4	0	3.26	4.25	0.00	0.00	0.00	0.00
8	310	311	10	6	4.12	4.90	2.08	3.26	3.59	6.32
11	294	302	14	10	4.90	4.90	4.11	5.62	4.38	6.35
14	272	283	18	12	4.59	4.90	5.79	7.58	5.79	7.84
15	244	258	20	14	4.90	4.90	7.06	8.68	7.06	8.68
16	214	229	20	16	4.90	4.90	7.59	9.68	7.59	9.68
17	182	198	20	16	4.90	4.90	9.87	10.00	9.87	10.00
18	148	165	20	18	4.90	4.90	9.85	10.00	9.85	10.00
19	112	130	24	18	4.90	4.90	10.00	10.00	10.00	10.00
18	74	93	24	18	4.90	4.90	10.00	10.00	10.00	10.00
19	38	56	24	20	4.90	4.90	10.00	10.00	10.00	10.00

Tube-count table for 25-in. inside dia. two-pass tubeside, fixed-tubesheet heat exchanger, TEMA Type L, M or N with 1-in. tubes on 1.25-in. triangular pitch. Shell nozzles of identical size, located at top and bottom of shell. Horizontal pass partition between the tube passes

EXPLANATION OF COLUMN HEADINGS FOR TABLES 1 AND 2

Shell ID, clear

Shell ID is shell inside diameter (25 in. here); Clear = clearance = (chord length of the row) minus (the number of tubes in the row times tube diameter).

NOTE: Flow area for the crossflow stream equals baffle pitch times clearance, when excessive bundle bypassing does not occur

(1) Row No.

= row number. The top row is designated #1 and the largest number is for the row adjacent to the horizontal centerline. Only rows above the centerline are examined in this table

(2) Vertical distance

= the row's distance from the shell's horizontal centerline (inches)

(3) Edge distance

= the distance from each of the outermost tubes in the row to the shell (inches)

(4) Tubes in row

= tube count in that row

(5) Total tube count (NoP)

= number of tubes in the exchanger without an impingement plate, with all upper rows deleted and symmetrical rows below centerline deleted

(6) Total tube count (IMP)

= total number of tubes in in exchanger with impingement plate and all upper rows deleted and one fewer row below the centerline deleted. In some exchangers there may be sufficient shell exit area with the additional row below the centerline

NOTE: No adjustment in the tube counts of (5) or (6) is made for possible omission of tubes to permit installation of tierods and spacers

(7) ZNB nozzle

= Nozzle size (inches) with no impingement plate. The largest nozzle complying with TEMA's limitations concerning velocity and density is listed in this column

(8) ZPB nozzle

= the largest nozzle usable when there is an impingement plate (having sides about 20% greater than the nozzle)

Columns (9) through (14) have maximum velocities of water, in feet per second at 68°F

(9) VNB velocity

= velocity with no impingement plate in the nozzle size of Column (7)

(10) VNS velocity

= velocity with the next smaller nozzle than the one in Column (7)

(11) VPB+ velocity

= velocity with an impingement plate, the biggest nozzle, and an extra row in the lower pass

(12) VPS+ velocity

= velocity with an impingement plate, the smaller nozzle, and an extra row in the lower pass

(13) VPB velocity

= velocity with an impingement plate, the biggest nozzle, and symmetrical rows of tubes

(14) VPS velocity

= velocity with an impingement plate, the smaller nozzle, and symmetrical rows of tubes

RIBBON FLOW IN A FOUR-PASS HEAT EXCHANGER

Table 2 describes the tube counts for a four-pass heat exchanger, where pass partitions are used to arrange the tubes for ribbon flow (shown in the diagram). The horizontal pass partitions are located on the horizontal centerline, and between Rows 6 and 7, both above and below the centerline.

Row 7 is the top row in Pass 2. Thus, in Table 2, Columns (5) through (14) have no values, since these rows pertain to nozzles, and there are no nozzles in Pass 2.

Tables 1 and 2 show tube counts for the top half of an exchanger. Because of the extra pass partition in the four-pass design, there are only 10 rows in upper pass of the four-pass heat exchanger (with tube counts listed in Table 2), while there are 11 rows in the 2-pass exchanger (Table 1).

TABLE 2. Four-pass tube-count table

Shell ID	Row no. (1)	Vertical distance (2)	Edge distance (3)
25 in.	1	11.285	1.75
	2	10.203	1.72
	3	9.120	1.17
	4	8.038	1.57
	5	6.955	0.51
	6	5.873	0.53
	7	4.123	0.68
	8	3.040	0.37
Clear	9	1.958	1.22
5.94 in.	10	0.875	0.72

Chemical Engineers' Handbook" [8] were prepared by a computer program of Nooter Corp. and made allowance for shell entrance area for nozzles that were one-fifth the shell diameter.

This question of entrance areas (and of shell exit areas) affects exchanger reliability. Tube failures due to vibration have risen during the latter half of the century; TEMA investigation shows that the majority of of the failures reported have been caused by inadequate space for shell entrance and shell exit areas.

When the shell nozzles are identical in size and located opposite to each other and when there is no impingement plate below the shell inlet nozzle, the shell inlet and exit areas required by the TEMA standard are equal. When rows of tubes are eliminated for shell inlet area, then identical rows are deleted below the centerline for shell exit area.

When there is an impingement plate, the shell inlet area is reduced. However, the method in the TEMA standards for determining the shell inlet area with impingement plates differs from the method when there is no

In dealing with tube counts, do not overlook any impingement plates in the designs

FIGURE 4. When the row of tubes on or adjacent to the shell's horizontal centerline straddles the vertical centerline, the unit is said to have an "off-centerline" tube layout

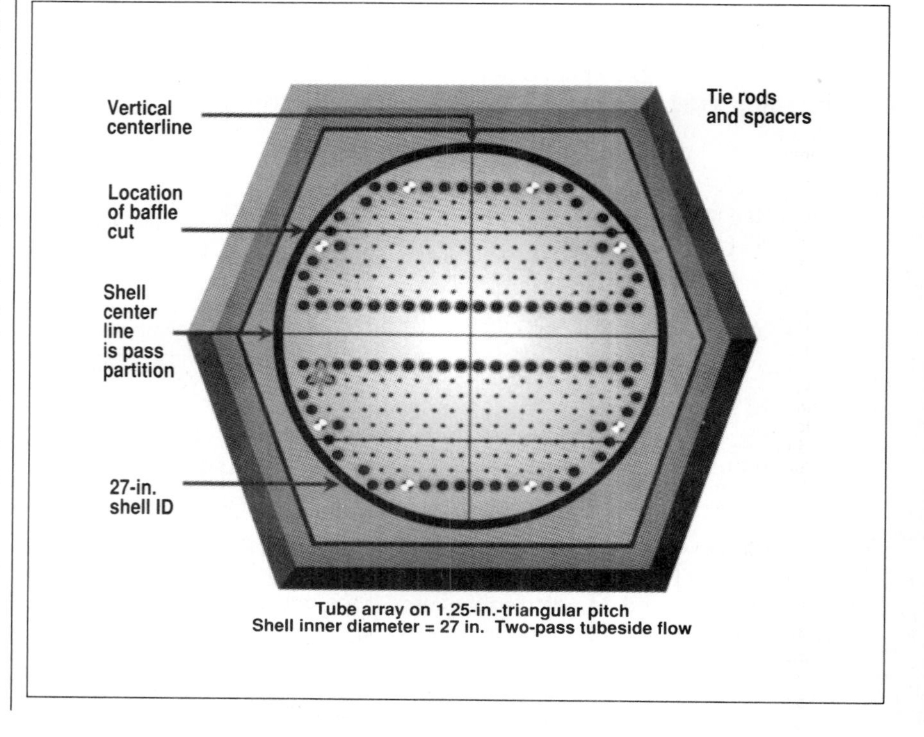

Tubes in row (4)	Total tube count NoP (5)	IMP (6)	Nozzle ZNB (7)	ZPB (8)	Velocities (ft/s) VNB (9)	VNS (10)	VPB+ (11)	VPS+ (12)	VPB (13)	VPS (14)
6	290	290	8	2	2.96	4.30	0.00	0.00	3.07	0.00
9	278	284	12	8	4.07	4.90	2.95	4.29	3.68	5.80
12	260	269	16	10	4.57	4.90	5.44	7.28	5.71	8.01
13	236	248	18	12	4.90	4.90	6.90	8.91	6.90	9.17
16	210	223	20	14	4.90	4.90	8.06	9.79	8.06	9.79
17	178	194	20	16	4.90	4.90	8.46	10.00	8.46	10.00
18	Pass 1 = 6 rows & 73 tubes									
19	Pass 2 = 4 rows & 72 tubes									
18										
19										

Tube-count table for 25-in. inside dia. four-pass tubeside, fixed-tubesheet heat exchanger, TEMA Type L, M or N with one-inch tubes on 1.25-in. triangular pitch. Shell nozzles of identical size and located at top and bottom of shell. Horizontal pass partition between the tube passes

TUBE COUNTS WHEN SHELL NOZZLES HAVE THE SAME ORIENTATION

Row number (1)	Tubes in row (2)	Total tube count (3)	Total count in Col. (5) of Table 1 (4)
1	1	312*	312
2	8	311†	310
3	11	303	294
4	14	292	272
5	15	278	244‡
6	16	263	214
7	17	247	182
8	18	230	148
9	19	212	112
10	18	193	74
11	19	175	38

*Duplicates the count of Col. (5), Table 1
†Duplicates the count of Col. (6), Table 1
Notes:
Column (2) above is the same as Column (4) of Table 1, and Column (4) above is the same as Column (5) of Table 1
Column (3) has all the tubes of the lower pass (156) plus some tubes of the upper pass
At Row 11, there are 156 plus 19, i.e., 175 tubes
At Row 10, there are 175 plus 18, i.e., 193 tubes
At Row 9, there are 193 plus 19, i.e., 212 tubes
Column (5) of Table 1 shows 11 different counts
Column (6) of Table 1 shows 10 different counts
Column (3) above shows 9 additional different counts
Thus, total number of different tube counts is 11 plus 10 plus 9, i.e., 30
Let N equal the number of tube rows (11 in this case); then the total number of different tube counts equals $3(N - 1)$

‡Random occurrence: This figure duplicates a tube count in Table 1. This random occurrence results in the total tube count being lower than predicted

TABLE 3. These tube counts relate to the heat exchanger featured in Table 1, when both inlet and outlet nozzles are located at the top of the shell, rather than 180 degrees apart

plate. In my calculations, I use the model for flow without impingement plate and then calculate a reduced shell entrance area with the impingement plate. Charles Hammack [9] advises that when there is an impingement plate, it may be possible to provide an additional row of tubes below the horizontal centerline and still have adequate shell exit area.

The most-extensive tube-count tables are to be published shortly in "Heat Exchangers, Economics and Management" [1]. These include maximum nozzle sizes at each row, with due allowance for shell inlet and exit areas. Fixed-tubesheet, U-tube, and internal-floating-head constructions are covered, as are two-pass and four-pass tube sides. The tables deal with 3/4-in. tubes on 15/16-in. triangular pitch and on 1-in. square pitch, as well as 1-in. tubes on 1.25-in. triangular and square pitch. The shell sizes go from 8 to 60 in.

The tube count of each row is listed. So is the total tube count with identical rows above and below the centerline, and with one additional row below the centerline. Also included are the maximum nozzle sizes with and without impingement plate, and the maximum velocity of water in these nozzles. Maximum velocities for nozzles that are one size smaller are likewise tabulated.

Most of the tube counts are for on-centerline tube arrangements. In cases where the off-centerline arrangement results in a greater count, it instead is used and "OFF CL" appears in the table. The clearance at the limiting row near the centerline is also given.

An example extracted from this book appears in Table 1. It shows data for a 25-in.-inside-diameter, fixed-tubesheet exchanger with two-pass tube side and 1-in. tubes on 1.25-in. equilateral triangular pitch, prepared in accordance with the tube study of Figure 3. Table 2, taken from the same source, pertains to four-pass tube counts.

The four-pass heat exchanger of Table 2, and the relevant tube layout, are based upon ribbon flow, which is illustrated on p. 80. The four-pass table includes a statement of the tube count per pass, and the number of rows in Pass 1 and in Pass 2. Passes 2 and 3 have identical tube counts. Passes 1 and 4 have identical tube counts when there is no impingement plate. The velocities are in columns (13) and (14).

Since there is no statement of "OFF CL" the tube layouts are on centerline. The row adjacent to the horizontal centerline (with the highest row number) has a tube located on the vertical centerline. When "OFF CL" appears in the left column at row 2, then the tubes of the bottom row straddle the vertical centerline as shown in Figure 4 for a 27-in.-inside-dia. U-tube heat exchanger with two tube passes.

How many tube counts?

Consider a horizontal, straight-tube heat exchanger having two tube passes, with its pass partition on the horizontal centerline. The shell nozzles are identical in size and are on the vertical centerline. There are three possible conditions, each leading to a number of tube counts:

A: The most-common is illustrated in Figure 2, with shell nozzles 180 deg

TABLE 4. Delaware Method factors for shellside heat transfer and pressure drop for a Size 10-240 Type BEM exchanger

N (1)	BAFL (2)	Win (3)	Row (4)	H (5)	XFLO (6)	NFA (7)	%NFA (8)	RL (9)	RB (10)	JC (11)	JL (12)	JB (13)
5.00	2.00	1	2	3.12	4.9	7	152	0.138	* 0.318	1.15	0.43*	0.69
4.00	2.51	1	2	3.12	6.1	7	121	0.141	* 0.318	1.15	0.44*	0.69
3.00	3.34	2	3	2.31	8.1	11	131	0.278	* 0.318	1.06	0.57	0.69
2.00	5.01	2	4	1.50	12.2	14	119	0.424	0.318	0.92	0.68	0.69
1.00	10.02	2	4	1.50	24.3	14	59	0.550	0.318	0.92	0.79	0.69
0.80	12.53	2	4	1.50	30.4	14	47	0.597	0.318	0.92	0.82	0.69
0.60	16.70	2	4	1.50	40.6	14	36	0.658	0.318	0.92	0.85	0.69
0.40	25.05	2	4	1.50	60.9	14	24	0.760	0.318	0.92	0.90	0.69
0.33	30.00	2	4	1.50	72.9	14	20	0.786	0.318	0.92	0.90	0.69

The factors above are tabulated for fixed-tubesheet and U-tube constructions, for a Size 10-240 Type BEM heat exchanger with the following tubesheet configuration: 20-ft. tube length, 3/4-in. outer-diameter tubes arranged on 15/16-in. triangular pitch, two-pass tubeside design. The shell ID is 10.02 in.

TABLE 5. Delaware Method factors for shellside heat transfer and pressure drop for a Size 48-240 Type BEM exchanger

N (1)	BAFL (2)	Win (3)	Row (4)	H (5)	XFLO (6)	NFA (7)	%NFA (8)	RL (9)	RB (10)	JC (11)	JL (12)	JB (13)
5.00	9.60	3	7	17.80	99.8	86	86	0.114	0.641	1.15	0.33	0.87
4.00	12.00	4	8	16.99	124.8	98	78	0.238	0.641	1.15	0.51	0.87
3.00	16.00	5	11	14.55	166.4	135	81	0.372	0.641	1.09	0.62	0.87
2.00	24.00	8	15	11.30	249.6	189	76	0.552	0.641	0.98	0.78	0.87
1.60	30.00	9	18	8.87	312.0	235	75	0.600	0.641	0.91	0.82	0.87

apart (i.e., in an "opposite" orientation). When symmetrical rows of tubes exist above and below the centerline and there is no impingement plate, then shell entrance area and shell exit area are equal.

Shell inlet area is reduced by installing an impingement plate above the top row of tubes. When there is an impingement plate and the tube rows above and below the centerline are equal, there is always more shell exit area than shell inlet area.

The total tube counts of Table 1 are in Columns (5) and (6). The former column applies to identical rows of tubes in both passes and is applicable both with and without an impingement plate.

There are 11 rows of tubes in the heat exchanger of Table 1. There are 11 possible tube counts, which are in Column (5). The smallest count is 38, with only one row above the centerline and one row below it.

B: When an impingement plate is installed, there may be sufficient shell exit area so that an additional row of tubes can be installed below the horizontal centerline. The applicable tube counts are in Column (6) of Table 1.

Note that the tube counts at Row 1 are identical in Columns (5) and (6). Thus there are ten different tube counts when there is an additional row in the lower tube pass.

C: When the shell nozzles have the same orientation, the deleted tubes provide both shell-inlet and shell-outlet area. The tubes below the centerline remain in place. The possible tube counts are shown in Table 3.

There are 156 tubes in each pass when the maximum number of tubes is installed. The smallest tube count in the exchanger with nozzles that have the same orientation is 156 plus 19 (which is the tube count of Row 11) as shown in Column (4) of Table 1. Total tube count is 175.

There are 11 possible tube counts, as shown in Table 3. Since two are duplicates of previous counts, there are nine new tube counts.

The total tube counts for the three conditions just described are 11 plus 10 plus 9, i.e., 30. If N equals the number of tube rows in a heat exchanger, then 3 times $(N - 1)$ gives the number of different tube counts in the exchanger.

Shellside calculations

In the open literature, the best approach to determining shell-side heat transfer coefficients and pressure

EXPLANATION

(1) N

The inside shell diameter is divided by the factor N to obtain the baffle spacing. Minimum spacing permitted by TEMA is one-fifth of the shell diameter

(2) BAFL

= baffle spacing (inches)

(3) Win

= the number of the row ("window row") nearest to the middle of the net free area (Column 7), where shellside fluid flows beyond the baffle cut. The top row is number one

(4) Row

= the row number of the baffle cut, which is at the centerline of the tubes in this row

(5) H

= height (inches) from the row to the horizontal centerline

(6) XFLO

= crossflow area (square inches) equals clearance times the baffle spacing of Column (2). Clearance equals chord length minus the product of the tube count of the row and the outside diameter of the tubes

(7) NFA

= net free area (square inches) is the area above the baffle cut that is not obstructed by tubes

--Re >/= 100--		--Re = 60--		-----Re = 20------			--Turbulent--		--Viscous--		
Prod (14)	TURB (15)	JR (16)	Prod (17)	JR (18)	Prod (19)	VISC (20)	ReNo (21)	PDt (22)	ReNo (23)	PD (24)	Spaces (25)
0.34	1.00	0.81	0.27	0.62	0.20	1.000	9999	100.0	100	100.0	115
0.35	0.89	0.81	0.28	0.62	0.21	0.888	7999	57.3	80	63.5	92
0.41	0.88	0.81	0.33	0.62	0.25	0.881	5999	37.9	60	47.9	69
0.43	0.71	0.79	0.33	0.57	0.24	0.653	4000	11.7	40	20.9	46
0.50	0.56	0.82	0.41	0.65	0.31	0.578	2000	2.7	20	7.0	23
0.52	0.51	0.84	0.43	0.68	0.34	0.560	1600	1.7	16	4.9	18
0.54	0.45	0.86	0.46	0.73	0.38	0.529	1200	0.9	12	3.1	13
0.56	0.38	0.88	0.49	0.77	0.42	0.475	800	0.4	8	1.7	9
0.57	0.35	0.90	0.50	0.79	0.44	0.450	668	0.3	7	1.3	7

The factors below are tabulated for fixed-tubesheet construction, for a Size 48-240 Type BEM heat exchanger, with the following tubesheet configuration: 20-ft. tube length, 3/4-in. outer diameter tubes arranged on 15/16-in. triangular pitch. The shell ID is 48 in.

-Re >/= 100--		--Re = 60--		-----Re = 20------			--Turbulent--		--Viscous--		
Prod (14)	TURB (15)	JR (16)	Prod (17)	JR (18)	Prod (19)	VISC (20)	ReNo (21)	PDt (22)	ReNo (23)	PD (24)	Spaces (25)
0.33	1.00	0.86	0.29	0.72	0.33	1.00	9999	100.0	100	100.0	24
0.51	1.31	0.88	0.44	0.76	0.50	1.374	7999	81.6	80	96.3	19
0.59	1.27	0.90	0.52	0.80	0.58	1.401	5999	40.8	60	63.4	14
0.66	1.12	0.85	0.56	0.70	0.65	1.081	4000	12.6	40	29.3	9
0.64	0.95	0.83	0.53	0.65	0.63	0.858	3200	5.2	32	16.6	7

OF COLUMN HEADINGS FOR TABLES 4 AND 5

(8) % NFA
= percent of crossflow area available for flow area above the baffle cut

(9) RL
= baffle-leakage correction factor, for effect of baffle leakage on pressure drop

(10) RB
= bundle-bypassing correction factor, for bypass-flow effect on pressure drop

(11) JC
= baffle-configuration correction factor for heat transfer relative to the fraction of total tubes in crossflow

(12) JL
= baffle-leakage correction factor for heat transfer

(13) JB
= bundle-bypassing correction factor for heat transfer relative to the fraction of crossflow area available for bypass

(14) Prod (Re ≥ 100)
= product of the four heat-transfer factors (JC, JL, JB, JR) when the Reynolds Number (Re) is 100 or greater. At Re ≥ 100, JR equals unity. For a given row, this product does not change for all Re numbers of 100 or more

(15) TURB (Re ≥ 100)
= ratio of heat transfer coefficients in turbulent flow, using 1.000 at the minimum baffle spacing

(16) JR (Re = 60)
= correction factor for the shellside heat transfer coefficient (to account for the build-up of an adverse temperature gradient at Re = 60). Fluid is in streamline flow

(17) Prod (Re = 60)
= the product of the four heat-transfer factors at Re = 60, in this case taking the value for JR from Column 16

(18) JR (Re = 20)
= correction factor that is applied to the shellside heat transfer coefficient (to account for the build-up of an adverse temperature gradient† at Re = 20). The fluid is in streamline flow

(19) Prod (Re = 20)
= the product of the four heat-transfer factors at Re = 20, in this case taking the value for JR from Column 18

(20) VISC (Re = 20)
= ratio of the shellside heat transfer coefficients in deep streamline flow, where the ratio value is 1.000 at the minimum baffle spacing

(21) Turbulent ReNo
= Reynolds number for the fluid flow. Turbulent flow (Re ≥ 9,999) exists at all baffle spacings

(22) Turbulent PDt
= relative pressure drop in turbulent flow, at the Reynolds Number of Column (21). Relative pressure drop in turbulent flow is set at 100 for the minimum baffle spacing. The flow quantity is unchanged

(23) Viscous ReNo
= Reynolds Number. Streamline flow commences at a Reynolds number of 100. The flow quantity is unchanged at all baffle spacings

(24) Viscous PD
= relative presure drop in streamline flow at the Reynolds Number of Column (23). Relative pressure drop in streamline flow is set at 100 for the minimum baffle spacing

(25) Spaces
= number of times that the fluid crosses the tube bundle with allowance for larger terminal baffle spacings at the tubesheets

† Note: In laminar flow, the heat transfer coefficient decreases with increasing distance from the heating surface, due to the development of an adverse temperature gradient.

TURBULENT FLOW			
N (1)	BAFL (2)	TURB (15)	PDt (22)
5.00	9.60	1.00	100.00
4.00	12.00	1.31	81.60 (b)
3.00	16.00	1.27	40.80 (a)
2.00	24.00	1.12	12.60

(a) As the baffle spacing decreases from 24 in. to 16 in., the pressure-drop ratio increases from 12.60 to 40.80. Shellside pressure drop increases by 224%. Shellside heat-transfer ratio increases from 1.12 to 1.27, i.e., by 4%

(b) As the baffle spacing decreases from 16 in. to 12 in., the pressure-drop ratio increases from 40.80 to 81.60. Shellside pressure drop increases by 100%. Shellside heat-transfer ratio increases from 1.27 to 1.31, i.e., by 3%

TABLES 6a (left) and 6b (right). The Size 48-240 Type BEM heat exchanger of Table 5 is examined more closely with regard to relationships that involve baffle spacing, pressure drop and heat transfer. For details on use of the tables, see the notes at the bottom of 6a and the box to the right of 6b

Ratio of heat transfer coefficients shellside:tubeside (at smallest baffle pitch)	Shellside coefficient [Btu/(h)(ft2)(degF)]	Tubeside coefficient	Overall heat transfer rate for both coefficients with 0.002 fouling resistance
1.00	100.00	100.00	45.45
	131.00	100.00	50.93
	127.00	100.00	50.32
	112.00	100.00	47.78
2.00	100.00	50.00	31.25
	131.00	50.00	33.75
	127.00	50.00	33.47
	112.00	50.00	32.33
0.50	100.00	200.00	58.82
	131.00	200.00	68.34
	127.00	200.00	67.23
	112.00	200.00	62.78
0.25	100.00	400.00	68.97
	131.00	400.00	82.42
	127.00	400.00	80.81
	112.00	400.00	74.47

Baffle spacing affects not only heat transfer but also pressure drop. A moral: don't overlook pumping costs

Brown Fintube Co.

Exchanger tube bundles come in various configurations, including hairpin

drop is the Delaware method [10]. It employs four correction factors for heat transfer and two correction factors for pressure drop. These are related to figures [11] in "Perry's Chemical Engineers' Handbook", 6th edition:†

Figure 10-20: JC, the heat-transfer correction factor for baffle cut and spacing

Figure 10-21: JL, the heat-transfer correction factor for effects due to shell-to-baffle and tube-to-baffle leakage

Figure 10-22: JB, the heat-transfer correction factor for bundle bypass flow (of the C and F streams)

Figure 10-23: JR*, the basic heat-transfer correction factor for adverse temperature gradient at low Reynolds number

Figure 10-24: JR, the heat-transfer factor for adverse temperature gradient at intermediate Reynolds number

Figure 10-25: RL, the factor for baffle-leakage effect on pressure drop

Figure 10-26: RB, the factor on pressure drop for bypass flow, similar in form to the JB factor

The designations of these factors (e.g., JC, JL, JB) correspond to column

†The Delaware method also employs three correlations from Perry's:

Figure 10-19: correlation of heat-transfer factor JK for an ideal tube bank

Figure 10-25a: correlation of friction factors for ideal tube banks with triangular and rotated square arrays

Figure 10-25b: correlation of friction factors for ideal tube banks with inline square arrays

Baffle spacing changes from:	to:	Heat transfer rate changes by this %	and	Pressure drop increases by this %
12.00	9.60	-10.8		23 (c)
16.00	12.00	1.2		100 (b)
24.00	16.00	5.3		224 (a)
12.00	9.60	-7.4		23 (cc)
16.00	12.00	0.8		100 (bb)
24.00	16.00	3.5		224 (aa)
12.00	9.60	-13.9		23
16.00	12.00	1.6		100
24.00	16.00	7.1		224
12.00	9.60	-16.3		23
16.00	12.00	2.0		100
24.00	16.00	8.5		224

READING THE TABLE:

1. When the shellside and tubeside coeffients are the same (at the minimum baffle spacing)
(a) The heat transfer increases by 5.3% as the baffle spacing goes from 24 to 16 in., and the pressure drop increases by 224%
(b) The heat transfer increases by 1.2% as the baffle spacing goes from 16 to 12 in., and the pressure drop increases by 100%
(c) The heat transfer falls by 10.8% as the baffle spacing goes from 12 to 9.6 in., and the pressure drop increases by 23%

2. When the shellside coefficient is double the tubeside coefficient (at the minimum baffle spacing)
(aa) Heat transfer increases by 3.5% as pressure drop increases by 224%
(bb) Heat transfer increases by 0.8% as pressure drop increases by 100%
(cc) Overall transfer rate falls by 7.4% as pressure drop increases by 23%

Note:
The overall effect on heat transfer is the greatest when the shellside coefficient is smaller than the tubeside coefficient (when the ratio in the first column is less than 1).
Example (bb) illustrates that there comes a point of diminishing returns, when overdesign of a heat exchanger results in practically negligible process improvements.
When the overall heat transfer is sufficient at a 16-in. baffle spacing, one can see that by adding more baffles to decrease the baffle spacing to 12 in., the heat transfer rate only improves by 0.8%, while the pressure drop doubles

headings in Tables 4 and 5, extracted from Reference [1]. Table 4 presents the Delaware method factors for a Size 10-240 Type BEM (fixed tubesheet) exchanger having a shell of 10.02-in. ID, 3/4-in. tubes on 15/16-in. triangular pitch, a two-pass tube side and a fixed-tubesheet construction. Table 5 is for a larger unit, a Size 48-240 Type BEM (fixed tubesheet) exchanger.

In the factor tables, baffle spacings vary from the smallest to the largest permitted by the TEMA standards. For each spacing, the 25 items shown in Table 4 or Table 5 are presented.

With regard to Column 9, for effect of baffle leakage, note that in accordance with the design philosophy of Reference [1], the *most probable* dimensions are used, e.g., baffle holes at 0.007 in. beyond drill diameter. As discussed earlier, I believe it is impossible to fabricate in accordance with the TEMA standards and have all baffle holes 0.010 in. greater than the drill diameter.

Heat transfer and pressure drop

Earlier, we looked at the tradeoff between a favorable heat-transfer effect and an unfavorable pressure-drop effect as a result of adding baffles. To examine this matter more closely, heat-transfer and pressure-drop data from Table 5 are developed more thoroughly in Tables 6a and 6b.

These relationships are examined in Table 6b when the shell-side coefficient is equal to the tube side coefficient, and when shell coefficient is 25%, 50% and 200% of the tube-side efficient (at the minimum baffle spacing). When both coefficients are initially at 100 Btu/(h)(ft^2)(°F) and the pressure drop is doubled, (from 5 to 10 psi) the overall heat transfer coefficient increases by 1.2%. When the shell-side coefficient is twice the tube side coefficient, the corresponding increase in overall rate is less than 1%. ■

The author

Frank L. Rubin is president of Practical Heat Transfer Consultants (P.O. Box 8242, Scottsdale, AZ, 85252, Tel.: 602-947-5689). Throughout the course of his career, Rubin has been employed as an engineer by several manufacturers of shell-and-tube heat exchangers. He has served as chairman of the Technical Committee on Heat Transfer Equipment of the American Soc. of Mechanical Engineers and of the equivalent committee of AIChE. Rubin was the originating editor of the Heat Transfer Equipment Section of the Chemical Engineers' Handbook (Fourth Ed.) and then served as Section Editor for two successive editions. He has also authored 30 technical papers. This article was adapted from his upcoming book, "Heat Exchangers, Economics and Management" (Hemisphere Publishers, Bristol, Pa.), which is scheduled for publication in 1993.

References

1. Rubin, Frank L., "Heat Exchangers, Economics and Management," Hemisphere Publishing Co., New York, N.Y., projected 1993.
2. Kern Donald Q., "Process Heat Transfer," 1950, McGraw-Hill, New York, N.Y.
3. Standards of the Tubular Exchanger Manufacturers Association, Tarrytown, N.Y.
4. "Standard for Power Plant Heat Exchangers," First ed., 1980, Heat Exchanger Inst., Cleveland
5. ASME Boiler and Pressure Vessel Code, Section VIII, Pressure Vessels, Div. 1, 1989, Am. Soc. Mech. Engineers, New York, N.Y.
6. Catalog of the now-defunct Downingtown Iron Works, Inc., Downington, Pa, c. 1947.
7. Rubin, Frank L., How to simplify shell-side calculations, *Chem. Eng.*, February 1957, pp. 257–260.
8. Perry, Robert H., and Green, Don, "Chemical Engineers' Handbook," Sixth Ed., 1984, pp. 11–13 to 11–16, McGraw-Hill, New York, N.Y.
9. Hammack, Charles, private communication, Engineers and Fabricators Co., Houston, Tex.
10. "Final Report, Cooperative Research Program on Shell-and-Tube Heat Exchangers," *Univ. of Delaware Engineering Experiment Status Bulletin*, vol. 5, June 1963.
11. Perry, *op. cit.*, pp. 10–29 to 10–31.
12. 1991 Errata pages for reference [3], pp. 91 and 102.

The performance attractions of plate and spiral exchangers are steadily growing

DON'T COMPACT HEAT

James R. Burley
Alfa-Laval Thermal, Inc.

Compact heat exchangers, more commonly referred to as plate or spiral heat exchangers, have been considerably improved in the past few years. They outperform traditional shell-and-tube exchangers in a broad and growing range of situations. Even so, many process engineers remain unfamiliar with how the compact devices work, where they can be used, and what they offer.

There are three main reasons for compact heat exchangers' good performance:

- They are designed to induce turbulent flow and, therefore, high heat-transfer coefficients
- They resist fouling
- Their flowpath maximizes the temperature driving-force between the hot-side and cold-side fluids

Compact heat exchangers are of course governed by the same heat-transfer principles that apply to other exchangers. The three performance reasons just cited become more clear via a brief review of some of those principles.

Exchanger performance

All heat exchangers are designed by using the equation

$$Q = U \times A \times LMTD$$

where Q is the exchanger duty (e.g., Btus transferred per hour), U is the overall heat-transfer coefficient, A is the heat-exchanger area and $LMTD$ is the log mean temperature difference between the hot side and the cold side.

For a given heat-transfer duty, the goal is to minimize A by maximizing U and $LMTD$.

The value of U is set by the formula

$$1/U = 1/h_h + t/k + 1/h_c + R_f$$

where h_h and h_c are respectively the individual heat-transfer coefficients for the hot and cold sides, t is the thickness of the heat-exchanger surface, k is the thermal conductivity of the heat-exchanger material, and R_f represents the resistance due to fouling on the heat-transfer surface.

We can raise the overall heat transfer coefficient U by upping either of the individual coefficients. But the increase will not be significant unless both are raised and R_f is lowered.

Consider, for instance, a processing situation where h_h and h_c are each 500 Btu/(h)(ft^2)(°F), t is 0.078 ft., k is 350 Btu/(h)(ft)(°F), and R_f is 0.001 (h)(ft^2)(°F)/Btu. Then

$$1/U = 1/500 + 0.078/350 + 1/500 + 0.001 = 1/192$$
$$U = 192 \text{ Btu/(h)(ft}^2\text{)(°F)}$$

If we increase the cold-side film coefficient by 50% we have

$$1/U = 1/500 + 0.078/350 + 1/750 + 0.001 = 1/220$$

Originally published August 1991

Alfa-Laval Thermal, Inc.

OVERLOOK EXCHANGERS

$U = 220$

This is a rise of only 14.6%.

Increasing both film coefficients by 50% improves the situation:

$$1/U = 1/750 + 0.078/350 + 1/750 + 0.001 = 1/258$$

$$U = 258$$

Thus, the increase over the base case becomes 34.4%.

Finally, if we can somehow reduce by one-half the effects of fouling, we have:

$$1/U = 1/750 + 0.078/350 + 1/750 + 0.0005 = 1/296$$

$$U = 296 \text{ Btu/(h)(ft}^2\text{)(°F)}$$

This is an overall increase of 54.2% from the base value of 192.

Plate and spiral heat exchangers tend to have high values for h_h and h_c and to incur relatively little fouling, so they are associated with high values for U. This is discussed in more detail on the next two pages.

The other major parameter affecting heat transfer, the log mean temperature difference, is defined as the logarithmic average difference in the temperatures between the hot and cold fluids:

$$LMTD = (\Delta T_1 - \Delta T_2)/\ln(\Delta T_1/\Delta T_2)$$

where ΔT_1 and ΔT_2 are the temperature differences at the two terminals of the heat exchangers.

Consider two heat exchangers, one cocurrent and the other countercurrent. In each, the hot-side fluid enters at 300°F and leaves at 250°F whereas the cold-side fluid enters at 100°F and leaves at 200°F. The temperature profiles appear in Figure 1.

For the cocurrent heat exchanger,

$$LMTD = (200–50)/\ln(200/50) = 108.2$$

whereas for the countercurrent heat exchanger,

$$LMTD = (100–150)/\ln(100/150) = 123.3$$

The driving force for the countercurrent exchanger is thus 14% higher than that for the cocurrent.

Plate and spiral heat exchangers are normally configured with the two fluids in fully countercurrent flow. Within most shell-and-tube exchangers, there is instead a mixture of cocurrent, countercurrent and cross flow, so a correction factor of less than 1 is applied to the $LMTD$ that would prevail if the flow were fully countercurrent. Correction factors are available from graphs.*

In practice, shell-and-tube exchangers are designed to avoid crossing temperature profiles, and are configured such

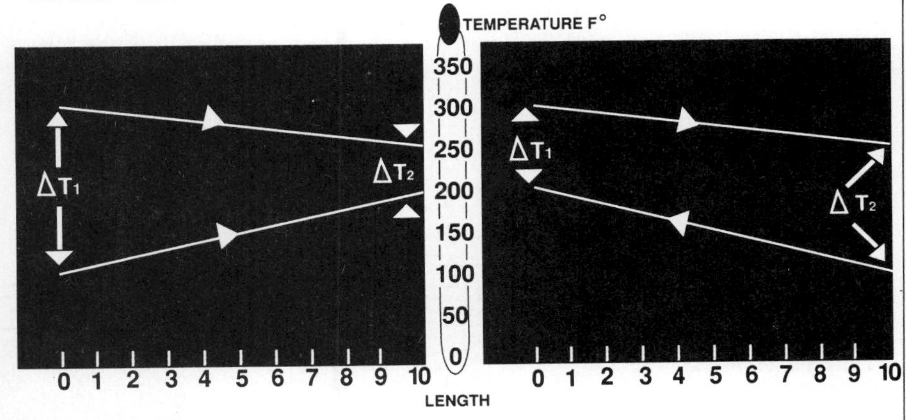

FIGURE 1. Countercurrent-exchanger temperature profile (right) yields bigger driving force than profile (left) for cocurrent exchanger with same inlet and exit temperatures

*For articles presenting such correction factors for a variety of shell-and-tube configurations, see "Quick Design and Evaluation : Heat Exchangers,", *CE*, July 1990, pp. 92-99, and "Charts for the Performance and Design of Heat Exchangers," *CE*, August 18, 1986, pp. 81-88. The same approach is used with double-pipe exchangers; for data, see "Correction Factors for Double-Pipe Heat Exchangers," *CE*, April 1991, pp. 163-164.

that in the extreme case the outlet temperatures are equal. In this situation the factor is 0.8.

Setting the stage

Shell-and-tube heat exchangers are by far the most common version found in process plants. Therefore, before discussing plate and spiral heat exchangers in more detail, it is worthwhile to look further at the shell-and-tube devices as a basis for comparison.

A shell-and-tube exchanger consists of a bundle of parallel tubes inserted inside a larger tube or shell. One fluid, typically a process fluid, flows inside the bank of tubes while the other flows through the shell around the bundle of tubes. The smallest of these exchangers have only a couple of tubes a few inches long, while very large ones may have thousands of tubes 30 to 40 ft. long. The exchangers are generally constructed totally of carbon steel, or with a carbon-steel shell and alloy tubes.

Consider the performance of these exchangers as a function of Reynolds number (Re). Well-documented studies of flow through a single tube reveal that for Re under 2,100 (i.e., laminar flow), the fluid builds up a continuous boundary layer that limits the heat transfer.

For intermediate values of Re ($2,100 < Re < 20,000$), the flow pattern is in transition and neither laminar nor turbulent. In this condition, the boundary layer separates from the tube wall and becomes unstable, forming a turbulent wake downstream. This condition produces a relatively high heat-transfer rate. But the latter is difficult to predict accurately, as the heat-transfer relationships become complex and depend greatly on the geometry of the system.

For Re greater than 20,000, fully developed turbulent flow is achieved and heat transfer rates are maximized as boundary-layer resistance is minimized. Heat transfer can be evaluated with reasonable accuracy using the well-established relationships involving Nusselt, Reynolds and Prandtl numbers (as discussed in, e.g., "Perry's Chemical Engineers' Handbook", 6th ed., pp. **10**-16 ff).

Flow patterns across the tube bundles (on the shell side) are complex, having both crossflow and longitudinal-flow components. The complexity becomes even greater if the shell has baffles to increase the cross-flow component.

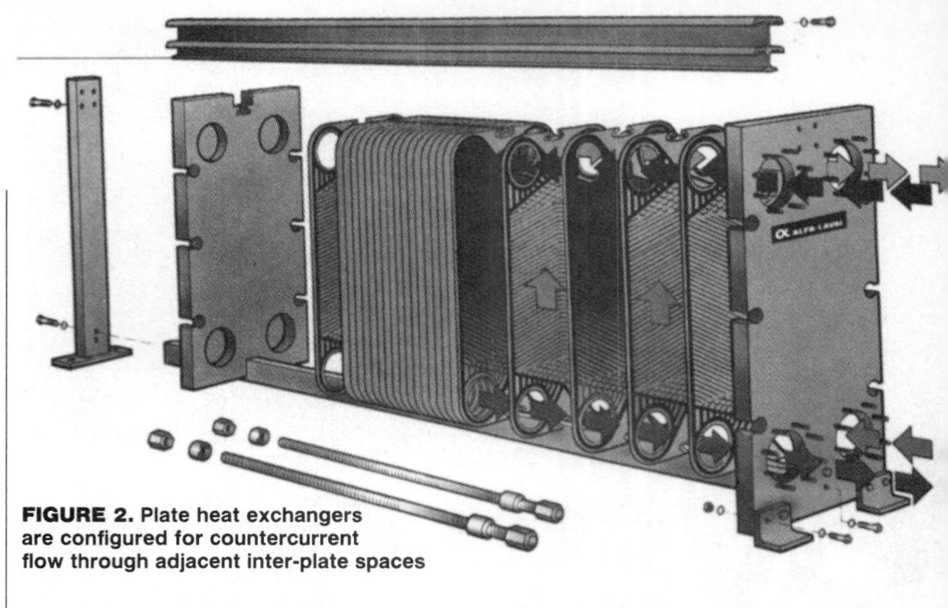

FIGURE 2. Plate heat exchangers are configured for countercurrent flow through adjacent inter-plate spaces

Plate heat exchangers

Plate heat exchangers are made of cold-pressed, corrugated-metal plates fitted into a frame. The frame consists of an upper and lower carrying bar that support and align the plates, plus two covers (also called pressure plates) that compress the corrugated plates. Gaskets seal the inter-plate spaces at the edges and direct the two fluids into their proper flow paths. Inlets and outlets for the fluids are provided at the corners of the plates. The fluids flow countercurrently between successive pairs of plates, as shown in Figure 2.

The corrugations not only increase the rigidity of the plate but also aid in inducing turbulence of the fluid flowing across it. The flow profiles are generally considered to be fully turbulent, except when viscosity is extremely high.

Individual plates range in size from 0.35 to over 30 ft^2. Each plate model has its own thermal-efficiency and pressure-drop characteristics, depending on plate length and surface area as well as the depth and angle of the corrugations. The size, number and type of plate depends on the flowrates, temperature profiles and physical properties of the fluids used, as well as such other parameters as allowable pressure drop and fouling tendency.

The gaskets are usually made of elastomers such as nitrile rubber, ethylene-propylene-diene-monomer rubber, butyl rubber, and copolymers of vinylidene fluoride and hexafluoropropylene (i.e., Du Pont's Viton), with the occasional use of graphite, polytetrafluoroethylene, or asbestos. The gaskets are designed with a double seal so that if gasket failure should occur, media intermixing is prevented.

A plate heat exchanger is the most efficient and cost-effective exchanger that exists. Due to the induced turbulence of the flow and the small hydraulic diameter* of the flowpath (which tends to increase the Reynolds number), the plate heat exchanger can obtain efficiencies of up to five times those achieved in a shell-and-tube version for a similar application.

Because it has no "tube side, shell side" distinction, the plate heat exchanger can be more appropriate than shell-and-tube exchangers when both the cold and hot sides are process fluids. Another attraction is that the heat-transfer area can be

*Hydraulic diameter is defined as cross-section area divided by twice the wetted perimeter.

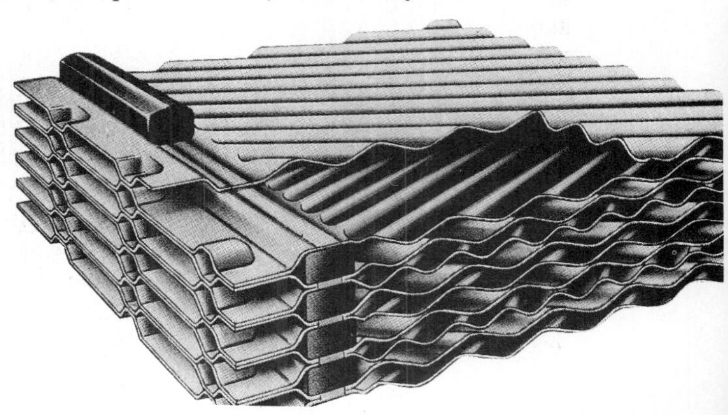

FIGURE 3. Use of welding instead of gasketing enables a plate exchanger to handle (in red regions at right) a fluid that would attack gasket elastomers

modified by adding or removing plates as the process conditions change.

Plate exchangers are used mainly in liquid-to-liquid applications. But they are also suitable for condensing or reboiling services, especially in light of their high rates of heat flux and their low holdup volume. Their limitation in this service is usually pressure drop, due to the small cross-sectional area through which the vapor phase passes.

Where fouling is a concern, the plate heat exchanger is often a good choice due to the high rates of shear and turbulence imparted upon the fluid. These tend to keep the heat-transfer surface notably free of deposits. Fouling factors are only about 1/10 those normally used for shell-and-tube exchangers in similar applications.

However, for services involving slurries that can cause particulate-type fouling, the tight hydraulic diameter of the plate exchanger may cause excessive plugging. Maximum acceptable solids loading is 40% by volume. Maximum acceptable particle size depends on the channel depth, but is typically about 0.1 in. If conditions exceeding these maxima are present, either a strainer should be placed upstream of the plate exchanger or else a spiral or shell-and-tube exchanger should be selected.

Plate exchangers require occasional service. Gaskets must be replaced, and plates thoroughly cleaned and inspected. All heat-transfer surfaces are accessible by removing the closure bolts and sliding the movable cover back. The higher-quality manufacturers have designs that allow for taking out any plate without removing the entire plate pack.

Due to the extremely thin (as low as 0.4 mm) sheets used in construction, stainless steel is the lowest-grade material suitable for plate heat exchanger construction. However, because the high efficiency of these exchangers enables them to be small, one made of stainless steel can cost less than a carbon-steel shell-and-tube for the same service. The plates are commonly pressed in stainless steel, titanium, nickel alloys, and several specialty materials such as tantalum.

The maximum design temperature for a plate exchanger is about 375°F. Maximum design pressure is about 365 psig.

So long as neither particulate fouling nor the service temperature or pressure pose a limitation, plate heat exchangers are especially preferable to shell-and-tube versions in a number of situations:

- A large change in temperature is required in one or both of the fluids (if the shell-and-tube option were used in this case, it is likely that multiple exchangers would be needed)
- There is a temperature cross (i.e., the outlet temperature of the hot-side fluid is colder than that of the cold-side fluid)
- The temperature approach is less than 10°F
- The installation space or weight is a consideration. (The plate units require less than 1/5 the floor space of the comparable shell-and-tube versions)
- High-alloy construction is required, particularly for both hot and cold fluids

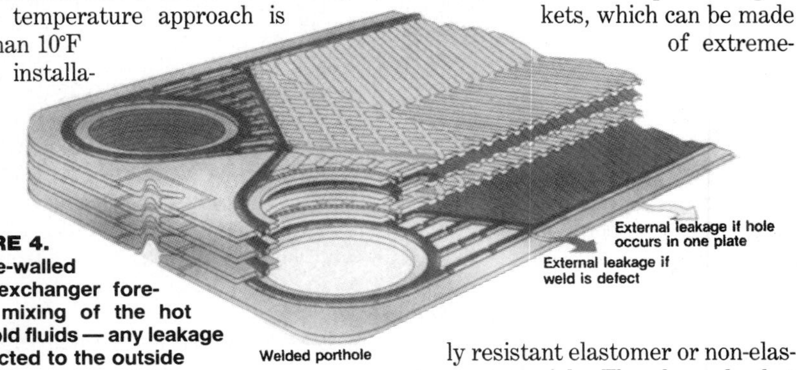

External leakage if hole occurs in one plate

External leakage if weld is defect

Welded porthole

FIGURE 4. Double-walled plate exchanger forestalls mixing of the hot and cold fluids — any leakage is directed to the outside

Advances in design

The plate heat exchanger is now 20 to 30% more efficient than it was ten years ago. Improvements in plate and gasket design and plate-pressing technique have increased their reliability and ease of use.

Gluefree gaskets: Gaskets had traditionally been glued to the plate, which made gasket replacement in the field difficult. New glue-free systems allow the gaskets to be attached to the individual plates by a clip-on or snap-on system. Glue-free gaskets eliminate up to 75% of the labor required to regasket a plate, thus reducing plant downtime and expense. They also add to the reliability of the repaired equipment, because they forestall new-gasket leakage problems due to improper cleaning of the old hardened glue. For an average-size heat exchanger, the snap-on gasketing procedure requires three hours, versus three days for a standard glued gasket.

Welded plates: Introduction of these has expanded the range of applications for the plate heat exchanger. Because the gaskets in contact with one medium are replaced by a laser weld (Figure 3), the device can handle aggressive media such as acids and solvents that attack elastomers.

The aggressive fluid is fed to and from the plates by flowing through thin channels formed by welding the reverse sides of two gasket grooves to each other. The only gaskets in contact with this fluid are two circular port-hole gaskets, which can be made of extremely resistant elastomer or non-elastomer materials. The channels that contain the non-aggressive medium are sealed by traditional elastomer gaskets.

Wide gaps: The wide-gap plate heat exchanger has been developed to effectively handle fluids containing fibers or coarse particles normally associated with, for instance, pulp-and-paper manufacture and sugar refining. It works on the same principle as the conventional plate heat exchanger but employs a plate pack with alternating wide-gap channels and conventional ones. Although this increased channel spacing means a lower overall heat transfer coefficient than those for traditional plate exchangers, the heat transfer is still three to four times more efficient than for a shell-and-tube. In practice, the wide-gap plate heat exchanger is being used to economically recover heat from difficult-to-handle waste streams.

Graphite: Plates made of graphite-containing composites give plate exchangers the added benefit of the inertness of that material. Such composites not only offer superior resistance to corrosion but also have good heat-transfer properties, low thermal expansion, high strength, and resistance to plastic deformation. This type of exchanger is especially suitable for handling acids or acid mixtures as found in pickling or plating operations where it is essential to forestall fluid intermixing.

Double walls: The double-wall plate heat exchanger is an attractive alternative to the double-wall shell-and-tube. The single plates of a normal exchanger are replaced by identical pairs of plates stacked together and laser welded around the port areas (Figure 4). In the unlikely event that a leak should occur, whether related to the plate or the gasket or the seal weld, the discharge would be to the atmosphere and thus visible on the outside of the heat exchanger.

Spiral heat exchangers

This version of compact heat exchangers is fabricated from two relatively long strips of metal sheet or plate, 6 to 72 in. wide, wrapped to form a pair of concentric spiral channels or passages (Figure 5). The edges of the channels are sealed, so each fluid flows through a single continuous pathway. The outer ends of the strips meet at headers on the periphery, and covers are fitted on each side of the spiral assembly to enclose the unit.

Various flow patterns can be developed by modifying the channel-welding patterns or fitting the channels with different covers or extensions. This allows the spiral to be used for a variety of heat transfer duties, as discussed later.

For most applications, each fluid flows in a single, curving spiral channel. This curvature imparts secondary flow effects which stimulate turbulence. This improves heat transfer, particularly at what would otherwise be low Reynolds numbers. Although the overall heat-transfer coefficients are generally not so high as with the plate exchanger, they are more attractive than those for a shell-and-tube.

The induced turbulence also keeps the fluid well mixed. This is important in applications with slurries.

Also, because there is a single flow channel rather than the many parallel channels on the tube side of shell-and-tube exchangers, flow-distribution problems that are a major source of fouling are all but eliminated. Whenever the spiral channel begins to foul, the decrease in cross-sectional area proportionally increases the fluid velocity, which in turn increases the turbulence and shear and thus opposes the fouling buildup.

By contrast, when a tube begins to foul, the pressure drop across that tube increases, inducing some of the flow to shunt to adjacent tubes. This leads to a decrease in fluid velocity, and hence in turbulence and shear, causing the tube to foul more readily and eventually to plug. Due to the ability of the spiral design to resist fouling, typical fouling factors used in spiral-heat-exchanger designs are 1/4 those used for the equivalent shell-and-tube designs.

Plate exchangers, of course, also present many parallel channels to the flow. But the turbulence effects in those exchangers effectively prevent fouling buildup, as discussed earlier.

The spiral may be the easiest of all heat-exchanger types to clean. The single-channel design combined with the high shear and turbulence make chemical cleaning very effective, and it alone is adequate in most cases. When mechanical cleaning is required, the short depth of the unit (72 in. maximum) makes high-pressure water washing very effective.

Materials of construction are typically carbon or stainless steels, but the spiral exchanger can be manufactured from any metal that can be cold rolled and welded. These include titanium,

FIGURE 5. Spiral heat exchanger provides a single continuous pathway for each fluid

nickel alloys and several newer materials, such as certain duplex stainless steels.

Like the plate heat exchanger, the spiral has no distinction between shell-side and tube-side flow patterns. Both fluids flow in an enhanced, tube-side-type channel. This makes the device especially useful in applications where process or fouling fluids (slurries or sludges in particular) are being used as both the hot and the cold media. The long channel length or fluid path, coupled with the fully countercurrent flow pattern, make it an excellent choice where temperature crosses are evident.

When the spiral exchanger is used as a condenser, the centrifugal action on the condensing vapor causes the liquid phase to separate from the vapor phase, minimizing entrained condensate in the noncondensible outlet stream. This allows for more complete condensing and condensate recovery.

By the addition of upper and lower extensions and changes in the welding pattern on the channels, a cross-flow spiral is formed. This kind of unit combines a large cross-sectional area of flow and a short flowpath with the higher-shear geometry of the spiral. The inherently low pressure drop of this arrangement, coupled with its high heat flux, makes this unit attractive as a compact low-pressure or vacuum condenser or a thermosiphon reboiler. Tests by Heat Transfer Research Inc. (College Station, Tex.) have confirmed that this type of spiral exchanger performs better than typical shell-and-tube thermosiphon reboilers.

These units are also effective as gas coolers or heaters. The short vapor path incurs very low pressure drops.

Improvements in rolling and welding technologies have allowed the spiral heat exchanger to achieve larger sizes and higher pressure ratings. And the use of continuous coil material reduces the amount of welding required in the internal portions of the spiral, significantly lowering its cost.

The limit on operating pressure is approximately 300 psig, which is lower than that for either the plate exchanger or the shell-and-tube. The temperature limit is 1,500°F.

SELECTION GUIDE FOR HEAT EXCHANGERS

Operational requirements	Shell-and-tube	Plate	Spiral
Max. design temperature, °C	High	150 to 175 (300 to 350°F)	815 (1,500°F)
Max. design pressure, bar (and psig)	High	25 (365)	21 (300)
DUTY			
Liquid-liquid	2	1	2
Gas-liquid	2	1-3*	1
Gas-gas	1	1-3*	1-3*
Condensation	1	1-3*	1
Vaporization	1	1-3*	1
MATERIALS OF CONSTRUCTION			
Mild steel	1	**	2
Stainless steel	2	1**	2
Titanium	4	1	3
Other exotic materials	3	1	2
Flexibility	--	1	--
MECHANICAL CLEANING			
One side	2	1	2
Both sides	3	1	2
Clogging, fouling	3-4	2***	1***
Low holdup volume	4	1	3

*Depending on operating pressure, vapor density, etc.

**Plate heat exchanger not produced in mild steel. For increasing length of thermal duty the plate heat exchanger (in stainless steel) tends to be more competitive.

***To forestall particulate-type clogging, a strainer should be placed upstream.

Code: 1 Usually best choice, economically or technically.
2 Often best choice
3 Sometimes best choice
4 Seldom best choice

Specification pointers

When writing a specification for either a plate or spiral heat exchanger, one must bear in mind the advantages that these offer over the shell-and-tube. For instance, it is normal in a shell-and-tube design to specify excess cooling fluid to compensate for the inefficiencies of the device. This is not necessary with the compact exchangers, which saves pumping costs.

It is also feasible to design for higher degrees of heat recovery. For example, a temperature approach as low as 2°F is practical, versus about 10° with shell-and-tube designs.

The costs of plate and spiral heat exchangers tend to be more sensitive to operating pressure than is the case with shell-and-tube units. Accordingly, these units should be specified for the design pressures actually needed for the application (instead of, as is common with shell-and-tubes, specifying for example a 150-psig design pressure when the operating pressure is only 75 psig). This is particularly important in a spiral design, where heavy-gauge materials are employed in the manufacturing.

Unfortunately, it has been common to include the standard Tubular Equipment Manufacturing Assn. (TEMA) fouling factors for shell-and-tube exchangers even when the specification is for a plate or a spiral exchanger. Because plate exchangers have such high efficiency, this practice leads to a strikingly great overdesign. For example, the effect of including an 0.001-(h)(ft²)/Btu fouling factor on a shell-and-tube exchanger with a clean (i.e., before allowing for fouling) U of 300 increases the surface area by 30%, whereas applying the same fouling factor to a plate heat exchanger with a clean U of 1,000 increases the area by 100%. The same is true for the spiral, although the difference is less dramatic. ■

Edited by Nicholas P. Chopey

The author

James R. Burley is marketing manager for the Key Accounts Group for plate and spiral heat exchangers for Alfa-Laval Thermal, Inc., 5400 International Trade Drive, P.O. Box C-32026, Richmond, VA 23231, Tel. (804) 236-1344. He joined the firm in 1979 as a design engineer and product manager for the spiral exchangers. Burley graduated summa cum laude from Rutgers University (New Brunswick, N.J.) with a B. S. in chemical engineering, and has taken graduate courses in business and marketing at Fairleigh Dickinson Univeristy (Rutherford, N.J.).

ENHANCE GAS PROCESSING WITH REFLUX HEAT-EXCHANGERS

These compact plate-fin units provide high thermodynamic efficiency in cryogenic separations

Adrian J. Finn
Costain Oil, Gas & Process Ltd.

Despite recent successes of membrane-based separations in low-throughput applications, cryogenic processing remains the best route for separating and purifying gas mixtures, especially when high recoveries are required. Now conventional units are being modified to yield even higher recoveries at lower costs. Throughout the chemical process industries (CPI), this is being accomplished with reflux or plate-fin exchangers, especially for processing of natural gas, and offgases from refineries and petrochemical facilities.

The concept of utilizing a heat exchanger as a multistage rectification device is not new [2]. However, only in the last fifteen years or so has accurate design of reflux exchangers become feasible. Also helpful have been the availability of prediction techniques for high-quality thermodynamic data, and process simulators that can rapidly solve the complex material, equilibrium and enthalpy relationships involved in simulating the performance of reflux exchangers.

Reflux exchangers are used for a variety of duties: air separation; recovery of hydrogen, ethylene, natural gas liquids (NGL) and liquefied petroleum gas (LPG); and purification of carbon dioxide. It is important to operate these cryogenic processes at high thermodynamic efficiency to minimize, or eliminate the need for, external refrigeration. The latest process integration and thermodynamic analysis techniques are often used to identify improvements in process efficiency, leading to savings in refrigeration.

The reflux exchanger is particularly suitable for recovery of heavy components from a light gas stream that contains only a few percent (up to 10%) of the heavier components. With these gas streams, the saving in refrigeration and the improvement in recovery give attractive returns on investment.

The relative volatility of the key components should ideally be two or more. If the relative volatility is much lower than two, then many stages may be required for good separation, and the required reflux rate may be too high for a reflux exchanger to be effective.

To design and size a reflux exchanger with confidence, one needs to predict the exchanger height by solving the heat- and mass-transfer-rate equations [3–5]. Also needed is a prediction of the

FIGURE 1. The brazed-aluminum plate-fin heat exchanger provides a high ratio of heat-transfer surface to volume

Originally published May 1994

unit's hydraulic performance for calculation of the incipient flooding point and minimum flow area.

The prediction of passage flooding can be difficult due to the complex geometry of plate-fin passages and because most reflux exchangers operate close to critical pressure. All these aspects of reflux exchanger design have received detailed attention from process designers, and extensive testing has been performed to ensure that design methods are valid for all commercial applications.

In recent years, there has been a growing trend toward the use of reflux exchangers not only in cryogenic temperatures, but also in hydrocarbon separations at warmer temperatures, and for control of hydrocarbon dewpoint (even in offshore applications). Their use in debottlenecking projects is also becoming popular.

Four projects that show the value and effectiveness of reflux exchangers are discussed below in more detail. The first example considers hydrogen recovery from demethanizer overheads; the second highlights a low energy process for NGL and LPG recovery from natural gas. The third is a simple process for recovery of ethylene from fluid-catalytic cracker (FCC) offgas; and the fourth is a similar process for olefin recovery from dehydrogenation-reactor offgas.

H_2 from demethanizer overheads

Hydrogen recovery by cryogenic processing has been practiced for many years and is still highly competitive for recovery of medium- to high-purity hydrogen at medium to high pressure. Hydrogen recovery is applied to gases resulting from either steam-reforming operations, such as those on ammonia plants, or offgas streams from refineries or petrochemical facilities [*6*].

Process flowsheets for 95–98% recovery of 90%-purity hydrogen are relatively simple because hydrogen has a high relative volatility. The refrigeration needed for the process can usually be provided by reduction in pressure of cold process streams and the resultant Joule-Thomson expansion.

However, high hydrogen purity can result in reduced hydrogen recovery. The use of a reflux exchanger at the

UNVEILING A REFLUX EXCHANGER

A reflux exchanger is a heat-exchange device in which both heat and mass transfer occur to effect a separation with high efficiency. In cryogenic processing, the reflux exchanger is usually a brazed-aluminium plate-fin heat exchanger (Figure 1). Due to its high ratio of surface area to volume, it is very compact and light, even when designed to operate with a temperature driving force of only 2–3°C.

Heat is removed from the feed vapor as it rises through the feed passages of the vertically mounted exchanger. The unit is so designed that the resulting condensed liquid runs back down the same passages as reflux, countercurrent to and in intimate contact with the rising vapor.

Other process or refrigerant streams flowing in adjacent passages provide the refrigeration needed to condense the upflowing vapor. In the feed passages, the downward-flowing liquid becomes enriched with the heavier components from the vapor stream while the more-volatile components are revaporized and stripped from the liquid by the upflowing vapor.

The reflux exchanger acts in much the same way as the upper part of a distillation column but a heat exchanger gives two important advantages over conventional distillation:

- The temperature difference between the condensing feed stream and the streams providing refrigeration is small. Large temperature-driving forces cause significant inefficiency in conventional distillation [1]
- The exchanger has in effect a large number of partial condensation stages so temperature and composition differences between vapor and liquid are small and separation is effected near to equilibrium conditions

As the reflux-exchanger process is performed with very small driving forces both for heat and material transfer, the refrigeration needed to condense the feed vapor is used effectively. Thus the overall refrigeration requirements are minimized.

To better understand the inherent advantages of the reflux exchanger, consider the analogy with the rectification section of a distillation column. In a column, all liquid reflux is normally produced at the condenser — the coldest point of the distillation process.

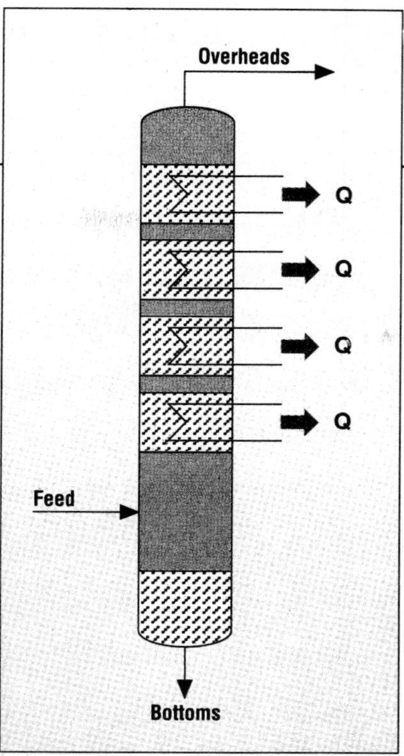

DENISE EARDMAN

FIGURE 2. A reflux exchanger is functionally equivalent to the multi-stage rectification section of a distillation column

However, the rectification section could theoretically be operated with a condenser on each stage of A column (Figure 2). Each condenser produces part of the total liquid reflux but at higher temperatures than with a single overhead condenser.

The reflux exchanger is a multistage rectification system, which is directly analogous to the multiple condenser concept. Reflux liquid is generated throughout the height of the exchanger and the whole of the exchanger length acts as a large number of separation stages.

The multistage rectification achieved by the reflux exchanger offers a few other benefits. In a partial condensation process, the feed is cooled to a sufficiently low temperature to ensure that most of the heavier components are condensed out and recovered. This can result in a relatively large amount of unwanted light components being dissolved in the liquid phase, and usually these must be removed by downstream fractionation.

In contrast, a reflux exchanger produces liquid at a higher temperature, with lesser amounts of dissolved light components. Consequently, the refrigeration load is reduced because of the reduction in the heat load for condensation. Alternatively, for the same refrigeration load, better recoveries can be achieved. ❐

'cold end' of the process can improve hydrogen recovery because it uses the available refrigeration more effectively. This has been shown to be the case in the following ethylene-plant project [7].

Recovery of hydrogen from the demethanizer overhead gas of an ethylene plant is conventionally performed using the process shown in Figure 3. In a typical installation, the gas sent for hydrogen recovery consists of hydrogen and methane in approximately equal proportions, and is available at about 35 bars and a temperature of −110°C. The hydrogen product is to be produced at essentially feed gas pressure and the methane-rich tail gas at as high a pressure as possible.

Most of the methane is condensed and collected in the first separator at −140°C. The condensate is expanded to 4.5 bars, evaporated in the first heat exchanger and is allowed to leave the plant as high-pressure tail gas. The second separator operates at −170°C. To reach this temperature, the condensate from the separator is expanded to low pressure and some hydrogen-rich vapor is injected into it. The condensate is reheated in the heat exchangers and permitted to leave the plant as low-pressure tail gas. The product stream from the second separator has a hydrogen purity of 98 vol. %.

It is possible for the low pressure tail gas to be brought out of the hydrogen-recovery unit at the same pressure as the high-pressure tail gas (that is, at 4.5 bars), but only by injecting hydrogen. Hydrogen injection dilutes the condensate to the extent that it evaporates at lower temperatures, and thermodynamic pinches are avoided in the heat exchangers. Thermodynamic analysis shows that the amount of hydrogen injected into the condensate could be reduced by use of a reflux exchanger (Figure 4), and thereby a higher hydrogen recovery can be achieved.

When the low-pressure tail gas is brought to the same pressure as the high-pressure tail gas, the hydrogen recovery is increased by about 3% using the reflux exchanger. Furthermore, since this route results in an increase in the mean temperature difference in the second exchanger, less surface area is required and there is a corresponding reduction in capital cost. By condensing the methane at relatively higher temperatures, the reflux exchanger eliminates any possibility of freezing, thus significantly increasing plant reliability (*CE*, January, pp. 37–41).

A reflux exchanger essentially performs the same duty as the rectification section of a distillation column

The reflux exchanger provides the most cost-effective way of ensuring high hydrogen recovery with minimal modification to an existing process flowsheet. In the present example, the reflux exchanger produces a small but worthwhile improvement in plant performance. In the next three examples, the reflux exchanger is crucial to overall plant performance.

NGL and LPG from natural gas

The conventional method of extracting natural-gas liquids from natural gas is by the use of a cryogenic turboexpander process (Figure 5). The desired components, either ethane- or propane-plus, are recovered by cryogenic partial condensation of the feed gas.

Refrigeration is generated by work-expansion of high-pressure feed gas, sometimes supplemented by mechanical refrigeration using propane. Liquids condensed from the feed are separated and passed to a low-temperature stabilizer column, which fractionates off the lighter components.

The expander exhaust normally contains some liquid that is fed as reflux to

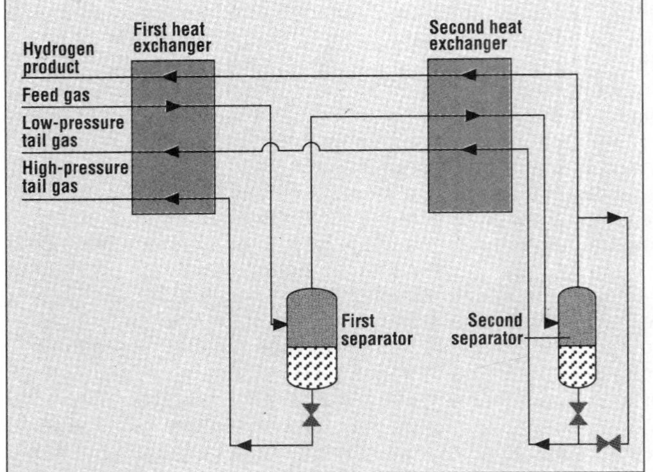

FIGURES 3 (left) AND 4 (below). Compared to a conventional system (left) for hydrogen recovery from demethanizer overheads, a unit equipped with a reflux exchanger (below) can boost hydrogen recovery by about 3%

FIGURES 5 (top right) AND 6 (bottom right). A reflux exchanger (bottom) offers lower operating and capital costs than those for a turboexpander (top)

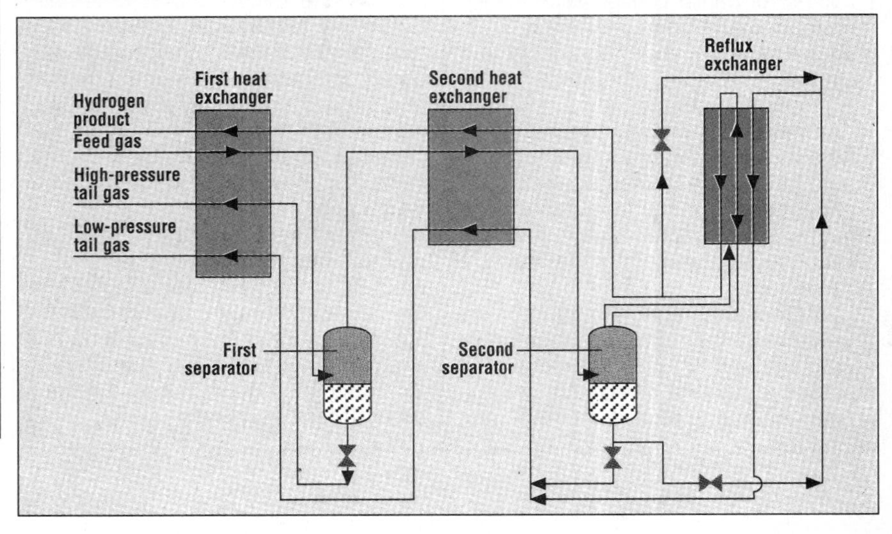

the top of the stabilizer. The expander outlet vapor and stabilizer overheads are reheated in a countercurrent fashion with feed gas, and this residue gas is then recompressed in the brake of the turboexpander. It is then boosted to the required pipeline pressure.

The turboexpander plant utilizes the pressure of the feed gas to generate the refrigeration needed for the process to operate. This is a disadvantage when the feed gas is not available at particularly high pressure, or when the residue gas must be returned to a pipeline at high pressure because the capital and energy costs for additional gas compression can be high, especially for high liquids recovery. To overcome this deficiency, a simple yet efficient alternative to the turboexpander process has been developed, based on a reflux exchanger.

The first NGL-recovery plant using the new reflux-exchanger process (Figure 6) was started up in 1985. It can achieve 97% propane recovery with a much lower power consumption than competing processes [*8–11*]. The low power consumption is due to the fact that no compression of the product gas is required. The overall capital cost is also less than that for a turboexpander plant of comparable capacity.

The feed gas leaving the dehydration unit passes through a heat exchanger where it is cooled and partially condensed before entering a separator.

COMPARATIVE LOSS ANALYSIS

	Turbo-expander	Reflux exchanger
Minimum theoretical work of separation, kW	67	67
Power requirement, kW		
Compression	194	134
Warm exchangers	351	134
Cryogenic exchangers	246	149
Expander and compressor	328	-
Distillation	157	31
Valves, mixers, pumps and so on	-	37
Total	1,276	485

Basis: 25 million std. ft^3/d feed; 95% propane recovery

TABLE 1. Thermodynamic comparison reveals why the total power requirement for the reflux-exchanger process is so much less than for the conventional turboexpander route

The vapor stream leaving the separator is then partially condensed in the reflux exchanger.

Liquid condensate from the bottom of the reflux exchanger is combined with the liquid hydrocarbons from partial condensation. This condensate is pumped and reheated through the feed gas cooler to near ambient temperature before being stabilized in a distillation column.

Overhead gas from the column is returned to the feed gas to ensure a high overall product recovery. Alternatively, the column could be fitted with an overhead condenser to reduce gas recycling. Stabilized liquid product from the column is reboiled with hot oil or low-pressure steam. Vapor from the top of the reflux exchanger, typically at a temperature of about –70°C, is reheated, first in the reflux exchanger to provide refrigeration and then with incoming feed gas. The resulting product gas is then delivered at about the same pressure as the feed gas to the plant.

The refrigeration for the process is typically provided by an external refrigeration cycle. A cascade cycle based on propane and ethylene or ethane is conventional, although a mixed-refrigerant cycle may also be used.

The reflux exchanger essentially performs the same duty as the rectification section of the stabilizer column (Figure 5) but it does the job so efficiently that the process needs much less refrigeration. The reflux-exchanger process operates at higher pressure and temperature than a turboexpander process, thereby further improving efficiency and minimizing temperature differences in the unit.

The turboexpander process has a wide temperature difference at the "cold end" of the process (Figure 7A), which contributes to its high power consumption. In the reflux-exchanger process, close temperature approach is achieved throughout the entire cooling range (Figure 7B).

There are two other important benefits over the turboexpander process:

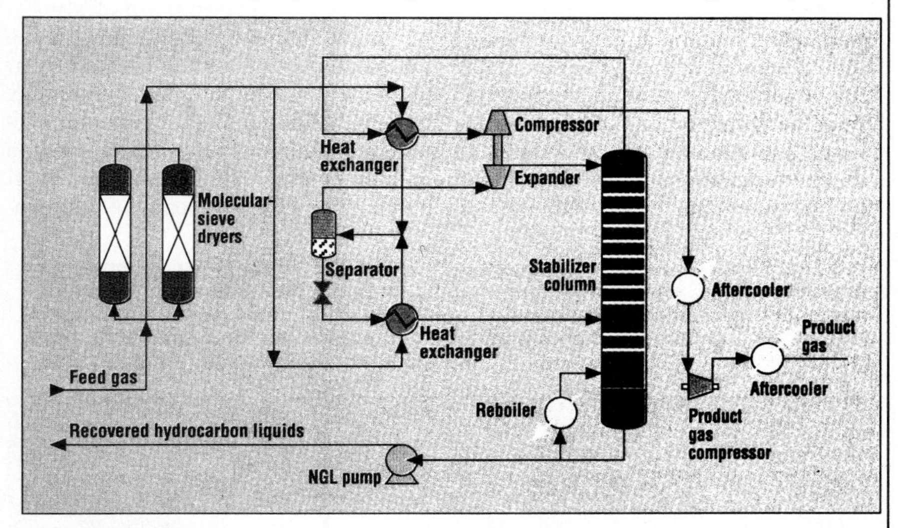

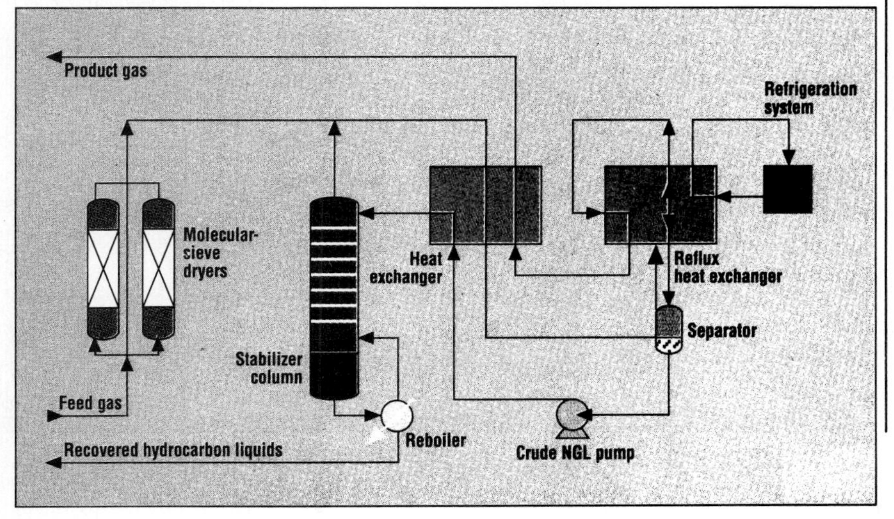

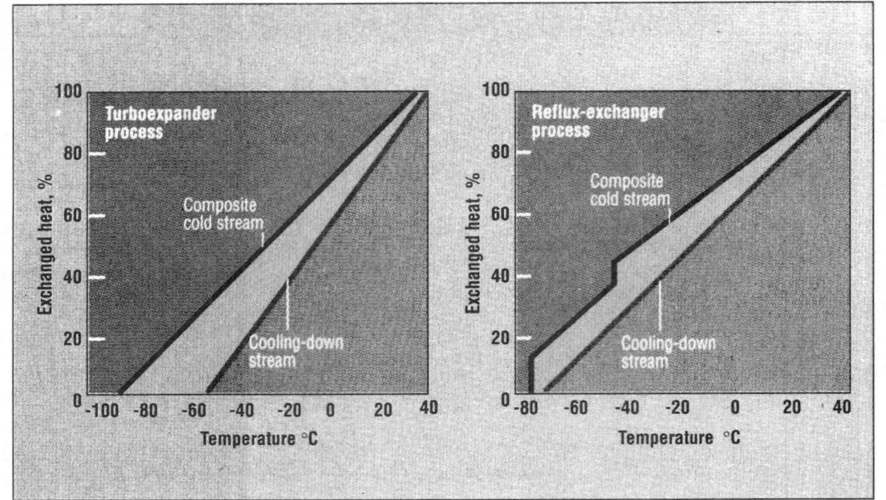

FIGURES 7A (left)AND 7B (right). Unlike the turboexpander process, which offers much of its refrigeration at the coldest section (left), a reflux exchanger allows cooling throughout a wide range of temperature (right)

With the present overcapacity in world ethylene production, less efficient plants need revamping and debottlenecking to maintain competitiveness in today's open and global marketplace. Under these circumstances, integration of one or more reflux exchangers can prove highly attractive.

In the recovery of C_{2+} from fluid-catalytic cracker (FCC) offgas, efficiency is very important because ethylene is used as a refrigerant at –100°C. The recovered liquid is pumped to an ethylene-plant chilling train to boost overall recovery of C_{2+} from the demethanizer. The very low capital investment for this project has resulted in a payback time of well under one year.

- Turndown and flexibility are greater than that usually obtained with a turboexpander plant
- Because the process is operated at relatively high temperature and pressure there is greater tolerance to carbon dioxide in the feed gas

Thermodynamic analysis of the two processes [*9*] shown in Table 1 reveals the power requirement of each part of each process. It is also clear why the total power requirement for the reflux-exchanger process is so much less than the turboexpander route.

The reflux-exchanger process is cost-effective for dewpoint control and liquid extraction (especially for gas rates below 100 million std. ft^3/d). Gas streams that cannot be economically processed in conventional cryogenic plants can be processed at low cost. For high-pressure feed gas, a modified process flowsheet has been developed, which also requires much less power than a conventional turboexpander process [*12, 13*].

There is increasing interest in using reflux exchangers for dew-point control during offshore hydrocarbon processing, and NGL recovery. This allows one to exploit the dual advantages of the reflux-exchanger process: efficient processing in light, compact heat-exchange equipment. The reduction in space and weight by using reflux exchangers can give tremendous cost savings. Evaluations have shown not only major capital and operating savings [*14*] but also improved reliability, especially where the feed gas can undergo substantial changes in composition (Figure 8).

Olefins from FCC offgas...

Reflux exchanger technology is particularly appropriate where very high recovery, typically 98% or greater, of high-value hydrocarbon components is required. Recovery of propylene and isobutylene from refinery offgases is becoming increasingly popular as new cracking and conversion processes are installed on refineries. Isobutylene in particular is a valuable feedstock for gasoline production. Ethylene recovery can be attractive, too. Recovery processes based on a reflux exchanger are very competitive for these important CPI applications.

Combining a reflux exchanger with a distillation column, either upstream of the column or acting as an auxiliary condenser, leads to high recovery of valuable hydrocarbons. Such schemes have been used on ethylene plants to improve ethylene yield with low incremental energy and capital cost [15].

... and dehydrogenation offgas

Catalytic dehydrogenation of propane or butane to form olefins is steadily increasing in importance, especially where olefins act as feedstocks for production of high-value products, such as polypropylene and methyl-*tert*-butyl ether. The effluent from the dehydro-

FIGURE 8 (below). Reflux exchangers reduce the weight and space requirements for hydrocarbon-processing and NGL-recovery units, and enhance plant reliability

FIGURE 9 (right). The reflux-exchanger process for olefins recovery from dehydrogenation offgas requires no external refrigeration. All the cooling is provided by the expansion of the refluxed vapor

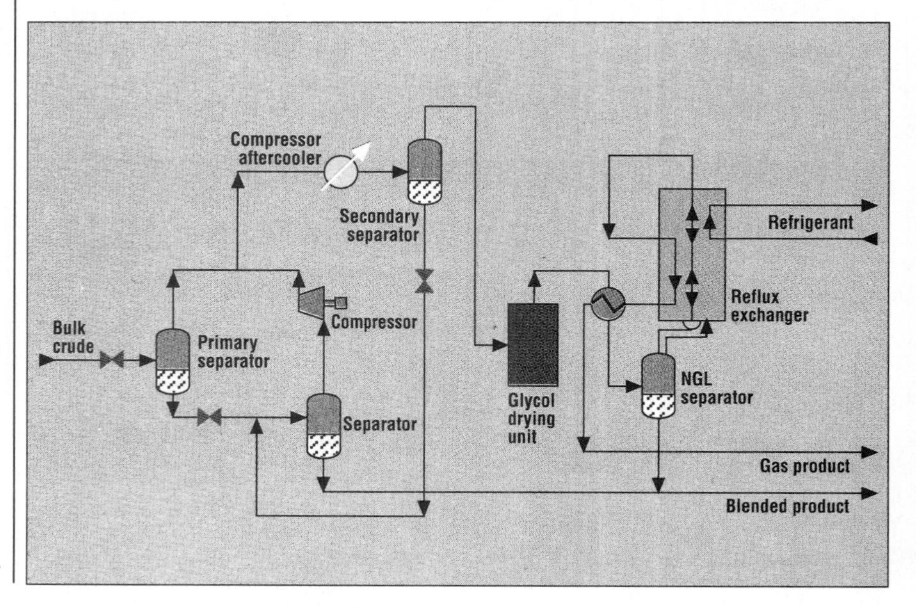

COMPARISON OF EQUILIBRIUM CONSTANTS

K values at 30 bars	Temperature, °C –75	+20
C_2H_4	0.17	1.21
C_2H_6	0.0084	0.90
C_3H_6	0.011	0.24
C_3H_8	0.008	0.17

TABLE 2. At the elevated temperatures of the reflux-exchanger process, the K values for the lighter components are much higher, allowing better recoveries of C_{4+}

genation reactor is a mixture of hydrogen, paraffins and olefins. It is conventional to use cryogenic technology to recover the desired propylene or butylene from this stream for fractionation, and the high product value justifies the cost of separating these products at high recovery levels.

The reflux-exchanger process shown in Figure 9 gives high recovery of hydrocarbons while cutting the level of lighter components in the liquid phase. This reduces energy requirements for downstream fractionation. No external refrigeration is necessary because all the refrigeration is provided by the expansion of the refluxed vapor in the turboexpander.

The amount of refrigeration available is limited by the pressure ratio across the expander. The reflux exchanger uses the available refrigeration as efficiently as possible to maximize olefin recovery. More than 99% recovery of the desired olefins can be achieved on a reflux-exchanger process of this type.

The level of dissolved light components in the liquid from a conventional partial-condensation process that relies on simple chilling, is much higher than that with the reflux-exchanger process. The reason is the difference in separator temperature between the two processes.

In C_{4+}-recovery applications, the reflux-exchanger process produces liquid at 20°C whereas the partial condensation process does so at –75°C. The equilibrium constants (the so-called K values) for the light components as given by the Peng-Robinson equation of state are much higher at the elevated temperature of the reflux-exchanger process (Table 2). The higher temperature leads to much less lighter components in the liquid phase, allowing easier fractionation of this stream and better recoveries of C_{4+} from hydrogenation offgas [*16*].

Also, the temperature difference between the top and bottom of the reflux exchanger is large. Supply of refrigeration at much higher temperatures in the reflux-exchanger process results in an extremely cost-effective route to liquids recovery. Not surprisingly, the reflux-exchanger process is well on its way to becoming a "conventional route" for olefin recovery from dehydrogenator effluents. ■

Edited by Gulam Samdani

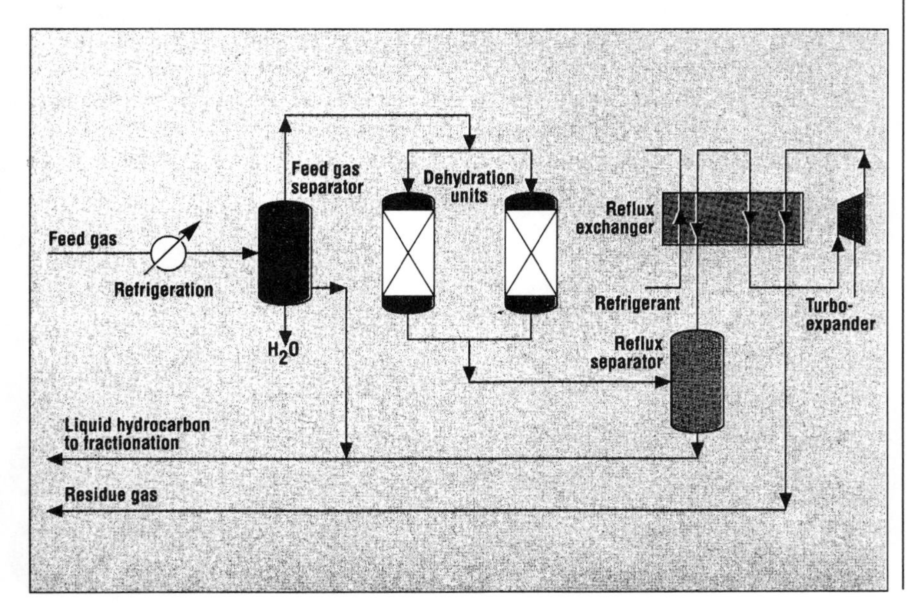

Acknowledgements

The author gratefully acknowledges the permission of the management of Costain Oil, Gas & Process Ltd. to publish this article. He respectfully recognizes the outstanding contribution of the late Martin Ruhemann (1903–1993) to the development of reflux exchanger technology.

The author

Adrian J. Finn is principal process engineer, responsible for proposals and detailed engineering on projects at the process contracting div. of Costain Oil, Gas and Process Ltd. (Costain House, Styal Road, Wythenshawe, Manchester M22 5WN, U.K.; tel: (44) 061-910-3227; fax: (44) 061-499-9218). He is also responsible for conceptual and feasibility studies in natural gas liquefaction, nitrogen rejection and natural gas liquids extraction as well as hydrogen, ethylene and liquefied petroleum gas processing. He is a member of the Institution of Chemical Engineers (U.K.), and the European Chapter of Gas Processors Assn. He holds a B.Sc.Tech. in chemical engineering and fuel technology from Sheffield University, and an M.Sc. from Leeds University (both in U.K.).

References

1. Steinmeyer, D., Save energy, without entropy, Hydrocarbon Processing, Vol. 71, No. 10, p. 55, October 1992.
2. Ruhemann, M., "Separation of Gases," 2nd ed., Oxford University Press, London, 1949.
3. Rohm, H. J., The Simulation of steady State Behaviour of the Dephlegmation of Multi-Component Mixed Vapours," Int. J. Heat Mass Transfer, Vol. 23, p. 141, 1980.
4. Davis, J. F., and others, Fractionation with Condensation and Evaporation in Wetted-Wall Columns, AIChE J., Vol. 30, No. 2, p. 328, 1984.
5. Tung, H. H., and others, Fractionating Condensation and Evaporation in Plate-Fin Devices, AIChE J. Vol. 32, No. 7, p. 1,116, 1986.
6. Banks, R., and Isalski, W. H., Excess fuel gas ? Recover H2/LPG, Hydrocarbon Processing, Vol. 66, No. 10, p. 47, October 1987.
7. Ruhemann, M., and Tichar, K., "Proceedings of CryoTech 73," p. 61, 1973.
8. Tomlinson, T. R. and Cummings, D. R., US Patent 4,675,030.
9. Tomlinson, T. R. and Banks, R., LPG extraction process cuts energy needs, Oil and Gas Journal, p. 81, July 15, 1985.
10. Limb, D. I., and Czarnecki, B. A., Reflux-exchanger process lifts propane recovery at Aussie site, Oil and Gas Journal, p. 35, December 14, 1987.
11. McKetta, J. J., ed., "Encyclopedia of Chemical Processing and Design," Vol. 45, p. 43, Marcel Dekker Inc., New York, 1993.
12. Tomlinson, T. R. and Czarnecki, B. A., US Patent 4,846,863.
13. Tomlinson, T. R., and others, Exergy Analysis in Process Development, The Chemical Engineer, No. 483 and 484, pp. 25 and 39, October 11 and 25, 1990.
14. Czarnecki, B. A., UK Patent 2,224,036.
15. Chiu, C-H, Advances in gas separation, Hydrocarbon Processing, Vol. 69, No. 1, p. 69, January 1990.
16. Isalski, W. H., "Separation of Gases," Oxford University Press, London, 1989.

EVAPORATORS EVAPORATORS EVAPORATORS EVAPORATORS

Spell out all the requirements of the system. Then choose the evaporator type and configuration

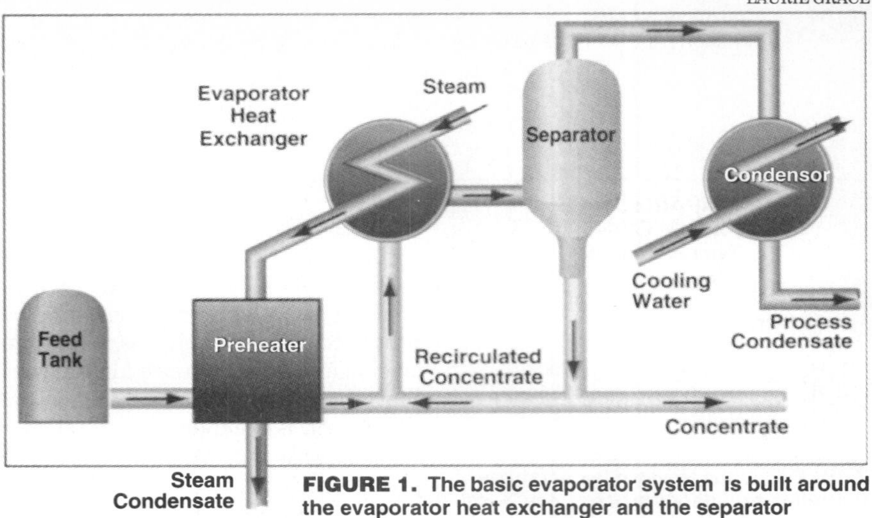

FIGURE 1. The basic evaporator system is built around the evaporator heat exchanger and the separator

An engineer selecting and specifying evaporation equipment faces numerous decisions. He or she must decide among the several types of evaporators, each having advantages and limitations. Once the type is selected, other choices must be made, notably the operating conditions and the number and order of evaporation effects.

Usually, an evaporator serves to concentrate a solution by boiling off water. It can also remove organic solvents. Two liquids can be separated by evaporation, if there is a large difference in boiling points and if complete separation is not necessary.

Most evaporators are steam driven, but systems can be designed using hot oil, hot water, or other convenient heat source. And the evaporator itself is ordinarily served by a variety of auxiliaries, as in the general scheme shown in Figure 1.

The feed tank is the starting point for material to be introduced into the system. Preheaters may be included if the feed is below its evaporating temperature. The evaporator heat exchanger can be one of several different types, determined by the nature of the liquid being concentrated.

The separator disengages the droplets of concentrated liquid from the vapor stream. Usually, this takes place in a relatively large-diameter vessel that lowers the upward velocity of the stream and lets the droplets settle. In some designs, the inlet stream enters tangentially, and the separation is aided by centrifugal force. A mesh or vane type mist remover is sometimes needed.

A portion of the concentrated product may recirculate to the evaporator heat exchanger; the rest is drawn off as product. Meanwhile, the vapor generated in the evaporator goes to a condenser. Direct-contact (spray) condensers have been popular in the past, but most new installations use surface condensers, usually shell-and-tube units and sometimes plate exchangers.

Evaporator systems commonly operate under vacuum. When using cooling-tower water as coolant in the condenser, it is usually practical to condense vapor at about 120°F. This requires an absolute operating pressure of about 3 in. Hg if pressure losses in the ducting and condenser are taken into account. This level of vacuum is fairly easy to obtain, using a two-stage steam jet ejector or a liquid-ring vacuum pump. The latter are usually preferable; they consume less energy.

The evaporator system also requires pumps, piping, and controls. Some applications require a cooler for the concentrated liquid.

Multiple effects

As discussed later, it is often advantageous to operate evaporators in combination. Figure 2 shows, for instance, a triple-effect arrangement. Vapor removed from the feed in the first effect becomes the heat source for the second effect, and that from the second effect is the heat source for the third effect.

In this example, the third effect is split into two stages: The second-effect vapor goes in parallel to the two third-effect heat exchangers whereas the process liquid passes through them in series. Each boiling-condensing cycle is an "effect" and each product pass is a "stage," so this design is a three-effect, four-stage evaporator.

The "economy" of an evaporator system equals the amount of evaporation divided by the amount of steam supplied. The system in this example has an economy of roughly 3, because 1 lb of supplied steam produces 1 lb of evaporation in each of the three effects. Due to preheat requirements, vent losses, and the difference in latent heat at dif-

Originally published April 1994

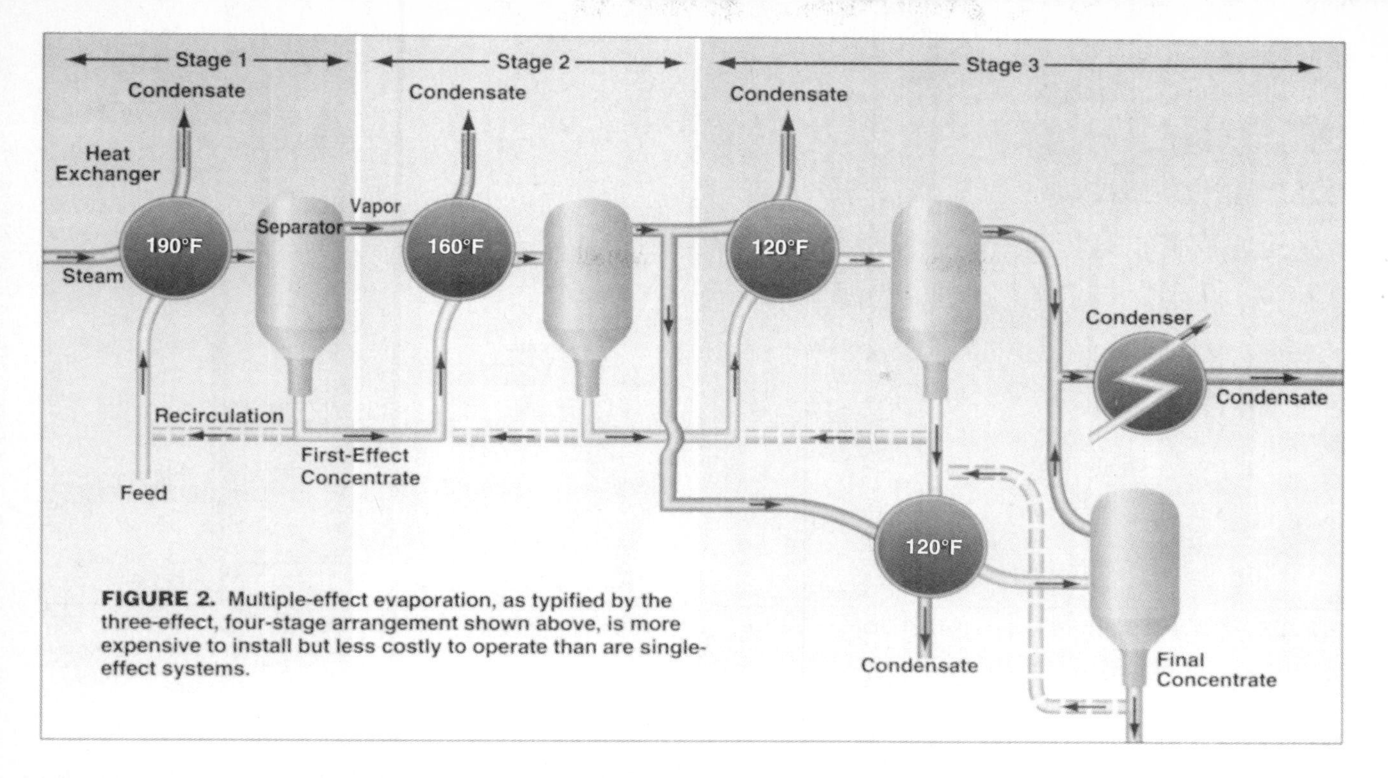

FIGURE 2. Multiple-effect evaporation, as typified by the three-effect, four-stage arrangement shown above, is more expensive to install but less costly to operate than are single-effect systems.

ferent temperatures, the actual economy is likely to be somewhat lower.

The lower steam requirement of the multiple-effect system is accompanied by an higher equipment cost. The available temperature difference at any single heat exchanger is considerably lower than that available in a single-effect evaporator, so the total surface area will be greater. And, the multiple-effect evaporator needs more vessels, pumps and piping. This tradeoff between capital and operating costs is an important consideration in selecting the best evaporator arrangement.

Evaporator types

An evaporators is classified according to the design of its heat exchanger. Most purchases today are for the film evaporators, with suppressed-boiling (forced-circulation) evaporators ranking next.

Film evaporators: In these, the process liquid is distributed as a film on the heat transfer surface. As evaporation takes place, vapor fills the core of the tube (or plate) passage. This vapor thins and thus accelerates the film.

When they can be used, film evaporators are the most cost-effective. They can usually be designed for low temperature differences, yielding high steam economy, by using several effects and vapor recompression (discussed later).

Film-evaporator heat exchangers tend to be less expensive per unit surface area than those for other evaporators. For instance, tubular film evaporators commonly employ tubes of large diameter (e.g., 2 in.) and lengths around 20 to 50 ft. This costs less than the narrower, shorter tubes in forced-circulation designs.

The process liquid occupies only a thin film on the tube wall, so liquid holdup is low. The need for cleaning chemicals is thus minimal.

Because of the thin film, residence time tends to be short. This and the low temperature differences make the film evaporator attractive for heat-sensitive materials.

Film evaporators are limited to relatively fluid products, because high viscosity thickens the film and lowers the heat-transfer coefficient. The practical upper limit is around 100 to 500 cP.

Accordingly, heavily fouling substances and those with significant suspended solids may better be handled in forced-circulation evaporators. For moderate fouling, film evaporators are acceptable if cleaned often.

All film evaporators have a minimum liquid wetting rate, to assure that the surface gets coated. If the required rate is above the product flowrate, one must recirculate the material back to the heat exchanger inlet, or else split the heat exchanger into multiple stages and operate them in series, roughly doubling the liquid rate per tube.

Heat-transfer coefficients for film evaporators are based on operating experience or pilot plant testing. Overall clean coefficients (assuming that the heat source is condensing steam) range from about 500 Btu/(h)(ft^2)(°F) when processing waterlike materials to 100 Btu/(h)(ft^2)(°F) or even less for products near the upper viscosity limits.

Film evaporators come in rising-, falling-, or rising-falling-film configurations. They are available as tubular or plate systems.

Rising-film tubular evaporators: This

HOW TO MAKE THE RIGHT CHOICE

Greg Lavis
APV Crepaco, Inc.

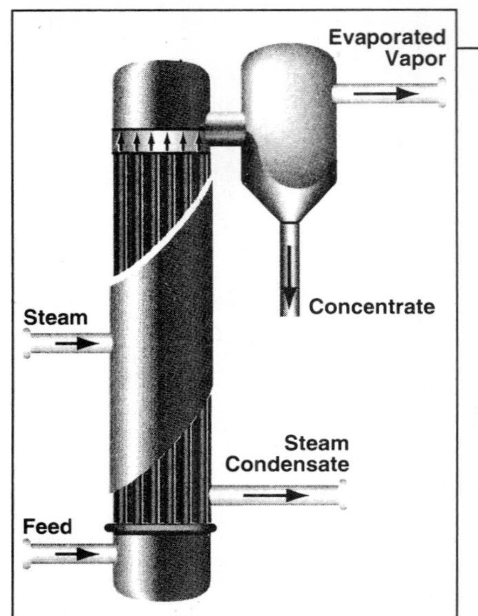

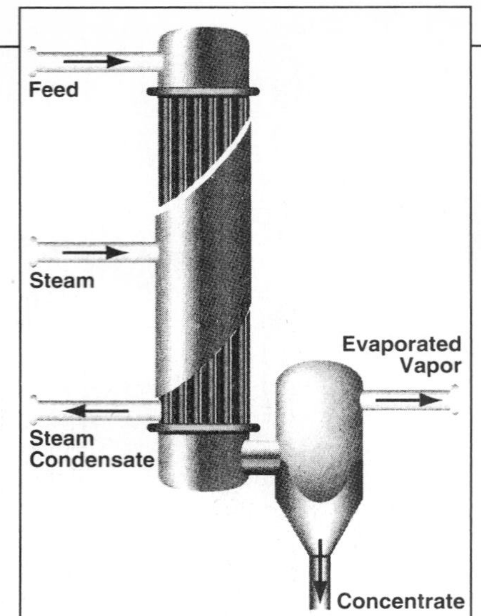

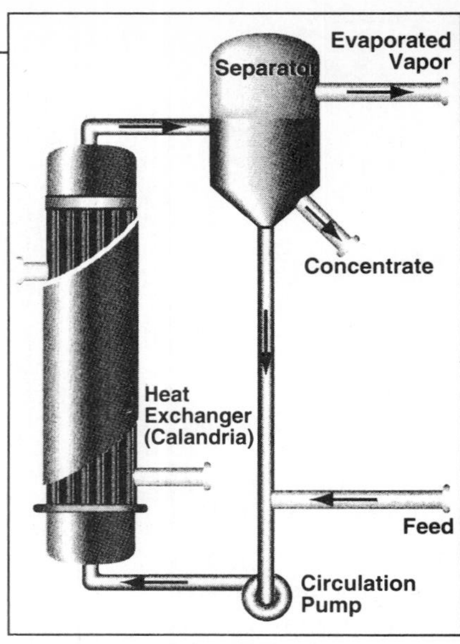

FIGURES 3, 4 and 5. Three workhorses of evaporation are (left to right) rising- and falling-film tubular evaporators and suppressed-boiling evaporators. The latter, also called forced-circulation evaporators, can use other boiling-suppression strategies instead of the static head shown here. Film units tend to be economical; forced-circulation units tolerate fouling products better

was the first film evaporator. The feed liquid (Figure 3) enters the bottom bonnet of the heat exchanger, flooding the space below the bottom tubesheet. Flow distribution is thus no problem.

As the liquid is heated and evaporation begins, vapor first appears as bubbles in the liquid pool. Farther up the tubes, the bubbles become larger until the fluid changes over to the slug-flow regime, wherein the vaporized gas is in the form of large plugs. Still farther up, the liquid film is formed, along with a vapor core.

If the feed is not adequately preheated or the evaporation rate per tube is low, the film develops slowly. This decreases the heat-transfer coefficient and increases the residence time.

The weight of the liquid in the tube creates a hydrostatic head. This increases the pressure and therefore the boiling temperature at the tube bottom. Accordingly, rising-film evaporators cannot operate reliably at temperature differences smaller than about 25°F.

Product recirculation (if needed) does not require a circulation pump. Merely maintain a high enough level in the separator to overcome the hydrostatic head in the tube.

Falling-film tubular evaporators: In this design (Figure 4), feed enters the top of the heat exchanger, and coats the wall of the tubes. The liquid film forms immediately due to gravity. As evaporation takes place, the vapor in the tube core accelerates, thinning the film further.

The falling-film evaporator produces a thinner, faster moving film than the rising film evaporator, so its heat-transfer coefficient is higher. The contact time, or residence time within the heat exchanger, is also short, typically about 15–30 s per stage.

Because evaporation is not needed to form the film and there is no hydrostatic head to overcome, one can design falling film evaporators for temperature differences as small as 6°F. This adds system flexibility, as illustrated later. With low temperature differences, however, the effect of pressure drop, and of the buildup of non-condensables, can be significant.

One correlation* that has been used to estimate the minimum wetting rate for a falling film in a tube is

$$\Gamma = 19.5\,(\mu_l\, S_l \sigma^3)^{0.2}$$

where Γ is the wetting rate in lb/(h)(ft), μ_l is the liquid viscosity in centipoise, S_l is the specific gravity, and σ is the surface tension in dyne/cm. In practice, the selected wetting rate varies with the product. For heat-sensitive products it is often desirable to keep wetting rates fairly low, to limit the residence time. High wetting rates help limit fouling. With this evaporator, recirculation requires a pump.

Liquid must be distributed evenly over the heat-transfer surface. This prevents localized overconcentration, dry spots, increased fouling, and reduced overall heat transfer coefficients.

*Minton, Paul E., "Handbook of Evaporator Technology," Noyes Publications, Park Ridge., N.J.,1986, p. 27.

†Perry and Chilton, "Chemical Engineers' Handbook," 5th ed., McGraw-Hill, New York, 1973, p. 5-57.

One distribution method is to extend the tubes above the tubesheet: Liquid is fed into the space below the top of the tubes, and overflows into them. Or, a distribution plate can be located above the tubesheet. A third method is to use a spray nozzle at the liquid inlet.

Some distribution schemes work better if the feed is preheated above the evaporation temperature. In these cases, the flashing of the liquid at a distribution orifice creates turbulence, which helps spread out the liquid.

Thickness of a falling film, without vapor shear, in laminar flow, is calculated† by:

$$z = \left[\frac{3\Gamma\mu_1}{g\rho_l(\rho_l - \rho_v)}\right]^{\frac{1}{3}}$$

where z is film thickness, Γ is wetting rate, μ_l is liquid viscosity, g is acceleration due to gravity, ρ_l is liquid density and ρ_v is vapor density.

The residence time in a falling-film evaporator stage is the holdup volume divided by the flow rate. If ρ_v is negligible, the residence time (assuming laminar flow) is:

$$t = L\left[\frac{3\mu_l\rho_l}{g\Gamma^2}\right]^{\frac{1}{3}}$$

where L is the tube length and t is time.

Vapor shearing thins and accelerates the liquid film. If the Reynolds number becomes such that the flow regime is no longer laminar, the two equations just cited do not apply. In practice, they tend to overestimate the film thickness and residence time.

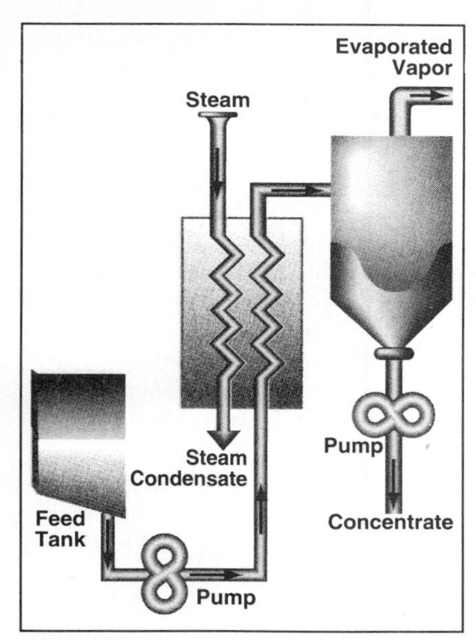

FIGURE 6. **The proprietary Paravap design is especially attractive for streams that are relatively clean but viscous**

Correlations to estimate heat-transfer coefficients in falling-film evaporators are available, but their accuracy is uncertain. In practice, overall coefficients based on experience or pilot plant data are usually preferred.

Rising-falling-film tubular evaporators: In this version, the falling-film pass is an extension of the rising-film one within the same heat-exchanger shell, so the effective tube length of this evaporator is twice the actual length.

This design is sometimes used where headroom is limited. Like the rising-film tubular evaporator, it also offers easy flow distribution.

Counterflow film evaporators: In a counterflow film evaporator, the liquid feed enters the top of the heat exchanger, and is distributed as a falling film on the evaporator tubes. The vapor flows upward in the center of the tubes, and exits at the top tubesheet.

When evaporating a liquid-liquid mixture, a counterflow evaporator increases the separation between the two liquid components. However, the vapor velocity in this evaporator must be low to avoid flooding.

Plate evaporators: In these units, as in plate heat exchangers,* the process fluids and the heating fluid flow separately between gasketed plates. In a rising-falling-film plate evaporator, the "plate units" each consist of steam plates, an upflow product-inlet plate and a downflow product-discharge plate; several plate units are assembled in the frame, and the incoming feed is distributed to them, in parallel. In yet a different plate-evaporator option, there are falling films solely.

Plate evaporators are flexible: Plates can be added or removed as needed for changes in load. And they are compact, needing less space than tubular evaporators for a given heat-transfer-surface area. Headroom requirements are much lower—a complete plate-evaporator installation will fit under a 20-ft ceiling, whereas tubular film evaporators might need more than 50 ft. Installation and building costs are significantly reduced.

In a falling-film plate evaporator, the residence-time correlation is the same as for a falling film tube. The length at a plate is much less than the length of the tube, so the residence time will be shorter, typically 5 to 15 s per stage.

This makes plate evaporators attractive for heat-sensitive products, such as foods. The ability to open the frame and inspect the heat-transfer surface is also beneficial in food service.

Suppressed-boiling evaporators: In these devices, also called forced-circulation evaporators, the process liquid circulates through the heat exchanger at a high rate. As the liquid is heated, boiling is suppressed by back pressure. In Figure 5, this is created by the static head of the piping at the heat exchanger exit. Boiling can instead be suppressed by a valve or orifice plate.

As the liquid leaves the heat exchanger and flows upward, the pressure is reduced, and the liquid flashes. Vapor is removed in the separator. Most of the liquid is recirculated. If the evaporator is operated in a continuous mode, a product stream is removed and a feed stream introduced.

Suppressed-boiling designs usually tolerate fouling products better than film evaporators. And, the heat exchanger can have a high liquid velocity, which helps keep it clean. Product viscosity is limited mainly by the type of circulation pump, and the pressure drop in the heat exchanger.

The required circulation rate is calculated by

$$m = q / C_p \Delta t$$

where m is the circulation rate, q is the heat exchanger heat load, Δt is the liquid temperature rise, and C_p is the liquid specific heat. If we ignore sensible heat, the heat load is the evaporation rate times the latent heat, and the above formula becomes

$$m = e\lambda / C_p \Delta t$$

where e is the evaporation rate and λ the latent heat.

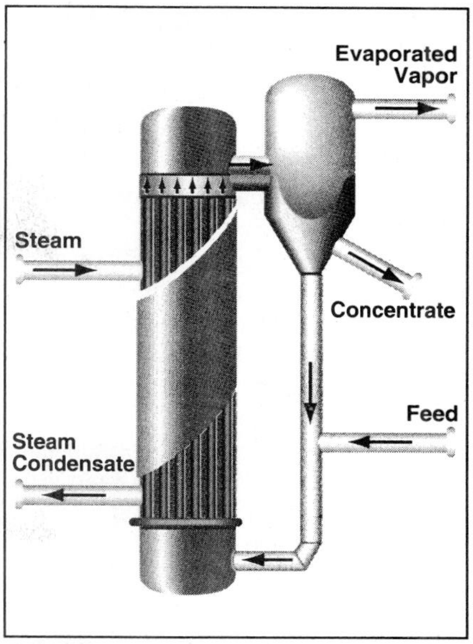

FIGURE 7. **The natural-circulation evaporator, shown here, is often employed as a component of a distillation system**

The high circulation rate means a low liquid-temperature rise, allowing a larger mean temperature difference over the heat exchanger. However, the high rate can complicate pump selection—the circulating liquid is at its boiling point, so pump net-positive-suction-head (NPSH) requirements must be kept in mind. Axial-flow, mixed-flow or standard centrifugal circulating pumps are usually selected.

When evaporating a viscous liquid, a positive-displacement pump is used. The heat-exchanger pressure drop is high in this situation, and the evaporator is frequently designed for a relatively low circulating rate and high temperature rise. These evaporator systems are usually limited to single- or two-effect configurations.

High velocities in the heat exchanger increase the heat-transfer coefficient and help limit fouling, but raise the pressure drop. Design of a forced-circulation evaporator stage is accordingly optimized by balancing heat-exchanger and pumping requirements.

Heat exchangers for suppressed-boiling evaporators are usually either shell-and-tube or plate designs. The former can be horizontal or vertical.

*For details on plate heat exchangers, see "Don't overlook compact heat exchangers," pp. 84–89, and "Tips to maximize heat transfer," pp. 104–110.

The product is nearly always on the tube side.

Plate heat exchangers offer higher heat-transfer coefficients and a more compact design. Operating pressures of up to 400 psi and temperatures of up to 400°F are possible, depending on the frame, plate, and gasket.

Other evaporators: The versions already discussed cover most chemical-process-industries (CPI) applications. However, there are other options.

Wiped-film evaporators: In a wiped-film or "thin film" evaporator, the heat exchanger consists of a jacketed cylinder, with wiper blades that distribute the product over its inner surface. The jacket contains the heating medium, usually high-pressure steam.*

Achieving a reasonable evaporation rate over the small heating surface requires large temperature differences. For this reason, multiple-effect configurations are not practical.

The cylinder wall is constantly scraped so the liquid film remains thin, even with viscous products. This prevents fouling. These evaporators are thus especially suitable for highly viscous materials (above 5,000 cP), and ones that tend to crystallize or foul.

Residence time in a wiped-film evaporator is short, since the liquid occupies only a thin film on a small surface. With heat-sensitive products, the advantage of a short residence time may offset the disadvantage of the high-temperature heating medium.

This evaporator incurs high capital cost per unit of evaporation. Rotor maintenance is an additional cost.

Paravap: The "Paravap" evaporator (Figure 6) is a proprietary APV design. Boiling takes place in a plate heat exchanger. High vapor velocities, and turbulence due to corrugations in the exchanger plates, cause the liquid to atomize. Even with viscous liquids, apparent viscosity at the plates is low.

This device is effective for viscous, relatively clean liquids. It offers cost savings over a wiped-film evaporator in several applications. One common use is concentration of corn or sugar syrup.

Natural-circulation evaporators: In these short-tube evaporators (Figure 7), the tube bundle can be outside (as shown) or inside the vessel. As evaporation occurs in the tubes, vapor displaces some of the liquid, and the static head in the tube bundle decreases.

The driving force for liquid recirculation is the difference between the static head of the column of pure liquid in the vessel and that of the liquid-vapor mixture in the heat exchanger. The most common use of this evaporator is as a thermosyphon reboiler for distillation.

*For an article on agitated thin-film evaporators, see p. 104.

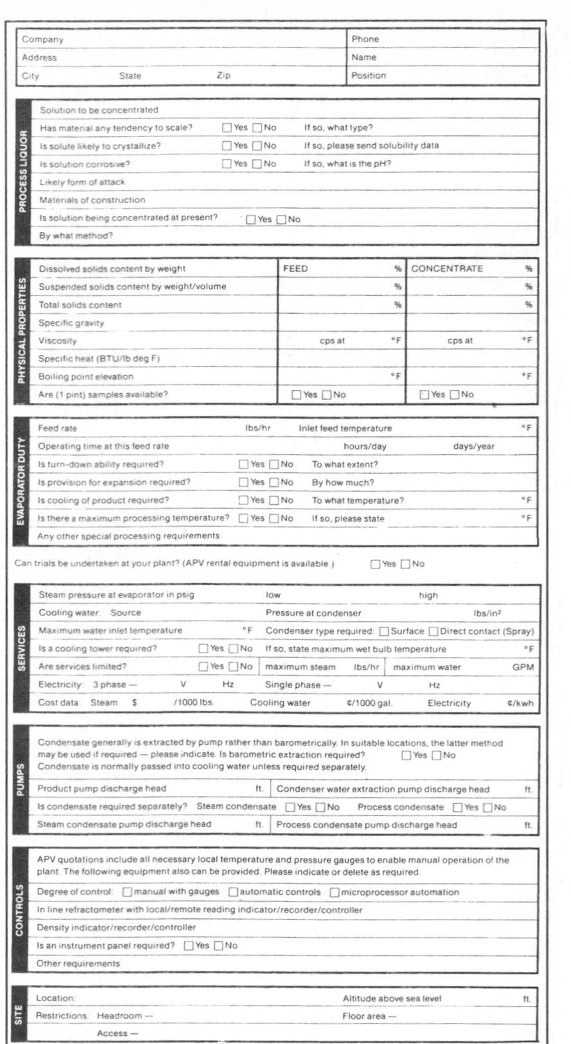

Company | Phone
Address | Name
City State Zip | Position

PROCESS LIQUOR
Solution to be concentrated
Has material any tendency to scale? ☐ Yes ☐ No If so, what type?
Is solute likely to crystallize? ☐ Yes ☐ No If so, please send solubility data
Is solution corrosive? ☐ Yes ☐ No If so, what is the pH?
Likely form of attack
Materials of construction
Is solution being concentrated at present? ☐ Yes ☐ No
By what method?

PHYSICAL PROPERTIES

	FEED	CONCENTRATE
Dissolved solids content by weight	%	%
Suspended solids content by weight/volume	%	%
Total solids content	%	%
Specific gravity		
Viscosity	cps at °F	cps at °F
Specific heat (BTU/lb deg F)		
Boiling point elevation	°F	°F
Are (1 pint) samples available?	☐ Yes ☐ No	☐ Yes ☐ No

EVAPORATOR DUTY
Feed rate lbs/hr Inlet feed temperature °F
Operating time at this feed rate hours/day days/year
Is turn-down ability required? ☐ Yes ☐ No To what extent?
Is provision for expansion required? ☐ Yes ☐ No By how much?
Is cooling of product required? ☐ Yes ☐ No To what temperature? °F
Is there a maximum processing temperature? ☐ Yes ☐ No If so, please state °F
Any other special processing requirements

Can trials be undertaken at your plant? (APV rental equipment is available.) ☐ Yes ☐ No

SERVICES
Steam pressure at evaporator in psig low high
Cooling water: Source Pressure at condenser lbs/in²
Maximum water inlet temperature °F Condenser type required: ☐ Surface ☐ Direct contact (Spray)
Is a cooling tower required? ☐ Yes ☐ No If so, state maximum wet bulb temperature °F
Are services limited? ☐ Yes ☐ No maximum steam lbs/hr maximum water GPM
Electricity: 3 phase — V Hz Single phase — V Hz
Cost data: Steam $ /1000 lbs. Cooling water ¢/1000 gal. Electricity ¢/kwh

PUMPS
Condensate generally is extracted by pump rather than barometrically. In suitable locations, the latter method may be used if required — please indicate. Is barometric extraction required? ☐ Yes ☐ No
Condensate is normally passed into cooling water unless required separately.
Product pump discharge head ft. | Condenser water extraction pump discharge head ft.
Is condensate required separately? Steam condensate ☐ Yes ☐ No Process condensate ☐ Yes ☐ No
Steam condensate pump discharge head ft. | Process condensate pump discharge head ft.

CONTROLS
APV quotations include all necessary local temperature and pressure gauges to enable manual operation of the plant. The following equipment also can be provided. Please indicate or delete as required.
Degree of control: ☐ manual with gauges ☐ automatic controls ☐ microprocessor automation
In line refractometer with local/remote reading indicator/recorder/controller
Density indicator/recorder/controller
Is an instrument panel required? ☐ Yes ☐ No
Other requirements

SITE
Location: Altitude above sea level ft.
Restrictions: Headroom — Floor area —
Access —

FIGURE 8. Questionnaire, courtesy of APV Crepaco, shows the information an evaporator maker should be given before starting a customer design

Define the requirements

The first step in specifying an evaporator is to spell out the requirements of the system. The nature of the product, the throughput and the available utilities all need to be specified. Any control-system requirements, limitations on equipment size or layout, and other special requirements must be stated.

It is convenient to structure this task around a design questionnaire, such as the one in Figure 8. The first sections deal with the nature of the feed. As already indicated, fouling tendencies usually dictate the type of evaporation selected, and may define the operating temperatures. Nature of the liquid also governs the materials of construction.

The properties of the process stream are sometimes not well known, so one must rely on testing, or on data from previous installations. The nature of the feed stream may vary due to fluctuations in upstream processing.

Three particularly important properties for specifying an evaporator are viscosity, boiling-point elevation, and presence of suspended solids. Any of these can cause problems.

In many cases, the viscosity will change more with temperature and concentration than do other properties—it is not uncommon to start with a waterlike feed and end with several thousand centipoise. Change in viscosity will usually be the main cause of difference in heat transfer coefficients between stages in the evaporator.

Complicating the picture, viscosity is often one of the the least-well-defined characteristics of the product. Knowing the viscosity at the feed and final concentrations does not usually define it at the intermediate concentrations. Furthermore, viscosity can be affected by changes in feed-stream composition or in upstream processing.

Boiling-point elevation is significant if the concentration of solute is high For instance, water under 3.72 psia boils at 150°F; if an aqueous solution at that pressure has a concentration such that its boiling-point elevation is 5°F, it boils at 155°F but the evaporated water vapor condenses (at that pressure) at 150°F, and the 5-degree difference is not available for heat transfer. In a multiple-effect setup, the effective temperature difference over the system is

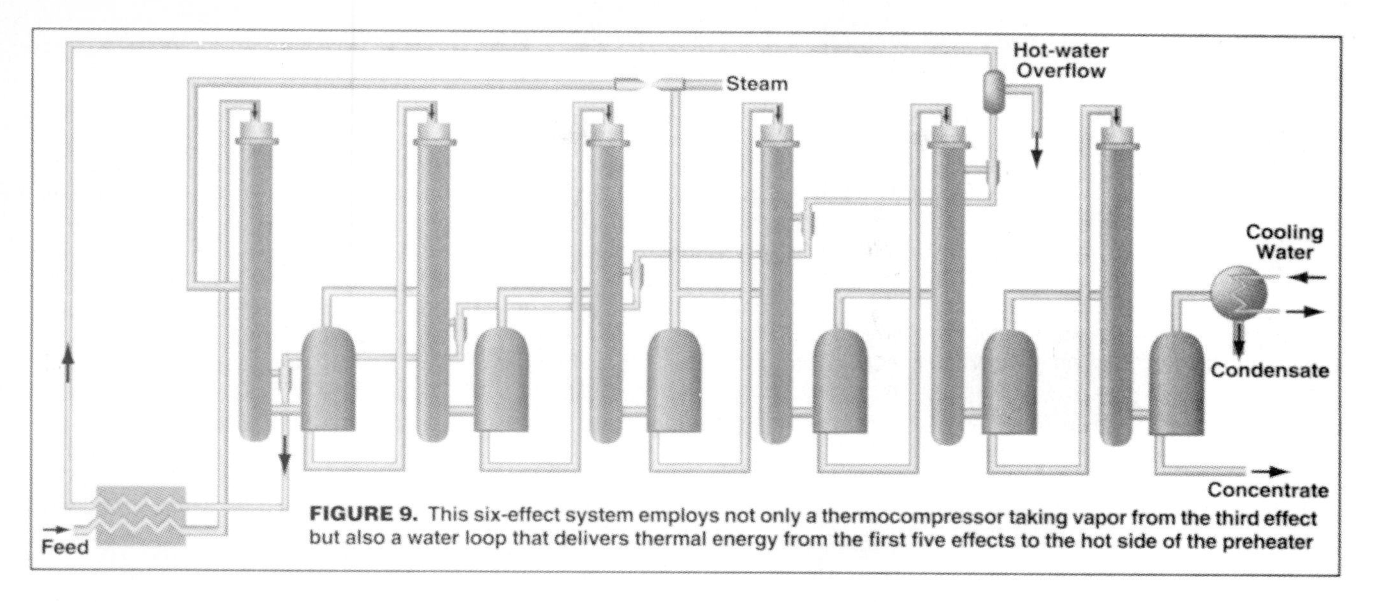

FIGURE 9. This six-effect system employs not only a thermocompressor taking vapor from the third effect but also a water loop that delivers thermal energy from the first five effects to the hot side of the preheater

reduced by the combined boiling-point elevations of all of the effects.

The ratio of suspended to dissolved solids can significant. Suspended solids tend to foul. In general, their presence also increases product viscosity notably.

Heat-sensitive products usually have an upper limit on processing temperature. Residence time must be short.

The sections of the questionnaire dealing with evaporator duty and services are used to define the capacity of the evaporator. Utility cost information is useful when capital and operating costs of different designs are compared.

The final sections of the questionnaire cover detail on pumping requirements, controls and site restrictions. The pump and controls information plays a role in the design process.

Determine the configuration

Once the duty is defined, the best configuration can be selected. This includes choosing the evaporator and the number of effects and stages, and deciding whether to use either thermal or mechanical vapor recompression.

When dealing with an evaporator supplier, it is wise not to specify any particular configuration, at least initially. This lets the supplier recommend alternatives, which can then be refined into the optimum design.

As regards the number of effects and stages, a major tradeoff is the lower operating cost but higher capital cost of increased steam economy. As with most process equipment, the price of an evaporator does not rise linearly with capacity. So, additional capital for more steam economy is easier justified at relatively high evaporation rates.

Because film evaporators (especially falling-film versions) can employ close temperature differences and are generally inexpensive per unit surface area, they are especially attractive for high-economy configurations. It often happens that while film evaporators are suitable for a given product at low concentration, a different type (often, forced circulation) becomes better as the concentration increases. In such cases, it is common to concentrate with a film evaporator as far as practical, then use a forced-circulation finishing evaporator for the final concentration. The finishing evaporator can be either an independent system, or combined with the main evaporator.

In one setup employing a three-effect evaporator plus a finisher, vapor from the first effect goes in parallel to the second effect and to the finisher. This allows a relatively large temperature difference for the finisher while still maintaining multiple-effect steam economy for most of the process.

In a multiple-effect array, the liquid can be sent to the effects in any order. The most common arrangement, called forward feed, sends the incoming feed to the first effect (i.e., the one heated by incoming steam), and the concentrated product from each effect becomes the feed to the next. This arrangement does the least damage to heat- sensitive products, as the most dilute stream is exposed to the highest temperature. Preheating to the boiling point is necessary only at the first effect, because the feed to the other effects is always above boiling temperature.

The reverse-feed arrangement, sending the incoming feed to the final effect and passing the successively concentrated product toward the first effect countercurrent to the vapor flow, is sometimes desirable. The more-concentrated solutions are at the higher temperatures, where solubility is often higher and viscosity almost always lower. This may allow reaching a higher concentration than in a forward-feed arrangement. However, heating is required at each effect.

Other, "mixed" feed arrangements play special roles. For example, it might be necessary to keep the final concentrate temperature above the final-effect boiling temperature but below the first-effect boiling temperature.

It may be advantageous to provide multiple stages within a single effect. Splitting an effect into two or more stages allows some stages to operate at lower product concentrations, at improved heat transfer coefficients. The overall surface area required decreases.

In film evaporators, multiple product stages can be built into a single heat exchanger, using a single separator. This provides the advantages of multiple-stage evaporation without the cost of additional separators and ducting. Such a configuration is available in both plate and tubular systems.

As more effects are used, the temperature differences across the heat exchangers are smaller. Pressure losses in the separators, ductwork and exchangers become more significant.

Boost from a thermocompressor

In plant situations where surplus high-pressure steam is available, a thermocompressor* can inexpensively raise steam economy. A thermocompressor is

*For more about thermocompressors, see "Pump up your energy savings," pp. 203–209.

a jet device in which high-pressure steam entrains low-pressure vapor.

When employed with evaporation, the thermocompressor usually takes vapor from the first-effect separator and returns it to the steam side of the first-effect heater (for an exception, however, see below). This entrained vapor produces more evaporation than it otherwise would, because it is supplying heat to an earlier effect. The improvement in steam economy comes without lowering the temperature difference across any effect.

Adding a thermocompressor increase the surface area of the effects receiving the entrained vapor but decrease that of the other effects. The changes approximately cancel out.

A thermocompressor is designed for a particular set of conditions. If the flows or pressures move away from these, performance will decline. Thus, a thermocompressor can hinder the flexibility of an evaporator system. In some cases, one can get around this by using multiple thermocompressors, or by having the supplier test the device at different operating conditions.

A six-effect tubular evaporator with a thermocompressor appears in Figure 9. In this case, the entrained vapor comes from the third effect, and brings about the equivalent of an additional three effects of evaporation.

Note also an added source of energy economy in this setup—the hot water loop that supplies the preheater. This circulating water picks up heat by condensing some of the vapor produced in each of the first five effects. Using inter-effect vapor for preheat is more efficient than using steam. The savings are especially notable in high-economy evaporation systems, where the energy requirement for preheating is a significant part of the overall energy use. Multiple-section preheaters, using condensate, and vapor from different effects, are commonly employed.

Mechanical vapor recompression

Another source of energy savings is mechanical vapor recompression (MVR). In a single-effect evaporator outfitted with MVR, the vapor leaving the separator is compressed to the pressure that corresponds thermodynamically to the saturation temperature required on the steam side of the heat exchanger. In most cases, no steam input to the evaporator is required once the system is running.

Potential energy savings with MVR are significant. A steam boiler needs enough energy to satisfy the latent heat of vaporization of water, and to offset losses and meet preheating requirements. In MVR systems, the compressor needs only enough energy to boost the vapor to the higher pressure.

FIGURE 10. This rising-falling-film installation illustrates that evaporator's virtue of requiring relatively little headroom in the plant

MVR evaporation was especially popular in the late 1970s and early 1980s, when energy was costly. Although lower energy costs have for the moment made new installations less common, engineers involved with evaporation should be familiar with the operating principles involved.

Compressor selection is critical. The compressor must overcome the pressure difference between the condensing and boiling sides of the evaporator, as well as losses in the separator and ductwork. The compressor is usually centrifugal. MVR evaporators with single-stage units can operate with compression ratios up to about 1.8.

Because a centrifugal compressor operates at high speeds, one must protect it from impingement of droplets. This usually requires a mist eliminator in the separator. The mistfree vapor is superheated to protect against condensation in the suction ducting.

Much of the compressor energy input is converted to heat, so the vapor temperature rises considerably. Mechanical inefficiency of the compressor adds to this rise.

When this vapor is desuperheated, by spraying hot condensate into it, additional vapor is generated. This can be employed in several ways: to make up for vapor that escapes with noncondensable gases through the evaporator vents; to compensate for heat losses; to serve for preheating. And, some can be removed to maintain the energy balance of the MVR evaporator system (or a small amount of steam can be added).

The condenser in the MVR system is much smaller than those in steam-driven systems. It needs only condense vent and excess vapor streams.

The available temperature difference across an MVR system is limited by the pressure-boost capability of the compressor. MVR systems have a lower available temperature difference than steam-driven systems, so products with high boiling point elevations, or suppressed-boiling evaporators with high liquid temperature rises, are not attractive for MVR designs.

The engineer must take care to avoid unstable compressor operation (surging) at low flow rates. If fouling decreases the heat-transfer coefficients, the throughput of an MVR system falls, and surging might occur.

High efficiencies can be achieved via multiple-effect MVR systems. And by designing for close temperature differences, equivalent steam rconomies in the range of 50 are not beyond reach. When the required concentration ratio of a product is high, an MVR main evaporator followed by a small steam driven finisher is a possibility.

Rather than using a compressor, an MVR system can be driven by a fan, at a compression ratio of about 1.2. This usually provide about 6°F of temperature difference for the evaporator.

Since the compressor ratio is low, the horsepower required per pound at inlet flow is low as well.

Fans operate at lower speeds than compressors, and are therefore less susceptible to damage from droplets. They are also less likely to surge.

Evaporator calculations

To design or specify an evaporator system, one must make mass and heat balances around the major components. Detailed review of this procedure is beyond the scope of this article. However, it is instructive to analyze a simple example, pertaining to the single-effect evaporator setup of Figure 1.

Assume that the evaporator handles 40,000 lb/h of 100°F feed containing 10% solids* and having a specific heat

It is a good idea to compare the attractiveness of several alternative configurations

of 0.95 Btu/(lb)(°F). Evaporation (under vacuum) is at 120°F, at which temperature the latent heat of evaporation is 1,026 Btu/h. The concentrate stream contains 40% solids; its specific heat is 0.8 Btu/(lb)(°F). Assume that the boiling-point elevation can be ignored.

The preheater, operating on utility-steam condensate from the evaporator heater, heats the feed to 125°F so that it enters the evaporator above its boiling point. The hot side of the evaporator heater employs 14.2-psia saturated utility steam condensing at 210°F, at which temperature its latent heat of evaporation is 972 Btu/lb.

The condenser condenses the evaporated vapor at 120°F (temperature losses during ducting are thus ignored), and subcools it to 100°F to lessen the vapor load on the vacuum system. The heat goes into cooling water that enters the vessel at 85°F and discharges at 105°F.

Mass balance: The first step is the product mass balance, establishing the flowrates of concentrated product and process condensate. Overall, $F = C + E$ where F is feed rate, C is concentrate rate and E is condensate rate. The solids leave with the concentrate, so the solids balance is $Fx_f = Cx_c$, where x_f and x_c are solids concentrations in feed and concentrate, respectively. In our case, (40,000 lb/h)(0.10) = (C lb/h)(0.40), so C is 10,000 lb/h. From the overall balance, E is 30,000 lb/h.

Heat balance, evaporator heater: Now, make a heat balance over the evaporator heater, starting with the product side. Because the feed enters at 125°F, 5 degrees higher than the evaporation temperature, it *lessens* the vessel's heat requirement by [40,000 Btu/h][0.95 Btu/(lb)(°F)][125 − 120], i.e., by 190,000 Btu/h. The heat needed for evaporation is (30,000 lb/h)(1,026 Btu/lb), i.e., 30,780,000 Btu/h. The net heat that must come from the hot side is thus 30,780,000 − 190,000, i.e., 30,590,000 Btu/h.

Accordingly, the amount of utility steam that must condense is (30,590,000 Btu/h)/(972 Btu/lb), or 31,471 lb/h, ignoring any heat available from subcooling of the condensate. Assuming that 1% of the incoming steam is vented, the total required steam flow is 31,471/0.99 or 31,789 lb/h. The amount vented is 318 lb/h.

Heat balance, preheater: Knowing the amount of condensed utility steam enables one to make a heat balance around the preheater, and thus determine the temperature drop this condensate must undergo. The heat needed to preheat the feed is [40,000 lb/h][.95 Btu/(lb)(°F)][125 − 100°F], or 950,000 Btu/h. Thus, the condensate will cool by [950,000 Btu/h]/[31,471 lb/h][1 Btu/(lb)(°F); i.e., by about 30°F.

Among other things, this calculation confirms that the utility-steam condensate is a feasible heat source. Even if this stream cools to, say, 180°F before reaching the preheater, its temperature leaving the preheater is 150°F, which is above the feed supply temperature.

Heat balance, condenser: This condenses the 120°F evaporated vapor and the 210°F vented steam, and cools the resulting water to 100°F. From steam tables, the difference in enthaply between 210°F saturated steam and 120°F water is 1,062 Btu/ lb.

Condensing the evaporated vapor releases (30,000 lb/h)(1,026 Btu/lb), i.e., 30,780,000 Btu/h. Condensing the vent steam and cooling it to 120°F releases (318 lb/h)(1,062 Btu/lb), or 337,716 Btu/h. And cooling the 120°F water to 100°F releases [30,000 + 318 lb/h][1 Btu/(lb)(°F)][120 − 100°F], or 606,360 Btu/h. Thus, the total condenser heat load is (30,780,000 + 337,716 + 606,360), i.e., 31,724,076 Btu/h.

Since the cooling-water temperature rises by 20°F, the cooling-water requirement is [31,724,076 Btu/h]/[20°F][1 Btu/(lb)(°F)], that is, 1,586,204 lb/h or 3,172 gal/min.

Multiple-effect calculations

Calculation for multiple-effect evaporators requires solving the above balances for each effect. Product from one stage becomes the feed to another stage, and evaporation from one effect becomes steam to the next effect. If a thermocompressor is used, or if vapor is drawn off for preheating, these factors must also be taken into account.

These material and heat balances form a series of simultaneous equations. Solution is usually by computer.

In almost all cases, it is worthwhile to investigate, several alternative configurations. In the present example, for instance, steam and electricity costs for the single-effect approach are about $190/h—but the figure drops to about $50/h for a triple-effect system, or $30/h for a six-effect system with thermal vapor recompression, or only $15/h for a fan-driven MVR. Of course, the rise in capital cost for the various cases must also be taken into account. ■

Edited by Nicholas P. Chopey

*In most applications, the feed material is a mixture of dissolved and suspended solids in water. For simplicity, all nonvolatile components are usually grouped together as "solids."

The author

Greg Lavis is a Senior Process Engineer in the Separations Technology Group of APV Crepaco (395 Fillmore Ave., Tonawanda, NY 14150). He has been employed at the company as a Process Engineer and an Automation Engineer since 1978. He holds a B.S. in chemical engineering from the State University of New York at Buffalo.

SPECIFICATION TIPS TO MAXIMIZE HEAT

John Boyer and
Gary Trumpfheller
ITT Standard

Last year in the U.S., the chemical process industries (CPI) invested more than $700 million in capital equipment related to heat transfer. Much of that investment was driven by a growing body of environmental legislation, such as the U.S. Clean Air Act Amendments. The use of vent condensers, for example, which use heat exchangers to reduce the volume of stack emissions, is on the rise.

Heat exchanger makers have also responded to growing environmental concerns over fugitive emissions by developing a new breed of leaktight heat exchangers, designed to keep process fluids from leaking and volatile organic compounds from escaping to the atmosphere. Gasketed exchangers are benefitting from improvements in the quality and diversity of elastomer materials and gasket designs. The use of exchangers with welded connections, rather than gaskets, is also reducing the likelihood of process-fluid escape.

Over the last decade, the use of heat exchangers has expanded into non-traditional applications. This, coupled with a variety of design innovations, has given engineers a wider variety of heat exchanger options to choose from than ever before.

Operating conditions, ease of access for inspection and maintenance, and compatability with process fluids are just some of the variables CPI engineers must consider when assessing heat exchanger options. Other factors include:

- Maximum design pressure and temperature
- Heating or cooling applications
- Maintenance requirements
- Material compatibility with process fluids
- Gasket compatibility with process fluids
- Cleanliness of the streams
- Temperature approach

The temperature approach is the difference between the inlet temperature of the cold fluid and the outlet temperature of the hot fluid. Certain exchanger designs operate better at different temperature approaches. Plate-and-frame exchangers, for example, work well at a very close temperature approach, on the order of 2°F. For shell-and-tube exchangers, however, the lowest possible temperature approach is 10°F.

As for cleanliness of the process fluids, shell-and-tube exchangers have tube diameters that can accommodate a certain amount of particulate matter without clogging or fouling. Plate-and-frame exchangers, however, have narrow passageways, making them more susceptible to damage from precipitation or particulate fouling.

Operators in the CPI have three general heat exchanger categories to consider. The most common, the shell-and-tube, can be found in almost every industrial application. In recent years, the plate-and-frame has emerged as a viable alternative to the shell-and-tube.

Air-cooled exchangers have gained popularity because of their ability to reduce water consumption.

SHELL-AND-TUBE EXCHANGERS

The shell-and-tube exchanger's flexible design, high pressure and temperature capabilities, and its ability to handle a high levels of particulate material make it the most common heat exchanger used in the CPI. Mechanically simple in design and relatively unchanged for more than 60 years, the shell-and-tube offers a low-cost method of heat exchange for many process operations. Following is a brief description of each of the most common shell-and-tube configurations:

Straight-tube, fixed-tubesheet exchangers

The fixed-tubesheet exchanger is the most common, and typically has the lowest capital cost per ft^2 of heat-transfer surface area. Fixed-tubesheet exchangers consist of a series of straight tubes sealed between flat, perforated

Originally published May 1993

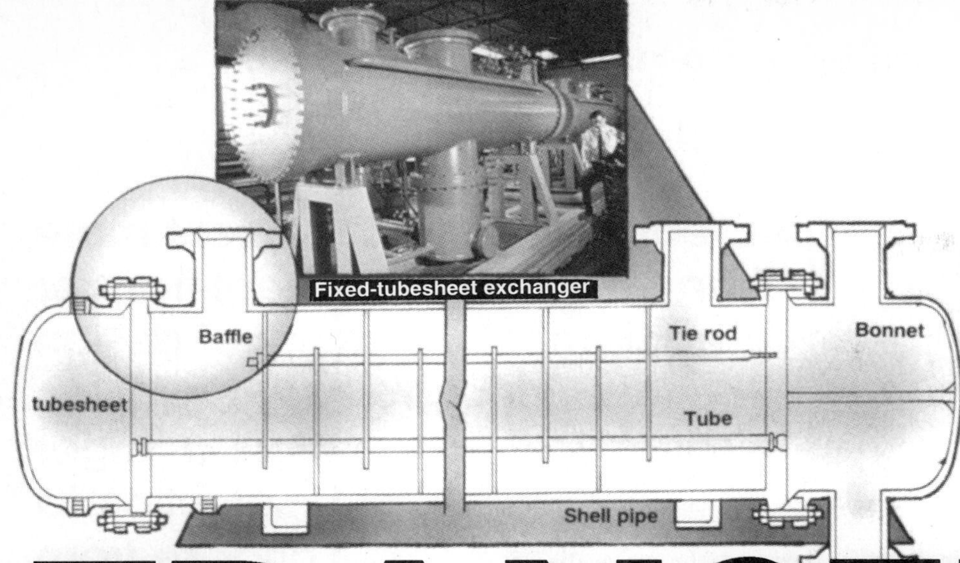

Figure 1 (left). In a fixed-tubesheet exchanger, straight runs of tubing are attached to two perforated tubesheets. The design has no shellside gasket or packed joints. This minimizes the potential for leakage, and makes this exchanger ideal for high-pressure operations, or those handling potentially lethal fluids

TRANSFER

A host of parameters must be evaluated when choosing among shell-and-tube, plate-and-frame and air-cooled heat exchangers

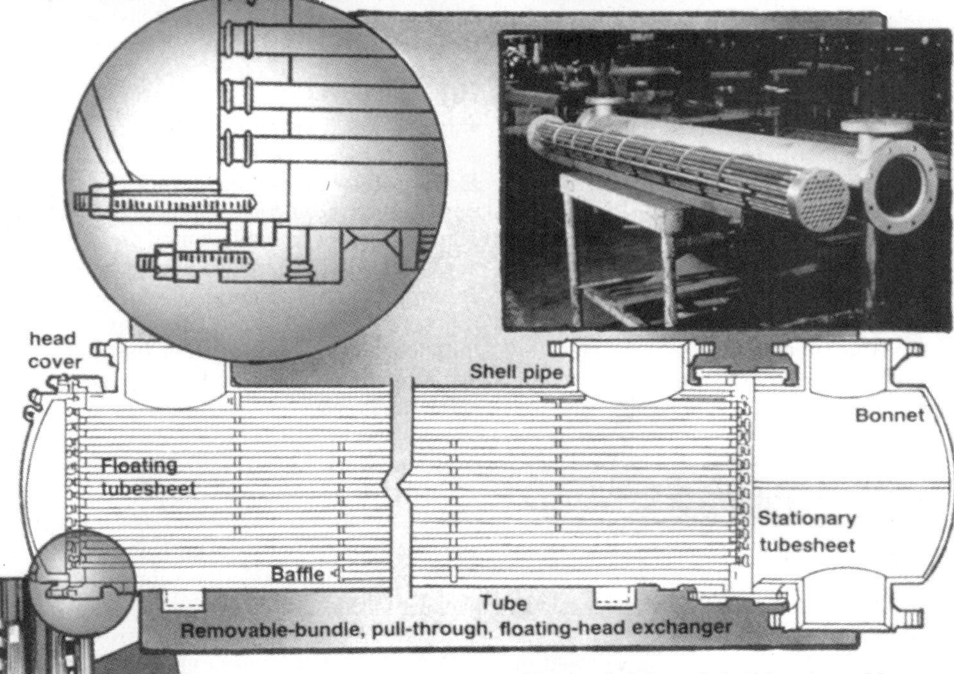

Figure 2 (above). In this removable-bundle exchanger, straight runs of pipe are attached to one fixed (or stationary) head, and one floating head, which allows the entire assembly to be removed for cleaning and repair. Also, the floating head accommodates differential thermal expansion during operation

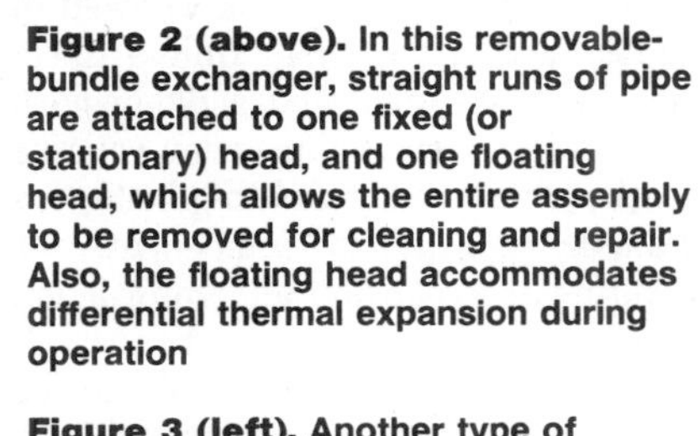

U-tube exchanger

Stationary bonnet

Baffle

U-bend

Tube

Stationary tubesheet

Shell pipe

Figure 3 (left). Another type of removable-bundle exchanger, the U-tube design has only one tubesheet, which allows for maximum differential expansion between the shell and the tubes during operation

metal tubesheets (Figure 1).

Because there are neither flanges, nor packed or gasketed joints inside the shell, potential leak points are eliminated, making the design suitable for higher-pressure or potentially lethal service. However, because the tube bundle cannot be removed, the shellside of the exchanger (outside the tubes) can only be cleaned by chemical means. The inside surfaces of the individual tubes can be cleaned mechanically, after the channel covers have been removed. The fixed-tubesheet exchanger is limited to applications where the shellside fluid is non-fouling; fouling fluids must be routed through the tubes.

The Tubular Exchanger Manufacturers Assn. (TEMA; Tarrytown, N.Y.) has established heat exchanger standards and nomenclature. Every shell-and-tube device has a three-letter designation; the letters refer to the specific type of stationary head at the front end, the shell type, and the rear-end head type, respectively (a fully illustrated description can be found in the TEMA standards). Common TEMA designations for the straight-tube, fixed-tubesheet exchangers are BEM, AEM, NEN. Common applications include:

- Vapor condensers
- Liquid-liquid exchangers
- Reboilers
- Gas coolers

Removable-bundle, externally sealed, floating-head exchanger

Floating-head exchangers are so named because they have one tubesheet that is fixed relative to the shell, and another that attached to the tubes, but not to the shell, so it is allowed to "float" within the shell. Unlike fixed-tubesheet designs, whose dimensions are fixed at a given dimension relative to the shell wall, floating-head exchangers are able to compensate for differential expansion and contraction between the shell and the tubes.

Since the entire tube bundle can be removed, maintenance is easy and inexpensive. The shellside surface can be cleaned by either steam or mechanical means. In addition to accommodating differential expansion between the shell and tubes, the floating tubesheet keeps shellside and tubeside process fluids from intermixing.

Last year, U.S. operators spent more than $700 million on heat-transfer equipment. A variety of design innovations have given operators a larger selection to choose from than ever before

Although the externally sealed, floating-head design is less costly than the full, internal-floating-head exchanger, it has some design limitations: both shellside and tubeside fluids must be non-volatile or non-toxic, and tubeside arrangements are limited to one or two passes. In addition, the packing used in this exchanger limits design pressure and temperature to 300 psig and 300°F.

Common TEMA designations are AEW and BEW. Applications include exchangers handling:

- Inter- and after-coolers
- Oil coolers
- Jacket water coolers

Removable-bundle, outside-packed, floating head exchanger

This design is especially suited for applications where corrosive liquids, gases or vapors are circulated through the tubes, and for air, gases or vapors in the shell. Its design also allows for easy inspection, cleaning and tube replacement, and provides large bundle entrance areas without the need for domes or vapor belts (Figure 2).

Unlike the previous design, only shellside fluids are exposed to packing, allowing high-pressure, volatile or toxic fluids to be used inside the tubes. The packing in the head does, however, limit design pressure and temperatures.

Common TEMA designations are BEP and AEP. Typical applications include:

- Oxygen coolers
- Volatile or toxic fluids
- Gas processing

Removable-bundle, internal-clamp-ring, floating-head exchanger

This design is useful for applications where high-fouling fluids require frequent inspection and cleaning. And, because the exchanger allows for differential thermal expansion between the shell and tubes, it readily accommodates large temperature differentials between the shellside and tubeside fluids.

This design has added versatility, however, since multi-pass arrangements are possible. However, since the shell cover, clamp ring and floating-head cover must be removed before the tube bundle can be removed, service and maintenance costs are higher than in "pull through" designs (discussed below).

Common TEMA designations are AES and BES. Typical applications include:

- Process-plant condensers
- Inter- and after-cooler designs
- Gas coolers and heaters
- General-purpose industrial heat exchangers

Removable-bundle, pull-through, floating-head exchangers

In the pull-through, floating-head design, the floating-head cover is bolted directly to the floating tubesheet. This allows the bundle to be removed from the shell without removing the shell or floating-head covers, which eases inspection and maintenance.

This is ideal for applications that require frequent cleaning. However, it is among the most expensive designs. And, the pull-through design accommodates a smaller number of tubes in a given shell diameter, so it offers less surface area than other removable-bundle exchangers.

Common TEMA designations are AET and BET, and typical applications include:

- Exchangers handling chemical fluids
- Hydrocarbon fluid condensers
- Air or gas compressors
- Inter- and after-coolers

Removable-bundle, U-tube exchangers

In the U-tube exchanger, a bundle of nested tubes, each bent in a series of concentrically tighter U-shapes, is attached to a single tubesheet (Figure 3). Each tube is free to move relative to the shell, and relative to one another, so the design is ideal for situations that accommodate large differential temperatures between the shellside and tubeside fluids during service. Such flexibility makes the U-tube exchanger ideal for applications that are prone to thermal shock or intermittent service.

As with other removable-bundle exchangers, the U-tube bundle can be withdrawn to provide access to the inside of the shell, and to the outside of the tubes. However, unlike the straight-tube exchanger, whose tube internals can be mechanically cleaned, there is no way to physically access the U-bend region inside each tube, so chemical methods are required for tubeside maintenance. As a rule of thumb, non-fouling fluids should be routed through the tubes, while fouling fluids should be reserved for shellside duty.

This inexpensive exchanger allows for multi-tube pass arrangements. However, because the U-tube cannot be made single pass on the tubeside, true countercurrent flow is not possible.

Common TEMA designations are BEU and AEU, and typical applications include:

- Oil cooling
- Chemical condensing
- Steam heating

Special designs

For applications with high vapor-flow and high-pressure conditions, a specially designed shell-and-tube exchanger is often the only viable solution. Special designs may also be called for when applications have close temperature crossings, meaning the outlet temperature of the warmed fluid exceeds that of the cooled fluid. Following are several examples:

- TEMA K-type shells, which allow for proper liquid disengagement for reboilers
- TEMA J-type shells, which accommodate high vapor flows by allowing for divided flow in the shellside
- Two-pass TEMA F-type shells, which can be used for applications when a temperature cross exists (below)
- TEMA D-type front head designs, which are often the answer for high-pressure tubeside applications

While these specially designed exchangers may be the solution to a process problem, construction costs tend to be higher than those of "standard" engineered shell-and-tube equipment.

Common TEMA designations include BKU, BJM, BFM and DED. Specially designed exchangers are often called for in:

- Reboilers
- Steam heaters
- Vapor condensers
- Feedwater heaters

Choosing off-the-shelf exchangers

Fixed-tubesheet and U-tube shell-and-tube exchangers are the most common types of off-the-shelf heat exchangers available today. Such stock models are typically used as components in vapor condensers, liquid-liquid exchangers, reboilers and gas coolers.

Standard fixed-tubesheet units, the most common shell-and-tube heat exchangers, range in size from 2 to 8 in. dia. Materials of construction include brass or copper, carbon steel, and stainless steel. Even though this exchanger is one of the least expensive available, it is still generally constructed to standards specified by the manufacturer. If the user desires, stock exchangers can be constructed to American Society of Mechanical Engineers (ASME) codes.

U-tube heat exchangers are commonly used in steam heating applications, or heating and cooling applications that handle chemical fluids as opposed to water. While the U-tube is generally the lowest-priced heat exchanger available, higher service and maintenance costs tend to be higher than other exchangers, since the nested, U-bend design makes individual tube replacement difficult.

Custom-designed heat exchangers, though more expensive than their off-the-shelf counterparts, are generally made to higher design standards than stock exchangers. Many manufacturers follow the TEMA standards for design, fabrication and material selection:

Table (right). To reduce the risk of fugitive emissions from a plate-and-frame exchanger, the gasket materials must be selected with the process fluids in mind

- TEMA B is the most common TEMA designation, and provides design specifications for exchangers used in chemical process service
- TEMA C guidelines provide specifications for units used in commercial and general process applications
- TEMA R guidelines provide specifications for exchangers used in petroleum refining and related process operations

Each of these classes are applicable to shell-and-tube heat exchangers with the following limitations:

- Shell diameter does not exceed 60 in.
- Pressure does not exceed 3,000 psi
- The product of shell diameter (in.) times pressure (psi) does not exceed 60,000

Standards set by the American Petroleum Institute (API; Washington, D.C.) are also generally accepted throughout the heat exchanger industry. These standards (API 614, 660 and 661) specify the mechanical design of the exchanger and list specific materials that can be used in construction of both water- and air-cooled exchangers.

While there are obvious advantages to purchasing a custom-designed exchanger that meets either TEMA or API manufacturing guidelines, these specifications add to the cost of the exchanger and may slow delivery time.

PLATE-AND-FRAME EXCHANGERS

In recent years, the plate-and-frame heat exchanger has emerged as a viable alternative to shell-and-tube exchangers for many CPI applications. Such units comprise a series of plates, mounted in a frame and clamped together. Space between adjacent plates form flow channels, and the system is arranged so that hot and cold fluids enter and exit through flow channels at the four corners (Figure 4).

Within the exchanger, an alternating gasket arrangement diverts the hot

TYPICAL GASKET MATERIALS	
MATERIAL and Maximum Operating Temperatures	COMPATIBLE FLUIDS
NITRILE 230°F	Mineral oils Most aqueous solutions Aliphatic hydrocarbons Inorganic acids (at low concentrations and temperaures)
EPDM (Ethylene propylene diene monomer) 320°F	Steam Higher-temperature aqueous solutions Inorganic and organic acids or bases
VITON* 212°F	Mineral oils Aliphatic and aromatic hydrocarbons Sulfur carbon carbons Trichloroethylene Perchloroethylene

Note: Gaskets are also available in other materials such as hydrogenated nitrile, butyl rubber neoprene, hypalon and silicon rubber to meet various application requirements.

* Viton is a Du Pont Co. trademark for a series of fluoroelastomers based on the copolymer of vinylidene fluoride and hexafluoropropylene

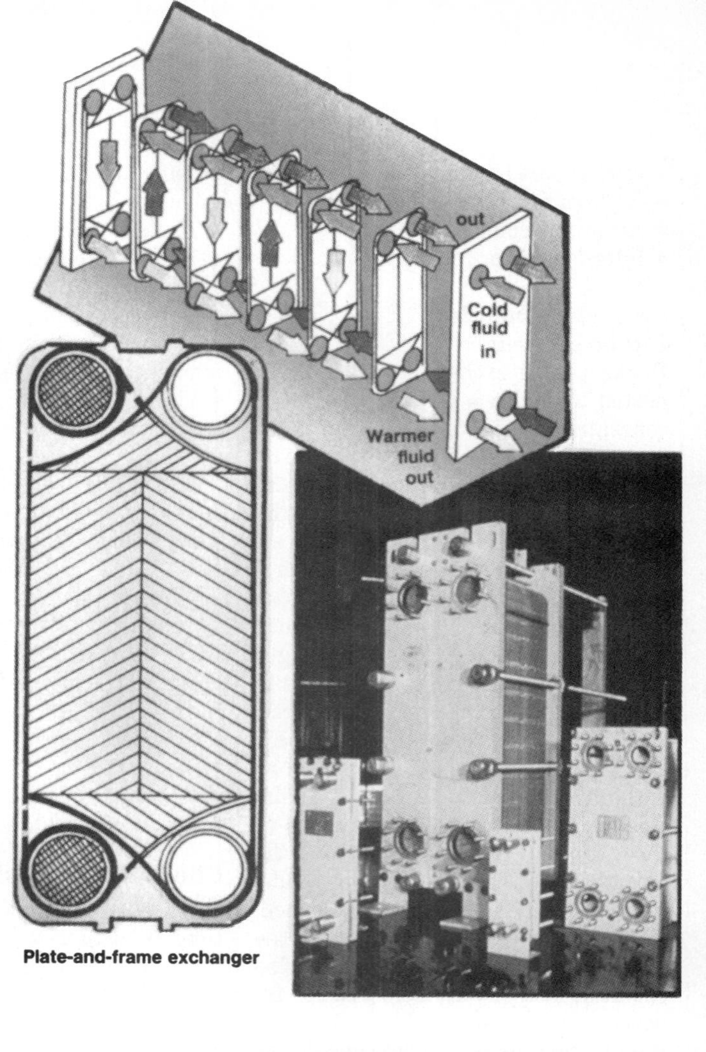

Figure 4 (above). When the plates of a plate-and-frame exchanger are assembled, the holes in the corners form a continuous channel. Alternating gasket patterns direct the hot and cold fluids into alternating passes. Heat transfer then takes place across the plates

and cold fluids from each inlet into an alternating sequence of flow channels. In this arrangement, each cell of heat transfer media is separated by a thin metal wall, allowing heat to transfer easily from one media to the other.

The plate-and-frame's highly efficient countercurrent flow typically yields heat transfer coefficients three to five times greater than other types of exchangers. As a result, a more-compact design is possible for a given heat-exchange capacity, relative to other exchanger styles.

A corrugated chevron or herringbone pattern is pressed into each plate for several reasons. First, the pattern gives the entire exchanger strength and rigidity. It also extends the effective surface area of plates and increase turbulence in the flow channels. Taken together, these effects boost heat transfer.

Depending on the applications, plate selection is optimized to yield the fewest total number of channel plates. Because the plates can be easily removed, service and maintenance costs are typically lower than that of shell-and-tube exchangers.

Although the plate-and-frame heat exchanger can be used in almost any application, the following selection criteria must be reconciled:

- Maximum design or working pressure is limited to 300 psi
- Temperature limits and fluids must be compatible with gasket materials (Table above)
- Plate materials must be compatible with process media
- The narrow passageways in the plate-and-frame can cause high pressure drops, making the exchanger incompatible with low-pressure, high-volume gas applications
- Rapid fluctuations in steam pressures and temperatures can be detrimental to gasket life. For this reason, applications that use steam favor shell-and-tube exchangers
- In applications where process media contain particulate matter, or when large amounts of scaling can occur, careful consideration should be given to the free channel space between adjacent plates
- The plate-and-frame design is best suited for applications with a large temperature cross or small temperature approach

Until recently, a major limitation to the plate-and-frame exchanger was the gluing method used to attach the gaskets to the plates during construction. The glue was often applied unevenly, greatly increasing the chance of process fluid leaking through the gasket groove of the plate and either intermingling with other fluids or escaping to the atmosphere.

Today, most exchanger manufacturers offer a new glueless gasket system. The plate construction uses clips and studs to secure gaskets to the plates. This method eliminates irregularities in the gasket groove and results in better sealing of the plate pack.

The new glueless system also cuts service and maintenance costs, since the plates can be cleaned or regasketed without removing them from the frame. However, for high-fouling applications where plates must be opened, removed and cleaned frequently, the glued gasket system may be the better choice.

Double-wall-plate exchangers

Another recent advance in plate-and-frame design is the double-wall plate heat exchanger, which offer greater protection against gasket failure. In traditional plate-and-frame exchangers, the process fluids are contained by gaskets and thin, metal plates. In double-wall exchangers, two standard plates are welded together at the port holes to form one assembly, with an air space between the plates. Any leaking fluid is thus allowed to collect in this interstitial space, instead of entering an adjacent fluid passageway and contaminating the other process stream.

Operating conditions, ease of access for inspection and maintenance, and compatibility with process fluids are critical selection criteria

Typical applications include:
- Domestic water heaters
- Hydraulic oil cooling
- Any service where cross contamination of process fluids cannot be tolerated

Welded-plate exchangers

In this design, the field gasket that normally contains the process fluid is replaced by a welded joint. When plates are welded together at the periphery, leakage to the atmosphere is prevented, so this design is suitable for hazardous or aggressive fluids.

The welded plates form a closed compartment or "cassette." Similar to gasketed designs, alternating flow channels are created to divert the flow of hot and cold fluids into adjacent channels. Aggressive fluids pass from one cassette to the next through an elastomer or Teflon ring gasket, while nonaggressive fluids are contained by standard elastomer gaskets. The use of welded joints can reduce total gasket area by 90% on the aggressive-fluid side. Typical applications include exchangers handling:
- Vaporizing and condensing refrigerants
- Corrosive solvents
- Amine solutions

Wide-gap-plate exchangers

Compared with traditional plate-and-frame exchangers, this design relies on a more loosely corrugated chevron pattern, which provides exceptional resistance to clogging. The plates are designed with few, if any, contact points between adjacent plates to trap fibers or solids. Some styles of this exchanger use wide-gap plates on the process side and conventional chevron patterns on the coolant side, to enhance heat transfer.

Typical applications include exchangers handling:
- White water in pulp-and-paper operations
- Slurries

Brazed-plate exchangers

In this design — the latest technological advance in plate heat exchangers — the elastomer gaskets found in most plate exchangers are replaced with a brazed joint, which greatly reduces the possibility of leakage. The heat transfer plates, which are only available in Type 316 stainless steel, are brazed together using either a copper or nickel brazing material.

These exchangers are built to manufacturers standards and are often offered as stock items. The brazed-plate exchangers are typically ASME rated to 450 psi. Temperature ratings vary from 375°F for copper brazing to 500°F for nickel brazing. As with other plate-and-frame exchangers, high heat transfer rates translate to compact designs.

Typical applications include:
- Units that vaporize and condense refrigerants
- Applications requiring high alloys
- Heat-recovery applications
- Brine exchangers
- Applications involving liquid ammonia, chlorine solutions, alcohols or acids

AIR-COOLED EXCHANGERS

In the air-cooled exchanger, a motor and fan assembly forces ambient air over a series of tubes, to cool or condense the process fluids carried within (Figure 5). The tubes are typically assembled in a coiled configuration.

Air is cheap and abundant, but it is a relatively poor heat transfer medium. To increase the heat transfer rates of the system, the tubes in air-cooled exchangers are typically given fins, which extend the surface area, increase heat transfer, and give such systems the nickname fin-tube coils.

Air-cooled exchangers are typically found in such applications such as:
- Heating and air conditioning
- Process heating and cooling
- Air-cooled process equipment
- Energy and solvent recovery
- Combustion air preheating
- Fluegas reheating

The diameter and materials specified for the tubes and fins depend on system requirements. The fins are commonly made from aluminum or copper, but may be fabricated of stainless or carbon steel. Tubes are generally copper, but can be made from almost any material, and they range in size from 5/8- to 1-in. outer diameter. The design of the air-cooled exchanger is such that individual coils can be removed independently for easy cleaning and maintenance.

Aluminum brazed-fin exchangers

In this design, corrugated plates and fins are added to a brazed-composite core, to create alternating air and fluid passages. This compact, lightweight design is considered the most cost-effective air-cooled unit available. Turbulence created in the fluid channels boosts efficiency. Typical applications include:

Figure 5. In the air-cooled exchangers shown below, tapered fins are added to the tubes, to extend the surface area and maximize heat transfer

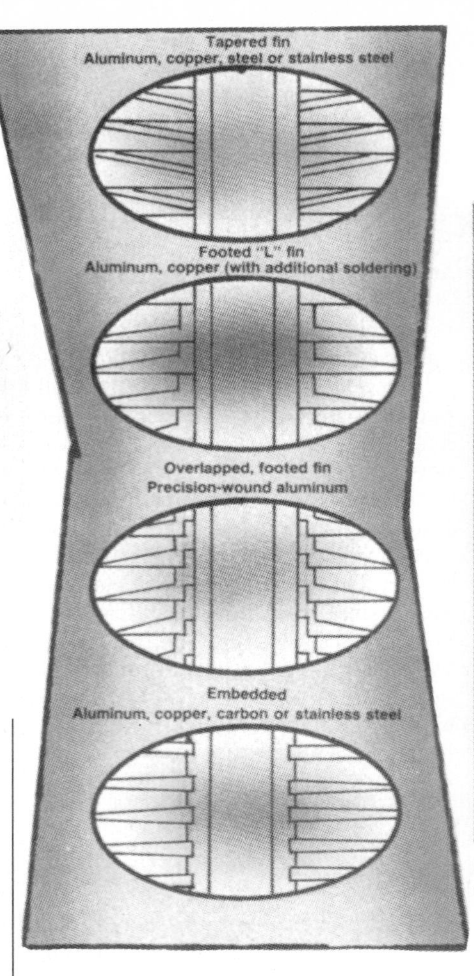

Figure 6 (center). Typically, the fins are wound helically under tension, to achieve a fin spacing between 4 and 14 fins per in. The geometry of the fin-tube connection dictates the heat transfer rates and temperature ratings. Fin-tube units are illustrated above

- Cooling lube oil for power equipment
- Cooling fluids for hydraulic equipment
- Cooling gear box fluids

Aluminum plate-fin exchangers

This type of exchanger is constructed with traditional heat exchanger tubing. Stacked, die-formed aluminum plates extend the surface to maximize air-side heat transfer. Like the brazed-fin exchanger, this unit is also used for oil and glycol cooling, but its higher flowrate expands its capabilities. Built from standard components, aluminum-fin exchangers are designed with a more solid construction than their brazed-fin counterparts. Typical applications include:

- Oil cooling
- Compressed-air cooling
- Water cooling with air

Fin-tube exchangers

In this design, one continuous fin is wrapped spirally around a series of individual tubes. Often referred to as a "heavy duty coil," this air exchanger has fin-tube attachments that can be built either to ASME and API standards, or to customer specifications. Often used in air-heating applications, the heavy-duty coil is available with several different fin variations, including the tapered fin, footed "L" fin, overlapped-footed fin and the embedded fin, which describe the geometries at the fin-tube interface (Figure 6). The method of attaching the fin to the tube is critical, since the loosening of this bond may hinder heat exchange.

Typical applications include those that:

- Heat air with high-pressure or high-temperature steam
- Heat or cool with high liquid flows
- Cannot tolerate condensate freezing, such as steam applications
- Heat air with hot water

Clearly, for most applications, and heat-exchanger types, there are a multitude of choices and options available; these guidelines should provide a basis for comparison. No matter what configuration is ultimately implemented, the emphasis on clean, efficient heat recovery throughout the CPI ensures that the heat exchanger will remain one of the most critical components in the manufacturing process. ■

Edited by Suzanne Shelley

Equipment sources

For information on suppliers of heat exchangers, consult Section 5 (Processing Equipment) of the 1993 *Chemical Engineering Buyer's Guide.*

The authors

John W. Boyer is market manager for chemical and allied products for ITT Standard (175 Standard Parkway, P.O. Box 1102, Buffalo, NY 14240; Tel.: 716-897-2800). He is responsible for marketing and placement of all heat transfer products for petrochemical, organic-chemical, agricultural-chemical, pharmaceutical and bioprocessing applications. During his four years at ITT Standard, Boyer has also worked as an application specialist and senior applications engineer. He holds an A.S. degree in chemistry, and earned a B.S. degree in chemical engineering from the State University of New York (Buffalo) in 1982.

Gary W. Trumpfheller is a market manager for general and process industries for ITT Standard (Tel.: 716) 897-2800). He is responsible for marketing and applications for all heat transfer products used in general markets, including pulp-and-paper production, gas processing and general manufacturing. Trumpfheller has also served as an application engineer and product manager during his 12 years with ITT Standard. He earned a bachelor's degree in mechanical engineering from the University of Buffalo in 1980.

MAXIMIZE TOWER POWER

John Hensley
The Marley Cooling Tower Co.

Industrial processes produce excess heat, usually in the form of hot water that must be cooled and re-used. Cooling towers dissipate this heat quickly, by circulating hot process water in the presence of air to maximize evaporation.

While the world's total fresh water supply is abundant, some areas have usage patterns that are heavily out of balance with the pace of natural replenishment. The proper handling and efficient re-use of this vital resource is mandatory, and does not limit itself to arid regions.

In fact, industrial water prepared for discharge must also be cooled to preserve the delicate balance of nature. Discharging hot industrial water into any estuary will impact the inhabitants, which are accustomed to a narrow range of moderate temperatures.

Cooling towers are designed to expose the largest surface area of transient water to the maximum air flow, for the longest period of time. Process water is delivered to hot water inlets located at the top of the towers. It is allowed to flow down through the heat-transfer media in the tower by either gravity (through a metering orifice) or by pressurized nozzles.

The type of air flow varies with tower design. In an unmechanized tower, air flow comes from wind or natural draft. For tighter control, air flow can be mechanically induced by a series of motor-driven fans.

As with any other piece of equipment, cooling tower performance can be optimized if the operating parameters are rigorously specified. The key considerations are discussed below.

Tower heat load

Cooling tower size is established by the total amount of heat to be removed from the system per unit time. Figure 1 shows the elements of the heat load configuration.

The *range* refers to the number of degrees (Fahrenheit, in this case) through which the circulating water is cooled (the temperature of the hot incoming water, minus the desired exit temperature of the cooled water).

Originally published February 1992

BILL FINEWOOD

Range is strictly a function of the heat load and the amount of water circulating through the tower.

The heat load, expressed as Btu/min, is calculated by multiplying the range by the mass flow of water (lb/min of water to the tower; which is calculated as follows: gal/min of water x 8.33 lb/gal of water).

The *approach* refers to the number of degrees separating the cold water temperature from the wet-bulb temperature of the air entering the tower. The inlet air temperature is given as the *wet-bulb temperature,* because it is measured with a wet-bulb thermometer (whose bulb, encased in a sleeve of wetted muslin fabric, provides the temperature of the air at 100% relative humidity).

The approach reflects the difficulty of the required thermal performance, and directly affects the size of the tower. As demonstrated in Figure 2, increasing the approach from 10°F to 15°F (accepting cooled water that is 5°F warmer than that originally specified), for a given heat load, range, and wet-bulb temperature reduces the tower size by almost 30%.

The specifier may not always be able to take advantage of this effect, however, due to the demands of the process and the ambient temperature conditions at site. Within a very limited margin, for example, the required cooled water temperature is dictated by the needs of the process plant, and the design wet-bulb temperature is a function of local (and historical) weather conditions. This may leave relatively little freedom of choice as regards approach temperatures.

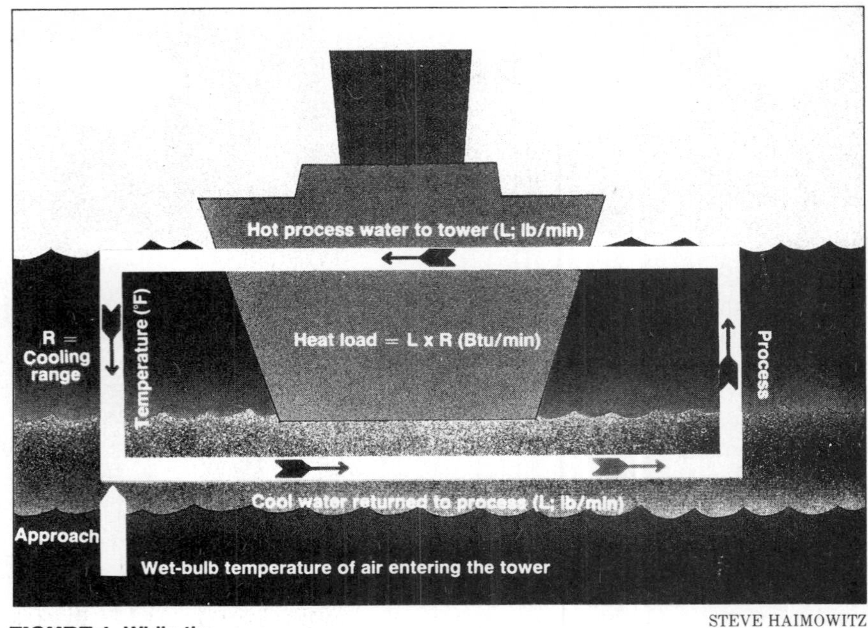

STEVE HAIMOWITZ

FIGURE 1. While the formula for heat load (top) suggests an infinite number of combinations, a relatively narrow band of flowrates and ranges is dictated by design limitations and other constraints

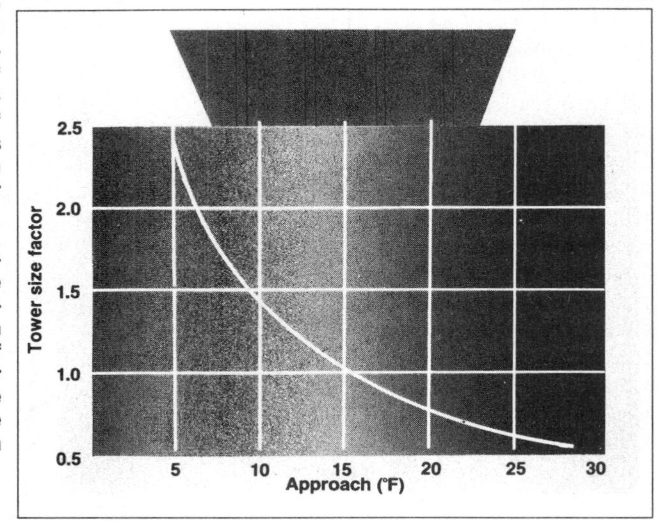

FIGURE 2. For a given heat load, range and wet-bulb temperature, the graph (right) shows that if you can accept water that is just a little warmer, you may be able to specify a smaller tower

Design wet-bulb temperature

Accurate determination of the wet-bulb temperature is vital if the cooling tower is to perform as planned. If the temperature of the air is consistently higher than that anticipated by design, then warmer-than-desired water will leave the tower. Conversely, if the actual wet-bulb temperature is consistently lower than expected, chances are good that the cooling tower is larger than required.

The precise wet-bulb temperature must be specified on the basis of conditions anticipated at the site proposed for a cooling tower. Analyses at industrial installations have shown that the best performance occurs when design wet-bulb temperatures are exceeded for no more than 5% of the total operating hours during a normal summer (generally considered the time of peak load demand for a cooling tower).

Even when peak wet-bulb levels exceed design temperatures, the gap is rarely more than 5°F. And, performance analyses show that, in general, the hours in which actual temperatures exceed the design temperatures are seldom consecutive. The residual effect on the total system is normally sufficient to carry through the above-average periods without detrimental results.

Air wet-bulb temperatures are routinely measured and recorded worldwide by the U.S Weather Bureau, U.S. military installations and airports. These data are an invaluable source of information to both users and designers of cooling towers*.

* The compiled weather statistics of the U.S. Depts. of the Army, Navy and Air Force are available through the U.S. Printing Office, Washington, D.C. A document entitled "Engineering weather data" contains excerpts from these and other key sources, and is available at no charge from The Marley Cooling Tower Co., 5800 Foxridge Dr., Mission, KS 66202

Proper siting, design and materials of construction will boost heat dissipation

Peruse your tower options

Cooling towers are broadly categorized by the following considerations: the mechanism used to provide the required airflow; the relative flow paths of air and water; the primary materials of construction; the type of heat transfer media applied; and the tower's physical shape. For purposes of this article, only those types and configurations that are most commonly used in the chemical process industries (CPI) are discussed.

Unmechanized air flow

Two types of towers are available that use no mechanical devices to aid the flow of air: atmospheric towers, and hyperbolic natural-draft towers.

Atmospheric towers use pressure sprayers to dissipate the hot water into fine droplets. This water-distribution system induces air currents through openings in the tower., either parallel slats, or garage-door type louvers.

The lack of automation makes these systems susceptible to variations in prevailing wind patterns, which causes system performance to fluctuate. In general, atmospheric towers are recommended only for relatively small size towers (on the order of 10–100 gal/min), and only for those processes where highly specific cooling is not critical.

Hyperbolic natural-draft towers, named for their shape, provide extremely dependable and predictable thermal performances. Air flow is produced by a density differential between the heated (less dense) air inside the tower, and the relatively cool (more dense) ambient air outside the tower. Performance is favored in climates with high relative humidity.

Such towers, often as tall as 500 ft., are suitable for high process water throughput (to 600,000 gal/min). They are common in the field of electric power generation.

Mechanized air flow

In mechanized cooling towers, fans are used to provide air flow of a known volume through the tower. Such automation improves thermal performance and stability, since the operator can regulate air flow (by cycling the fan, or manipulating its capacity) to compensate for changing atmospheric and load conditions.

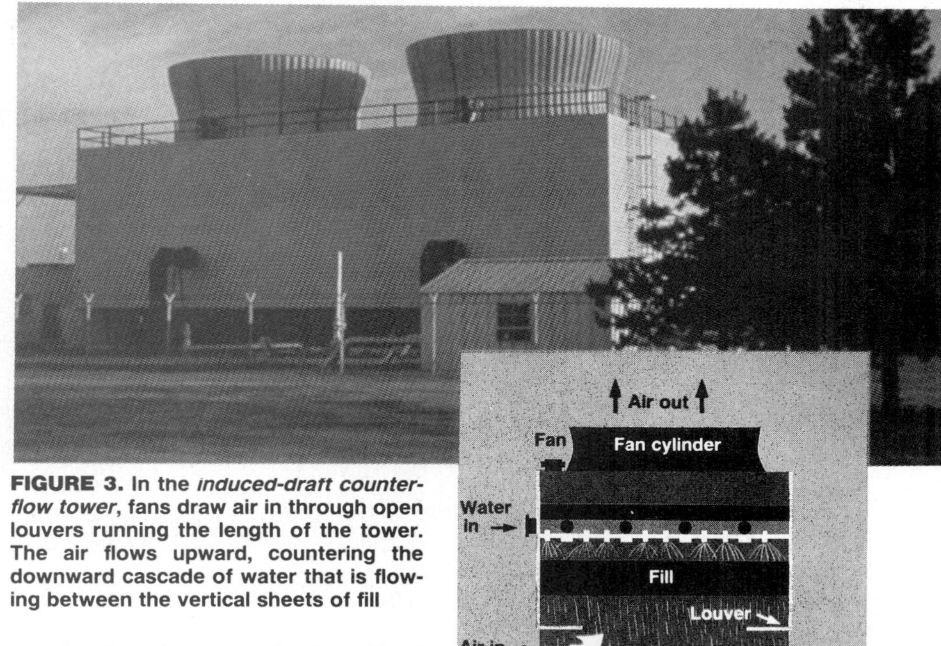

FIGURE 3. In the ***induced-draft counterflow tower***, **fans draw air in through open louvers running the length of the tower. The air flows upward, countering the downward cascade of water that is flowing between the vertical sheets of fill**

Mechanical-draft towers are favored in the chemical and petroleum-refining industries. They use one or more motor-driven fans to provide airflow through the tower. In the *induced-draft* configuration, fans are located in the exiting air stream at the top of the tower. By blowing exhaust air out the top, fresh air is sucked into the air inlets at the base of the tower.

One configuration of the induced-draft tower is the *counterflow* tower, because the upward flow of air counters the downward cascade of water (Figure 3). Such towers are built in modular cells. Additional cells can be added end to end, as increased capacity is needed. This makes the counterflow tower very versatile, operating efficiently with throughput

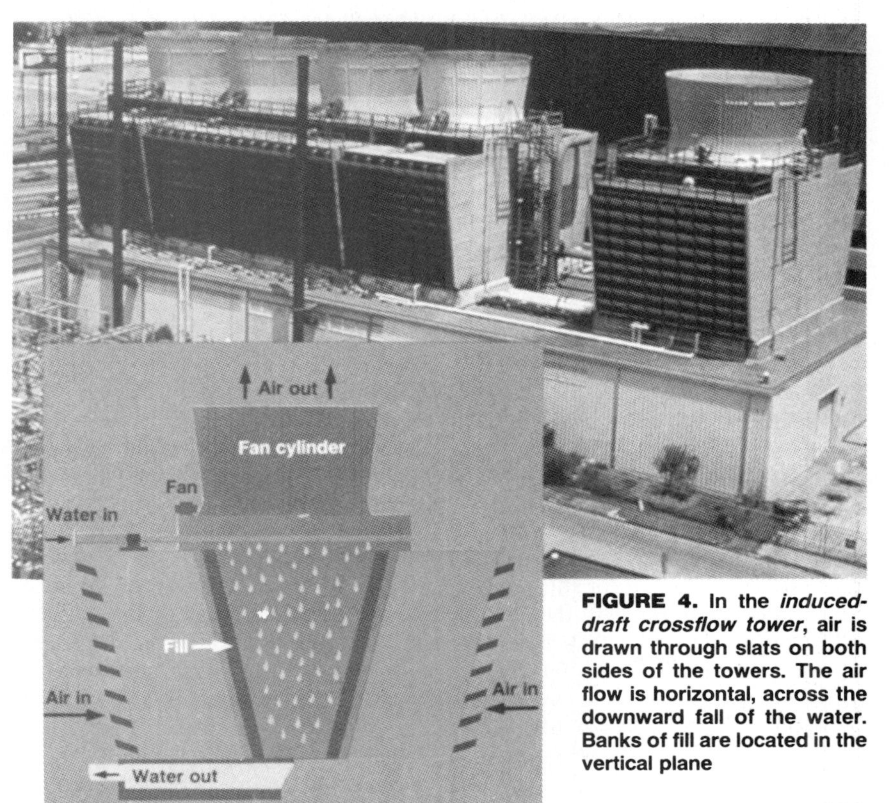

FIGURE 4. In the ***induced-draft crossflow tower***, **air is drawn through slats on both sides of the towers. The air flow is horizontal, across the downward fall of the water. Banks of fill are located in the vertical plane**

from 15 to 10,000 gal/min per cell.

Counterflow towers are less open than crossflow towers, so the warm water moving through the tower is exposed to relatively little sunlight. This minimizes algae growth. However, this enclosed nature impedes access to the unit for servicing.

Figure 4 shows an induced-draft *crossflow* tower. In this configuration, air is drawn in from both sides of the towers, so that water cascades down through the horizontal flow of air.

Both crossflow and counterflow towers are equally efficient, but each offers advantages. Thanks to the emergence of high-efficiency film fill (defined below), counterflow towers have established a position as the lowest-cost, most space-saving type, and are the current design of choice under normal operating circumstances.

However, maintenance personnel prefer crossflow towers, whose key components are open to view and easily accessible for servicing. While the fans, motors, speed reducers and drive shafts are accessible in both designs, the crossflow tower provides ready access to other components, including the water-distribution system, fill, drift eliminators, and internal tower structure.

Tower fill maximizes contact

The counterflow tower shown in Figure 5 contains no fill, relying instead upon a pressurized spray system to disperse the water into fine droplets that fall through a mechanically induced, upward-moving stream of air. To optimize cooling, the falling distance should ideally be maximized to increase contact time. While that goal is easily established by building ever-taller cooling towers, at some point, it becomes economically infeasible to increase the tower size.

A better way to prolong contact time between air and water to promote evaporation is to interrupt the progress of the falling water with one of two types of tower "fill": film-type or splash-type. Counterflow towers lend themselves ideally to film-type fill, while crossflow towers use both with equal facility.

Film-type fill consists of corrugated or rippled sheets (Figure 6). Water flowing over them is spread into a thin film, which maximizes the liquid surface area. Also, these sheets promote water flow over a large vertical area, which maximizes both the exposure to the air flow, and the contact time.

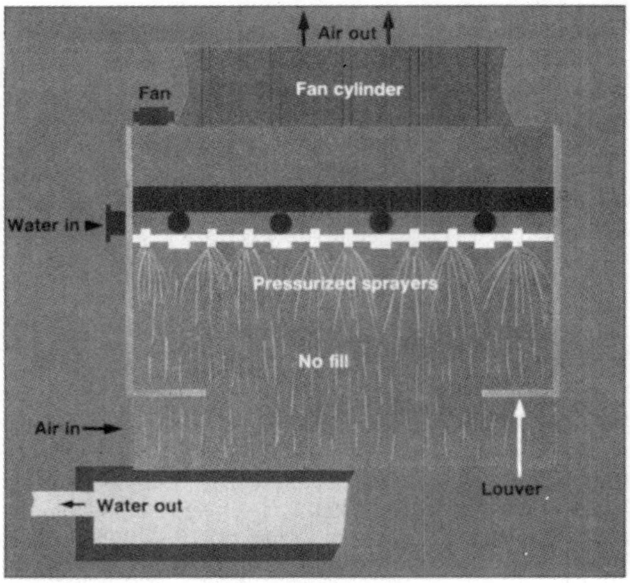

FIGURE 5. Towers without fill are used in applications such as food, steel and paper processing, where high product carryover can build up on the fill. In this no-fill counterflow tower, a pressurized sprayer disperses the water into tiny droplets, which fall through a mechanically induced, upward flow of air. Without the benefit of the flow retardation guaranteed by the fill, towers without fill must be larger for a given heat load than those with fill

Applied properly, film-type fill has the capability to provide significantly more-effective cooling capacity than splash-type fill for a given tower size. However, film-type fill is susceptible to poor air and water distribution, if the sheets are not evenly spaced. This will reduce cooling efficiency and defeat the purpose of the fill.

The fill — which maximizes the water's surface area and its contact time with the flow of air — is considered the most important element of a cooling tower

Film-type fill should not be specified in situations where the circulating water is contaminated with product carry-over, debris, silt or mud, since such contaminants can plug the passages. In those cases, splash-type fill is recommended.

Splash-type fill uses staggered rows of splash bars to interrupt the water's vertical descent (Figure 7). The successive offset levels, or parallel splash bars, are placed horizontally, below the water-distribution system. By repeatedly arresting the water's fall so that it splashes into small droplets, the exposure of the water surface to the passing airstream is maximized.

Because the levels of the splash fill are relatively widely spaced, clogging in this scenario is not a problem, and the unit can be easily cleaned. However, unless adequately supported, the splash bars are susceptible to sagging. If the parallel levels are allowed to sag or warp, the set spacing will be altered, and the water and air will "channel" through the fill in separate flow paths causing the tower's thermal performance to degrade.

The support grids that are used to hold the fill in place are typically manufactured of fiber-reinforced polyester (FRP), and the splash bars are typically of extruded PVC, polypropylene, polyethylene or treated wood.

Materials of construction

In general, the description of a cooling tower as being wood, steel, fiberglass or concrete refers to the material used in its structure. The remaining components may or may not be of the same material.

Wood. Because of its availability, workability, relative low cost, and durability under the very severe operating conditions encountered in cooling towers, wood remains the predominant structural material. The basic structural standards are specified in the "National design specification for wood construction (NDS)," which is published by the National Forest Products Assn. (Madison, Wis.).

Because cooling towers are subject to a harsh operating environment at elevated temperatures, extra structural concerns — beyond accepted norms — are dictated. To augment the NDS standards, the Cooling Tower Institute (CTI; Houston, Tex.) has issued its own standard specifications: STD-114 pertains to the design of cooling towers with Douglas Fir lumber; STD-103 for the design of Redwood towers; STD-119 for structural joints and timber connectors; and WMS-112 for wood preservative treatments.

Recognizing that wood suffers a reduction in structural characteristics in a prolonged wet environment under elevated temperatures, these standards prescribe appropriate reductions in allowable structural loads.

They also require greater attention to the placement and fit of structural connectors for withstanding the long-term dynamic loads imposed by wind and weather, by the continuous operation of the tower itself, and even by earthquakes. Specifiers of wood towers should require the strictest adherence to the CTI standards.

California Redwood has been a long-time favorite for cooling tower construction. However, the harvesting of ancient Redwoods — with their well-known structural characteristics — is now forbidden, and the modern-day Redwoods don't boast the same structural strength. Consequently, Douglas Fir is normally considered the preferred alternative.

Regardless of type, wood must be treated with a reliable preservative to prevent fungal attack and retard decay. Chromated copper arsenate (CCA) and acid copper chromate (ACC) are the most common. Such preservatives are diffused into the wood by total immersion in a pressure vessel, with pressure being maintained either until a prescribed amount of preservative is retained by the wood, or until the wood refuses to accept further treatment.

While such treatment helps to extend the service life of wood towers, these chemicals are the subject of growing environmental concern. During operation, small concentrations of CCA and ACC may be present in the water leaving the tower during blowdown; at the end of the tower's service life, the disposal of chemically impregnated wood may call for special handling and permitting.

Although wood is relatively insensitive to chlorides, sulfates and hydrogen sulfide in the process water, it can be damaged by excessive levels of free chlorine, and wood is sensitive to prolonged exposure to excessively hot water. Design temperatures of the hot water should be limited to 140°F, or should be controlled to that level with the use of an initial cold-water by-pass step.

While these concerns have not significantly lessened the use of structural wood for cooling towers, they have caused some users to consider alternative materials of construction:

Plastics. Plastics use is on the rise as a structural material for cooling towers, and a material for components, as well. Contributing factors include plastic's inherent resistance to microbiological attack, corrosion and erosion; its compatibility with other materials; formability; good strength-to-weight ratio; and acceptable cost level.

For many years, fiber-reinforced plastic (FRP) has been used in such components as structural connectors, fan blades, fan cylinders, fill supports, piping, casing, louvers, and louver supports. More recently, sophisticated molding and pultrusion manufacturing techniques have made its use viable in structural applications, as well. Polypropylene has virtually become the standard material for spray nozzles, while polyvinyl chloride (PVC) has become the principal material of construction for fill bars and sheets.

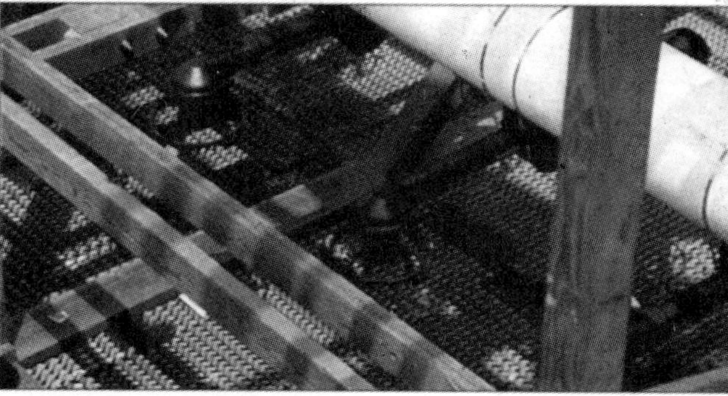

In general, plastic components have the best resistance to harsh operating conditions. Certain plastics, such as PVC, are inherently fire resistant; where required, others may be formulated for fire retardancy, the most common formulation being fire-retardant FRP.

Steel and stainless steel. Many of the smaller towers designed for factory assembly are primarily of steel construction. Local building and fire codes may also dictate that larger field-erected towers be constructed of steel.

Steel towers are growing in popularity, thanks to their consistent structural strength, building and fire code requirements, and environmental acceptability. In these cases, galvanized steel is the predominant choice for structural support, basins, partitions, decking, and many other major components.

Where long-term corrosion resistance with minimal maintenance impact is a high priority, stainless steel works well. Although the cost impact of stainless steel is significant compared to wood, it often compares favorably to the cost of FRP towers.

Concrete. Concrete has been used for many years to build hyperbolic cooling towers, and the technology has also been applied in large mechanical draft towers. Currently, however, concrete is being considered as a viable alternative material on the smaller mid-range (25,000 to 100,000 gal/min) industrial towers. In many cases, the higher initial cost of concrete construction is justified by its reduced fire risk and higher load-carrying capability.

Basically, design philosphies for concrete cooling tower construction coincide with those espoused by the American Concrete Institute (ACI; Detroit,

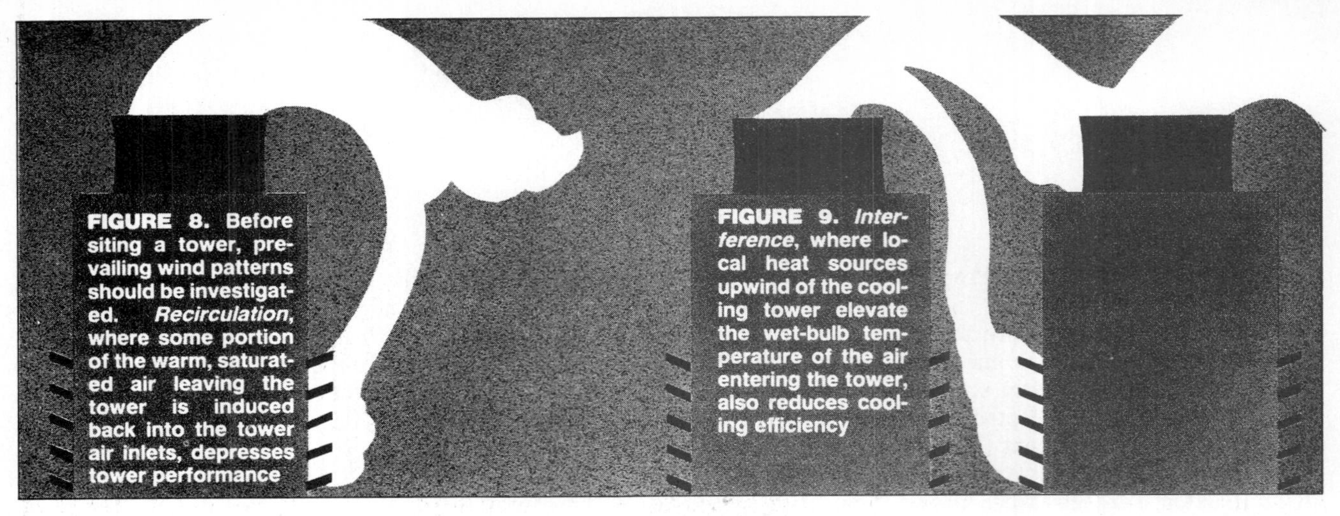

FIGURE 8. Before siting a tower, prevailing wind patterns should be investigated. *Recirculation*, where some portion of the warm, saturated air leaving the tower is induced back into the tower air inlets, depresses tower performance

FIGURE 9. *Interference*, where local heat sources upwind of the cooling tower elevate the wet-bulb temperature of the air entering the tower, also reduces cooling efficiency

Mich.), except denser mixes and lower water-cement ratios are called for, compared with more commercial (less industrial) construction. Typically, ACI's Type II cement is preferred.

Component materials

Regardless of the primary structural material, steel is used for many components of the cooling tower where high strength is required. This includes fan hubs for larger diameter fans, unitized supports for stabilization of the mechanical equipment, fan guards and driveshafts guards, as well as the tie rods, bolts, nuts and washers.

Cast iron is used for gear cases, anchor castings, flow-control valve bodies, and fan hubs for intermediate sized fans. Although cast iron enjoys good corrosion resistance, it is usually either galvanized or coated with a high grade enamel for cooling tower use. For severe water service, it may be sandblasted and coated with several layers of an epoxy compound.

FIGURE 6. *Film-type fill* (left) consists of corrugated sheets. Water flowing over the fill is spread to a thin layer, which maximizes the water's surface area and its contact time with air flowing through the tower

FIGURE 7. *Splash-type fill* (right) consists of staggered rows of horizontal splash bars. Water flowing over them splashes into small droplets, which increases surface area and contact time with the air

Bolts, nuts and washers do not lend themselves to a protective coating, other than galvanization or cadmium plating. Severe water conditions normally dictate a change in materials, and an appropriate grade of stainless steel is the popular choice because of excellent corrosion resistance in the aerated conditions existing in cooling towers. Stainless steel is also the most-used material for driveshafts, although driveshafts of mandrel-wound carbon fiber are becoming popular.

For special conditions, such as salt or brackish water service, copper alloys are sometimes used for the driveshaft. Silicon bronze fasteners are also suitable for service in salt water, but they are relatively soft and must be protected against erosion. The use of Naval brass is normally discouraged because of its susceptibility to stress-corrosion cracking. The use of more-sophisticated metals, such as monel and titanium, is usually precluded by cost considerations.

Selection of aluminum alloys for use in cooling towers is done sparingly. Only the more corrosion-resistant alloys are used, and only for specific components of significant cross section. Among these components are fan blades, fan hubs for smaller size fans, some ladder assemblies, and handrail fittings for steel-framed cooling towers.

Tower siting and orientation

The orientation of a cooling tower is important, since local heat sources upwind of the cooling tower can elevate the wet-bulb temperature of the air entering the tower and interfere with the tower performance. A detailed site-plan drawing must be included with the specifications. It should show not only the intended location of the cooling tower, but the placement of any towers anticipated in the future. This drawing should indicate the elevations of all process facility structures at the location.

A contour map (or rose diagram) of average winds should be worked up, emphasizing the seasonal wind force and direction. Cooling towers can impact their own performance through recirculation (Figure 8) and the performance of other towers (present or future) through interference (Figure 9). It behooves the specifier to discuss the potential interaction of the tower with other process operations onsite. It may be wise to elevate the design wet-bulb temperature by one or more degrees to more accurately reflect the anticipated temperature of the entering air at an industrial site.

Avoid poor performance

Poorly designed cooling towers will be costly in more ways than just inefficient thermal performance. To rectify performance deficiencies, the manufacturer may alter the existing tower, install additional cooling tower capacity, or refund a percentage of the contract price, proportional to the performance deficiency.

In any case, the additional materials and labor required to achieve the desired performance is limited to the scope of the original contract. The cost of any additional foundation, basin, piping, electrical wiring, or control mechanisms required would typically fall to the owner.

These potential costs are obvious and significant, yet there are hidden operat-

KEEP AN EYE ON THE WATER

Water serves to constantly wash the air flowing through a cooling tower. Contaminants stripped from air compound the problem of contaminants already in the makeup water. Add to this the fact that constant evaporation serves to concentrate contaminants in the water, and you can see how trouble multiplies.

For the best performance of standard construction materials, the following "normal water" conditions have been arbitrarily defined, and are accepted by industry:

- Circulating water should have: a pH between 6 and 8.5; a chloride content (as NaCl) below 750 ppm; a sulfate content (SO_4) below 1,200 ppm; a sodium bicarbonate ($NaHCO_3$) content below 200 ppm; a maximum temperature of 120°F; no significant contamination; and adequate treatment to minimize corrosion and scaling
- Chlorine, if used, should be added intermittently as needed, so that free residual chlorine does not exceed 1 ppm
- The surrounding atmosphere should be no worse than "moderate industrial" where rainfall and fog are only slightly acid, and do not contain significant chlorides or hydrogen sulfide (H_2S)

Water conditions outside these limits should be investigated, so that contributing factors can be rectified. In general, wood and plastic components are very tolerant of chemical excursions far beyond these limits. Conversely, carbon steel items are relatively unforgiving of all but the most limited variations.

The following controls can be put in place, to maintain the desired water quality:

Blowdown. Evaporated water exits the cooling tower in a pure vapor state, leaving behind its total dissolved solids (TDS) to concentrate in the recirculating mass of water. The TDS content is controlled via blowdown, where a portion of the circulating water is continuously evacuated and replenished with relatively pure makeup water.

The rate of blowdown — different for each cooling tower — is determined by the following equation:

$$B = E - [(C-1) \times D] / (C-1)$$

where:

B = Rate of blowdown (gal/min)

E = Rate of evaporation (gal/min). Evaporation can be approximated by multiplying total water flowrate (gal/min) by cooling range (°F) by 0.0008 (a factor)

D = Rate of drift loss. Drift is the amount of circulating water lost as droplets are entrained in the exhaust air stream. It can be approximated by multiplying total water flowrate in gpm by 0.0002 (a factor)

C = The approximate level to which contaminants concentrate in the circulating water;

where: $C = (E + D + B)/(D + B)$

Scale prevention. Calcium carbonate, which has a solubility of about 15 ppm, is the principal scale-forming ingredient in cooling water. It results from the the decomposition of calcium bicarbonate.

If agents (such as sulfuric acid) are added to the system to convert a portion of the calcium bicarbonate to calcium sulfate, the resultant calcium sulfate concentration should be kept below 1,200 ppm. Otherwise, sulfate scale will form (which is denser, and even harder to remove than calcium carbonate scale). A regime of treatment chemicals can be tailored for a given system, to keep scale-forming solids in solution at the tower's operating conditions.

Corrosion control. As the water circulating through a cooling tower (water that originally met the "normal water" standards listed above) becomes progressively contaminated, tower corrosion can occur, as a result of high oxygen or carbon dioxide content, or low pH. If the problem cannot be rectified in the source water, various treatment compounds, used as inhibitors, build and maintain a protective film on the metal parts.

Electrolytic action — where increasing dissolved solids increase the conductivity — also promotes corrosion. This is particularly endemic of systems whose water is rich in chloride and sulfate ions. Adjusting the blowdown rate is usually sufficient to minimize this phenomenon.

Biological growth reduction. Algae and slime (a gelatinous organic growth) typically develop in cooling towers, and reduce cooling effectiveness. Chlorine and chlorine-bearing compounds are effective algicides and slimicides, but excess chlorine can damage wood and other organic materials of construction. If used, chlorine should be added intermittently (as a shock treatment), and only as necessary to control slime and algae. Concentration must not exceed 1 ppm, and care should be taken to avoid concentration at the point of entry, since a localized reduction in pH will hasten corrosion there.

Foaming and discoloration. Persistent foaming, resulting from certain combinations of dissolved solids or foam-causing compounds, can often be controlled by increasing the blowdown rate. In extreme cases, foam-suppressing chemicals can be added to the system.

Control of foreign materials. Suspended materials, brought into the system from the air, are best removed by continuous filtration. Oils and fats should be removed from the recirculating water by a skimmer, since they reduce the thermal performance of the system. □

ing costs, too. In most cases, tower alterations will call for increased fan power — the additional operating cost of which accrues throughout the life of the cooling tower.

While the vendor's bid price is important, operational parameters are of paramount concern during cooling tower specification. These include design features and functional benefits to the process, the quality of tower design and workmanship, the vendor's proven performance record, product reliability, warranty, ease of maintenance, and the availability of parts and service. All too often, specifiers go with the lowest bid in a competitive-bid situation — and come to regret it later.

The following questions must be satisfied when shopping around for a cooling tower: How much would it cost in terms of product output if the cooling tower were to miss its specified performance by even a couple of degrees (Figure 2)? How good is the warranty of the vendor whose tower is assembled from various commercial components? To whom would you appeal for warranty satisfaction? Who will be your source for future parts and service?

A thorough economic evaluation includes the cost impact of all operational and peripheral factors, added to the quoted price of the cooling tower system. Price enters the picture only as a tiebreaker, when two or more bidders have satisfied the primary concerns. The following factors are part of the total evaluated cost:

Pump head evaluation. Because there is a head loss attributable to cooling tower operation, pump energy must be expended throughout the life of the plant. It is expressed as (current) dollars per foot of head.

Fan power evaluation. On mechanical-draft cooling towers, a similar evaluation of the extrapolated cost of fan operation should be considered. Fan power evaluation is normally expressed as dollars per horsepower, or dollars per kilowatt.

Basin evaluation. Space considerations are important, too. Real estate that is occupied by a cooling tower is unavailable for other use. Costs are also incurred during site preparation and during the laying of the foundation. These costs are expressed as dollars per ft^2 of basin area.

Electrical and wiring evaluation. Cooling tower motors are individually wired. Therefore, a portion of the owner's overall electrical installation cost will vary with the number of fans on the tower. This evaluated cost is expressed as dollars per fan.

Piping evaluation. Unless the location and relative size of the tower is fixed by the plans and specifications, variations in cooling tower configuration and placement can result in significantly different amounts of piping required. The installed cost of piping required beyond a specified point should be evaluated in terms of dollars per foot. ■

Edited by Suzanne Shelley

The author

John Hensley is manager of product services for The Marley Cooling Tower Co. (5800 Foxridge Dr., Mission, KS 66202, Tel.: 913-362-1818). Hensley, who attended the U.S. Naval Academy (Annapolis, MD), has been with the company for 40 years. For 24 years, he was located at the company's New York regional office, where he worked with owners, engineers, and contractors to design cooling towers for specific industrial applications, including power generation. In 1981, he returned to Marley's general office in Kansas, where he wrote a comprehensive book, entitled "Cooling Tower Fundamentals," which is available at no charge from the company. Currently, Hensley is involved in technical writing and marketing for the company.

Section III

Operation and Maintenance of Heat-Transfer Equipment

GETTING TOP PERFORMANCE FROM HEAT EXCHANGERS

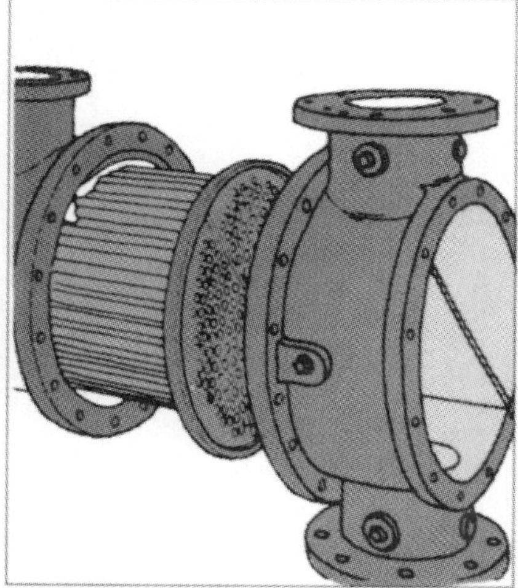

FIGURE 1. Make sure there is enough room for removing a tube bundle

Proper installation, operation and maintenance will ensure long, successful service

David F. Fijas, ITT Standard

A shell-and-tube heat exchanger is a pressure vessel that is designed to operate within certain pressure and temperature limits. It will give trouble-free performance if it is well-designed thermally and mechanically, stored carefully before installation, installed correctly, operated within its design limitations, and its surfaces are cleaned periodically.

The process system of which the heat exchanger is a part must be safeguarded by safety relief valves and instrumentation so that the exchanger's design limitations are not exceeded. Operating personnel should be aware of these limitations.

Storage tips

Upon the delivery of a heat exchanger, check the protective covers for shipping damage. If the exchanger has been damaged extensively, notify the carrier promptly. If it is all right but will not be installed immediately, take precautions to keep it from becoming contaminated or rusted.

Remove any accumulations of dirt, water, ice or snow from the exchanger, then wipe it dry. Store it indoors, if possible. Prevent rust by coating the exchanger with a preservative. If, when in service, the exchanger will handle an oil, avoid both both rust and contamination by filling it with an oil. If, for delivery, the exchanger was not filled with oil or another preservative, open its drain plugs to draw off any accumulated moisture, then close the plugs. An accumulation of moisture usually indicates that rusting has already started, and steps should be taken to restore the surfaces.

Installation rules

If the exchanger's tube bundle is removable, install the exchanger so

FIGURE 2. Tubes may have to be replaced on site if the bundle is fixed

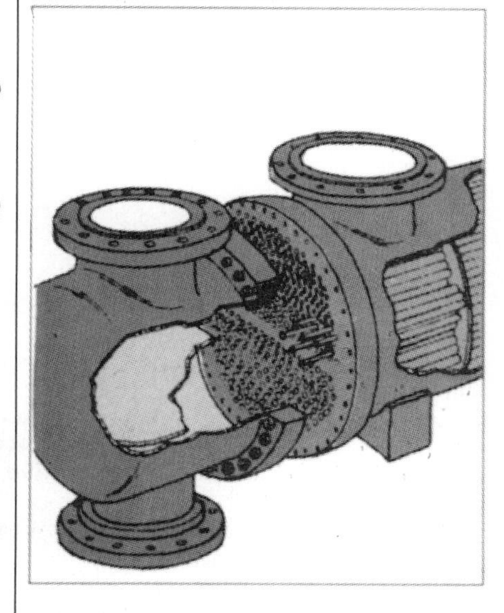

FIGURE 3. Let the tubesheets or baffles, not the tubes, support a tube bundle

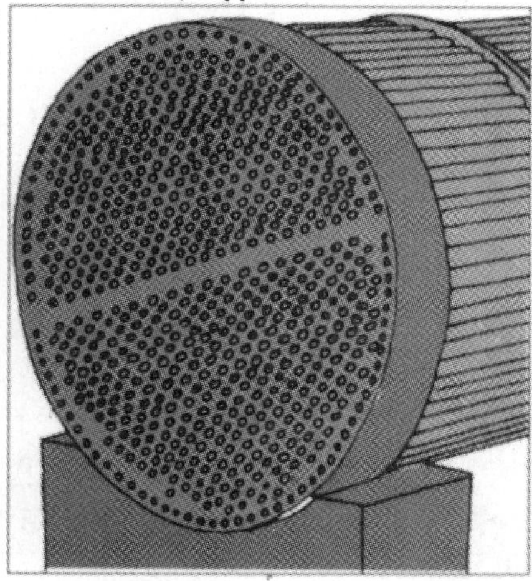

Originally published December 1989

Shellside Fluid type	Shellside Temperature	Tubeside Fluid type	Tubeside Temperature	Startup	Shutdown
Fixed tubesheet* (bundle not removable)					
Liquid	Hot	Liquid	Cold	Start both fluid flows gradually and simultaneously	Shut off both fluid flows gradually and simultaneously
Condensing gas (e.g., steam)	Hot	Liquid or gas	Cold	Start hot fluid first, then slowly start cold fluid, avoiding temperature shock†	Shut off cold fluid first, then hot fluid
Gas	Hot	Liquid	Cold	Start cold fluid first, then hot fluid	Shut off cold fluid gradually, then hot fluid
Liquid	Cold	Liquid	Hot	Start both flows gradually and simultaneously	Shut off both flows gradually and simultaneously
Liquid	Cold	Gas	Hot	Start cold fluid first	Shut off hot fluid first
Liquid	Hot	Liquid	Cold	Start cold fluid first, then hot fluid gradually	Shut off hot fluid first
U-Tube: packed floating head, packed floating tubesheet and internal floating head (all these types have removable bundles)					
Condensing gas (e.g., steam)	Hot	Liquid or gas	Cold	Start cold fluid first, then hot fluid gradually	Shut off cold fluid first, then hot fluid gradually
Gas	Hot	Liquid	Cold	Start cold fluid first, then hot fluid gradually	Shut off hot fluid first
Liquid	Cold	Liquid	Hot	Start cold fluid first, then hot fluid gradually	Shut off hot fluid first
Liquid	Cold	Gas	Hot	Start cold fluid first, then hot fluid gradually	Shut off hot fluid first

(Column group heading: Fluid location and relative temperature)

*If the tubeside flow cannot be shut off with the fixed-tubesheet exchanger, a bypass should be installed to shunt this flow; in which case, the tubeside fluid should be bypassed before the shellside fluid is shut off.

†In all startups and shutdowns, regulate fluid flows so as to avoid thermal shock, regardless of whether the bundle is removable or not. When stopping or starting fluid flows, observe extreme caution with insulated exchangers, because the metal parts could stay hot for a long time, causing thermal shock.

TABLE 1. Recommended procedures for starting up and shutting down heat exchangers

that there will be sufficient clearance at its stationary end for pulling out the bundle (Figure 1). If the bundle is fixed, provide enough clearance at one end to permit removing removing and replacing tubes, and enough room at the other end for tube rolling (Figure 2).

The piping should have valves and bypasses arranged so that the exchanger can be isolated safely for inspection, cleaning and repairs. Temperature-sensing wells and pressure-gage taps should be placed in the piping to and from, and as close as possible to, the exchanger. Vent valves should be located so that gases can be purged from both the tubeside and shellside.

The mounting supports should keep the exchanger from settling and causing piping strains. The foundation bolts should be set accurately. In concrete footings, pipe sleeves of at least one pipe size larger than the bolt diameter should be slipped over the bolts and cast in place. The larger sleeves make it possible to adjust the bolt centers after the foundation has set, facilitating installation.

In addition to the usual liquid-level controls, relief valves and level and temperature alarms, gage glasses or liquid-level alarms should be installed in all the gas spaces to alert operators to any failures in the condensate-drain system that could flood the exchanger. A surge drum upstream from the exchanger may also be necessary to dampen fluid pulsations caused by a pump or compressor. Such pulsations result in vibrations that reduce exchanger life. Flowrates higher than those specified in the design also can cause vibrations that can severely damage the tube bundle.

A heat exchanger that is taken out of service for a long time should be protected against corrosion as has been recommended for storage. If removed from operation for a short period, an exchanger should be drained and, Pipe drains should not be connected to a common manifold because this arrangement makes it difficult to determine if the exchanger has been thoroughly drained.

Set the exchanger level and square

so that pipes can be connected without being forced. Before piping up the exchanger, check all of its openings for foreign materials, including plugs, bags of desiccants and shipping covers. Avoid exposing the insides of exchanger to the atmosphere, because moisture or contaminants may enter it, and cause damage from freezing or corrosion.

If support cradles or feet are fixed to the exchanger, loosen the foundations bolts at one end of the exchanger, after the piping has been completed, to allow it to move freely. The holes in the support cradle or feet are oversized to permit the movement.

Operational keys

Before putting the exchanger into service, be sure all the connected piping and equipment is clean, to prevent plugging the tubes and shellside passages with refuse. Installing a strainer or a settling tank in the piping to the exchanger is a sensible precaution.

Open the vents and gradually allow the fluids to enter the exchanger, according to the procedure outlined in Table 1, in which are tabulated recommended startup and shutdown procedures. After the system has been filled completely with the operating fluids and all the air has been vented, close all the manual vent valves.

After the exchanger has been brought up to its operating temperature, retighten the bolts on all gaskets or packed joints, to prevent leaks and gasket failures. To guard against water hammer, drain the condensate from a steam-heated exchanger, both when starting up and shutting down the exchanger. To keep the fluids from freezing or from corroding surfaces, drain the fluids when shutting down the exchanger.

Maintenance guidelines

Most exchangers become fouled (e.g., contaminates form scale and sludges, which are deposited on surfaces). Even a light coating of scale or sludge on either side of the tubes can significantly reduce heat transfer.

The need for cleaning is generally signaled by a marked rise in pressure drop or drop in heat transfer. Such cleanings should not be long delayed because cleaning becomes more difficult as the layer of scale or deposit becomes thicker. Indeed, if the tube cleaning is put off too long, so many of the tubes could become plugged that flow through them could be reduced so drastically that the consequent drop in heat-transfer rate could become intolerable. In addition, flow through some tubes and not through others could cause variations in differential thermal expansion of metals, and this could result in tube leakage.

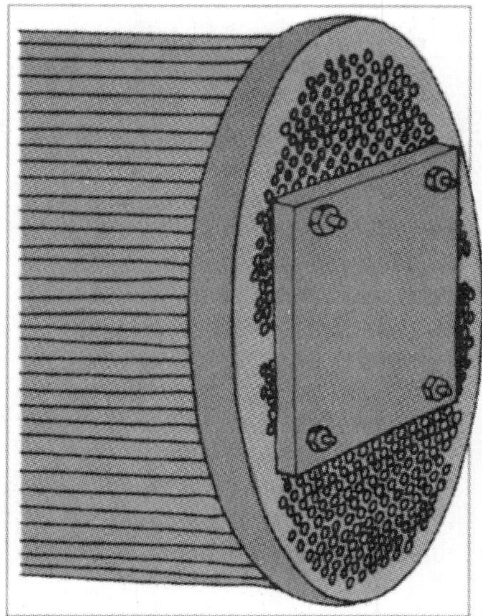

FIGURE 4. A steel bearing plate will protect tubes during bundle handling

FIGURE 5. Put a handle on a bundle by running a cable through a pair of tubes

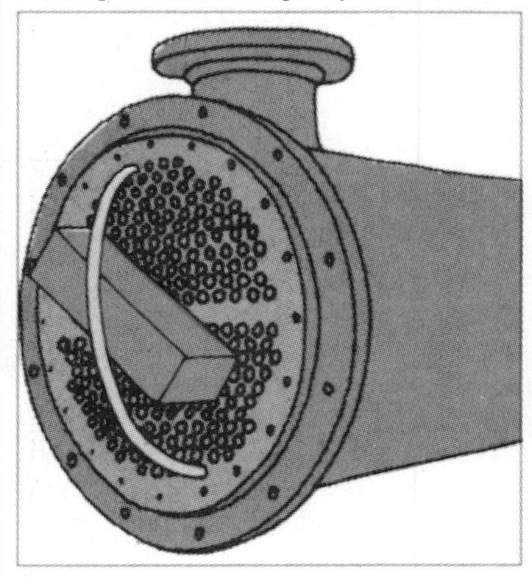

FIGURE 6. Do not drag a tube bundle out of its shell — lift it out

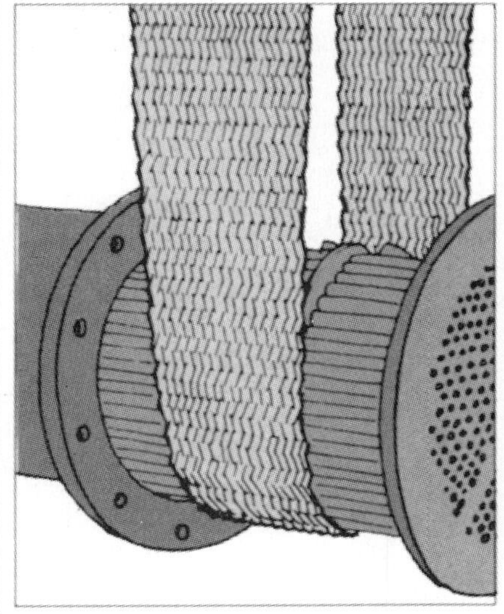

Depending on the exchanger, remove tubeside covers and bonnets as necessary to gain access for inspecting and cleaning the tubes. It may be necessary to remove the tube bundle (bundles in fixed-tubesheet exchangers cannot be removed) to inspect or clean the outside of the tubes. If the exchanger is equipped with sacrificial anodes or plates, replace these, if required, when an exchanger has been opened.

Care must be exercised when removing tube bundles to ensure that the tubes are not damaged. When out of the shell, a bundle should not be supported by the tubes but should rest on the tubesheets, baffle plates or blocks contoured to the periphery of the bundle (Figure 3). Bundles should not be handled with hooks or other tools that could damage the tubes.

To withdraw a bundle, pass rods through two or more tubes, and take the load on the floating tubesheet. A steel bearing plate should be attached to each end of the bundle by means of nuts threaded on the rods (Figure 4). A soft wood filler board inserted between the bearing plate and the tubesheet will keep the tube ends from being damaged. The bundle can then be pulled or lifted by means of forged steel eyebolts screwed into the bearing plates.

An alternative to the foregoing arrangement is to thread a steel cable through one tube and return it through another, with a hardwood spreader block inserted between the cable and each tubesheet to keep the tube ends from being damaged (Fig-

STEEL RODS			STEEL EYEBOLTS	
Tube size, in.	Rod size, in.	Safe load/rod, lb	Size, in.	Safe load, lb
5/8	3/8	1,000	3/4	4,000
3/4	1/2	2,000	1	6,000
1 and larger	5/8	3,000	1¼	10,000
			1½	15,000

TABLE 2. Safe load for steel rods and eyebolts in bundle removals

ure 5). Safe loads for steel rods and eyebolts are listed in Table 2.

When removing the bundle, it may be necessary to break the bundle free from the shell by means of a jack on the floating tubesheet. In such a case, the tube ends should be protected by a large bearing plate and filler board. When pulling the bundle out of the shell, lift it by means of a cradle formed by bending a light-gage plate into a U-shape (Figure 6). Dragging the bundle could bend baffles and support plates.

Sludge and similar soft deposits can usually be removed by hot wash oil or light distillate circulated through the tubeside or shellside. Soft salt deposits may be washed out by circulated hot fresh water. There are commercial products that effectively remove more-stubborn deposits. Do not attempt to clean tubes by blowing steam through them. This could overheat them, causing expansion strains that could result in leaks at tubesheet joints.

Chemical cleaning is generally preferred to mechanical cleaning, especially if the tubes have inserts or longitudinal fins. It may, however, be necessary to resort to mechanical cleaning if the scale is hard and cannot be removed chemically. Neither the inside nor the outside of the tubes should be hammered with a metallic tool. If the tubes must be scraped, the tool should not be sharp enough to cut them.

For locating ruptured or corroded tubes or leaking joints between tubes and tubesheets, the following procedure is recommended: (1) remove the tubeside channel covers or bonnets; (2) pressure up the shellside with a cold fluid, preferably water; and (3) check the tube joints and ends for indications of leaks. For testing certain types of exchangers this way, it may be necessary to seal off the space between the floating tubesheet and the inside shell diameter may means of a test ring, which can be purchased or made.

Tighten a leaking tube joint by means of a parallel roller tube expander. Do not roll a tube beyond the back face of the tubesheet. The rolling depth should not exceed the tubesheet thickness minus ⅛ inch. Tubes that are not leaking should not be rolled because this needlessly thins tubewalls.

New gaskets should be installed when a exchanger is reassembled. Composition gaskets become dried out and brittle in service. When compressed, metal or metal-jacketed gaskets match the contact surface. They tend to become work-hardened with service and do not seal adequately if recompressed. ■

The author

David F. Fijas is an engineering manager with ITT Standard (175 Standard Parkway, Buffalo, NY 14240-1102) responsible for the mechanical and thermal-hydraulic design of heat exchangers, product development and contract administration. He has long been involved with Heat Transfer Research, Inc. (HTRI), a cooperative industry-sponsored and supported research organization (whose 150 members include representatives from heat exchanger manufacturers, engineering contractors and processing companies), having chaired a shell and tube subcommittee for many years. A recipient of B.S. and M.S. degrees in mechanical engineering from the State University of New York at Buffalo, he is a member of the American Soc. of Mechanical Engineers.

HOW TO BOOST THE PERFORMANCE OF FIRED HEATERS

Selective revampings can hike thermal efficiency and capacity

A. Garg, Engineers India Ltd.

A fired heater is an insulated enclosure in which heat liberated by the combustion of fuel is transferred to a process fluid flowing through tubular coils. Typically, the coils are arranged along the walls and roof of the combustion chamber, where heat is transferred primarily by radiation. Heat can also be transferred via convection in a separate tubebank.

Revamping a fired heater to recover energy from its fluegas can be justified at today's low fuel prices, even if its heat duty is as low as 1-million kcal/h. A heater — particularly one that is overfired — can also be changed to make it operate at a higher heat duty. The upper limit of fired-heater thermal efficiency is about 92% (LHV).*

Several revamping schemes can improve fired heater performance. The major ones are installing a convection section in an all-radiant heater, enlarging the heat transfer area of the convection section, converting a natural-draft heater to a forced-draft one, and adding air preheating or steam-generation equipment.

*Lower Heating Value; Higher Heating Value (HHV) is the total heat obtained from the combustion of fuel at 15°C. LHV is HHV minus the latent heat of vaporization of the water formed by the combustion of the hydrogen in the fuel. The net thermal efficiency of fired heaters is based on LHV, that of boilers on HHV.

Installing a convection section

Most heaters with a heat duty of up to 3 million kcal/h were built entirely as radiant heaters. They generally have a net thermal efficiency of from 55% to 65%, and their fluegas temperatures range upwards from 700°C. The installation of a convection section in these heaters could recover additional heat and bring down their fluegas temperatures to within 50–100°C of their inlet feed temperatures. Doing this could boost their thermal efficiencies up to 80%, and sometimes even higher.

If the revamped heater is operated at the same heat duty, the radiant heat flux is reduced or additional heat duty is extracted in the convection section. Such an investment will normally be paid out in two to three years.

This alteration should be preceded by a careful checking of the heater's existing foundation and structure to ensure that the additional loading can be safely borne. More space will be taken up if an outboard convection bank (one mounted on an independent external structure) is added. Normally, the height of the stack must be raised. Sometimes, an induced-draft fan has to be installed to overcome the extra draft loss. Also, the pressure of the process fluid may have to be hiked to offset the additional tube pressure drop. A typical all-radiant fired heater to which a convection section has been

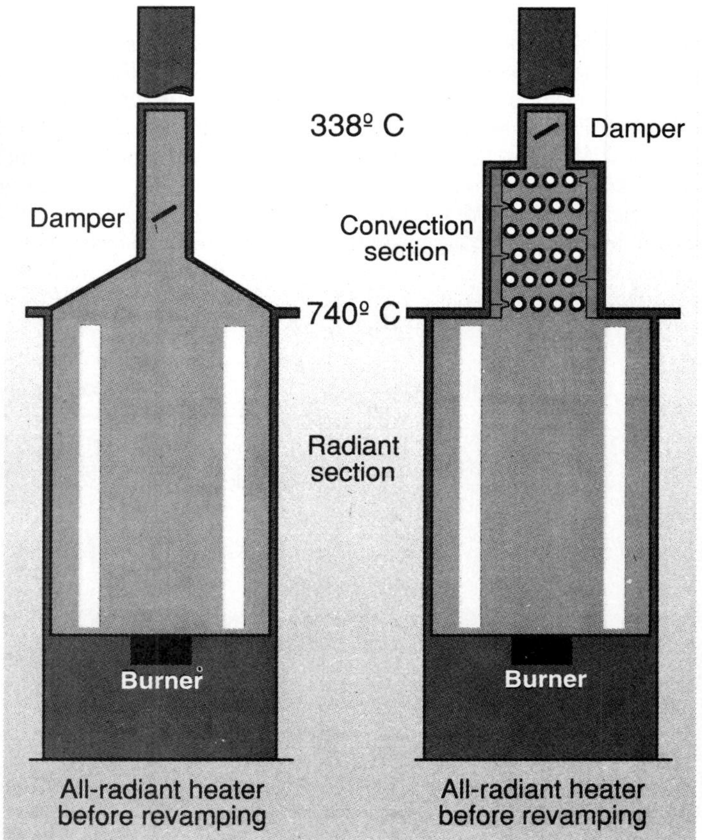

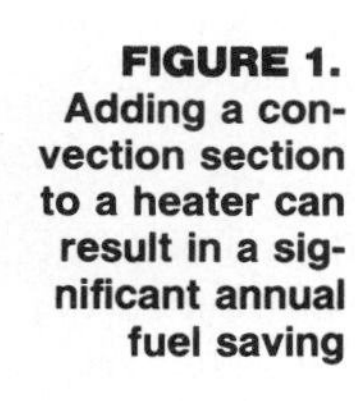
FIGURE 1. Adding a convection section to a heater can result in a significant annual fuel saving

Originally published November 1989

HEATER OPERATING PARAMETERS BEFORE AND AFTER CONVECTION RETROFITTING

Parameter	Before	After
Heater duty, million kcal/h	1.86	1.86
Thermal efficiency, %	60	81
Fuel fired, million kcal/h	3.10	2.30
Fluegas temperature at stack, °C	740	338
Fuel saving, metric ton/yr	—	630
Radiant flux, kcal/(h)(m²)	28,604	22,347
Fluid pressure drop, kg/cm²	1.4	1.8
Stack height, m	21	18.5
Inlet fluid temperature, °C	290	290

TABLE 1. Adding a convection section to this heater hiked its thermal efficiency by 35%

added is shown in Figure 1. Listed in Table 1 are design and operating parameters of a vertical, cylindrical, all-radiant, forced-draft fired heater before and after it was retrofitted with a convection section.

Increasing convection surface

The heat transferred to the process fluid passing through the convection section can be boosted by the addition of heat-transfer surface, to reduce the fluegas temperature to the stack to within 50–100°C of the fluid inlet temperature. This can be done by:

1. Adding tubes. Two additional rows of tubes can be installed in the convection section of most heaters without making a major change, except for relocating the inlet piping terminal. If space for adding tubes has not been provided, the convection section can be extended into the breeching or offtake* to make space.

2. Replacing bare tubes with extended-surface tubes, to gain more heat-transfer area. A typical studded tube provides 2 to 3 times more heat-transfer area than a bare tube, and a finned tube 4 to 5 times as much. Care must be exercised in the revamping because extended-surface tubes of the same size as the bare tubes will not fit in the existing tubesheets of the convection section. A convection section with studded or finned tubes will be compact in height but may require the installation of soot blowers if heavy fuel oil is fired.

3. Substituting finned tubes for studded tubes. Even when heavy fuel oil is fired, low density finned tubes (118 fins/m, 2.54-mm thick and 19-mm high) have performed satisfactorily. Improved burners have significantly reduced the formation of soot and ash, and soot blowers can keep finned surfaces clean. Finned tubes provide larger heat-transfer surface than studded tubes, and cause much less pressure drop. The characteristics of the tubes are compared in Table 2.

*Breeching is the enclosure in which the fluegas is collected (after the last convection coil) for transmission to the stack or outlet duct; an offtake is the duct piece that connects the breaching to the stack.

The additional heat-transfer surface increases the pressure drop, and the lower stack gas temperature reduces the draft. The obvious remedy to the lower draft is to resort to an induced-draft fan or make the stack taller, or do both. However, a load limitation may not allow either option. In such a case, one possibility would be to install a grade-mounted stack or to place the convection section and stack on separate foundations.

Natural to forced draft

The burners of natural-draft heaters require high levels of excess air and have long flame lengths. The air pressure drop available across their registers (which control the supply of combustion air) is limited. Because the combustion air is induced at very low velocities, good mixing of air and fuel is difficult. This leads to excess air levels close to 30–40% for fuel oil and 20–25% for fuel gas.

In forced-draft burners, air pressure energy promotes intimate mixing of fuel and air, with excess air limited to 10-15% for fuel oil and 5-10% for fuel gas. Forced-draft burners also offer the following advantages: (1) more-efficient combustion of heavy fuel oils, (2)

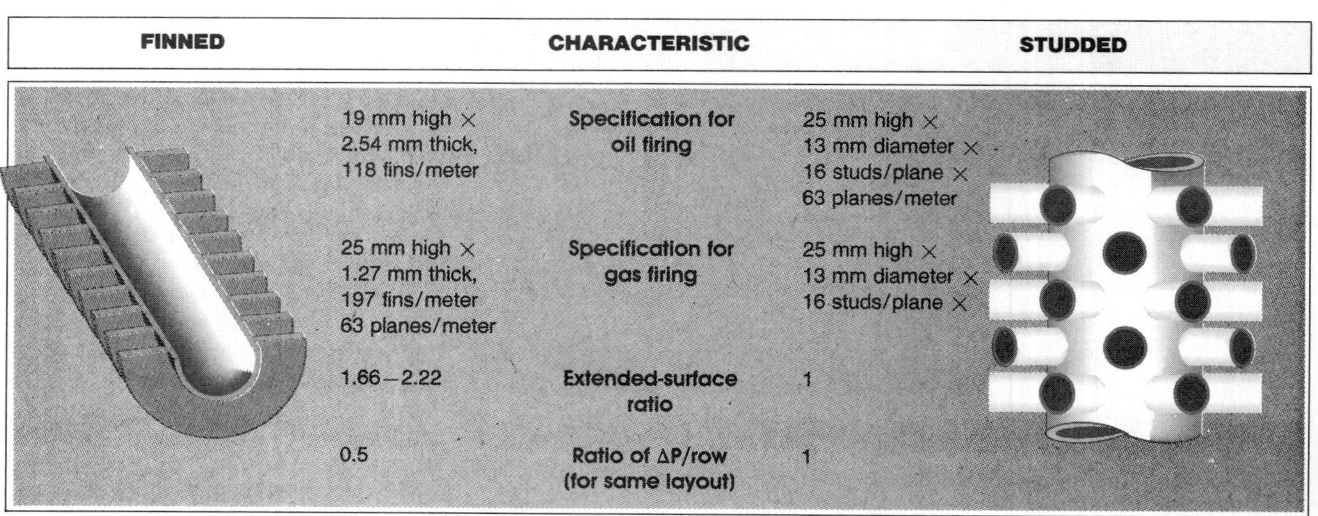

FINNED	CHARACTERISTIC	STUDDED
19 mm high × 2.54 mm thick, 118 fins/meter	Specification for oil firing	25 mm high × 13 mm diameter × 16 studs/plane × 63 planes/meter
25 mm high × 1.27 mm thick, 197 fins/meter 63 planes/meter	Specification for gas firing	25 mm high × 13 mm diameter × 16 studs/plane ×
1.66–2.22	Extended-surface ratio	1
0.5	Ratio of ΔP/row (for same layout)	1

TABLE 2. Finned tubes provide larger heat-transfer surface per unit length than do studded tubes but are more difficult to keep clean

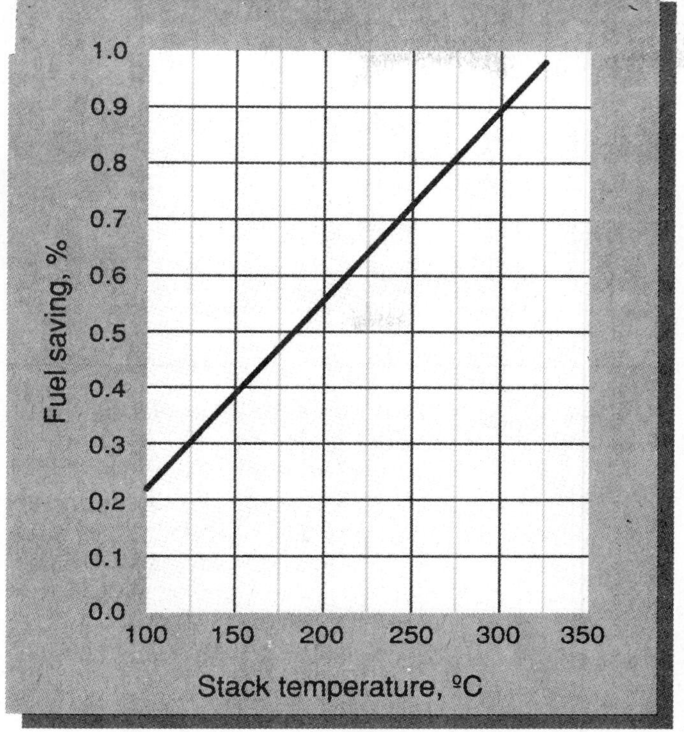

FIGURE 2. Percent efficiency improvement per 10% reduction in excess air

FIRED-HEATER PARAMETERS BEFORE AND AFTER CHANGING BURNERS

Parameter	Before	After
Heater duty, million kcals/h	10.0	10.0
Thermal efficiency, %	79	84
Excess air, %	40	10
Fluegas temperature at stack, °C	340	320
Atomizing steam consumption, kg/kg fuel	0.50	0.15
Flame dimensions, height × diameter, m	4.0 × 1.0	2.0 × 0.6
Fuel saving, metric ton/yr	—	603
Number of burners	8	8
Heat release/burner, million kcal/h	1.6	1.5

TABLE 3. Replacing natural-draft with forced-draft burners reduced fuel usage

FIGURE 3. Replacing burners may require raising the heater floor

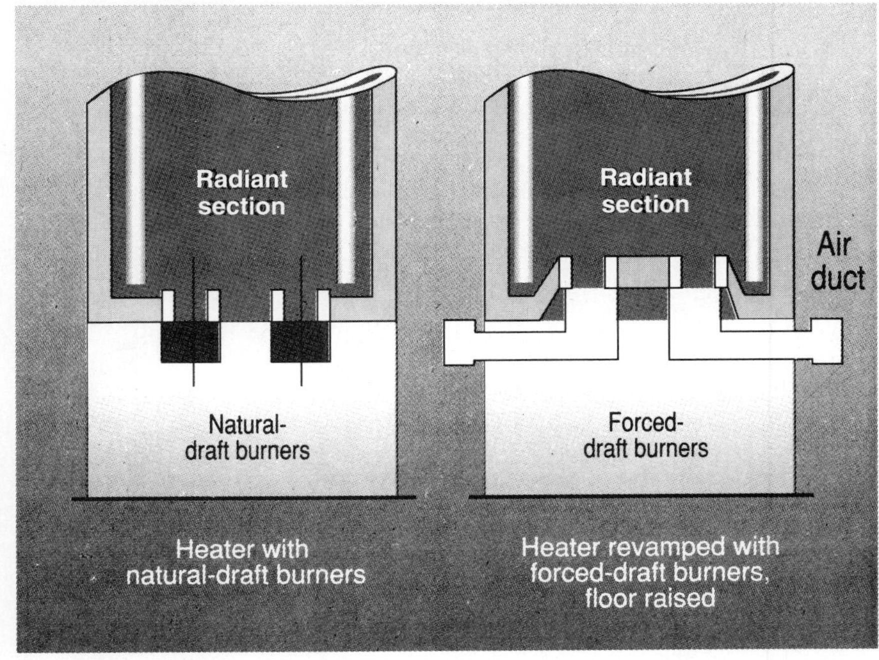

reduced particle emission, (3) lower consumption of atomizing steam, (4) better control of flame shape and stability, (5) less oil dripback, (6) quieter operation and (7) the possibility of preheating the combustion air.

A 10% reduction in excess air means a 0.5 to 1.0% fuel saving, depending on the fluegas temperature (Figure 2). Excess-air reduction normally results in a 2–3% fuel saving, as well as in better heat transfer. A 1 to 2% fuel saving in atomizing steam can also be realized.

The total energy saving only from replacing natural-draft with forced-draft burners will not provide an economic return. However, the change can be justified when combined with such benefits as higher capacity and the elimination of flame impingement.

A vacuum heater having natural-draft burners was plagued with short run lengths, chiefly due to flame impingement problems and tube failures. Replacing the burners with forced-draft ones lengthened the runs substantially (Table 3).

Before replacing burners, check the heater floor elevations, because forced-draft burners require ductwork and deeper windboxes. The floor may have to be raised to accommodate the burners (Figure 3). Space should also be available for ducts and fans.

Combustion air preheating

Adding an air preheater has remained the most popular way of revamping fired heaters. Every 20°C drop in the exit fluegas temperature boosts efficiency by 1%. Total savings range from 8% to 18%. An air preheater is economically attractive if the temperature of the fluegas is higher than 350°C. An economic justification must take into account the cost for fan power, as well as the capital cost. Heat is recovered with an existing heater by means of a combustion air preheater installed between the convection section and the stack (Figure 4).

A revamping entails installing forced-draft burners and a forced-draft or induced-draft fan, or both, as well as the air preheater. Space must be available for the preheater, fan, ducts and dampers. The heater should be able to operate with the heat-recovery system bypassed. Among the disadvantages of

For a comprehensive exposition of fired heaters, refer to the four-part series by Herbert L. Berman that appeared in the following issues of *Chemical Engineering*: Part I — Finding the basic design for your application, June 19, 1978, pp. 99-104; Part II — Construction materials, mechanical features, performance monitoring, July 31, 1978, pp. 87-96; Part III — How combustion conditions influence design and operation, Aug. 14, 1978, pp. 129-140; and Part IV — How to reduce your fuel bill, Sept. 11, 1978, pp. 165-169. Also see Good Heater Specifications Pay Off, by A. Garg and H. Ghosh, *Chem. Eng.*, July 18, 1988, pp. 77-80.

a preheater is the formation of an acid mist and, thus, faster corrosion and more frequent maintenance, and a higher concentration of NO_X in the fluegas.

Two types of air preheaters are currently used with fired heaters: the regenerative and the recuperative. Although the recuperative preheater is larger and costlier than the regenerative type, it is simpler, requires less maintenance, resists corrosion and needs no power.

Another preheater is the circulating-liquid type (Figure 5), in which a transfer fluid extracts heat from the heater's convection section and heats the combustion air passing through an exchanger. The heater needs only a forced-draft fan. Because the fluid is pumped from a tank through a closed loop, this arrangement is attractive if a hot-oil circulation system already exists in a plant. For low temperatures, boiler feedwater can serve as the transfer fluid.

With the addition of a preheater, the heater must be rerated because preheating boosts the radiant heat absorption. This raises the radiant heat flux and tubewall temperatures. Because the flames are shorter with forced-draft preheated-air burners, this type of fired heater can generally be operated (this should be checked) at about a 10% to 25% higher duty. The parameters of a vertical cylindrical heater before and after being fitted with an air preheater are listed in Table 4.

Steam generation

Waste-heat boilers and boiler feedwater preheaters can be an economical solution to recovering heat from heaters that

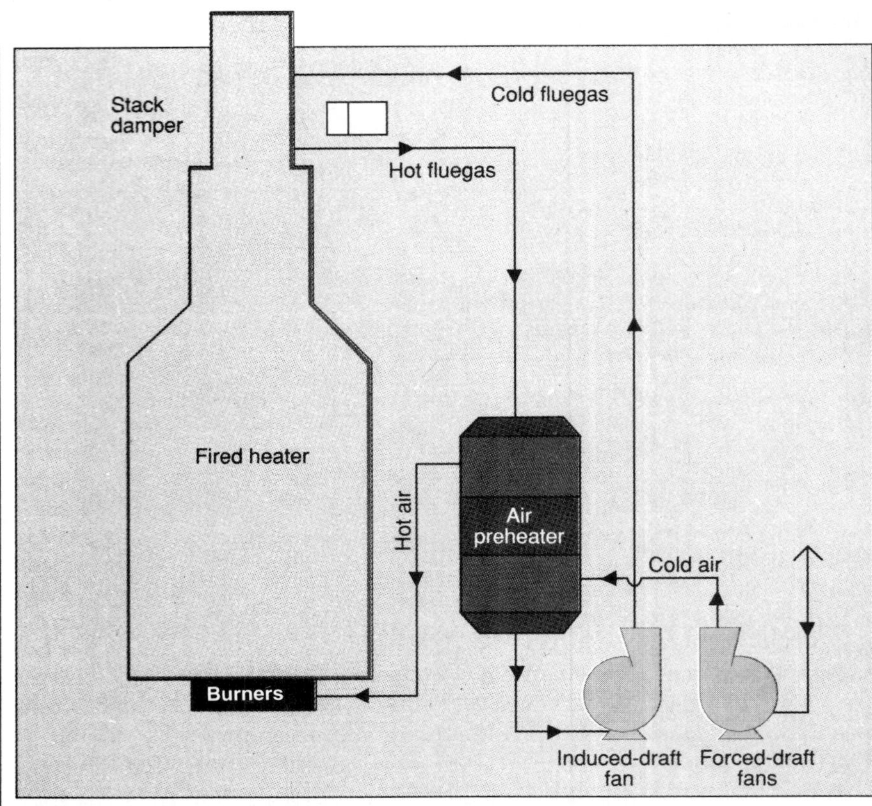

FIGURE 4. A preheater can salvage heat that would otherwise be lost into the atmosphere

FIGURE 5. A preheater can also get its heat from oil circulated in a closed loop

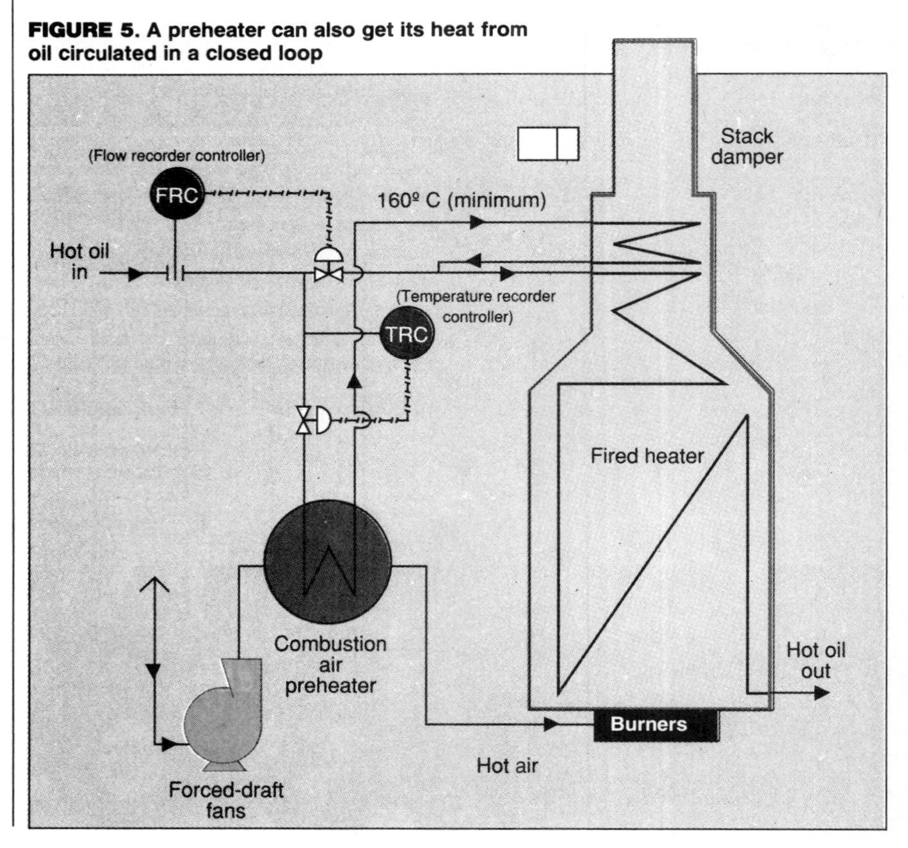

HEATER PARAMETERS AFTER THE ADDITION OF AN AIR PREHEATER

Parameter	Before	After
Heater duty, million kcals/h	4.30	4.73
Thermal efficiency, %	70	87
Excess air, %	40	15
Fuel fired, million kcal/h	6.14	5.44
Fuel savings, metric ton/yr	—	1,050
Radient heat flux, kcal/(h)(m²)	27,700	30,810
Electricity consumption, kW		5
Combustion air temperature	40	145
Fluegas temperature at stack, °C	370	228

TABLE 4. **Adding a preheater raises radiant-heat absorption, making tubewalls hotter**

would otherwise be wasted. For this revision, neither a forced-draft fan nor new burners and controls are necessary. The fluegas can be cooled to within 50°C and 100°C of the inlet boiler feedwater temperature generating medium- or low-pressure steam. Cold-end corrosion limits bringing down the the stack temperature, and the inlet temperature of the boiler feedwater must be controlled to avoid the condensation of acids.

Most heaters that have a convection section can be fitted with boiler feedwater preheaters in an extended convection section without the need for major modification. Frequently, an outboard convection system with an induced-draft fan and ducting is required for waste-heat boilers. With pyrolysis heaters and steam-naptha-reforming heaters, waste heat boilers are preferred because waste heat represents a good steam source for both.

In one instance, a revamping hiked heater duty by 15%. The addition of a boiler feedwater preheater raised the thermal efficiency from 85% to 90%. The convection section before and after revamping is shown in Figure 6.

Recovery constraints

The presence of sulfur in a fuel imposes a serious constraint on the extent to which heat can be extracted from stack gases. About 6% to 10% by weight of the sulfur burned in a fuel appears in the fluegas as SO_3, which condenses as sulfuric acid. The greater the SO_3 content in the stack gas, the higher the dewpoint temperature.

The formation of SO_3 drops sharply with the reduction of excess air, and rises if vanadium is present in the fuel oil. The tubewall temperature should be kept at least 15°C higher than the sulfur dewpoint. A minimum temperature of 135°C is recommended with fuels having a sulfur content of less than 1%, and 150°C with fuels having 4% to 5% sulfur.

Some steps typically taken to avoid flue dewpoint corrosion are to preheat the combustion air with low-pressure steam or hot water, to recycle part of the hot air from the air preheater outlet to maintain a higher air inlet temperature, and to use low-alloy corrosion-resistant steel, or a nonmetal. ■

155º C
Boiler feedwater preheater, 5 rows studded
250º C
Soot blowers, 8 per row
Steam generator coil
377º C
Soot blowers, 8 per row
Steam supply heating coil, 2 rows studded
419º C
Hydrocarbon preheating coil, 9 rows studded + 2 rows bare
Soot blowers, 8 per row
843º C
#1
#2
Convection section

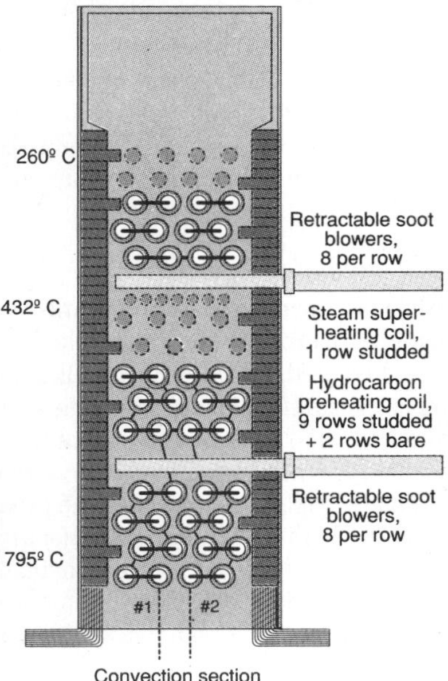

FIGURE 6. **Convection section before and after the addition of preheater**

References

1. Bonnet, C., Save Energy in Fired Heaters," *Hydro. Proc.*, March 1982.

2. Cherrington, D. C., and Michelson, H. D., How to Save Refinery Furnace Fuel," *Oil & Gas J.*, Sept. 2, 1974

3. "Refinery Energy Conservation," Noyes Data Corp., Park Ridge, N.J.

4. Rijwijk, A. G., "Forced Draught Process Burners — A Case History," Duiker Projkten BV, the Netherlands.

5. Seabold, J. G., How to Conserve Fuel in Furnaces and Flares, *Hydro. Proc.*, November 1982.

Acknowledgment

The author is thankful to M. Ghosh and the management of Engineers India Ltd. for granting permission to publish information in this article.

The author

A. Garg is deputy manager, Heat & Mass Transfer Div., Engineers India Ltd. (5th floor, P.T.I. Bldg.., 4, Sansad Marg, New Delhi — 110 001, India). His career responsibilities have included the design, sales and commissioning of fired heaters and furnaces. He is a chemical engineering graduate of the Indian Institute of Technology.

Both operating temperature and the quality of contact between product and belt affect capacity

GET MORE OUT OF

Per Lindén, Sandvik Process Systems, Inc.

Debottlenecking of existing plants remains an important option for increasing a firm's production capacity. But in planning a debottlenecking project, engineers sometimes focus on the key reaction or separation steps in the process, overlooking the impact of thus-gained capacity increases on steps at the end of the processing line.

To help forestall problems this could cause, here are guidelines for increasing the capacity of a widely used solidification-and forming device in plants producing solids, the belt cooler. The principles can be readily adapted for use with drum flakers.

A belt cooler is basically a conveyor, with a belt made of stainless steel. Typically the desired product in viscous-liquid form (for instance, molten sulfur) is fed onto the belt so as to form either a continuous layer, continuous strips, or drop-formed pastilles. Cold water sprayed against the underside of the moving belt carries away heat, and the product cools and solidifies as it approaches the discharge end of the conveyor. The solid product is then either packaged directly (as is commonly the case with pastilles) or broken up into smaller pieces, such as flakes.

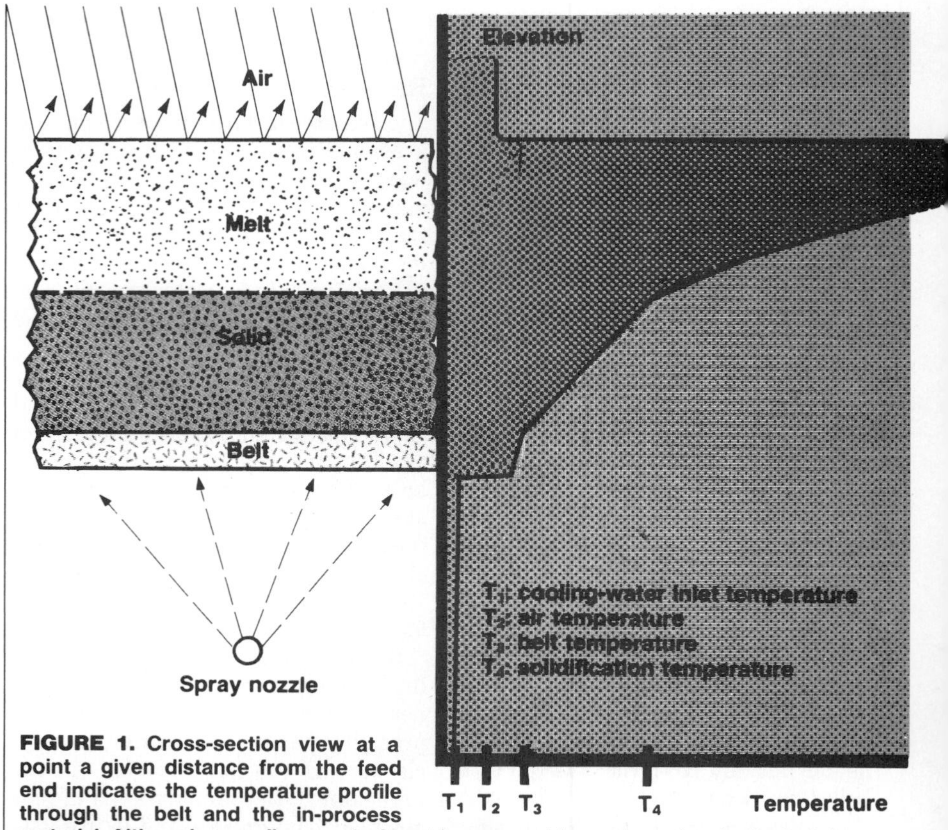

FIGURE 1. Cross-section view at a point a given distance from the feed end indicates the temperature profile through the belt and the in-process material. Although a small amount of heat is removed from above by air, the bulk is instead taken out from below by the spray of cooling water

Heat-transfer scenario

Ordinarily, three kinds of heat must be extracted from the melt: sensible heat from the liquid, latent heat of solidification or crystallization, and sensible heat from the solid. Only a minor amount of this heat is transferred to the air above the belt, except in cases where the product layer is thick, the air is unusually cold, or the air is blown by fans. Instead, the heat is mainly picked up by the cooling water. This can take place at heat-transfer coefficients as high as 500 kcal/(m^2)(h)(K) if the system is well designed.

Four heat-conduction mechanisms are involved, occurring simultaneously in series: conduction through the molten layer to the boundary between still-liquid and already-solidified phases inside the material; through the solidified layer to the belt; through the belt; and into the cooling water. The temperature profile for the system, at a point that is a given distance from the feed end of the belt, appears in Figure 1.

As heat removal takes place, the boundary between molten and solidified phases within a given portion of the in-process material — for instance, within a given liquid drop that is becoming a pastille — moves steadily upward within the material, away from the belt.

In principle, it is possible to analyze such solidification and cooling rigorously because the differential equations are known and can be solved numerically. However, this is seldom done

Originally published September 1989

YOUR BELT COOLER

for the sizing of belt coolers, because in most cases not enough data on the material (e.g., thermal conductivities and heat capacities of the melt and the solid) are known. In practice, rather than measure all the needed data, it is easier to make empirical cooling tests on a pilot-scale belt cooler.

First, the cooling water

When seeking to improve performance of a given belt cooler, the first thing to consider is the cooling-water temperature. Traditionally, belt coolers are built fairly long, and designed to be used with water at ambient temperature (the alternative of using refrigerated water and a shorter belt usually cannot be justified economically). For an already installed system, then, it may well be feasible to increase capacity by chilling the water further.

However, bear in mind that the speed of the solidification process is limited by the rate of heat diffusion within the product itself, and that the diffusion rate through the solid portion is slower than that through the still molten portion. What's more, as in any heat transfer situation, the driving force is the temperature difference between the in-process material and the cooling water. Because of both of those factors, the rate of heat loss from a given portion of in-process material (e.g., a given drop of molten material) tends to continually lessen as the melt cools and begins to solidify, and as the phase boundary between solid and liquid moves inward within the material, and as the solidified material itself cools.

Thus, the maximum heat transfer tends to take place in the upstream segments or zones of the belt (near the feed end), and it is uneconomical to allocate the coldest available cooling water to those zones. Instead, the temperature of the cooling water sprayed against a given zone should be cooler the closer that zone is to the discharge end, so that the temperature gradient between material and water is approximately the same along the entire belt.

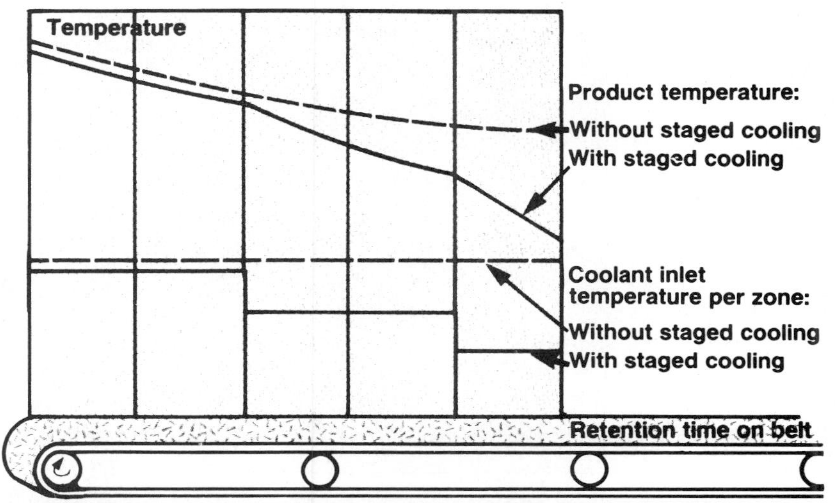

FIGURE 2. Heat removal for the belt system as a whole is enhanced if colder water is used toward the discharge end, rather than using water with a uniform inlet temperature throughout

The point is brought out in Figure 2, in which the ordinate is temperature and the abscissa is retention time on the belt for a give portion of material. Note that the abscissa can also be visualized as a profile of the belt cooler itself, with the upstream end at the origin. The dotted lines for temperature of product and of coolant assume that each zone receives coolant at the same temperature; the solid lines show the faster cooling possible — and, therefore, increased system capacity possible — if the downstream zones instead receive cooler water. The trick is to find the optimum gradient of coolant-water inlet-temperature for the successive zones.

FIGURE 3. Solidified material may sit properly on belt (at top) — or, undesirably, it may curl away from the belt (lower two views)

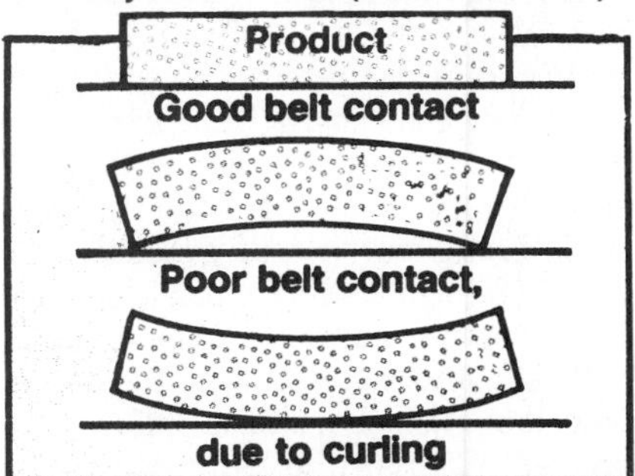

There is another important reason for not using the coldest available water near the upstream end of the belt. Many products shrink or expand substantially when going from liquid to solid state, which can cause the material to curl as it solidifies (Figure 3). The greater the temperature difference between molten product and the cooling water, the greater the curling tendency. Once a portion of material has curled, the cooling rate for it can fall off drastically because the resulting air pocket between material and belt is an efficient insulator.

Assess the other temperatures

Lowering the feed temperature also increases capacity. Often this temperature is governed by the previous steps in the process. In such cases, it may be economical to install a heat exchanger to cool the stream before it feeds onto the belt cooler.

Similarly, it is not uncommon for the piping that brings the product to the belt cooler to be jacketed and heated for ease of pumping the material. In those situations, the segment of jacket on the piping just upstream of the belt might instead be fed with coolant.

Another parameter that can sometimes be changed is the discharge temperature: the higher that temperature, the higher the belt capacity. Admittedly, there is also a greater risk of too-soft material jamming breakers and other size-reduction equipment downstream of the belt cooler; but most coolers are designed with a safety factor in that regard, and it is not imprudent to infringe upon that factor somewhat.

In particular, it is often possible to locate the point of completed solidification by feeling with the hand the point where the product is becoming brittle on the cooler. The belt speed might then be increased to, in essence, move that point closer to the discharge end. This should be done with caution, however, to forestall jamming downstream.

The recommendations to this point can be summarized as follows: (1) Optimize the cooling-water temperature so that the temperature gradient between product and water is fairly constant along the belt and the product does not curl away from the belt; (2) Minimize the amount of heat that must be removed from the material while it is on the belt, by feeding at the lowest practical temperature and discharging at the highest. These recommendations may increase the capacity of a given installation by 5 to 50%, at very modest cost.

Switching from a single- to a double-belt cooler can provide a sizable increase in the capacity of the system

Product	A	B	C	D
Viscosity, cP	500	1,500	20,000	4,000
Feed temperature, °F	350	350	365	510
Coolant temperature, °F	86	86	86	85
Discharge temperature, °F	92	95	95	125
Flake thickness, in.	0.07	0.07	0.07	0.12
Belt load, lb/ft^2	0.4	0.4	0.4	0.5
Specific capacity with single belt, $lb/(ft^2)(h)$	18.5	38	19.1	14
Specific capacity with double belt, $lb/(ft^2)(h)$	69	86	67	32.6
Capacity of single-belt cooler, lb/h	2,072	4,256	2,140	1,570
Capacity of double-belt cooler, lb/h	7,728	9,632	7,504	3,650
Factor of increase	3.7	2.3	3.5	2.3

More steps are possible

If the capacity has to be further increased, there are four additional alternatives:

The first is to add air hoods. By blowing chilled air over the top of the belt and its contents, the capacity can be increased by as much as 10-35%.

This is particularly useful for a line that handles dropformed pastilles because they have relatively much surface area available for air contact. However, such an increase in belt speed must be made with caution. Chilled air can indeed solidify the surface of a pastille quickly; but if not enough heat has been removed (by the air and by the cooling water) from the entire pastille before it leaves the cooler and goes to storage, the temperature throughout it will gradually become uniform and the particles may become soft enough to agglomerate.

It is appropriate to look further at how dropformed pastilles differ from layered material (the latter for subsequent flaking at the belt discharge) as regards capacity of a belt cooler. On the one hand, dropforming leads to about 10-20% less capacity because less of the belt is covered. On the other hand, the uncovered portions of belt adjacent to the individual pastilles serve to some extent as "cooling flanges." More significantly, since pastilles undergo no flaking or other size reduction after they leave the belt, they need not be solidified to the point of brittleness.

Another capacity-increase option, suitable for layered material but not pastilles, is to replace the typical weir-based feed system with cooled double rollers. As the incoming material passes between the rollers, it becomes in essence a partially solid strip, with both upper and lower surfaces already solidified before the material reaches the belt.

In a similar debottlenecking modification for layered material, one can mount a set of cooled, chrome-plated rollers along and above the cooler itself, to press the product down against the belt. This method is often used for products with high viscosities with tendency to curl.

Finally, there is the option of converting the existing cooler to a double-belt configuration, by simply adding another cooler on top of the existing one. This second belt, cooled from above by water, presses downward upon the in-process material.

For a material that tends to contact the single belt well, the addition of the second belt increases capacity by a factor of about 2 to 2.5. For a product that instead tends to curl, as discussed earlier, the factor can be even higher. The capacity increase possible by going from single to double belt is illustrated by actual test data in the table. Columns A and C pertain to materials normally offering poor belt contact, Columns B and D to materials where the contact is good. ■

The author

Per Lindén is Marketing Services Manager for Sandvik Process Systems, Inc., 21 Campus Rd., Totowa, NJ 07512; Tel. (201) 790-1600. He earned his M.S. in mechanical and industrial engineering at the Royal Institute of Technology, Stockholm, Sweden. He has a background in research and product development.

LAURIE GRACE

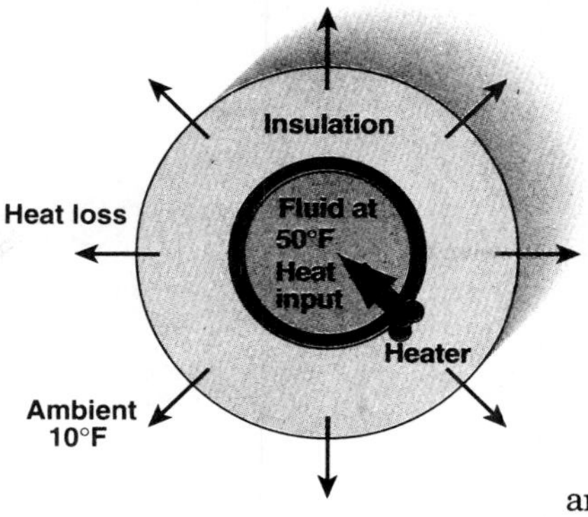

PROVEN TOOLS AND TECHNIQUES FOR ELECTRIC HEAT TRACING

Dinesh Gupta
Raychem Corp.
and **Bob Rafferty**
E.I. du Pont de Nemours and Co.

Piping systems equipped with poor insulation, and such heat sinks as valves, pipe supports and instruments, tend to lose or gain a large amount of heat whenever there is a significant difference between the pipe and ambient temperatures. Variations in ambient temperatures from freezing winter nights to scorching summer days often affect the extent of heat loss or gain as well. Consequently, the fluid inside a pipe may undergo phase changes (such as freezing or vaporization), and the viscosity can change to adversely affect the fluid's flow patterns.

Heat tracing is a way of maintaining the temperature variation in a pipe within an acceptable range. Figure 1 shows a typical insulated pipe with a desired fluid temperature of 50°F (10°C). When the ambient temperature is, say 10°F (–12°C), the insulated pipe loses heat through thermal insulation. The amount of heat lost, of course, depends on the difference in the pipe and ambient temperatures, pipe diameter, and the type and thickness of thermal insulation. In order to hold the pipe temperature at 50°F, an external heater is required on the pipe under the insulation to compensate for the heat loss. This external heater is typically called a tracer.

When the fluids don't move

Stagnant fluids pose a different heat-tracing problem from that of flowing fluids. When the fluid is flowing, the residence time for the fluid is normally so short that the temperature variation due to heat loss via the usual heat sinks is negligible. Although the heat losses are not uniform, the temperature variation in the system is negligible due to the low residence time (Figure 2).

However, when the fluid stops flowing, the variation in temperature is directly proportional to the localized heat losses at various points. The heat losses through such heat sinks as valves, piping supports and instruments are difficult to estimate accurately because the losses are sensitive to the particular layout of the installed systems. Figure 3 shows typical temperature variations at the piping supports and valves relative to the main pipe.

Thermal insulation typically prevents 80–90% of the heat loss from a pipe, and external heaters compensate for the remaining 10–20% to maintain the desired temperature. Figure 4 compares typical net heat losses for bare and insulated pipes.

Further, minor deterioration in the thickness of the insulation or the so-called k-factor of the insulation can have a large impact on the heating or cooling requirements. For example, a 10–20% deterioration in insulation performance may call for almost double the amount of external heating or cooling to maintain the pipe's temperature. Therefore, the importance of proper design and installation of thermal insulation in the overall system performance cannot be over-emphasized.

Attending to smaller pipes

Because of their lower thermal inertia, small-diameter pipes cool down or heat up much more quickly than large-diameter pipes. Figure 5 shows the typical relative cooldown times for pipes with various diameters.

Another key variable is the prevailing temperature of the pipe. Because of the greater temperature differential, high-temperature pipes cool down much more quickly than low-temperature pipes (Figure 6). Therefore, high-temperature pipes must be heat-traced more accurately and provided with higher safety margins for heat losses and quality of thermal insulation.

How to prevent freezing

Freeze protection of water lines has been the most common application of electric heat tracing. The volumetric expansion during the change of phase from liquid to ice can cause bursting of pipes and instruments. Without proper heat tracing

Originally published May 1995

and thermal insulation, freezing can occur at any point in a system, especially near a valve or a support. To allow sufficient safety margin, pipes are typically maintained at around 40°F (5°C).

Minimizing temperature swings

The first step in managing temperature variations is to calculate the heat loss for the pipes and heat sinks within a reasonable tolerance. The second step is to compensate for these heat losses by installing an adequate amount of thermal insulation. These steps are particularly critical near the heat sinks, such as valves and supports, because they have parts that are always exposed to the atmosphere and can have high heat losses due to fin effect.

Next, one needs to select an electric heater cable that can minimize temperature variation. Self-regulating heaters are electrically parallel in construction (Figure 7) and continuously vary their heat output in response to variation in the temperature of the pipe, support or a valve.

If the temperature drops below a set level, the heater automatically increases its heat output, thereby reducing overall temperature variation in the system (Figures 8 and 9). If an extra amount of heat is required near the heat sinks, the self-regulating heaters (Figure 9) can be overlapped without causing problems of overheating or burnout. Further, they can be cut to length in the field. Thus, they provide extra flexibility and safety margins for the overall system performance. As a rule, self-regulating heaters are preferred because they can compensate for heat loss more rapidly, compared with constant-wattage units (Figure 10).

Reliability means 'less is more'

Generally, more components in a system lead to less reliability because overall reliability of a system is a cumulative multiplier of the individual reliability of each component. For example, if there are six components in a system operating in series, each offering a 90% reliability, then the cumulative reliability of the system is $(0.9)^6$ or 53%. If the number of components is reduced to 3, the reliability improves to $(0.9)^3$ or 73%.

(Continued)

Controlling fluid temperature in piping networks minimizes flow problems

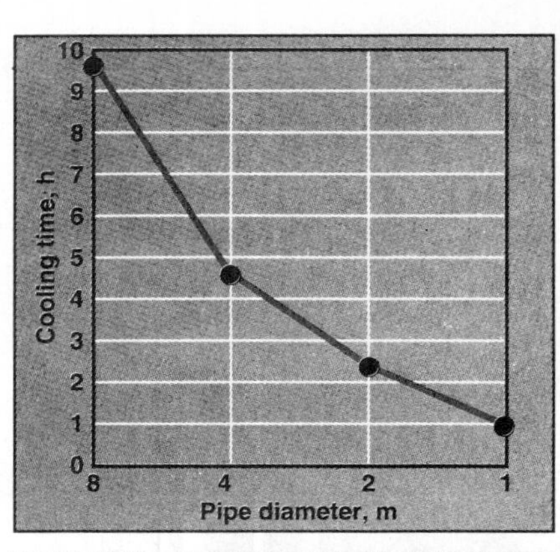

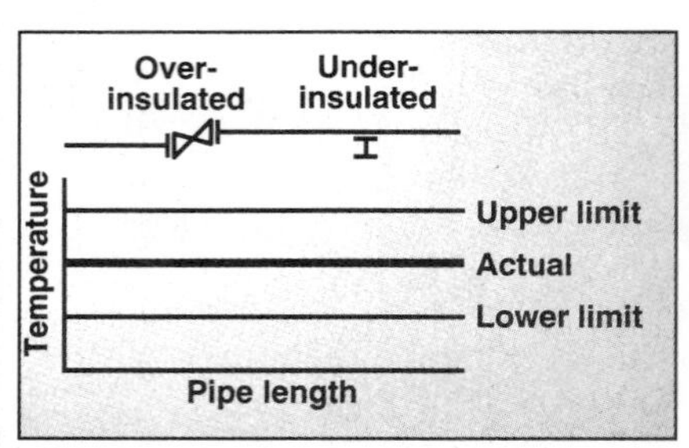

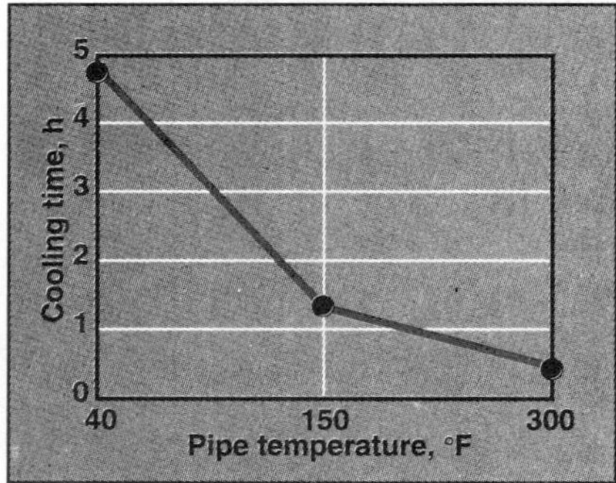

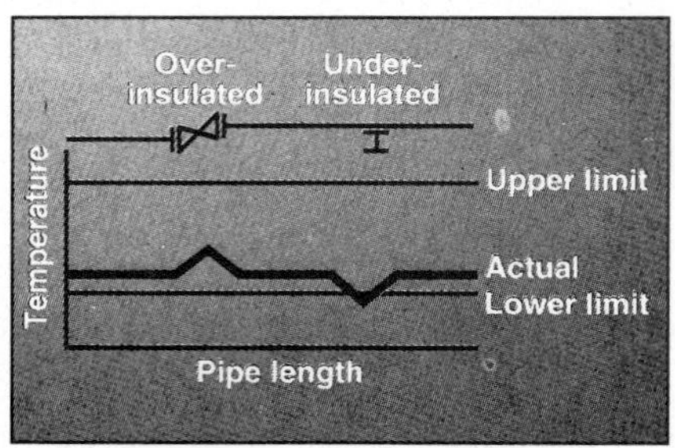

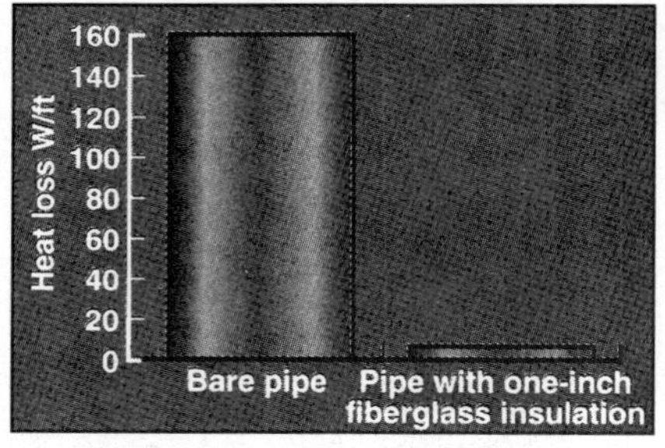

FIGURE 1 (facing page). Insulation is critical to success with heat tracing

FIGURE 2 (left top). A flowing pipe sustains a relatively flat temperature profile

FIGURE 3 left middle). Without heat tracing, a stagnant pipe undergoes temperature variation

FIGURE 4 (left). Heat loss is significant without adequate insulation

FIGURE 5 (right, upper). Heat tracing must be carefully designed to accommodate the fast cooling of small-dia. pipes

FIGURE 6 (right, lower). Heat tracing is more challenging for hotter pipes, which cool faster than cooler pipes

FIGURE 7 (below). Self-regulating heaters adjust their heat inputs in response to their heating requirements

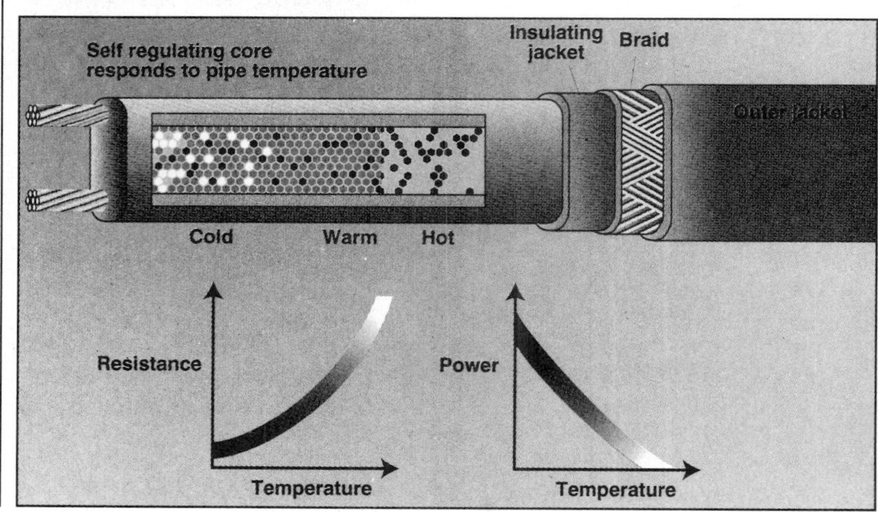

Point sensing and control

When designing for on-off temperature control using what is known as the point-sensing method, the different flow paths in a complex piping network must be considered separately, and individual heater circuits must be provided for each flow path (Figure 11). Point sensing is the most common technique used for control of the temperature of fluids in pipes. It involves separating the flow versus no-flow segments of the piping system, installing a sensor for each segment, and using the signal from each sensor to turn the heater on or off for individual circuits (Figure 12).

The sensor should be located on the straight section of the pipe away from the heat sinks. The frequency of sensors is generally one per control segment because additional sensors typically do not improve the overall reliability.

The method is effective for simple, long runs of pipes but increases the number of circuits substantially for complex piping systems. This attribute of the point-sensing method not only leads to much higher installed costs but also reduced reliability for complex systems. However, this method is more energy efficient, compared with the other two methods elaborated below.

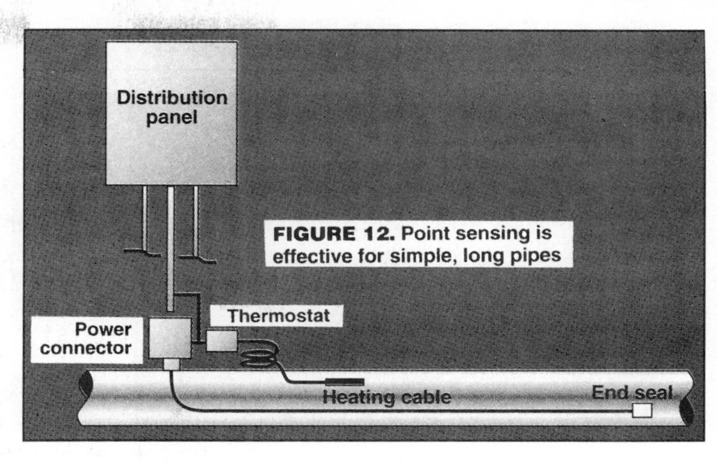

FIGURE 12. Point sensing is effective for simple, long pipes

Ambient sensing and control

Ambient sensing is the most common technique used for protection of piping systems against freezing. The method involves the use of an ambient sensor that turns all heat tracing circuits on or off. All heaters operate simultaneously as soon as the ambient temperature is below the set point. The amount of heat provided to each pipe is proportional to its heat loss, regardless of whether the pipe contains flowing or stagnant fluids (Figure 13).

The ambient sensing method minimizes the number of circuits as well as the installed cost. However, energy consumption is greater, relative to the point-sensing method, unless self-regulating heaters are used.

Dead-leg sensing and control

The dead-leg technique is a relatively new technique. It is simple to design and leads to fewer circuits, producing higher reliability and lower installed costs. Temperature ranges obtained with this technique can approach that of point sensing. This is perhaps the best method for freeze protection as well as broad-range temperature control of complex piping systems.

The dead-leg sensing technique involves selecting the pipe with the smallest diameter in a piping network that must be maintained at a given temperature. This smallest-diameter pipe must be in permanent, no-flow condition. This can be a non-flowing section of a piping network or a separately constructed piece of pipe. The pipe length is typically called a "dead leg" because of its no-flow condition.

A sensor and a controller are used to control the desired temperature on the dead leg. Then, all heater circuits installed on other pipes with different diameters in the network are turned on or off through a contactor based on the signal and the set point of the controller on the dead leg. All heating cables remain on until the dead leg meets its temperature requirements, regardless of the flow conditions of the different pipe sections (Figures 14 and 15).

The smallest-diameter pipe is selected because it offers the highest frequency of cycling. If properly designed, the dead-leg control method provides fairly uniform temperature within the required temperature range for the selected portion of a network (Figure 16). To minimize wide temperature variation among various pipes, self-regulating heaters are recommended.

Estimating the total installed cost

For a point-sensing thermostatic control system, the typical costs include materials and labor to install the heating cable, terminations for the heater, conduit and wiring for power supply, distribution panels, and control and monitoring equipment (Figure 17). The

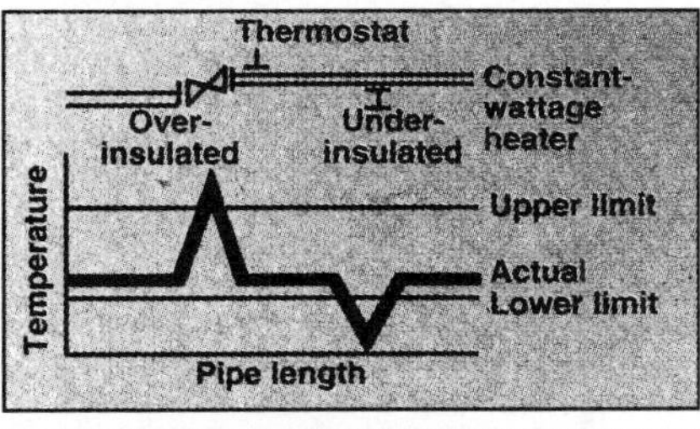

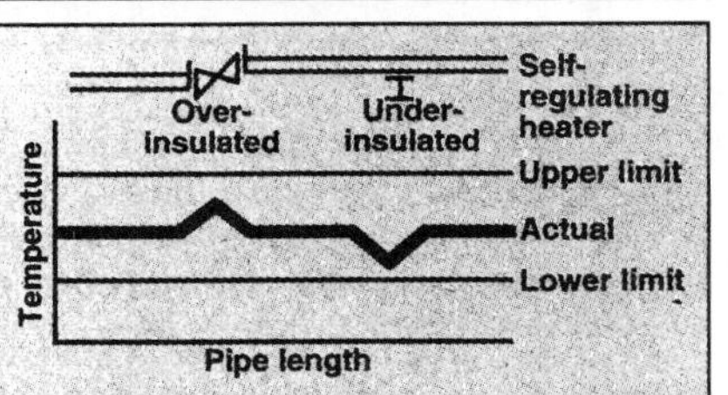

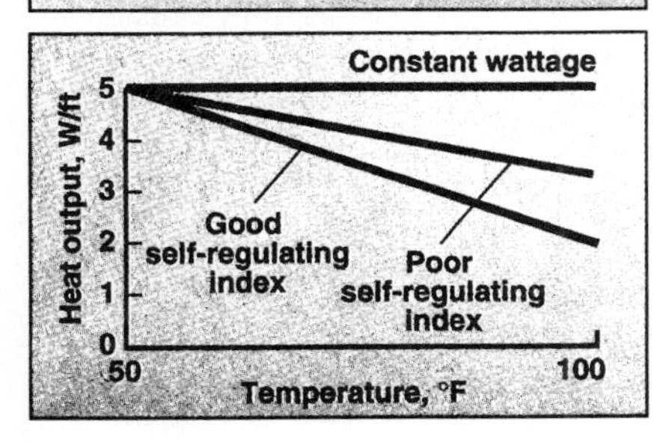

FIGURE 8 (left, top). There are significant temperature swings with constant-wattage heaters

FIGURE 9 (left, middle). Self-regulating heaters minimize temperature swings

FIGURE 10 (left, bottom). With a higher self-regulating index, a heat tracer can handle wider temperature variations

FIGURE 11 (below). The point-sensing method uses thermostatic control at various points in a piping network

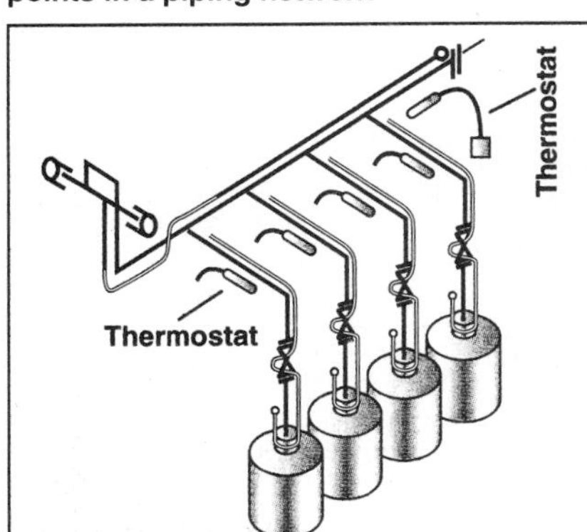

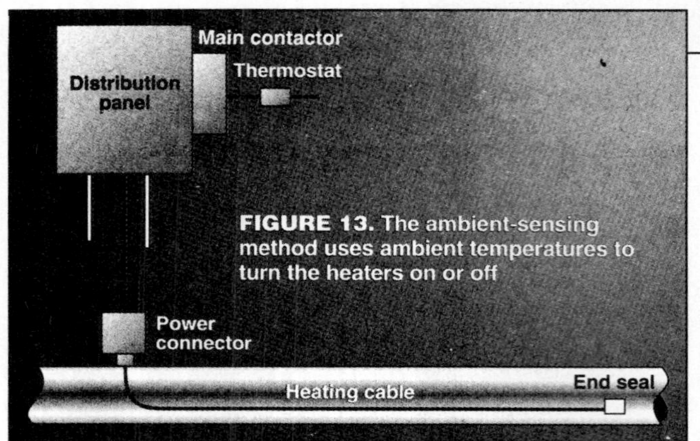

FIGURE 13. The ambient-sensing method uses ambient temperatures to turn the heaters on or off

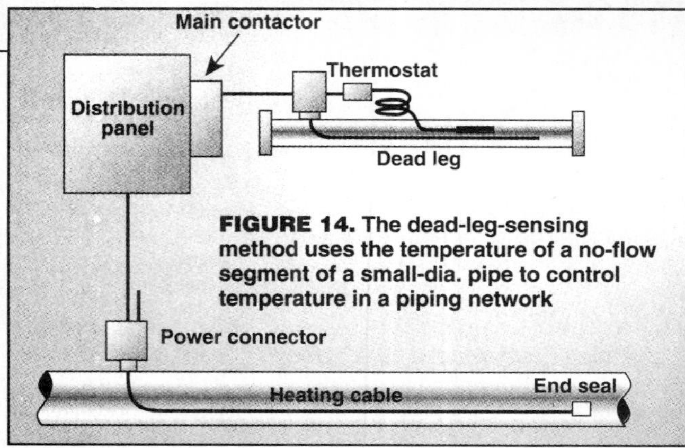

FIGURE 14. The dead-leg-sensing method uses the temperature of a no-flow segment of a small-dia. pipe to control temperature in a piping network

heater costs include the heating cable itself and installation labor. This cost is proportional to the length of pipe and amount of heat loss. A typical way to optimize this cost is by reducing the heat loss and using high-power heating cables if multiple cables are required.

Electrical distribution, terminations and monitoring costs depend on the number of power points or circuits required to design the system. To optimize this part of the cost, the number of circuits must be minimized.

One must also include the cost of design. Heat-tracing systems are typically designed on a per-circuit basis. Therefore, this cost is also proportional to the number of electrical circuits. A simple breakdown of costs shows that the bulk of the total installed cost is directly proportional to the number of heat-tracing circuits used in the system.

Comparing typical costs

For complex piping systems, the total installed cost of electrical heat-tracing systems largely depends on the choice of the control method. The point-sensing method of control requires the separation of flow paths and thereby increases the number of power-supply points, number of components, wires and conduits and number of controllers. This leads to much higher installed cost, compared with the ambient-sensing and dead-leg methods (Figure 18).

Next, one must consider the two components of operating cost: the cost of energy consumption; and the cost of maintenance. The cost of maintenance for systems utilizing ambient sensing or dead-leg sensing methods is much lower than the point-sensing method due to fewer components and the simplicity of the system. However, the energy consumption can be higher, depending on the amount of time the pipes carry flowing or stagnant fluids.

Dead-leg sensing systems save substantial amounts on the installed cost. They can also offer greater energy efficiency for freeze-protection applications, compared with ambient-sensing systems. Overall, reliability of heat tracing is much higher with dead-leg sensing systems due to the simplification of design, reduced number of components, flexibility of centralized monitoring of pipe temperatures, and electrical integrity. ■

Edited by G. Sam Samdani

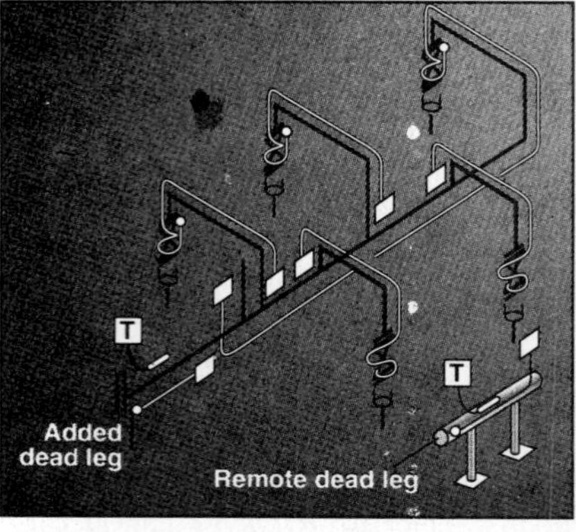

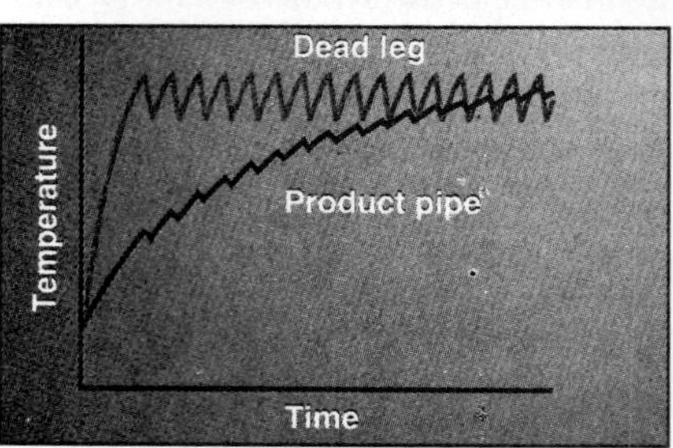

FIGURE 15 (left).
All heaters stay turned on until the dead leg reaches the required temperature

FIGURE 16 (left, below).
With dead-leg sensing, a pipe quickly approaches the control temperature

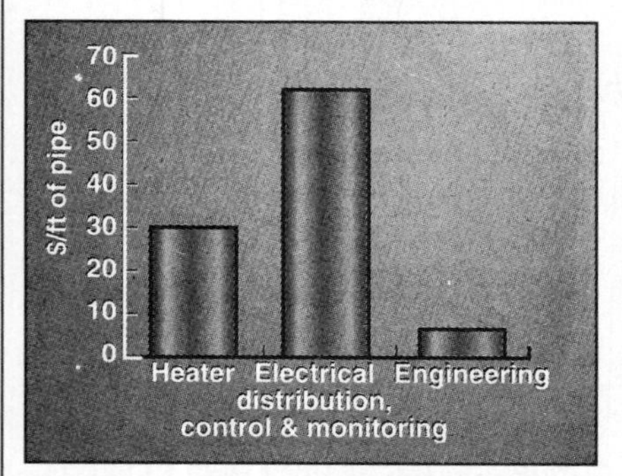

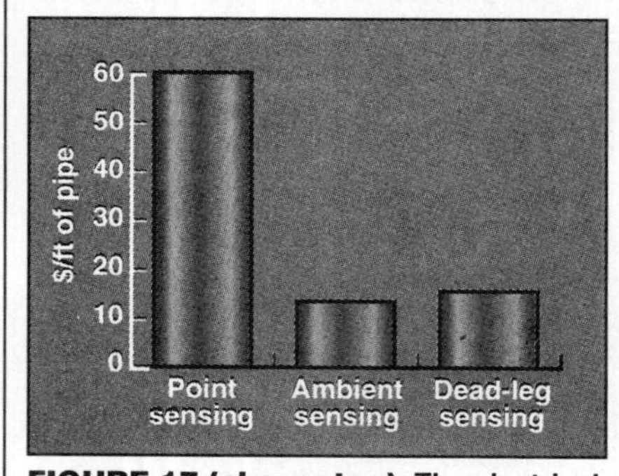

FIGURE 17 (above, top). The electrical distribution, control and monitoring elements make up a major portion of the total cost for a point-sensing system

FIGURE 18 (above, bottom). Installation of the point-sensing method is the most expensive among the three electrical heat-tracing alternatives

Authors

Dinesh C. Gupta is a senior marketing manager with the heat-tracing products division of Raychem Corp. (300 Constitution Drive, Menlo Park, CA 94025-1164; tel: (415) 361-2722; fax: (415) 361-6036). He received a bachelor's degree in mechanical engineering from the University of Delhi, India, and an MBA from the University of Santa Clara, California. He was plant manager for a large chemical facility before joining Raychem in 1978 as engineering manager. He has been involved with the design and operation of complex heat-tracing systems for over 15 years.

Nicholas (Bob) Rafferty is an engineering technical associate with E.I. du Pont de Nemours and Co.'s Engineering Div. in Newark, Del. He is the electrical resource for the corporate Electrical Technology Consulting Group and has over 25 years experience across a broad range of electrical heating assignments in construction, plant maintenance, project engineering and consulting.

HOT TIPS TO REJUVENATE OLD BOILERS

New refractories extend boiler life and increase efficiency by cutting leaks

A. L. Brandstatter
and Howard Sawatzki

Many steam boilers built before 1960 were constructed so well that they not only remain in service today, but give little indication that they might be nearing the end of their useful life. Many of the pre-1960 water-wall boiler designs were built with pressure components rated so conservatively that tubing and drum wall thicknesses easily remain within original specifications today.

But, as with all process equipment, age takes its toll on steam boilers. The efficiency of these aging boilers drops, resulting in higher emissions that can be of concern to regulators.

In these boilers, the drop in efficiency primarily results from air leaking into the furnace and heat leaking out of the furnace. In turn, both of these leaks are related to the condition of the refractories inside the furnace.

The role of refractories as protective linings is well understood. As such, the potential for catastrophic failure of refractory materials gets the immediate attention of plant engineers.

However, the deterioration of refractories over time is not often viewed as a factor in a resulting drop in furnace efficiency. In fact, even when engineers are aware of the link, the losses in efficiency tend to creep up so gradually that they seldom seem to warrant immediate attention or to draw funding from other types of plant improvement that are more directly related to increasing production.

Fortunately, the efficiency losses of older boilers can be corrected at a much lower cost than by replacing the units with newer, more-efficient models. In fact, by correcting age-related boiler problems engineers can often raise furnace efficiency to the point that emissions of certain gases are also minimized. At the same time, this rejuvenation restores the boiler's original steam-generating capacity, enabling it to support further production increases without requiring any expansion of the power plant.

Excess air is critical

As concerns with fuel economy and the environment continue to grow, engineers are paying more and more attention to the level of excess air in a furnace. That's because each fuel combusts most efficiently at a specific concentration of excess air: For example, coal-fired boilers need approximately 30 to 35% excess air, while oil-based units need about 15%; and natural-gas ones about 10%.

Too much excess air wastes heat because the furnace has more air to heat up. High air levels also increase nitrogen-based emissions (because the primary source of nitrogen is the air).

In contrast, too little excess air can lead to incomplete combustion, which increases carbon losses either through the ash pit or through undesirable levels of particulate entrained in the fluegas. Aside from complying with stack-opacity regulations, the control of particulates is essential to prevent undue erosion and clogging of critical areas between the furnace and the stack.

The presence of too much excess air often is first revealed by simple fluegas temperature readings. The temperature of the gas at the furnace outlet – before the gas reaches air preheaters, economizer, scrubbers and precipitators – should be close to 550°F.

Temperature readings significantly lower than 550°F can reflect the cooling effect of high levels of excess air. Gas temperatures significantly higher can indicate failure or shrinkage of refractory baffles, which can result in the hot furnace gases bypassing some of the heat-transfer surfaces. High temperatures also signal the buildup of scale, slag or particulates on boiler tubes to the extent of inhibiting heat transfer.

The actual level of excess air is determined with an Orsat or other type of gas analyzer that draws samples of the fluegas and measures the percentage by volume of oxygen and carbon diox-

FIGURE 1. Aging boilers can be rejuvenated with new refractories. Here, shredded plastics materials are propelled by compressed air and blown onto the host surface, where they build up snugly around prefired refractory anchors to a uniform thickness

Originally published September 1994

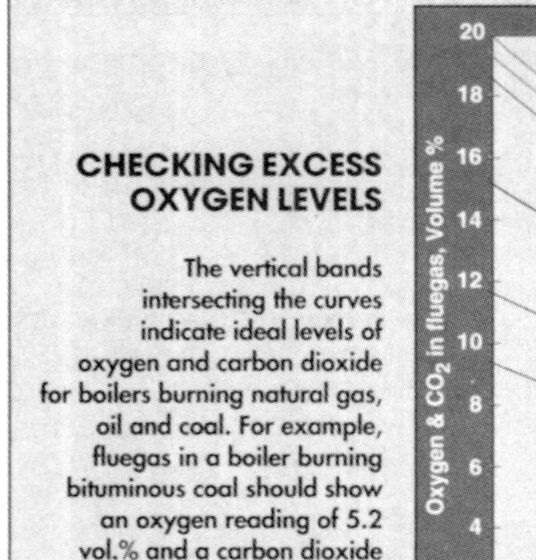

CHECKING EXCESS OXYGEN LEVELS

The vertical bands intersecting the curves indicate ideal levels of oxygen and carbon dioxide for boilers burning natural gas, oil and coal. For example, fluegas in a boiler burning bituminous coal should show an oxygen reading of 5.2 vol.% and a carbon dioxide reading of 14 vol.%; these readings indicate excess air levels between 30 and 35%

STEVE HAIMOWITZ

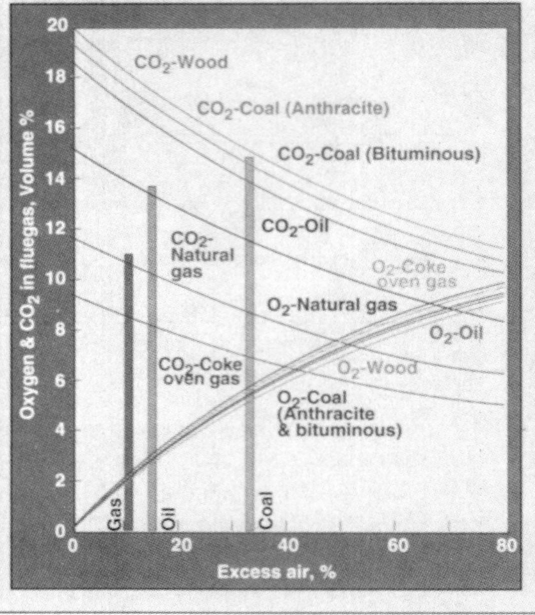

ide. Those measured values are then compared with established curves for different kinds of fuels to pinpoint the percent of excess air (box, above).

The presence of too little excess air (and, sometimes, the presence of too much) can be corrected by adjusting and balancing the forced draft (FD) and induced draft (ID) levels of a furnace. Thus, the first response to an excess-air problem should always be to make sure that the FD and ID fans and dampers are properly adjusted.

Cracks admit tramp air

If, however, the excess air level is still high after the furnace's FD and ID adjustments are verified, it is very often a signal that "tramp" air is leaking into the furnace. In older boiler furnaces, the presence of tramp air usually indicates that the refractories are cracked or deteriorating.

Typically, cracks appear in brick or tile refractories along mortar lines and expansion joints. In rammed plastic refractories, cracks form along lamination lines between slabs of material that were not rammed thoroughly enough into a true monolithic mass.

Actually, cracks do more than let in excess air – they also permit flyash to infiltrate the lining and buildup behind the refractory. This buildup of flyash can eventually grow sufficiently to generate a pressure strong enough to distort the furnace's exterior plate and collapse the refractory into the furnace.

Since the cracks open a path to the exterior steel, their location can be identified as hot spots whose temperatures can be measured on the furnace's exterior. The spots not only add to heat losses, but can become a safety hazard for operating personnel.

Refractories can cure leakage

New refractory construction techniques offer several ways to counteract excess-air problems. One method is to use monolithic linings that have no mortar or lamination lines to initiate cracking, and which require no expansion joints. Such monolithic refractories require significantly less installation downtime than is needed for traditional bricklaying or plastic-ramming methods.

One monolithic construction alternative is the installation of plastic refractories by pneumatic gunning (Figure 1). Plastic gunning material is a soft, somewhat sticky mixture of clay, aggregates and binders (but no cement). It is shipped either in factory-granulated form or as extruded slugs that are shredded at the job site.

The granular or shredded material is loaded into the gun-feed system and propelled by compressed air through a hose across distances of up to 250 ft. At the host surface, the material is blown out of the hose and built up to the desired lining thickness.

The host surface – sometimes this is the structural steel of the furnace – is fitted with pre-fired refractory anchors that are typically spaced on 12- to 18-in. centers. Any open areas behind the anchor may be backed with a temporary lattice of wood strips or a plywood sheet. The gunned material then builds up snugly around and between these anchors, and uses their length as a guide for uniform lining thickness.

Another monolithic construction uses a castable material that is mixed like concrete and poured into forms, similar to the pouring of a building foundation (Figure 2). This approach is most often specified where gunned plastics cannot be used, for example when replacing the lining between the water-wall tubes and the outer insulation layer, or when replacing linings where the workspace is severely restricted. (Note, however, that if the boiler's outer steel casing and insulation are to be replaced, and if there are no workspace limitations, then plastic refractories can be gunned against the tubes from the outside before the insulation and steel plate are replaced.)

Before installing castable linings,

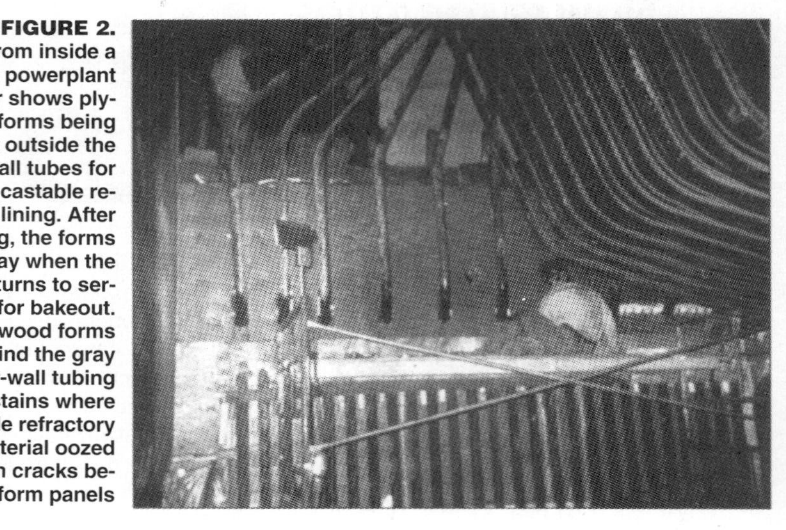

FIGURE 2. A view from inside a large powerplant boiler shows plywood forms being built outside the water-wall tubes for pouring castable refractory lining. After curing, the forms burn away when the boiler returns to service for bakeout. The plywood forms behind the gray water-wall tubing show stains where castable refractory material oozed through cracks between form panels

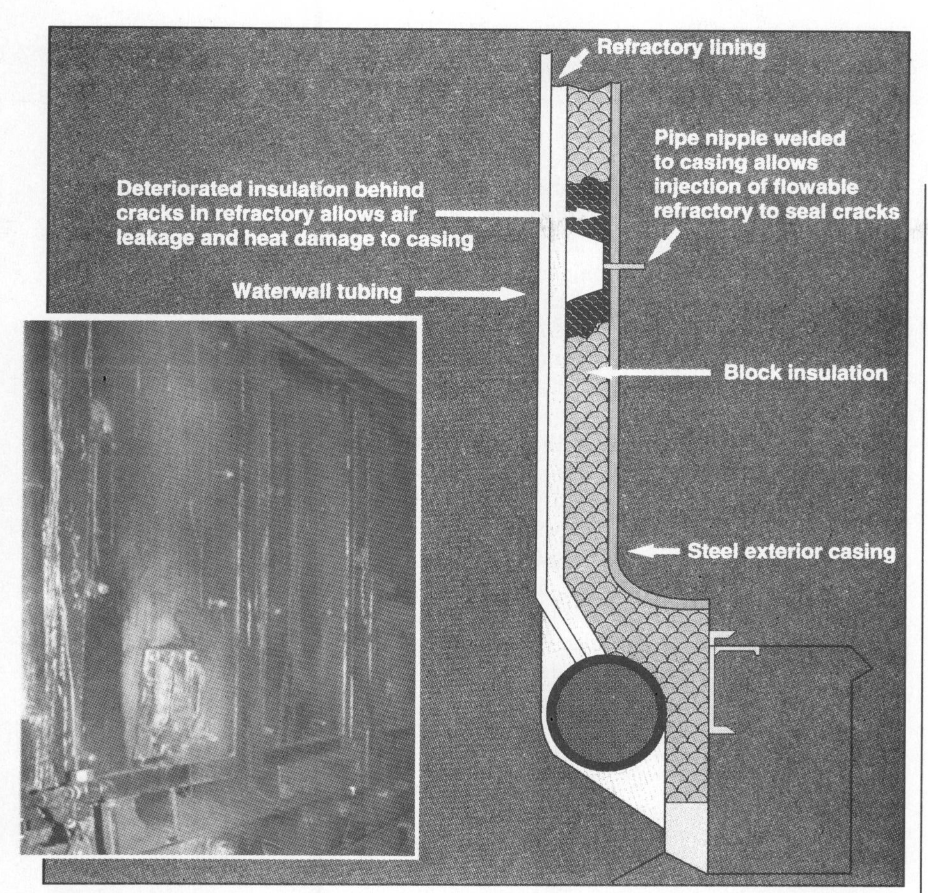

FIGURE 3. The exterior of an old boiler wall with nipples welded to the outer steel-plate casing on 18-in. centers. The nipples enable the flowable refractory to be injected into the inner cracks and voids, to seal air leaks without shutting down the boiler. The cross-section shows the relative positions of the deteriorated refractory and pipe nipples

wooden mold forms must first be built at the site. After pouring, the material is allowed to cure for at least 24 h. Then the forms are removed and the lining is dried by bringing the boiler up to steam load at a slow, controlled rate. In a water-wall installation, the wood cannot be removed from between the refractory and the tubes, and so is burned away during bakeout.

Plastic gunning methods require less downtime than a rammed plastic installation by allowing easier delivery of refractory materials into the confined spaces that are typical of furnaces and boilers. The technique is also far less labor-intensive than ramming, and less sensitive to fatigue-related errors.

Plastic gunning refractories generally install faster than castables (and don't require the forms needed for castables), and are easier to apply on a furnace's contoured shapes, such as ceilings and arches. Most importantly, plastic refractories allow bake-out to begin immediately after the gunning is finished (thus minimizing downtime), and they offer better thermal shock resistance than castable materials.

Certain castable mixes also can be applied by pneumatic gunning, saving time for building and removing forms. However, they require longer bakeout schedules, and they have a low thermal shock tolerance, typical of poured castables. With both castables and gunned plastics, the shrinkage that occurs during bake-out is nullified by the material's thermal expansion in operation, thus eliminating expansion joints.

Minigun systems

Plastic gunning has traditionally been restricted to relatively high-volume refractory installations. But, recently, "mini" systems to make gunned plastic refractories feasible for smaller-scale projects – usually those installations requiring less than 15 tons of refractory – are becoming available.

As a downsized system, the hoses cannot reach as far as those of the bigger systems, but the more-compact support machinery can, in many cases, move closer to the application site. In addition, mini systems make gunning more competitive for repairing walls, ignition arches and baffles of power-plant boilers, where routine maintenance may require masons to ram plastic in selective locations.

A third and relatively new alternative for quickly repairing and sealing cracked, delaminated and leaking refractory walls is injection of refractories. Since the service can be performed from the outside of the furnace, the boiler remains online (Figure 3).

First, the locations of hotspots are detected on the outer steel casing by infrared or contact sensing. The hotspots locations guide the positioning of pipe nipples, which are welded directly to the casing. Holes then are drilled through the casing inside each nipple. Finally, a hose attached to the nipple pumps a soft, flowable refractory material into the void behind the casing where it fills the cracks and hardens into an air-tight seal. The effectiveness of the method can be monitored by taking measurements of the oxygen content of the furnace as the leaks plug up.

Injectable refractories provide permanent sealing of air-leaking cracks in otherwise sound linings. Further, they can be used as a low-cost way of temporarily staving off costly replacement of deteriorating furnace walls. ■

Edited by Jayadev Chowdhury

The authors

A.L. Brandstatter has devoted his entire career to refractories, having joined Plibrico Co. (1800 North Kingsbury St., Chicago, IL 60614; Tel: (312-549-7014) upon graduation in 1951 from the Illinois Institute of Technology. In 1974, he was named to his present position of market manager of boilers and incinerators, where he coordinates all boiler and incinerator activities, guides development of new refractory products, and works with the construction and engineering departments in the installation of refractories. He is a member of the American Soc. of Mechanical Engineers (ASME).

Howard R. Sawatzki is an independent consulting engineer with his own firm, Howard Engineering Co. (Glenwood, Ill.), since 1986. Previously, he held managerial, design and production engineering positions with Wickes Boiler Co. (Saginaw, Mich.), Combustion Engineering Corp. (Saginaw), Bigelow Co. (New Haven, Conn.), Lasker Boiler and Engineering Corp. (Chicago) and Solid Fuels, Inc. (Hinsdale, Ill.). He holds a B.S. in mechanical engineering from Michigan State University, and was elected to Tau Beta Pi, the National Engineering Soc. He is a registered professional engineer in Connecticut, and a member of ASME.

BEWARE OF CONDENSER

Brad Buecker
Burns & McDonnell

Many chemical process plants generate steam for power production and process use. Recovering this steam as condensate, and returning it to the boiler, is an economical way to recycle heat. This is usually done in a water-cooled, steam-surface condenser located at the exhaust of a turbine. Poor performance of such a condenser — which is really a large heat exchanger — can significantly decrease a plant's heat-recycling efficiency.

The most-common causes of condenser inefficiency are: microbiological growth on the water side, scale formation on the water side, tube pluggage by debris and air in-leakage on the steam side. These problems can cost a plant dearly. Well-planned treatment programs to combat these causes, however, pay for themselves many times over.

What happens in a condenser

Ideally, when steam leaves the turbine of a condenser, it has used all of its available heat for work and is saturated. The condenser uses cooling water to remove the steam's latent heat and convert the steam to condensate.

Condenser pressure depends on cooling-water temperature, which, under normal circumstances, is low enough to produce a strong vacuum. As the following hypothetical example shows, this is important thermodynamically:

1. Steam at a pressure of 1,000 psig (1,014.7 psia) and temperature of 950°F enters a turbine
2. The turbine is adiabiatic and reversible, and has no bleed lines for feedwater heating

Steam tables [1] show that the enthalpy of the turbine inlet steam is 1,477 Btu/lb and the entropy is 1.6325 Btu/(lb)(°R). Because the turbine is adiabatic and reversible, note that the entropy of the exhaust steam is the same as that of the throttle steam, since no heat is transferred during the reversible expansion.

Accordingly, we can calculate the enthalpy of the exhaust steam for various conditions. In the hypothetical turbine, if the steam is taken from the turbine exhaust at atmospheric pressure (14.7 psia), its enthalpy is about 1,081 Btu/lb. Thus, 396 Btu/lb of heat is available for work. The enthalpy of the condensate at these conditions is 180 Btu/lb. Turbine efficiency is given by

$$\frac{h_{\text{turbine inlet}} - h_{\text{turbine outlet}}}{h_{\text{turbine inlet}} - h_{\text{condensate}}} \times 100$$

Thus, (1,477–1,081)/(1,477 – 180) × 100 = 33%.

If, however, all incoming steam exhausts into a vacuum, the situation changes noticeably. With relatively cool circulating water, absolute condenser pressures can reach as low as 0.5 in. Hg (0.3 psia). The enthalpy of steam exhausting into these conditions is about 843 Btu/lb, and the condensate enthalpy is reduced to 27 Btu/lb. The amount of heat available for work is 1,477 – 843 = 634 Btu/lb. Thus, the turbine efficiency would equal (1,477 – 843)/(1,477 – 27) × 100 = 44%, a drastic increase over the previous illustration.*

This example is exaggerated, as no utility turbine is reversible or designed to exhaust to atmospheric conditions. Also, some steam is typically extracted from the turbine for feedwater heating, so an increase in condensate temperature would decrease the quantity of steam needed for feedwater heating. Thus, higher condensate temperatures would alter the heat transfer in feedwater heaters.

*Too much cooling, or "condensate subcooling," also decreases efficiency. However, this phenomenon is usually less serious than tube fouling.

However, the quantity of steam extracted for feedwater heating is much smaller than that passing through the condenser. So, for the purpose of our discussion, the difference between this hypothetical situation and an actual turbine is negligible.

Consider a condenser in which tube fouling has increased the backpressure from 0.5 in. Hg (0.3 psia) to 2.5 in. Hg (1.5 psia). The enthalpy of the exhaust steam at 2.5 in. is 922 Btu/lb and the condensate enthalpy is 77 Btu/lb. The enthalpy at 0.5 in. Hg is 843 Btu/lb. The efficiency is (1,477 – 922)/(1,477 – 77) × 100 = 40%. Thus, almost 80 Btu/lb of work are lost, and the turbine efficiency drops from 44% to 40%.

Turbine efficiencies of 44% and 40% equate to turbine heat rates of, respectively, 7,750 and 8,530 Btu/kWh. For a power production of 250 MW, the increase in heat rate results in an increased heat input of 1.4×10^{11} Btu. If the fuel is coal with a heating content of 10,000 Btu/lb, a utility would have to fire an extra 7,000 tons of coal per month to offset the drop in efficiency. The costs of tube fouling can be considerable.

Microbiological fouling

Condensers, often a good location for microbial growth, are warm and provide a constant source of microbial introduction (Figure 1). In addition, they are usually located so far downstream from a point of biocide injection point that any biocide that actually reaches

Originally published April 1995

Bacteria, scale, debris and leakage are the usual suspects

organisms is too dilute to be effective.

Many bacteria colonies that form in condenser tubes secrete a gelatinous film that binds them — and a lot of other material, including silt — to the tube walls. These films not only retard heat transfer, but can promote corrosion of tube metal. Once formed, the colonies or their secretions can be difficult to remove chemically.

The author once shock-treated a condenser with chlorine to remove colonies that had coated virtually all of the tubes. The condenser only regained 50% of its lost efficiency, indicating that much material still remained. Only by mechanically scraping the tubes were plant personnel able to restore the condenser to normal performance. Analyses taken of the deposits

Detroit Edison

Nalco Chemical Co.

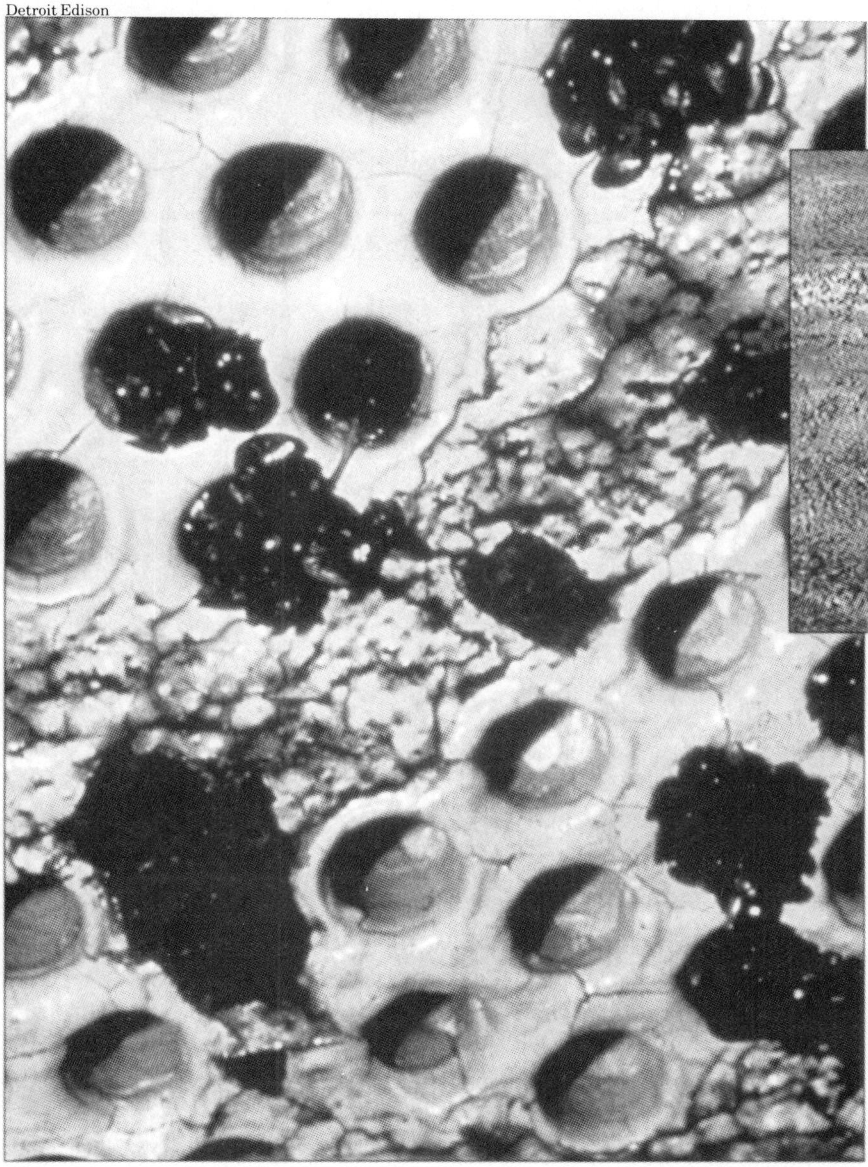

FIGURE 1. Closeup of a condenser tube (below) shows thick, dried microbiological growth and silt. **FIGURE 2.** Zebra mussels (left) on a condenser tubesheet are a fouling nightmare

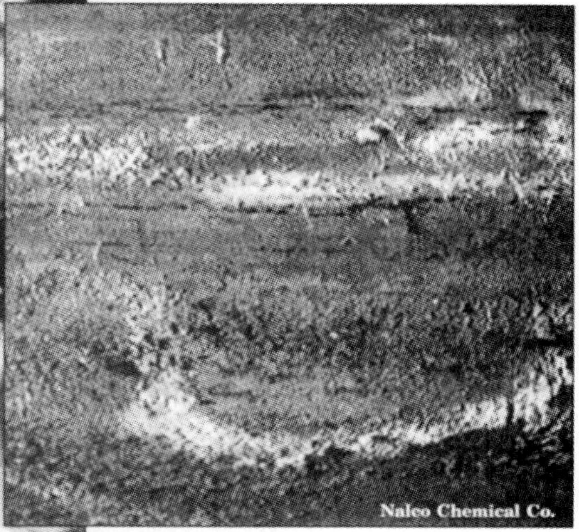

FOULING

before tube cleaning indicated organism counts as high as 100,000,000 in a few grams of sample.

Microbiological fouling is usually not difficult to detect. A rise in the condenser's terminal temperature difference (TTD) often indicates this type of fouling. The TTD is the difference between the temperature of the condensing steam and that of the circulating water at its outlet.

When condenser waterbox doors are opened for inspection, spongy buildups on the waterbox internals and the tubesheet typically indicate microbial growth. The tubes often have accumulated circumferential deposits of silt that are held in place by the secreted matter. The material looks and behaves much like mud.

Chemical treatment is the best method to prevent microbe buildups, but treatment programs must be set up carefully to be effective. For many years, chlorine was the top choice for treatment programs. However, chlorine's use is questioned because the substance reacts with organics to form chlorinated compounds, which include trichloromethane — a trihaloorganic (THO) that is a suspected carcinogen.

The U.S. Environmental Protection Agency (EPA), in revamping the Clean Water Act, may include provisions to restrict or ban the use of chlorine. Such regulations would force industry to use other chemicals, such as chlorine dioxide (ClO_2), ozone or nonoxidizing biocides, none of which form THOs. These chemicals, like chlorine, can be effective, though their costs are usually higher.

Often, however, the chemical is injected too far upstream of the condenser and has lost its strength by the time it's most needed. A method called targeted treatment — injecting chemical at strategic locations along the condenser inlet tubesheet — is proving useful. The biggest drawback of this technique is that it typically costs more than conventional systems.

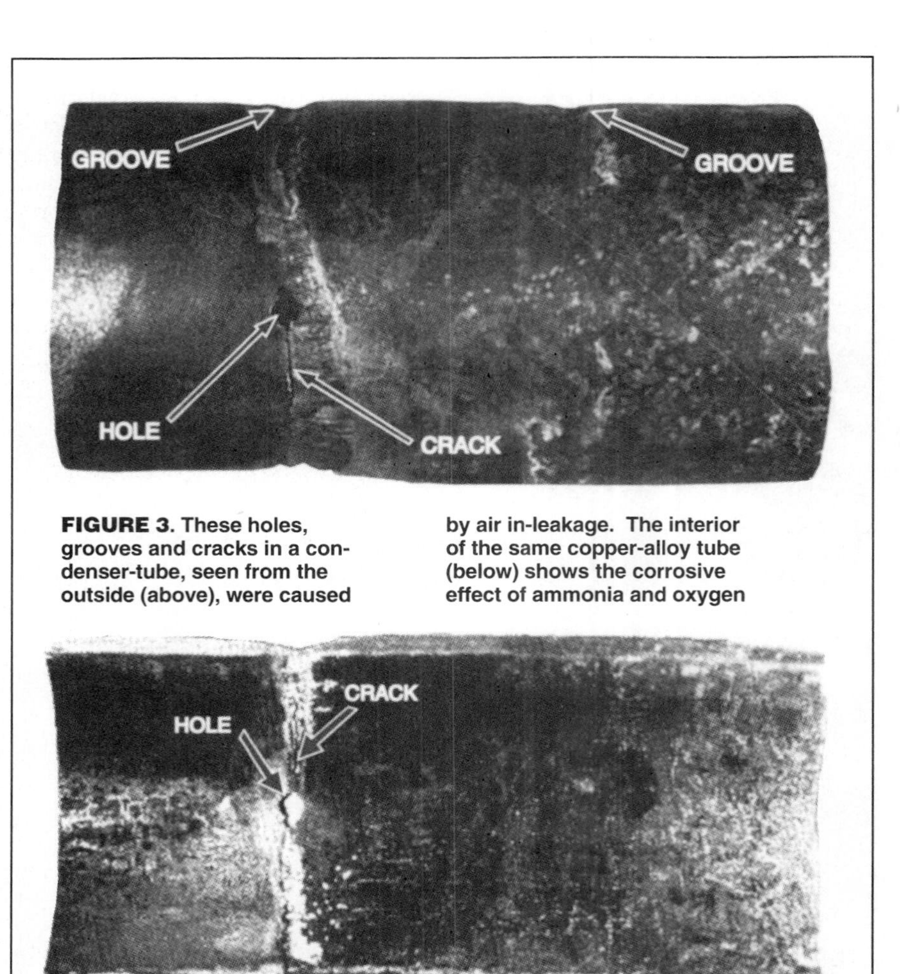

FIGURE 3. These holes, grooves and cracks in a condenser-tube, seen from the outside (above), were caused by air in-leakage. The interior of the same copper-alloy tube (below) shows the corrosive effect of ammonia and oxygen

City Water Light & Power

Scaling

This is a problem that is most frequently seen in recirculating systems. In recirculating systems, mineral compounds in cooling water concentrate several times above makeup levels, often approaching saturation levels. The water then passes through the condenser, becoming heated in the process. Many compounds, most notably calcium carbonate ($CaCO_3$), become less water-soluble as the temperature rises.

When a compound in recirculating water reaches its saturation point in a condenser, the compound precipitates and forms a layer of scale on the tubes. Calcium carbonate scaling is the most common type; however, calcium sulfate and phosphate, manganese compounds and silicates can also be a nuisance.

Although scaling is usually seen in recirculating systems, it sometimes occurs in once-through condensers. For example, an incidence of $CaCO_3$ condenser scaling was due to a drought. The condenser received its cooling water from a lake that lost much water to evaporation, which caused a large buildup in mineral content.

Scale looks different from microbiological buildups. Because the phenomenon involves crystal growth, the deposits are harder and often smoother. Calcium carbonate scale frequently appears as a brown layer because of impurities in the crystal structure. Manganese deposits are often black.

The effects of scaling are similar to those of microbe fouling: The condenser gradually drops off in performance. Scale deposits are harder to remove, however. Calcium carbonate and calcium sulfate can be removed by mechanical scraping, but usually some debris remains. For manganese and silica deposits, or deposits in tough-to-reach areas, chemical cleaning may be the only solution. Calcium carbonate will dissolve in acid, but other deposits require more-exotic treatments, such as ethylenediaminetetraacetic acid (EDTA) or acids containing fluorides.

A number of chemicals are used for on-line scale treatment. Sometimes adding sulfuric acid to cooling water will keep the pH low enough to prevent calcium carbonate deposition. Often, other chemicals must be added to cooling water. For example, phosphates and phosphonates tie up calcium ions and prevent them from forming scale products, and some polymers modify scaling compounds' crystal structures, preventing them from depositing on the tube walls. No generic treatment can be recommended. Each facility must be evaluated separately, and a custom treatment program developed for each.

Air in-leakage

Utility condensers generate a strong vacuum, which makes some air in-leakage inevitable. Thus, they are equipped with air-removal systems to remove gases. But, if the air in-leakage is too great, the equipment cannot handle the

FOUL PLAY: ZEBRA MUSSELS

The zebra mussel has two characteristics that make it particularly threatening to condensers. First, young mussels are microscopic, and can pass through conventional intake screens. Second, as the mussels develop, they produce anchors (byssal threads) that attach to surfaces. Unless the surrounding water velocity is very high, or tube surfaces are very smooth, mussels will attach to nearly anything, including other mussels. Colonies can become very thick. Since their introduction into the U.S. in 1986, the mussels have been spreading, and range from Michigan in the north to New Orleans in the South.

Utility and industrial personnel unprepared for this pest often discover their problem only when equipment begins to malfunction. By then, mechanical removal of mussels (scraping or water blasting; Figure 4) is about the only viable option. Thus, engineers would be well advised to be prepared to deal with mussels before they arrive.

A chemical-treatment program may be the most economical measure. However, select any treatment with care. Zebra mussels sense some biocides, particularly chlorine, and close when they detect it (Figure 5). They can remain closed for up to three weeks. The most effective chemical-treatment programs are those that kill young zebra mussels (veligers), which are susceptible to biocides. Thus, if adult mussels can be prevented from entering a cooling system, periodic biocide treatment will control colony formation.

An alternative method, albeit an expensive one, is heat treatment. Zebra mussels cannot tolerate temperatures much higher than 100°F. If the inlet water can be heated to this temperature for just a few hours, the heat destroys both juvenile and adult mussels. Unfortunately, circulating water flow rates are usually high, so much effort is needed to heat water to lethal temperatures. The most effective means of heating inlet water is to recirculate condenser discharge.

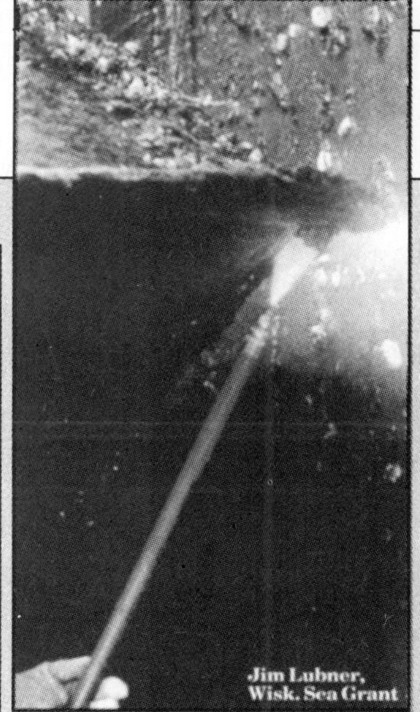

Jim Lubner, Wisk. Sea Grant

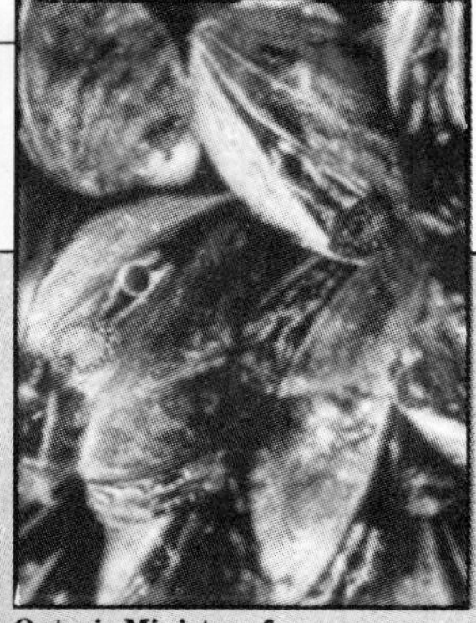

Ontario Ministry of Natural Resources

FIGURE 4. High-pressure waterblasting (left) is one way to get rid of zebra mussels. FIGURE 5. Mussels open to filter water for nutrients; however, they can stay closed for weeks if they detect biocides

Not all plants are equipped with the proper piping arrangement for recirculation. The capital costs to install this or other heating systems are usually high.

Research continues on methods to control zebra mussels. For instance, some reports indicate that carbon dioxide anesthetizes zebra mussels, upon which they become much more susceptible to biocides. Some paints are toxic to mussels — though also harmful to other aquatic life.

Also, a number of coatings are being developed to provide a surface too smooth for mussel attachment. The two organizations below can provide information on mussels and potential treatment programs. To receive this data, circle the appropriate number on the Reader Service card. ❒

Zebra Mussel Clearinghouse **625**
Michigan Sea Grant **626**

extra load. In this case, excess air coats the tubes and inhibits heat transfer.

This problem is quite common. Typical points of air in-leakage include cracks in an expansion joint between a turbine and condenser, cracks in a condenser shell at condensate-return lines, explosion diaphragms on a low-pressure turbine, low-pressure heater vents, and condensate pump seals.

The author observed a condenser that showed good efficiency at high load, and poor efficiency at low load. Because condenser efficiency changed so drastically depending on load, plant workers realized that the condition was unrelated to tube fouling, but was likely due to air in-leakage.

The problem was traced to a faulty trap on a condensate-return line from a gland steam exhauster. The trap had stuck open; but at high loads, enough condensate flowed through the line to minimize air leakage. At low loads, however, the line was only partially full and the condenser vacuum pulled air back through the exhaust vent. Once the trap was repaired, the efficiency problems vanished.

As the above example indicates, an air leak often causes a sudden change in condenser performance, which differentiates it from tube fouling or scale formation. For instance, if an explosion diaphragm fails, the TTD may rise a couple of degrees in a short time. Excess air in-leakage can frequently be verified by examining the discharge from the air-removal system.

A general rule of thumb is that the air-removal rate from a unit should be less than or equal to 1 scfm/100 MW. A condenser-shell failure or other type of air in-leakage problem will often increase this removal rate several fold.

Air leaks should be located and repaired as quickly as possible. Not only does in-leakage cause performance problems, but the excess oxygen entering the system can corrode the condensate-and-feedwater lines. In the presence of ammonia, oxygen can attack copper-based heater and condenser tubes (Figure 3).

Ammonia corrosion is most often prevalent in and below air-removal compartments. Ammonia concentrates in these areas and dissolves in the condensate. The condensate then runs down tube-support plates.

When excess air is present from in-leakage, ammonia and oxygen attack copper-bearing tubes at a tubesheet interface. This localized attack can thus cause condenser tubes to fail long before their expected tube life is reached.

A common technique used to locate air leaks is helium leak detection. A helium detector is installed at the discharge of the condenser's air exhauster. Helium is sprayed at various points around the condenser. A leak pulls the helium into the system, upon which it is collected by the air-removal equipment and eventually discharged through the air evacuator. The helium concentrations detected at exhaust vents help determine the size of the leak.

Tube pluggage

Another common problem occurs when leaves, vegetation and other kinds of aquatic life pass through worn trash screens and build up on the inlet tube sheet. Such problems can usually be cured with normal equipment repair.

More difficult to combat are aquatic organisms that are originally small enough to pass through screens, but grow to a problem size inside the circulating water system (Figure 2 and box, p. 141). Mollusks often grow to a size that fits neatly into condenser tubes. Removal of these pests is frequently time-consuming because some travel into the tubes before becoming lodged. They then must be removed physically.

Programs for condensers

Good monitoring techniques help detect the hazards listed above. One simple method is monitoring the TTD. As tubes begin to foul, less heat is transferred from the condensing steam to the cooling water. Thus, steam condensate temperatures grow warmer as outlet-water temperatures grow cooler. Plotted over time, the TTD can be a good indicator of tube fouling.

Computer programs are also useful. One program, developed by the author and co-workers with information supplied by the Heat Exchange Institute (Cleveland, Ohio) and the General Physics Corp. (Columbia, Md.), uses the TTD and other easily determined variables to calculate a condenser cleanliness factor. This program, in Basic, is effective in monitoring condenser performance, and has been used to detect microbiological fouling, tube scaling and air in-leakage. ■

Edited by Irene Kim

Reference

1. Steam Tables, Keenan, J. H., Keyes, F. G., Hill, P.G., Moore, J. G.; New York: John Wiley & Sons Inc., 1969

Author

Brad Buecker is a senior chemist at Burns & McDonnell Engineering (4800 E. 63rd St., Kansas City MO 64130; Tel. 816-333-4375). He formerly worked as an analytical chemist, flue-gas desulfurization (FGD) process and project engineer and results engineer for City Water, Light & Power (Springfield, Ill.). He has a B.S. in chemistry from Iowa State University and an Associate of Arts degree in engineering from Springfield College (Springfield). Buecker has written or coauthored 16 articles on boiler water chemistry, analytical and process chemistry, corrosion and computer programs for boiler calculations. He received an award from the Electric Power Research Institute (EPRI) for work on reducing scale buildups via gypsum recycling.

INNOVATIVE BAFFLES COUNTER HEAT-EXCHANGER VIBRATION

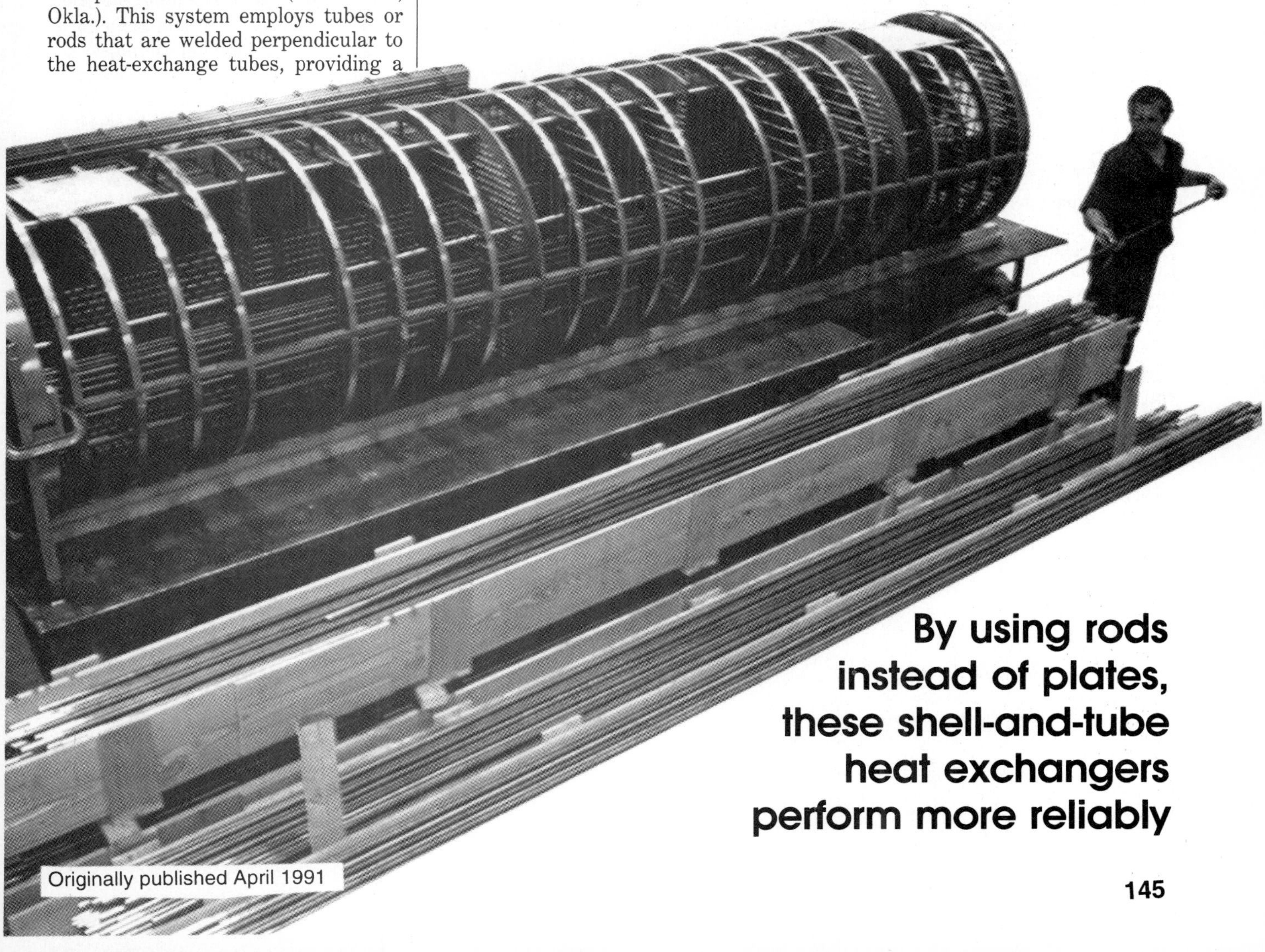

By using rods instead of plates, these shell-and-tube heat exchangers perform more reliably

In heavy-duty shell-and-tube heat exchangers, vibration can be a constant worry. With high volumes of fluids passing through rod bundles, and temperatures varying from one side of the exchanger to the other, vibration can set in, causing gaps to form between the plates and the rods. When materials that can react with each other are present that begin leaking through the gaps, unintended reactions can occur that can seriously harm the heat exchanger.

Both these problems existed at the heat exchanger trains of P. T. Badak LNG Co., whose natural gas liquefaction plant in Bontang, East Kalimantan, Borneo, is one of the world's largest. Starting in 1984, and winding up this year, most of the existing heat exchangers will have had their internal baffles replaced with the RODbaffle system, a technology developed by Phillips Petroleum Co. (Bartlesville, Okla.). This system employs tubes or rods that are welded perpendicular to the heat-exchange tubes, providing a

Originally published April 1991

Long titanium tubes are threaded between the stainless steel rods of the baffle support

high degree of stability to the exchanger while minimizing pressure drop or heat-transfer problems.

The Indonesian plant contains five trains of liquefaction units, each with four heat exchangers using sea water on the (cooling) tubeside and gas containing monoethanolamine on the shellside. The amine was present to sweeten the gas by removing carbon dioxide. Each exchanger is 16 ft long and 45 in. dia., containing 944 tubes. Originally, the tubes were constructed of a copper-nickel alloy.

When the exchangers were put into service initially, vibration set in, causing leaks so that the higher-pressure gas-amine stream would enter the tubeside. The resulting increase in pH caused calcium, magnesium and aluminum salts to drop out of solution, clogging the tubes with an extremely hard scale. Frequent shutdown and extensive cleaning were required.

The switch to the RODbaffle design virtually eliminates vibration. The rods are designed to lock tubes into place so that they cannot vibrate laterally. While changing the original tubes and baffles, Phillips and its equipment contractor, Titanium Ltd. (St. Laurent, Quebec) were instructed to switch to titanium tubes and stainless steel RODbaffle supports. The titanium provides better corrosion resistance under the harsh processing conditions.

Because the new tubes and baffles were a retrofit, Phillips was constrained by the existing design parameters, and chose to retain four of the original plate baffles in the tube bundle. Up to 20 RODbaffle units are used in each bundle. "Had we designed these as RODbaffle heat exchangers from scratch, we probably would have used longer, smaller-diameter bundles to capitalize on the RODBaffle design's low pressure drop characteristics," says C. C. Gentry, senior engineering associate at Phillips. In other applications where RODbaffles have been used, pressure drops have been cut by 50%, and overall heat transfer efficiency improved by 1.0%.

Working with titanium is not a trivial task; for example, all welding must be performed under an inert-gas atmosphere, because of the reactivity of the metal and oxygen. According to Phillips, Titanium Ltd. was able to provide excellent fabrication services, including in-house production of its own pipe, welding wire and small-diameter rods, all of titanium. —*Phillips Petroleum Co., Bartlesville, Okla* **463**
—*Titanium Ltd., St. Laurent, Quebec, Canada* **464 ■**

The proper condensate-removal system will make conventional steam-flow control work

OPERATING STEAM-HEATED EXCHANGERS AT HIGH DOWNTURN

STEPHEN WALSH

Albert Armer, Spirax Sarco Inc.

Shell-and-tube heat exchangers are common in the chemical process industries. Yet controlling their operation still presents problems at turndown operation — i.e., operation at the "ragged edge" of temperature control when flows have been cut back substantially.

Turndown limitations have led to recommendations that simple steam-inlet control and condensate removal through steam traps (Figure 1) be replaced with condensate-flow control (Figure 2) [*1,2*]. However, the following detailed analysis of condensate-flow control will show that it is not a panacea. A better approach to operating with steam-flow control is then presented that will not only enhance an exchanger's performance at turndown operation but also extend its service life.

The usual arrangement for condensate-flow control is shown in Fig. 2. If, however, the temperature controller (TIC) directly regulates Valve B, an increase in steam demand could open the valve wide even if only a little condensate were held back in the exchanger. Steam would then be lost to the condensate-return system.

A condensate receiver, with its pressure balanced against that of the supply steam and with a level controller regulating Valve B (as in Figure 1), would prevent steam from blowing through the exchanger and condensate receiver. If fluctuations in steam demand actuate Valve A and change the condensate flowrate, the condensate level in the exchanger's steam space will rise and fall. Full steam pressure in the exchanger would keep the condensate flowing even if the steam pressure in the exchanger approached the pressure in the condensate-return line.

However, the response of the temperature controller in this arrangement can be unacceptably sluggish because of the slowness of changing the condensate level in the exchanger. In one case, it took about 15 minutes to satisfy a three-fold increase in heat-load demand, and about 45 minutes to reach a comparable decrease [*1*].

If the Figure 2 exchanger has a shell of 17¼-in. I.D. and 124 1-in.-dia. tubes, the maximum rate of steam condensation in it would be about 3,280 lb/h. Valve A would have to be large enough to pass this flow with a head of h, which might provide a differential of 1 psi. A valve having a flow coefficient (C_v) of 7.4, which would pass 3,700 lb/h of steam, might be selected.

At half load (steam flow of 1,640 lb/h), a decrease in loading that would

Originally published December 1990

close Valve B would fill the exchanger with condensate at a rate of about 1,640 lb/h, flooding 5% of the tube area in about 18 seconds. An increase in load that would fully open Valve B would expose 5% of the tube area in about 22 seconds. At 90% load, however, closing Valve A raises the condensate level in the exchanger at the rate of 2,952 lb/h, and a 5% covering of the tubes would occur in about 10 seconds. Conversely, the wide opening of Valve B would lower the condensate level at about 748 lb/h, and the tube-surface area would increase 5% in about 42 seconds.

Obviously, the exchanger's performance depends considerably on the maximum capacity of Valve B and on exchanger geometry (e.g., tube diameter and arrangement) and, without taking instrument lag time into account. Clearly, the condensate-control scheme is likely to result in temperature control that, for at least part of the time, is going to be little better than that with steam-flow control. At high loads, when the steam pressure in the exchanger will be high enough to clear the condensate, the exchanger's performance would actually be poorer with condensate-flow control.

There are other disadvantages to condensate-flow control. Because of the changing condensate level in the exchanger, some tubes are alternately exposed to steam and and cooled by the fluid inside. The resulting expansion and contraction of the tubes will shorten tube service life and cause joints to leak sooner. Condensate leaving the exchanger may be at a temperature well below saturation; if it contacts uncondensable gases, it can quickly dissolve them and become aggressively corrosive. If the steam is in the tubes, water hammer usually occurs in them at the condensate surface.

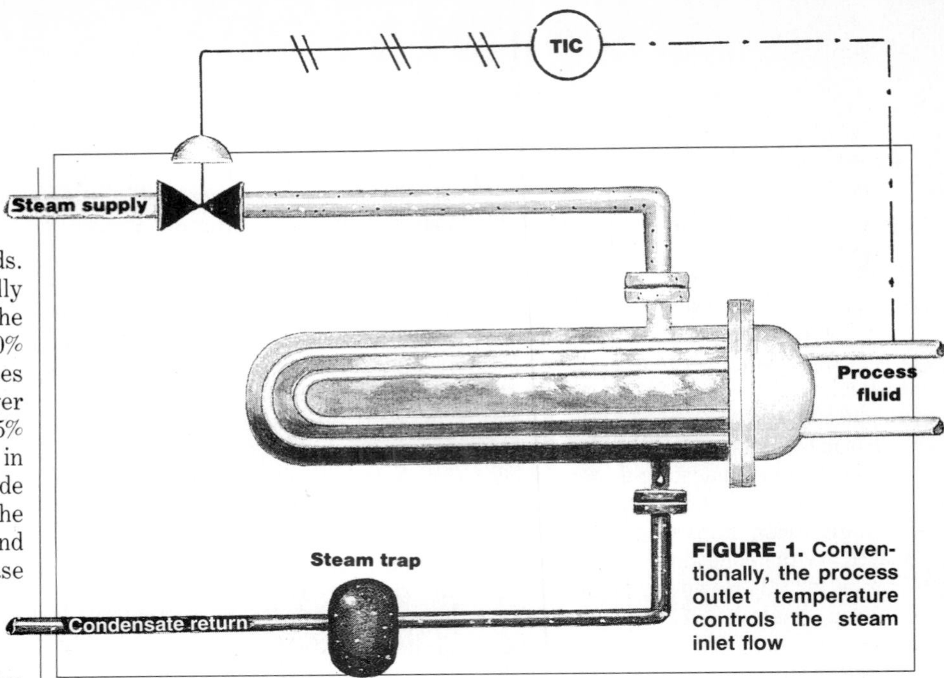

FIGURE 1. Conventionally, the process outlet temperature controls the steam inlet flow

An alternative

At maximum load, or at a reduced load within the useful turndown band, the steam pressure in an exchanger will usually be high enough to push condensate through a steam trap and into the condensate-recovery system. Only if the steam loading is very high and a trap of sufficient capacity is not available might it be advisable to consider installing a condensate receiver with a level-control valve.

Even in such an unusual circumstance, however, one ought to look at draining a receiver by means of a float-operated steam trap (Figure 3). The pressure in the receiver can be balanced via a line back to the steam space of the exchanger, which would effectively make the receiver a part of the

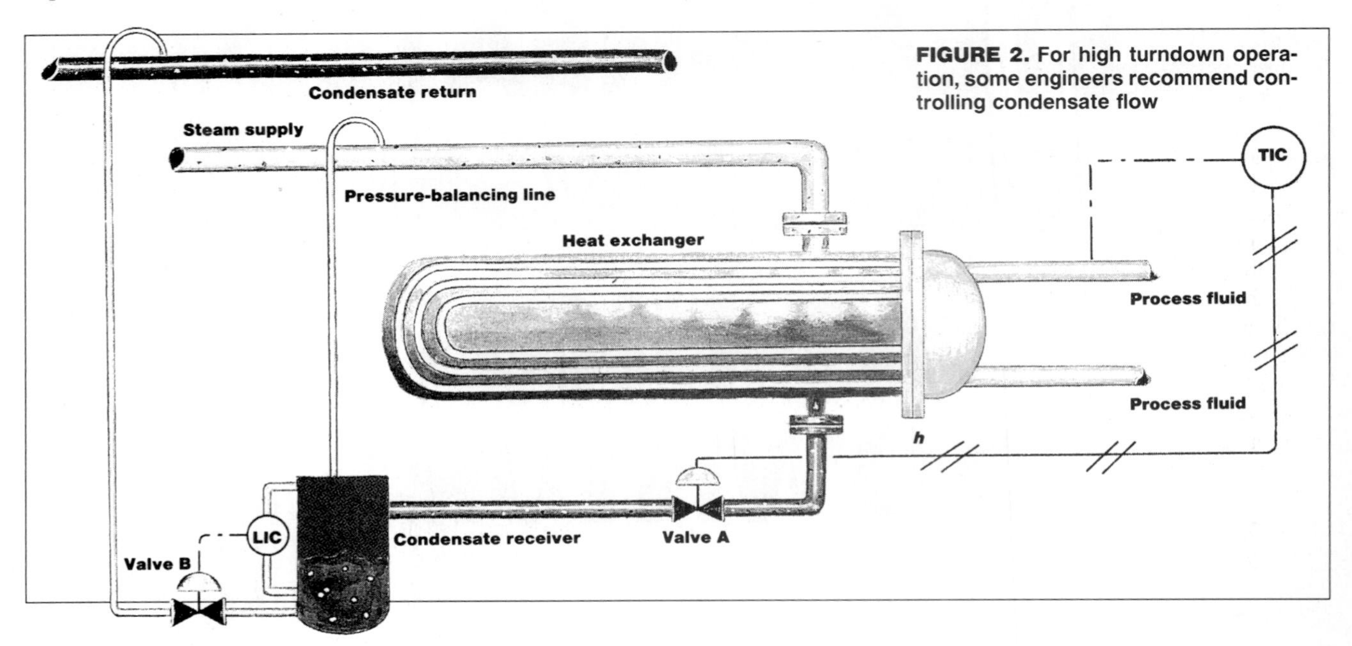

FIGURE 2. For high turndown operation, some engineers recommend controlling condensate flow

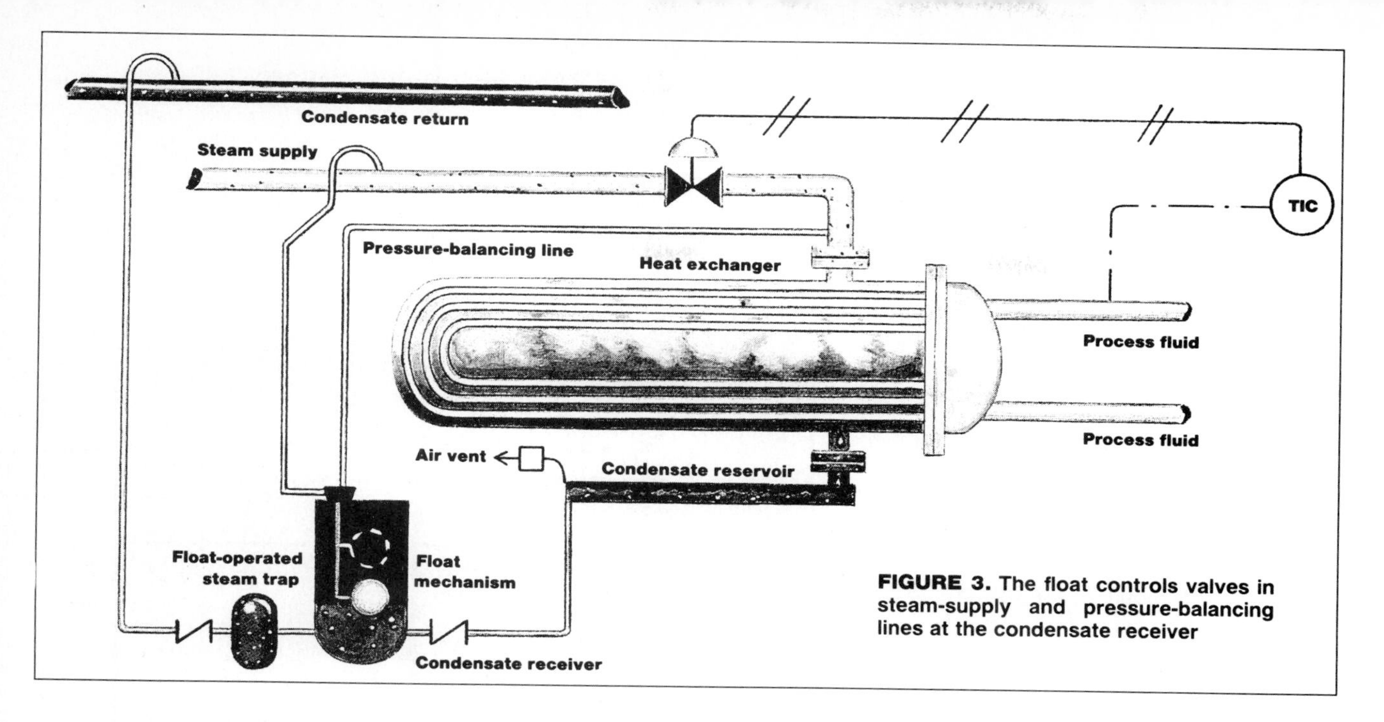

FIGURE 3. The float controls valves in steam-supply and pressure-balancing lines at the condensate receiver

exchanger's steam space. Condensate formed in the steam space would then readily fall to the lowest point in the system, which would now be the receiver. Draining the condensate through a steam trap will keep an exchanger clear of condensate and functioning optimally.

At light loading, the pressure in the steam space may approach or fall below the stalling pressure — i.e., the backpressure from the condensate-return system plus the differential head needed to push the lower condensate load through the steam trap. This problem can be countered by fitting a float-operated trip mechanism in the condensate receiver (Figure 3).

The mechanism remains inactive as long as condensate passes freely across the bottom of the receiver and through the steam trap. When, however, the exchanger load falls to the stalling point and the receiver becomes flooded, the mechanism trips, closing a valve to the balance line and opening a valve in the steam-supply line (Figure 3).

The steam flowing in via the supply line to the receiver raises the pressure in the receiver, pushing the condensate through the steam trap. After the condensate level drops to the tripping level, the mechanism trips in the opposite direction, closing the steam valve and opening the valve that discharges through the pressure-balancing line to the heat exchanger. During the discharge phase of the cycle, the check valve admitting condensate to the receiver stays closed. A reservoir between the exchanger and the receiver ensures that condensate does not temporarily flood the exchanger.

Such a receiver, with a tripping mechanism and steam and check valves, is standard equipment for pumping condensate with an atmospheric-venting control scheme. Used with a float trap in an unvented hookup, it ensures that condensate will be removed from the exchanger at all loads and operating pressures. The cost of pumping is virtually zero because the heat content of the motive steam is recovered in the exchanger.

Predict turndown availability

Condensate pumping is not necessary if the turndown required of an exchanger never lowers the steam pressure to the stalling point. The exchanger steam pressure simply pushes the condensate through a steam trap. Of course, if the sum of the differential pressure across the trap and the backpressure exceeds the pressure of the steam in the exchanger, a steam trap would serve no useful purpose even at maximum load. Only condensate pumping is required.

Between these extremes are a range of possibilities. At high loads, a steam trap prevents the loss of steam to the return system. At low loads, condensate must be pumped into the return system.

The temperature of the steam in an exchanger (T_{2m}) for any load factor M, can be calculated via Mathur's correlation [1]:

$$T_{2M} = [(T_3 + T_4)/2][(1 - M)/100] + = M\,T_2/100 \quad \textbf{(1)}$$

Equation (1) is similar to the correlation of Pathak and Rattan [2]:

$$T_{2f} = 0.5\,(T_3 + T_4)(1 - f) + fT \quad \textbf{(2)}$$

In Equation (2), f is the fraction of the design heat load (comparable to Mathur's M), and T is temperature. Both correlations are applicable when the process-side inlet and outlet temperatures are constant but the process flowrate varies.

Figure 4 provides the load factor M, and shows the performance of an exchanger whose steam flow is controlled. (Figure 4 is comparable to Mathur's Figure 3.) The steam temperature at 100% loading appears on the ordinate at T_2. It may be below the temperature of the supply steam by an amount that depends on the pressure drop through the control valve when it is passing 100% of the steam load.

The process inlet and outlet temperatures, T_3 and T_4, respectively, also appear on the ordinate, the first at 100% load and the second at 0% load. The midpoint of T_3 and T_4 also appears on the ordinate at the 50% loading — i.e., $(T_3 + T_4)/2$. A horizontal line from this point intersects the T_3-T_4 line at a load factor of 50%. A diagonal line to T_2 tracks the steam temperature at load factors less than 100%, with a cutoff line indicated at T_4. If the steam temperature is not at T_4, the process fluid can never reach that temperature.

A horizontal line from temperature

T_{bp} (the temperature at the sum of the trap differential pressure and backpressure intersects the steam temperature track at the stalling condition (i.e., normal flow through the steam trap stops). From Figure 4, steam temperature at the stalling condition is given by:

$$T_{bp} = (T_3 + T_4)/2 + (M/100) \times [T_2 - (T_3 + T_4)/2] \quad (3)$$

$$T_{bp} = (T_3 + T_4)/2 + M\,T_2/100 - (M/100)[(T_3 + T_4)/2] \quad (4)$$

$$T_{bp} = [1 - (M/100)][(T_3 + T_4)/2] + MT_2/100 \quad (5)$$

Note that Equation (5) is the same as Equation (1).

The load factor, M, is given by:

$$M\% = \frac{t_{bp} - [T_3 + T_4)/2}{T_2 - [T_3 + T_4)/2} (100) \quad (6)$$

If the process flowrate is constant but its inlet temperature varies, the steam temperature at varying loads is tracked by the line joining T_2 and T_4 in Figure 4. Many air heater coils operate in this way, with the inlet temperature changing with the season but with a fan providing a substantially constant flowrate. In such a case, the load factor M is given by:

$$M\% = [(T_{bp} - T_4)/(T_2 - T_4)]100 \quad (7)$$

Conventional arrangements

Heat exchangers are sometimes located high enough so that condensate can drain by gravity from an exchanger to a steam trap, and from the steam trap to a receiver vented to atmosphere. A condensate pump then transfers the condensate from the receiver and, usually, returns it to a deaerator. Figure 4 and Equations (6) and (7) are valid for such an arrangement.

In such a case, the backpressure on the steam trap obviously would be zero. Stalling would occur at the pressure at which the differential across the trap equals the steam pressure. If the trap were located low enough to provide a hydrostatic head equal to the trap differential pressure, the exchanger could function smoothly down to atmospheric pressure.

A further turndown will require a vacuum breaker, which must be carefully located in the system. It would be ineffective if fitted to the steam trap.

CARLA MAGAZINO

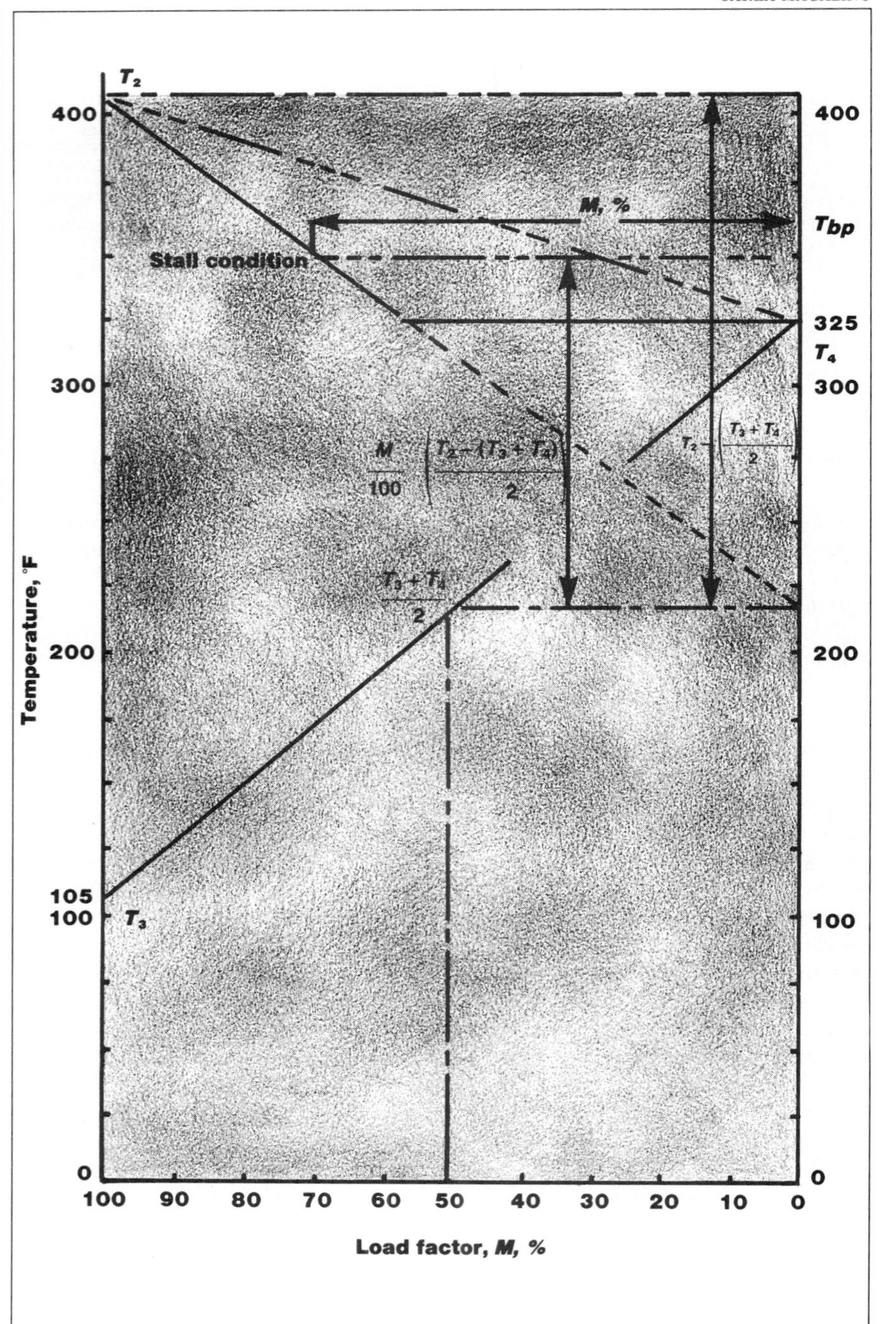

FIGURE 4. Predict the performance of exchangers operated by steam-inlet control

The pressure in the trap, due to the gravity dropleg from the exchanger, for pushing out condensate would also hold shut the vacuum breaker.

A vacuum breaker should be located above any likely condensate level. With steam in the shell of an exchanger, a breaker can be connected to the shell, or even to the steam inlet pipe. With steam in the tubes, the pressure drop may make this location less satisfactory. A better choice may be to connect the breaker to the exchanger's condensate header or to the condensate drainline immediately below the exchanger.

A breaker may admit air to eliminate a vacuum. The air that enters when a higher heat loading calls for a positive pressure in the exchanger can create problems, so an automatic vent must be installed to discharge it.

Control via steam flow

How would the Figure 2 exchanger perform with steam-flow control? At half load (as before, 1,640 lb/h), a decrease in load of 5%, if closed the steam valve, would result in the condensation of excess steam in the shell to the new equilibrium condition in about 1.4 seconds. A similar increase of load, if it opened wide the control valve, would bring on a new equilibrium pressure in about 1 second. In practice, the valve would be repositioned simultaneously with the pressure change in the shell, and the time lags in instrument responses to the change in shell pressure would not be significant.

The advantages conferred by steam-flow control with condensate removed by trap-and-pump combination include:

1. The exchanger output can match the applied load without limit, from 100% to 0 load
2. A steam trap can handle the condensate load because sufficient steam pressure to operate it is always available from the exchanger or a pump
3. Sized for full load, the steam control valve will meet all operating conditions
4. Water hammer, and metal differential expansion and corrosion are avoided
5. Exchanger performance is predictable, and the control response is immediate at all loads

References

1. Mathur, J., Performance of Steam Heat Exchangers, *Chem. Eng.*, Sept. 3, 1973, pp. 101-106.
2. Pathak, V. J., and Rattan, I. S., Turndown Limit sets Heater Control, *Chem. Eng.*, July 18, 1988, pp. 103-105.

The author

Albert Armer is an applications engineer with Spirax Sarco , Inc. (P.O. Box 119, Allentown, PA 18105; tel.: 215-797-5830), a manufacturer of steam traps, temperature and pressure controls and other heating and steam products. His experience includes more than 25 years of engineering steams systems in the food and pharmaceutical industries. He received his degree in mechanical engineering from Barrow College (England), and is a member of the British Institute of Plant Engineers. He has had several articles published, the previous one in *Chemical Engineering* being "The Right Steam Trap?", Feb. 15, 1988, pp. 68-75. ■

Section IV

Dealing with Heat-Transfer Fluids

HEAT TRANSFER FLUIDS — TOO EASY TO OVERLOOK

Richard L. Green and
Ronald C. Morris
Monsanto Co.

In various process situations, many engineers almost instinctively opt for steam. That tendency can be costly

For process heating, distributed systems that transfer heat indirectly from a central combustion unit are usually preferable to either electrical heating of individual process units or direct fuel-fired heating. Electrical heating is expensive; direct fuel firing entails the risk of high hot-spot surface temperatures that can cause process-side surface fouling and loss of product quality.

For indirect process heating, the engineer can choose either steam or heat-transfer fluids (HTFs). The most common HTFs for heating are a variety of organic compounds or mixtures, among them alkylated or benzylated aromatics, hydrogenated and unhydrogenated polyphenyls, modified terphenyls, mixtures of bi- and diphenyls and their oxides, and polymethyl siloxanes. Some of these fluids are stable at temperatures high as 750°F. Among other heat transfer fluids are mineral oils, molten nitrate salts, and polyalkylene glycols. Detailed descriptions of numerous individual fluids and discussion of their process applications appear on pp. 160–168 and in *CE*, May 1994, pp. 75–78.

Steam is inherently attractive. It is inexpensive, it displays a high heat-transfer coefficient as it condenses, and it can be generated economically in large boiler plants. On the other hand, its vapor pressure is an order of magnitude larger than that of most organic HTFs (Figure 1).

Accordingly, as the temperature requirements for a steam-heated process go up, the operating pressure of the steam system must rise considerably. This rise has implications for the piping and process equipment, as discussed later.

Organic and synthetic HTFs, by contrast, generally do not suffer vapor-pressure limitations. In most cases, their upper operating limit is instead set by thermal stability.

For instance, alkylated benzene [1], a typical single-medium HTF, can operate at temperatures up to 500°F. Mixtures of biphenyl and diphenyl oxides [2] are stable up to 750°F; this type of fluid, frequently used for vapor-phase heating above 500°F, is often referred to as "organic steam." Typical mineral-oil HTFs [3] can be an economic option in the range of 300-to-500°F, and the modified terphenyls [4] are employed at up to 650°F.

Two easy choices, one difficult

Accordingly, if heat is to be supplied to a process at below about 300°F and the expected temperature variations are not great, steam is widely (and usually rightly) recognized to be the better choice. Similarly, for temperatures above approximately 500°F, HTFs are the widely acknowledged correct option.

However, choosing between steam and an HTF for the intermediate, 300–500°F range requires more care — not only economic evaluation, but also awareness and consideration of a variety of process factors. In our experience, many of the latter are either unknown or overlooked. Such omission is likely to lead the engineer to opt for steam when an HTF is in fact the sounder choice.

Steam availability

The efficiency of steam as a heat carrier is due to not only its high heat of vaporization but also its good thermal conductivity. Indeed, heat fluxes in steam boilers can reach figures in the vicinity of 130,000 Btu/h/ft^2. However, proper evaluation of the steam-heating alternative must take into account a variety of other boiler-related factors, some of them not obvious.

In choosing between steam and an HTF, a key point to keep in mind is the availability of and *overall* need for

Originally published April 1995

steam at the site. When planning a grassroots, multi-process plant, the engineer can design the central steam generation and distribution facilities to current requirements throughout the facility, as well as likely future process expansion. For a single process user, on the other hand, including a steam boiler in the plant plan may be difficult to justify.

If the requirement for process heat is at an existing rather than a grassroots site, the distance from the new process facility to the boiler plant or an appropriate steam header must be considered. And if a header is indeed nearby, it must have sufficient capacity to supply the new use in addition to its existing users. The new process unit will lower the header pressure because of the increased pressure drop through the piping.

The introduction of the new, steam-heated process unit on an existing site may instead trigger the need for an additional boiler. If so, the engineer must make sure that the long-term steam requirements are adequate to assure efficient utilization of a boiler that is big enough to be economical.

When a new process unit leads to the need for a new boiler, a joint economic assessment of both is required. First the allocated cost of steam generation, in capital dollars per pound-per-hour of steam capacity, is determined by dividing the capital cost of the existing and the proposed new boilers by their total capacity. This capital cost must take into account the cost of steam piping from the boiler to the new process unit, as well as that for the condensate-return lines.

Then this number is multiplied by the steam requirements for the new process site (including a provision for losses) to determine the latter's cost allocation. The process project under assessment must earn an adequate return on not only the process-related capital outlay but also this allocated steam-system capital.

Piping considerations

When steam is to be the heat source, particular attention must be paid to the piping. Such scrutiny is especially important for heat that is to be delivered above 390°F, which is approximately the saturation temperature for 200-psig steam.

Above 200 psig, heavy-schedule pipe, fittings and valves may be required (Figure 2). For example, 250-psig saturated steam requires Schedule 80 extra-heavy pipe in sizes from 1 1/2 in. to 4 in., and double-extra-heavy pipe in larger sizes. Such requirements can increase piping cost by as much as 50% (Figure 3).

Steam systems also require steam traps. And for plant locations requiring cold-weather startup, these systems may require heat tracing on the piping and instrumentation lines.

By contrast, properly selected HTFs employ not only thinner-walled but also smaller-diameter pipe for a given temperature and heat load. And they flow easily even in the cold.

On the other hand, leakage is a bigger concern with HTFs than with steam-based systems. This is due to not only the cost of fluid replacement but also to concern for safety, because single-medium organic fluids are flammable. Accordingly, good sealing practice is especially important with an HTF-based system.

Process equipment

Many of these piping considerations also pertain to process equipment. For instance, the size and configuration of items such as jacketed vessels, plate

STEVE HAIMOWITZ

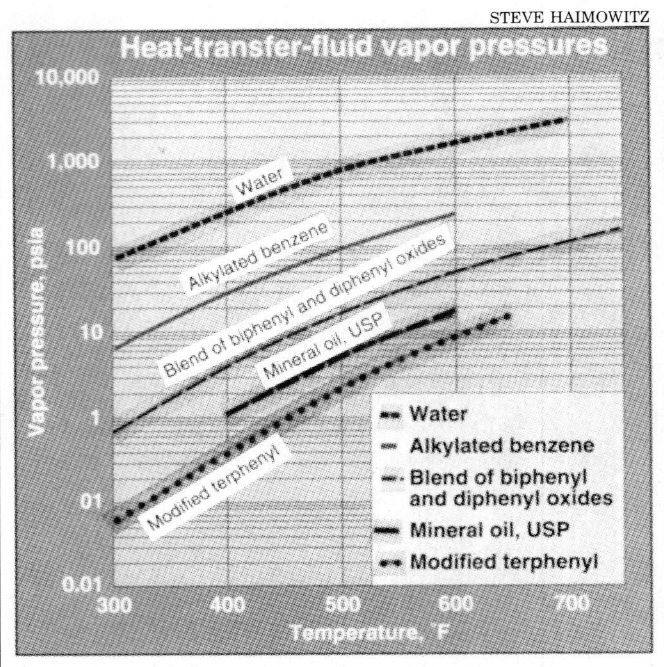

FIGURE 1 (left). Organic heat-transfer fluids' vapor pressures differ markedly from each other, but all are far lower than that of water (steam)

FIGURES 2 and 3 (below). Higher steam pressures and greater pipe diameters both tend to increase the required thickness of piping. In either case, the result is a higher cost. The costs shown are in 1994 dollars

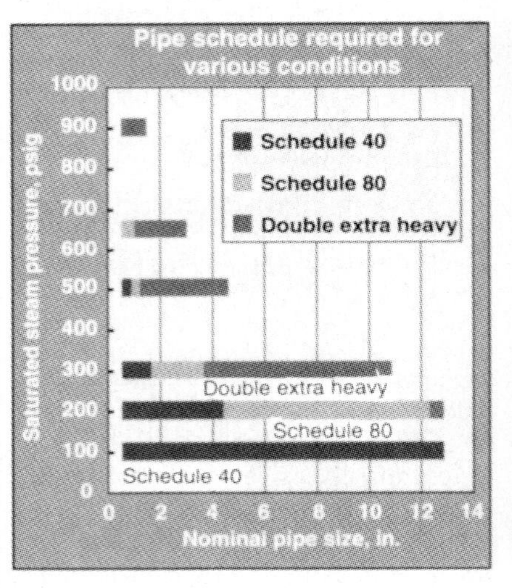

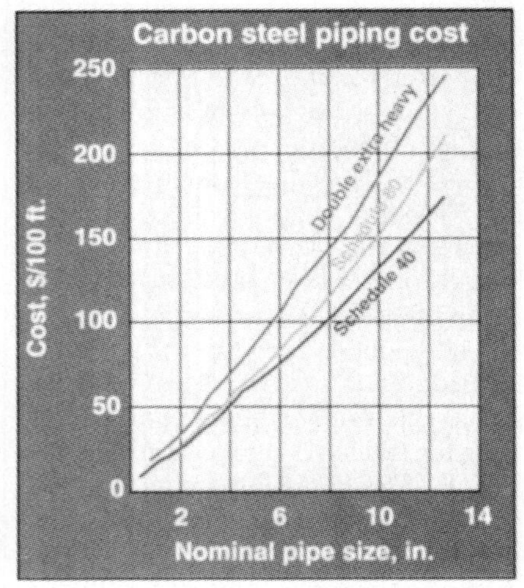

heat exchangers, and dryers may make high-pressure heat transfer surfaces expensive or impossible to fabricate.

If high-pressure steam is to be used with large-diameter vessels (whose size may be economically attractive for batch-size or throughput reasons), thick walls for the jacket and vessel must be provided. For example, the wall thickness for a jacket to surround an 8-ft-dia reactor with 600-psig steam (which condenses at about 490°F) would be approximately 2.7 in. Such an arrangement can also subject the reactor-vessel wall itself to significant buckling stress.

If the vessel is instead heated by interior coils in contact with the process fluid, these difficulties may become compounded by concern for fouling and corrosion, caused by trace salts and ion-containing droplets entrained in the steam. And if the process requires expensive materials of construction, the cost penalty for high-pressure, heavy-wall coils may become significant, as it does with jacketing.

Less-obvious considerations to keep in mind are the temperature-sensitivity of the process, the required closeness of temperature control, and the heat-transfer capability of the process fluid. All three of these factors mainly involve the film coefficient on the process side of the vessel or coil through which the heat enters.

When this film coefficient is low but the hot-side coefficient is high, the resulting high wall temperature on the process side may degrade the process fluid. This problem can in principle arise whether HTFs or condensing steam is on the hot side. However, modern systems employing HTFs can minimize or avoid it by controlling the heating-fluid flow. Such control is difficult or impossible with steam.

In fact, the problem rarely arises with HTFs. When they are used, the hot-side film coefficients are usually comparable in magnitude to those on the process side.

If agitation on the process side is desirable for process reasons, the process side coefficient may be significantly higher than that of the hot side. In this event, jacket baffles or nozzle inserts may be included in HTF-based systems to eliminate dead spots and provide turbulent flow, thus raising the hot-side coefficient.

In choosing between steam and heat-transfer fluids, consider design and operational factors alike

Operation and maintenance

Often-overlooked drawbacks to steam heating also arise with respect to operating and maintaining the system. For instance, it is not widely realized that higher-pressure steam systems can require a licensed operator at the location (usually an onsite central utility unit) where the steam is generated.

With the generation of steam there is an ongoing need for water-treating chemicals and equipment. Even with these provisions, there is also a risk of scaling and corrosion damage.

HTFs are non fouling and non corrosive by design, so systems employing them rarely experience those problems. On the other hand, flammability of the single-medium organics may call for extra firefighting protection, such as sprinklers. Flash-point fires occur only rarely with most fluids having flash points above 200°F, which takes in virtually all of the major HTFs.

Heating plus cooling

Perhaps the most clear-cut advantage of HTFs over steam comes in situations that require equipment (e.g., reactors) to be not only heated but also cooled during the operating cycle. Opting for steam in such a situation can create design, operating and maintenance problems, because the alternating use of the steam and either chilled water or brines poses an extreme corrosion risk, as well as thermal stresses.

If steam is nevertheless to be used for the heating portion of the cycle, the engineer has two alternatives. The first is to install completely separate systems for the heating and the cooling, which is expensive.

The second alternative, also expensive, is a closed system in which the cooling portion of the cycle employs a secondary-coolant brine that is indirectly cooled by water or refrigeration. Maintaining the proper concentration of the brine is difficult, due to the continual dilution by condensate at switchover from heating to cooling.

A properly designed thermal-liquid HTF system avoids these problems. A key element in its design is the selection of a fluid whose operating-temperature range covers both portions (hot and cold) of the process cycle.

Examples

Many of the points made in this article can be illuminated by considering two examples. Each requires a choice between steam or an HTF.

Example 1. It is necessary to supply 6.3 million Btu/h of 365°F heat to a 5,500- gal, pressurized, carbon-steel reactor in a new, grassroots facility. An ample supply of steam is available at 250 psig from a 750-million-Btu/h boiler, in a 14-in. header that passes about 300 ft from the planned reactor.

Fuel gas to heat an HTF is also available onsite. A fuel-gas-fired heater could be located about 150 ft from the gas line and about 175 ft from the reactor. The steam usage is charged at $5.75/million Btu. The fuel-gas cost is $5.10/million Btu. The efficiency of the fuel-fired heater for the HTF is assumed to be 85%.

System design and HTF selection: First, assume that steam will be used. The required steam flowrate dictates 6-in. piping, and its pressure calls for double-extra-heavy pipe. To deliver the heat to the process fluid, assume use of an internal coil made of 4-in. carbon steel pipe. The required surface area is calculated at 2,174 ft^2, and the diameter and steam pressure call for Schedule 80 thickness. The 750-million-Btu/h boiler costs $1.6 million; the allocated capital charge for it is calculated at $2,050 per million-Btu-per-hour.

Now, repeat the procedure but assume use of an HTF. The piping for fuel gas and the HTF itself becomes Schedule 40, as does the coil. The size of the coil is calculated to be about the same as in the steam option. Coil-inlet temperature for the HTF fluid inlet temperature is 500°F, and its required flowrate is found to be about 230 gal/min.

The results appear in Table 1. It is obvious that the piping and coil are significantly less expensive for the HTF-based system. And, this option incurs no allocation of steam plant capital.

However, these attractions are not sufficient to offset the cost of the fuel-fired heater, so the capital cost of the HTF is more than twice that of the steam system.* The annual costs for the HTF option also exceed those for the steam. In this case, the use of steam is much more attractive.

Example 2. Heat is required for a new residue-recovery system consisting of two 3,000-gal. still pots. One pot at a time is taken off line to concentrate its residue before draining.

The average heat requirement is 7.8 million Btu/h, with a maximum at 9 million Btu/h. The pot temperature rises from 350°F to 430°F during the concentration. Sufficient 500-psig steam is available at the power plant, 3,500 ft from the location. Elevated pipe supports and bridges are already in place between the power plant and the site, but there is no steam header on them. The cost of the steam is $7.5/million Btu.

Fuel gas is available 100 ft from a suitable heater location, 125 ft from the pot stills. Fuel gas cost is $5.05/million Btu. Other costs are as stated in Example 1.

System design and HTF selection: If steam is used, the 4-in. steam-supply line must be double-extra-heavy pipe. The steam system requires coils in the stills, because even a full jacket would not supply sufficient heat from 500-psig steam. Though acceptable, the coil is less desirable than a jacket because of possible problems in draining and cleaning the stills. The 3-in. coils also must be fabricated of double extra heavy pipe. Allocated capital cost of the 800-million-Btu/h, $2.2-million boiler is $2,750 per million-Btu-per-hour. If an HTF is used, with an inlet temperature of 600°F, it is found that a jacket can be employed rather than a coil.

COMPARISON OF HTF AND STEAM FOR A REACTOR
(all numbers in thousands of dollars)

	Steam	HTF
Reactor coil	20	15
Steam supply piping	38	
Gas supply piping		7
HTF circulating piping		12
HTF circulating pump		6
Surge tank		10
Heater		135
Steam plant allocated capital	13	
Heat Transfer Fluid		5
Total Capital	71	190
Annual steam cost	290	
Annual fuel gas cost		302
Other annual expenses	5	19
Total Expense	295	321

COMPARISON OF HTF AND STEAM FOR A RESIDUE STILL
(all figures in thousands of dollars)

	Steam	HTF
Coils in stills	30	
Jackets for stills		10
Steam supply piping	278	
Gas supply piping		5
HTF circulating piping		12
HTF circulating pump		5
Surge tank		10
Heater		190
Steam plant allocated capital	25	
Heat transfer fluid		6
Total Capital	333	238
Annual steam cost	468	
Annual Fuel Gas Cost		371
Other annual expense	30	26
Total Expense	498	397

TABLES 1 and 2 (above). Careful determination of the comparative costs is necessary when choosing between steam and a heat transfer fluid (HTF). Such care is especially important in the 300-500°F temperature range, into which both the above examples fall

A GLANCE AT MAJOR SECONDARY COOLANTS

SECONDARY COOLANT	Pumping[1] Limit, °F	Temperature for 10 cP Viscosity[2], °F	Needs Corrosion Inhibitors?[3]	Normal Boiling Point, °F
Aqueous Glycol Solutions:				
Propylene glycol: 30 wt. %	5 (FP)	17	Yes	216
Ethylene glycol: 30 wt. %	6 (FP)	<6 (FP)	Yes	216
Propylene glycol: 50 wt. %	−41	51	Yes	223
Ethylene glycol: 50 wt. %	−32	20	Yes	224
Other Coolants:				
Synthetic hydrocarbon[4]	<−120	−52	No	378
Silicone oil[4]	<−120	−85	No	347
Alkylated benzene[4]	−100 (FP)	−100	No	358
Water	32 (FP)	32	Yes	212

1. Fluidity limit, where coolant is at its freezing point (FP) or 2,000 cP viscosity.
2. Possible limit to efficient heat exchange, below which turbulent flow is not effectively achieved in economic heat exchangers.
3. When in contact with metal equipment.
4. Data shown are for typical commercial grades in coolant service.

TABLE 3. For coolant service in process situations that involve moderate temperatures, aqueous glycol solutions should not be overlooked. One key attraction they offer, not obvious in the table, is that the engineer need not be concerned with flammability

The results appear in Table 2. In this case, the long run of heavy steam-supply piping offsets the cost of the heater, the cost of the steam is more expensive than the fuel cost for the heater, and selection of the HTF option permits the use of the more-desirable jacket. Even if there were no foreseeable alternative use for the steam capacity at this site and the allocated steam-plant capital could be excluded, the economics would still favor the HTF.

Note that at the process conditions in this example, the steam system is approaching the limits of its practicality. Even if higher-pressure steam were available, long runs of large supply piping for extremely high pressures would not be feasible. The HTF system, by contrast, provides additional flexibility, because its operating temperatures could still be raised to the temperature use limit of the HTF.

Other HTFs

This article has focused mainly on organic or synthetic, single-medium HTFs, and has contrasted them mainly

*We have assumed once-through use of the steam. If the steam condensate is instead returned to the central boiler, the capital and operating costs associated with the condensate return system would narrow (but not close) the cost gap between the two options.

with steam. However, the engineer should also be aware of water-based, two-component HTFs such as aqueous solutions of propylene glycol or ethylene glycol (Table 3).

These fluids are often employed as coolants, and accordingly they do not frequently compete with steam. For applications falling within temperature ranges in which the organic-or-synthetic fluids and the aqueous solutions alike can operate, the latter transfer heat more efficiently. They also pose only minimal requirements regarding fire safety. ■

Edited by Nicholas P. Chopey

References

1. "Therminol LT Heat Transfer Fluid — Low Temperature Range, −100°F to 600°F," Pub. No. 9175, Monsanto Co.
2. "Therminol VP-1 Heat Transfer Fluid — Vapor Phase/Liquid Phase Heat Transfer Fluid," Pub. No. 9115, Monsanto Co.
3. "Therminol XP Heat Transfer Fluid — With FDA/USP/NF Status, 0°F to 600°F," Tech Bulletin 7239262, Monsanto Co.
4. "Therminol 66 Heat Transfer Fluid — High-Temperature, Low-pressure Fluid 20°F to 650°F," Tech. Bulletin 9146A, Monsanto Co.
5. Walas, S. M., "Chemical Process Equipment: Selection and Design," p. 213, Butterworths Publishing, London, p. 213, 1988.
6. Peters, M. S. and Timmerhaus, K. D., "Plant Design and Economics for Chemical Engineers," 4th ed., pp. 625, 809, McGraw-Hill, New York, 1991.
7. Walas, S. M., "Chemical Process Equipment: Selection and Design," p. 686, Butterworths Publishing, London, 1988.
8. Glasscock, D. A., and Hale, J. C., Process Simulation: the Art and Science of Modeling, *Chem. Eng.*, November 1994, pp. 82-89.
9. Green, R. L, Larsen, A. H., and Pauls, A. C., The Heat Transfer Fluid Spectrum, *Chem. Eng.*, February 1989, pp 90-98.
10. Seifert, W. F., Matching the Fluid to the Process, *Chem. Eng.*, February 1989, pp 99-104.

Authors

Richard L. Green, a Senior Research Specialist for the Monsanto Chemical Group of Monsanto Co., 800 N. Lindbergh Blvd., St. Louis, MO 63167, Tel. (314) 694-4615, currently provides worldwide technical services for Monsanto's Therminol heat transfer fluids. He has developed a wide range of heat transfer media operating from −100 to 750°F, as well as functional fluids for special lubricant applications and fire-resistant hydraulic fluids for commercial aircraft. He holds B.S.M.E. and M.S.Engr.Sci. degrees from Washington University, St. Louis.

Ronald C. Morris, 522 Arborwood Drive, Ballwin, MO 63021, Tel. (314) 227-1016, is a Consultant for Process Engineering and Computing. He does consulting in engineering computations, heat transfer systems, batch processing and project evaluation. His prior experience includes manufacturing and engineering technology with Monsanto Co. He holds a B.S. in chemical engineering from the University of Kansas, and an M.S. in engineering management from Northeastern University. He is a fellow of AIChE and a continuing-education lecturer on practical project economic evaluation.

GET FLUENT ABOUT HEAT TRANSFER FLUIDS

If you get involved with a process that operates below 32° or above, say, 350°F, you may need a working knowledge of heat-transfer fluids. The two articles following enable you to discuss these fluids knowledgeably with the firms making them. And they offer informa-

THE HEAT-TRANSFER-

Richard L. Green,
Alvin H. Larsen and
Allen C. Pauls,
Monsanto Chemical Co.

Over the past 30 years, a variety of organic heat-transfer fluids have been developed for indirect heating and cooling of processes. Here, we summarize the benefits of various heat-transfer methods, then take up the relationship of heat-transfer fluids to system design. Specific classes of heat-transfer fluids are reviewed, with emphasis on their properties that affect thermal efficiency and operational performance. We present equations to help quantify and rank the performance of specific fluids in terms of their properties. Finally, we discuss how the operating environment can affect the performance of heat-transfer fluids.

Heat-transfer fluids' niche

There are two basic kinds of process heating. One is direct firing, as in rotary kilns or submerged-combustion evaporators, where the flame or the combustion products directly contact the in-process material. The other is indirect firing, which employs a heat-exchange surface between the heat source and in-process material. Direct firing is confined to relatively few process situations, such as manufacture of cement or lime. Indirect firing is far more widely used in the chemical process industries.

Indirect firing, in turn, includes two versions: direct heating, as in fired heaters, with only one heat-exchange surface between the primary heat source and the in-process material; and indirect heating, whereby a heat-transfer fluid circulates continuously between the heat source and the material. In many process situations, either direct or indirect heating is feasible. The choice depends on process characteristics, system cost (including both capital and operating costs) and safety of operation.

Direct heating offers the highest thermal efficiency. It entails no need for secondary heat-transfer media, with their associated equipment and operational costs. High process temperatures — to about 1,800°F — are achievable. Unfortunately, if the material being heated is sensitive to the high surface temperatures associated with direct heat transfer, there can be severe fouling on the process side and deterioration of the process material, particularly at any local hot spots. Furthermore, process situations calling for multiple individual direct heaters incur a comparatively high risk of fire. Process heating of thermally stable materials such as gases and heat-transfer fluids is the most prominent use of direct heating.

Conversely, while indirect heating has lower thermal efficiency and requires a circulation loop, it can accommodate multiple processes safely with uniform heat-transfer-surface tem-

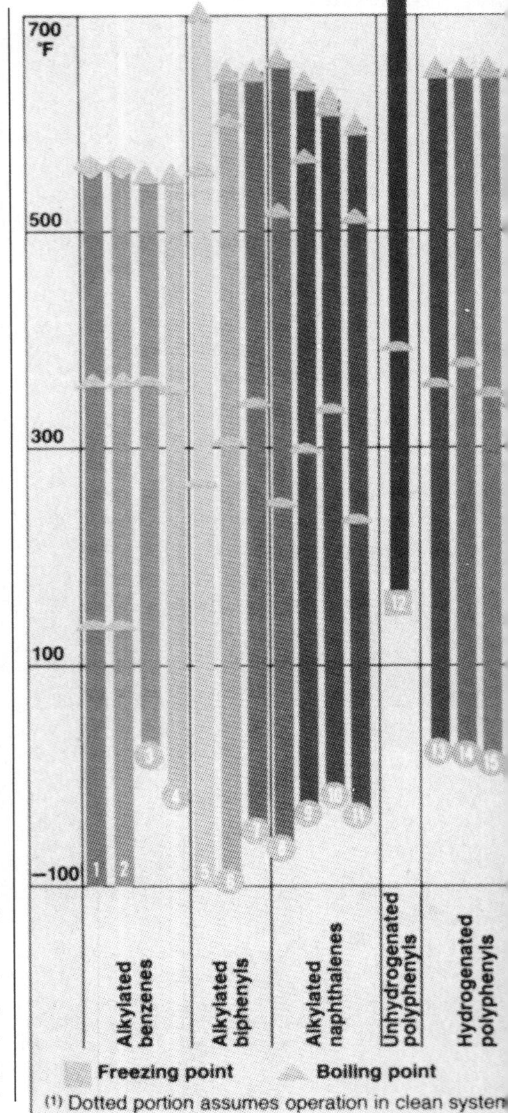

(1) Dotted portion assumes operation in clean system

Originally published February 1989

tion on selecting and using the fluids, as well as on taking them into account when designing the process system itself.

Written by engineers from two leading firms in this field, the articles each present an independent, self-contained point of view. The first takes a wide-ranging and conceptual look at heat transfer fluids, while also reviewing various types and their respective capabilities. The second emphasizes practical considerations to be noted when selecting a fluid for a particular situation.

Heat-transfer fluids are widely used with reactors. Thus, an apt prelude to this report was the article "Getting jacketed reactors to perform better" in our Dec. 19, 1988 issue, pp. 149-152. And for latest news developments about the fluids, see p. 39 in this issue.

FLUID SPECTRUM

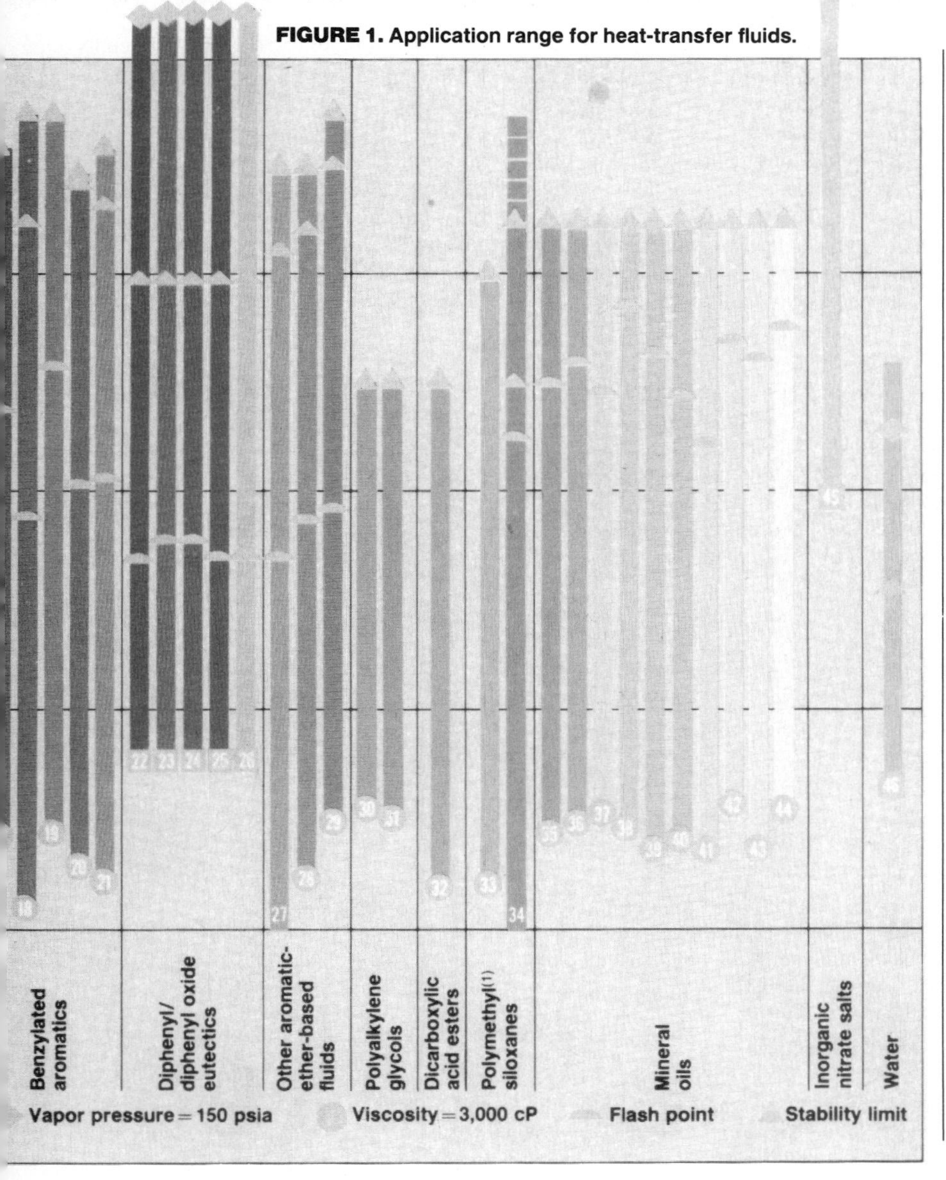

FIGURE 1. Application range for heat-transfer fluids.

perature: no hot spots, and precise, responsive control of the process temperature.

There are two modes of indirect heat transfer: in the vapor phase, where the transfer is by latent heat of vaporization and condensation; and in the liquid phase, where the transfer is sensible heat driven by temperature change. The advantages of each of these alternatives appear on the next page.

In some indirect-heating applications, temperature control is the primary function, and relatively small amounts of energy are transferred. Other cases entail large amounts of energy transfer, with significant temperature change. A sample listing of applications, including typical temperature levels, is likewise shown on the following page.

First, consider the process

The choice of a particular heat-transfer fluid and of liquid- or vapor-phase mode are governed by process needs: level and range of process temperature, system operating pressure, and required heat-transfer rate. The goal is to not only provide effective heat transfer but also avoid environmental, safety, corrosion, and fluid-degradation problems. A number of past review articles [1, 2, 3, 4, 5] cover these concepts.

In system design, a process temperature level with certain control tolerances is usually required. The heat-transfer system may have multiple users. And it may entail a heating-and-cooling cycle

VAPOR- AND LIQUID-PHASE HEAT TRANSFER EACH OFFER ADVANTAGES

Liquid-phase heat transfer

- Operating-temperature range can be wide.
- Response to process changes is quick.
- Piping is smaller, operates at lower pressure and costs less than that for vapor-phase systems.
- Flow control enables heat transfer to users that operate at multiple temperatures.
- Maintenance of piping systems is simpler than with vapor-phase systems.
- A given loop can be used for both process heating and cooling.
- Heat loss is less than with vapor phase.
- Fluid-makeup requirements are low.
- Operation is safer, due to low-pressure systems.
- Licensed operator is not required.
- Noncondensable-gas vents and condensate-return systems are not needed.

Vapor-phase heat transfer

- Temperature across exchanger surface is uniform.
- Temperature can be regulated by adjusting the pressure.
- Small free-convection system is practical, without need for pumps.
- Amount of heat-transfer fluid in system is relatively small.
- Heat-transfer coefficient is higher than with liquid-phase fluids.
- Pumping rate for condensate-return pumps is lower than that for pumps that handle liquid-phase fluids.
- Flow passage is typically more simple, and need not be uniform.

that requires the heat-transfer fluid to perform over a broad range of operating temperatures. Generally, low-temperature fluidity is desirable in a heat-transfer fluid, for an occasional cold startup and to allow effective heat transfer in cooling operations.

Vapor-phase heat-transfer fluids may be used in the liquid state, provided a static pressure is imposed to suppress boiling at the operating temperatures. The pressure limitations on such usage are especially likely to appear at system expansion tanks and process equipment having large shells, such as reactor vessels. Similarly, the vapor pressure-temperature relationship for the heat-transfer fluid becomes important in determining the system fluid-containment pressures. Liquid-phase systems can generally operate with ANSI B16.15 Class 250 components (for pressures to 150 psia) to 500°F and with Class 300 components (to 330 psia) at higher temperatures.

In liquid- or vapor-phase systems operating at up to about 750°F, carbon steel is generally used for the heater, pumps, valves and piping. Carbon steel is also satisfactory for the process units, unless the in-process substances dictate otherwise. Higher-alloyed steels are required for operation at higher temperatures.

The rates of heat transfer between the system heat sources and sinks depend on the heat-transfer fluid's viscosity, density, thermal conductivity and heat capacity, as well as the fluid velocity, geometry and temperature gradients associated with heat-transfer surfaces. Liquid-phase heat-transfer fluids should have moderately high density, high heat capacity, high thermal conductivity and low viscosity at the operating temperatures, to achieve a high heat-transfer coefficient. Fully developed turbulent flow at moderate velocity is generally desirable in liquid-phase systems, because it combines a high heat-transfer coefficient with economical pumping power. As for vapor-phase systems, a high heat-transfer coefficient is favored by high density, high thermal conductivity, high heat of vaporization and low viscosity. These systems operate most effectively at conditions favoring nucleate boiling and condensation. The effects of various fluids on heat-transfer performance are discussed later in this article.

The heat-transfer fluid should offer good thermal stability over the operating-temperature range, especially with regard to deposit formation on metal surfaces and debris formation in the fluid. It should also be able to tolerate small levels of contamination by process materials, air and water. Compatibility with the system material of construction should lead to long service life in the operating temperature range.

Since most heat-transfer fluids slow-

TYPICAL APPLICATIONS FOR HEAT-TRANSFER FLUIDS

Process	°F	Examples
Cogeneration	600-1000	
Distillation, other separations:	550-700	natural oils, crude oils, solvent recovery
Solar energy recovery:	400-1000	power generation, process heat
Textile processing:	400-630	spinning, finishing
Asphalt and pitch heating:	400-600	manufacture of roofing products, carbon electrodes
Plastic processing:	400-600	molding, extrusion, laminating, film processing
Petroleum and natural gas processing:	350-550	gas drying, purification, enhanced oil recovery, crude oil heating
Ship and barge cargo heating:	350-500	heavy fuels
Temperature control during synthesis:	300-700	acids, bases, alcohols, olefins, esters, others
Waste heat recovery:	200-650	from power plants and from incineration, for various process applications

ly degrade in use, the choice of heat-transfer fluid often hinges on whether the product is reclaimable or disposable. Can the fluid be reclaimed by distillation or simple filtration? Is economical disposal by incineration feasible, or is landfilling required?

Play it safe

Not only must the overall system be designed to operate safely; the heat-transfer fluid should also be environmentally and toxicologically acceptable should it leak. And it should not be reactive with or corrosive toward the outside medium or the process. The system design should provide shielding from potential sprays of hot heat-transfer fluid, including metal insulation covers and other devices around seals.

As regards fire safety of the heat-transfer fluid, bear in mind that combustion can be caused by either a direct ignition source or self-ignition at higher temperatures. Fires in the presence of an ignition source occur when a sufficient concentration of the fluid in air is above a minimum temperature; this is known as the flash point. A higher temperature, at which the ignition is sustained, is known as the fire point. And the minimum temperature of self-ignition without an ignition source, at which the heat-transfer fluid becomes air-ignited, is called the autoigniton temperature (AIT).

Autoignition and flash point fires alike can occur in fuel-fired heaters, where the heat-transfer fluid can leak into the combustion chamber. To provide for the quenching of such fires, the heater design should include inert gas or steam snuffing. Leaks into heat-transfer-system insulation can cause autoignition-related fires, due to decomposition of the heat-transfer fluid by oxidation, in combination with wetting of the extended surface of the insulation. In areas of high potential for leakage, such as flanges, pump seals and valve packing, cellular glass insulation can be employed to forestall insulation soaking, and various shieldings can be used to prevent contact with the insulation.

For selecting a heat-transfer fluid, a number of industry-standard tests are employed to measure flash, fire and autoignition temperatures [6, 7, 8]. However, the autoignition temperature given by these methods has proved optimistic: Recent findings indicate that in large, heated, metal systems, the true value can be as much as 200°F below the standard measured value [9].

WHAT TO LOOK FOR IN A HEAT TRANSFER FLUID

Thermal properties

- Broad fluidity range, to allow cold startup and process heating and cooling
- High density, high thermal conductivity, high heat capacity and low viscosity over the process operating-temperature range

Containment properties

- Compatibility with economic materials of construction (e.g., carbon steel); no corrosion of or other reactivity with system components or external materials
- Low vapor pressure, to allow low-pressure-system design and low external leakage; narrow boiling range, to prevent fluid loss during venting operations
- In liquid phase systems, low vapor pressure to prevent localized boiling and volatility problems, such as pump cavitation and two-phase flow

Stability properties

- Thermal stability, offering long fluid life at the operational temperatures
- High-boiling thermal-decomposition products soluble in the heat-transfer fluid; no deposits or sludge; low-boiling products ventable from the system expansion tank or from system vents
- Ability to tolerate small levels of contamination by process materials, air and moisture

Fire safety

- High flash-point temperature if ignition sources are present; minimal formation of low- boiling decomposition products, which lower the flash point
- Autoignition temperature greater than the maximum environmental temperature (including leakage into insulation and high-temperature air ducts)
- No formation of stable mist from external leakage

Environmental and toxicological properties

- Low toxicity, confirmed by MSDS with acute toxicity measured and estimates stated for other hazards (such as TLV, products of combustion)
- Low reactivity with in-process materials, in case of leakage to the process
- Economical reclamation or disposal

Potential leakage problems

Another hazard consists of a leaking heat-transfer fluid that forms a stable fog that can be ignited. Surface-energy effects make such a fog reactive and allow rapid flame propagation from the ignition source, causing a mist explosion-conflagration. Fire protection controls have been developed to sense such mists; they should be included in any systems using fluids that can mist when they leak [10, 11].

Fluid leakage from heat-transfer systems also poses toxicological and environmental concerns. Heat-transfer fluids should cause little irritation when in contact with skin and eyes, and minimal toxic effect when ingested or inhaled. In the U.S., the degree of toxicity of heat-transfer fluids should be judged by Material Safety Data Sheets (MSDS), in compliance with OSHA Hazard Communication Standard 29CFR 1900.1200.

Some fluids, especially the more-volatile ones, need ventilation to prevent annoying odors and to be at concentrations below their threshold limit values (TLVs). The TLV is the acceptable maximum concentration in air that is considered safe for daily inhalation; it should be known from experience, or estimated from testing.

In light of these risks of fire and environmental and toxicological problems due to leaks, the heat-transfer system must employ the most reliable technology for system control and containment of leakage. This should include explicit provision for collecting whatever leaks might in fact occur.

An overall listing of desired heat-transfer-fluid properties appears at left. And specifics on heat-transfer-system design with respect to components, layout and controls can be found in a number of references [5, 12, 13, 14].

Types of heat-transfer fluids

Water, gases, organic liquids, inorganic salts and liquid metals have all been employed in indirect heat transfer to

FIGURE 2. Vapor pressure differs markedly among the fluids.

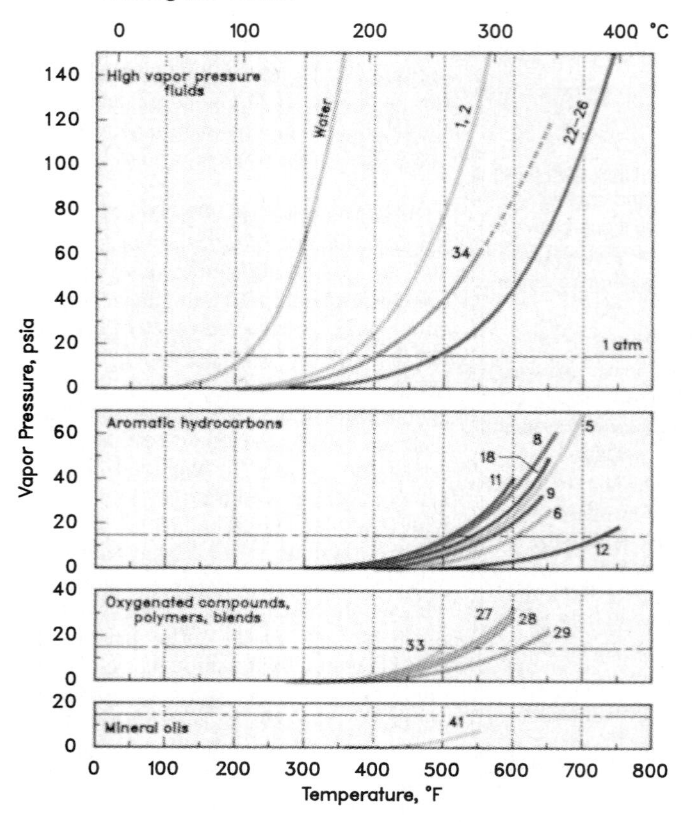

FIGURE 3. Densities of most fluids cluster around that of water.

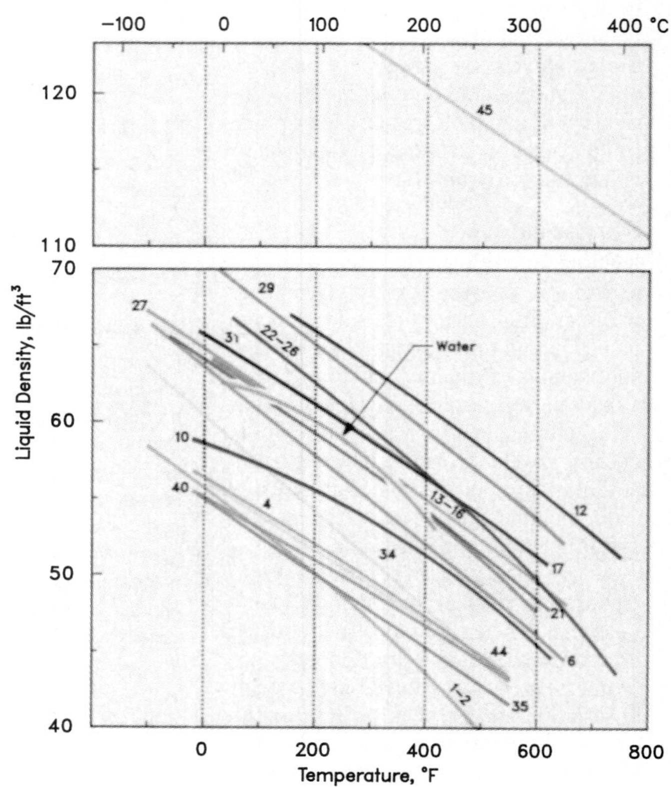

match process needs. However, some of these families of materials have significant limitations:

Although gases — air and nitrogen, in particular — are usable over a very wide temperature range and are stable, they require extreme pressure to achieve good heat transfer. Liquid metals such as sodium and mercury have high thermal stability, but pose environmental and safety problems and need costly containment systems. Brines and water-glycol solutions raise material-compatibility and stability problems at uses above about 300°F. Refrigerants are very effective at low process temperatures, but require high pressurization in higher-temperature processing.

Accordingly, the discussion here will focus on water, inorganic-salt eutectics, and a variety of organic and organic/inorganic materials employed as heat-transfer media. These substances are the ones most commonly used in the important 300 — 1,000°F temperature range.

A convenient way of classifying heat-transfer fluids and matching them to process needs is to consider their individual ranges of temperature applicability. Such ranges for a number of commercially available fluids appear in Fig. 1 on pp. 90 and 91.

Low-temperature applicability is usually limited by pumpability, related to a viscosity limit of approximately 3,000 centipoise in centrifugal pumps having bypass capability. Some fluids have a freezing-point restriction, where formation of crystalline materials would block flow.

The upper temperature limit is either the thermal stability limit or the temperature at which the vapor pressure reaches 150 psia. Vapor pressures of several heat-transfer fluids are shown in Fig. 2. As for thermal stability, the limits given in Fig. 1 were determined from laboratory thermal stability measurements and from operating experience at various temperature levels. The laboratory measurements were conducted in sealed stainless-steel ampoules with an inert-gas atmosphere for 1,000 h at a constant temperature. Experience substantiates that the temperature-stability limit in operating systems corresponds to a 5-10% decomposition in the ampoule test, as measured by thermal-cracking and condensation products formed outside the boiling range of the unused heat-transfer fluid.

The flash points indicated in Fig. 1 permit a relative comparison of fire safety, while the indicated normal boiling point provides an indication of fluid volatility.

Heat-transfer fluids can also be classified by chemical structure:

Alkylated benzenes of relatively low molecular weight (Fluids 1 and 2 in Fig. 1) are used for low-temperature cooling, and more generally in processes where the heat-transfer fluids stay below 500°F. With their narrow boiling range, they can be effectively employed in the vapor phase. Their low flash point requires extra fire-protection measures. As the molecular weight increases (Fluids 3 and 4), the flash point and boiling point rise to much higher levels, but the thermal stability decreases.

Alkylated biphenyls (Fluids 5-7) and *alkylated naphthalenes* (Fluids 8-11) have greater thermal stability than the higher-molecular-weight alkylated benzenes. Their flash points are higher,

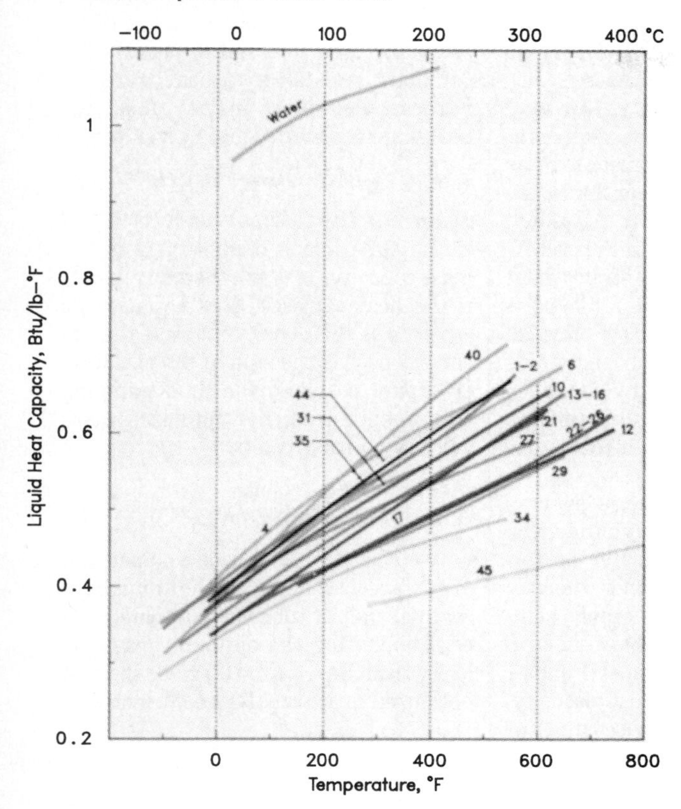

FIGURE 4. Except for water and nitrate salts, heat capacities differ little.

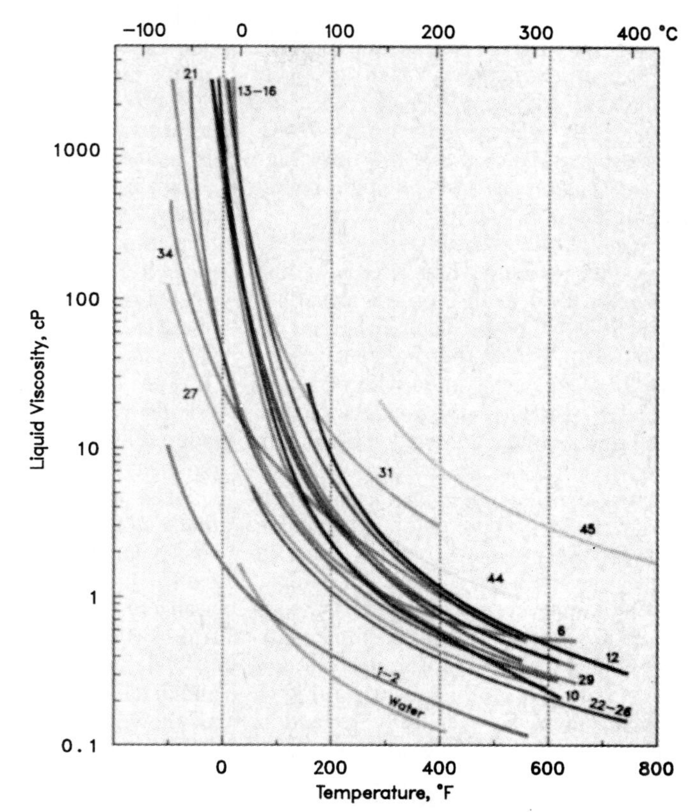

FIGURE 5. Viscosities vary greatly according to the temperature.

and the boiling points are nearer the stability limit than is the case for the low-molecular-weight alkylated benzenes. At the same degree of alkylation, the alkylated biphenyls are more stable than the alkylated naphthalenes. The 3- and 4-ring *hydrogenated polyphenyls* (Fluids 13-16) sacrifice some low-temperature fluidity, but they gain in the upper temperature region with good thermal stability and high boiling point. Operation below the boiling point substantially reduces leakage problems due to volatility. *Unhyudrogenated polyphenyls* (Fluid 12) require thermal tracing to perform at a 160°F startup temperature, but offer thermal stability to 750°F, a little above the boiling point.

Benzylated aromatics, such as those containing three rings (Fluids 17 and 19) and two rings (Fluids 18, 20 and 21), maintain good low-temperature startup capability and in some cases higher boiling points than the alkylated biphenyls and naphthalenes, but lower stability than the hydrogenated polyphenyls.

The other organic fluids stable up to 750°F are the *diphenyl/diphenyl oxide eutectics (DP:DPO)*. These (Fluids 22-26) have been used for over 50 years in both liquid- and vapor-phase systems; but because they operate well above their boiling point, fluid loss due to volatility is common. Heat tracing for startup is required in most climates, because the fluids have a 54°F freezing point. Low flash point and mist problems due to leakage require special fire-protection equipment. Fluids that contain DPO have a strong, characteristic aromatic-ether odor, which can lead to unsatisfactory working conditions should excessive leakage occur.

Other *aromatic-ether-based fluids* are a blend of DP:DPO and alkylated biphenyl (Fluid 27), alkylated DPO (Fluid 28), and a DPO blend with 3- and 4-ring aromatic ethers (Fluid 29). These fluids have generally good low-temperature fluidity and good stability, but at the expense of high volatility at typical operating temperatures and low flash points. Ether-based fluids generally form high-boiling condensation products when they are thermally stressed, requiring frequent fluid reclamation by vacuum distillation.

Polyalkylene glycols (Fluids 30 and 31) and *dicarboxylic acid esters* (Fluid 32) have high initial flash points but relatively low stability. These fluids often thermally decompose to low-boiling products, which can nullify the high-flash-point attraction of the fluid. These glycols and esters have the potential of more-rapid biodegradation than other organic fluids, which is an advantage in environmentally sensitive areas.

Polymethyl siloxanes (Fluids 33 and 34) are extremely sensitive to moisture and to other contaminants, a property that accelerates decomposition and solids formation. In ultraclean systems free of contaminants, any breakdown that occurs tends to form low-boiling compounds only. Additives can reduce the sensitivity to contamination (Fluid 34). Industrial experience indicates that the maximum useful stability is in the 500-600°F region. The close proximity of the flash point and the boiling point of Fluid 34 is an indication that its boiling range is broad. Fluid use also leads to a reduction in its flash point from 350°F to 130°F or lower, due to degradation.

Mineral oils (Fluids 35-44) have moderate stability, but are generally more sensitive to contamination by air and moisture than are the synthetic fluids. Their decomposition often produces sludges, and rust particles that are caused by mild corrosion in the system. The mineral oils generally have high flash points, as well as high boiling points.

Inorganic nitrate salts (Fluid 45) have an extremely high thermal stability, no flash point, and low volatility. Their 300°F melting point and large volume change upon freezing generally require a compact system with bath-type heaters. Nitrate salts are strong oxidizing agents and can form explosive mixtures with materials being processed.

Water (Fluid 46) is inexpensive, nontoxic, nondegrading and nonflammable. Its thermal properties provide by far the best heat-transfer efficiency in its operating temperature range. The temperature use range for water is unfortunately limited by its 32°F freezing point and its vapor pressure, as shown in Fig. 2. Water must be chemically treated to prevent system fouling and corrosion.

Heat-transfer performance

Fluid properties affecting heat-transfer performance include liquid density, heat capacity, viscosity and thermal conductivity. In addition, for vapor-phase systems, latent heat of vaporization is important. As noted in the box on p. 93, it is desirable that the density, thermal conductivity and heat capacity be high, and the viscosity low, over the temperature range of operation. Properties of selected fluids of each type are discussed below, and some are depicted in Figs. 3 — 5.*

Densities of most heat-transfer fluids are clustered about that of water, as shown in Fig. 3. The density of inorganic nitrate salts (Fluid 45) is about twice that of water, and the densities of polyphenyls (Fluid 12) and some aromatic-ether-based fluids (Fluids 28-29) are 10 to 20% higher than that of water.

Heat capacity of water, at low temperatures, is over twice that of any other heat-transfer fluid considered, as shown in Fig. 4. As for other fluids, the mineral oils (Fluids 35-44), alkylated benzenes (Fluids 1-4) and alkylated biphenyls (Fluids 5-7) have higher heat capacities than the average, while polyphenyls (Fluid 12), DP:DPO (Fluids 22-26), some aromatic-ether-based fluids (Fluids 28-29), polymethylsiloxanes (Fluids 33-34) and inorganic nitrate salts (Fluid 45) have lower-than-average heat capacities.

Since viscosities can range over as many as four orders of magnitude, viscosity can affect heat-transfer performance more than any other thermal property. Compared to other heat-transfer liquids, water and low-molecular-weight alkylated benzenes (Fluids 1 and 2) have by far the lowest viscosity, as shown in Fig. 5. The viscosities of DP:DPO (Fluids 22-26) and its blends (Fluid 27) are also somewhat lower than average over their range of applicability. Inorganic nitrate salts (Fluid 45) have a higher viscosity than organic fluids, and the viscosities of polyphenyls (Fluid 12), polyalkylene glycols (Fluids 30 and 31), and mineral oils (Fluids 35-44) are somewhat higher than average.

The thermal conductivity of water is four to six times that of organic fluids, and the thermal conductivity of inorganic nitrate salts (Fluid 45) is over three times that of organic fluids. Of the organic fluids, polyalkylene glycols (Fluids 30 and 31) and polyphenyls (Fluid 12) have thermal conductivities that are significantly higher than average, while low-molecular-weight alkylated benzenes (Fluids 1 and 2) and mineral oils (Fluids 35-44) have values lower than average.

The latent heat of vaporization of water is five times that of organic fluids used as vapor-phase media, which include low-molecular-weight alkylated benzenes (Fluids 1 and 2) and DP:DPO (Fluids 22-26).

The relative heat-transfer performance of heat-transfer fluids as a function of fluid temperature is shown in Fig. 6 for a cost-optimized exchanger. The heat-transfer fluid is assumed to be on the tube side and to provide the dominant resistance to heat transfer. For turbulent flow inside tubes, the heat-transfer coefficient is given by:

$$h = 0.023(k/D)(DG/\mu)^{0.8}(c\mu/k)^{0.4} \quad (1)$$

where k is the thermal conductivity, D is the tube inside diameter, G is the mass velocity, μ is the viscosity, and c is the heat capacity. G is equal to ρv where ρ is the density and v is the linear velocity. If the sum of the exchanger capital cost and the fluid pumping cost is optimized, the optimized mass velocity (15) is given by:

$$G = (1.6K_s\, g_c\, D^{0.2}\rho^2/(0.023\, K_p\mu^{0.2}))^{1/2.8} \quad (2)$$

where K_s is the annualized cost per unit of surface area and K_p is the annualized cost per unit of tubeside pumping power. Eliminating the optimum mass velocity from Eqs. (1) and (2) gives the cost optimized heat-transfer coefficient as:

$$h_o = (0.023/D^{0.143}) \times (1.6\, K_s\, g_c\, /0.023\, K_p)^{0.2857} \times (k^{0.6}\, c^{0.4}\, \rho^{0.571}\, /\mu^{0.457}) \quad (3)$$

Based on this equation, the relative heat-transfer-performance parameters of the heat-transfer fluids shown in Fig. 8 are defined as:

$$F = (h_o\, /h_{ref}) = (k/k_{ref})^{0.6} \times c/c_{ref})^{0.4}(\rho/\rho_{ref})^{0.571}\, (\mu_{ref}\, /\mu)^{0.457} \quad (4)$$

where $k_{ref} = 0.3$ Btu/h-ft-°F, $c_{ref} = 1$ Btu/lb-°F, $\rho_{ref} = 62$ lb/ft^3, $\mu_{ref} = 1$ cp, and h_{ref} is h_o using these property values.

The heat-transfer-performance factor of water increases from 1.0 at 60°F to 2.8 at 400°F; the heat-transfer performance of water is 3.5 to 5.5 times better than the best of the other heat-transfer fluids considered. Of the fluids shown in Fig. 6, low-molecular-weight alkylated benzenes (Fluids 1 and 2) provide the best thermal performance at temperatures to 550°F. At that temperature, following in approximately decreasing order of heat-transfer performance are DP:DPO (Fluids 22-26), DPO blend (Fluid 27), inorganic nitrate salt (Fluid 45), alkylated naphthalene (Fluid 10), benzyated aromatic (Fluid 21), aromatic-ether-based fluid (Fluid 29), benzylated aromatic (Fluid 17), high-molecular-weight alkylated benzene (Fluid 4), mineral oil (Fluid

*For organic heat-transfer fluids within a moderate temperature range from -40 to 430°F, densities are easy to measure, and the results given here should be accurate to within 1%. Viscosities below 10 cP are relatively easy to measure, and results should be accurate to within 10%. Vapor pressures and heat capacities are somewhat more difficult to measure, but are likely to be accurate to within 5%. Most difficult to measure is thermal conductivity, because of free-convection effects in low-viscosity fluids; errors and biases may be 15% or more. Except for water, we have estimated the latent heat of vaporization from other properties, using corresponding-states behavior. Extrapolation of density, vapor pressure, and heat capacity outside measured ranges is subject to greater uncertainty, but has been adjusted for thermodynamic consistency to the extent possible, treating each fluid as a pure pseudo-component. Extrapolation of measured viscosity data to 3,000 cP requires an extended empirical viscosity model, and is subject to considerable uncertainty.

FIGURE 6. Heat-transfer-performance curves summarize effects of properties.

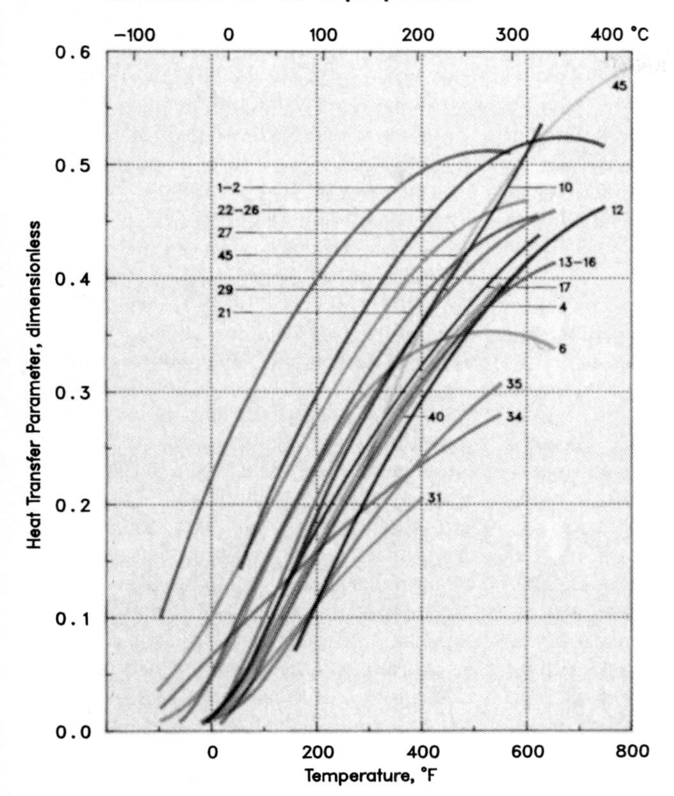

FIGURE 7. Relative pumping power proves similar to heat-transfer performance.

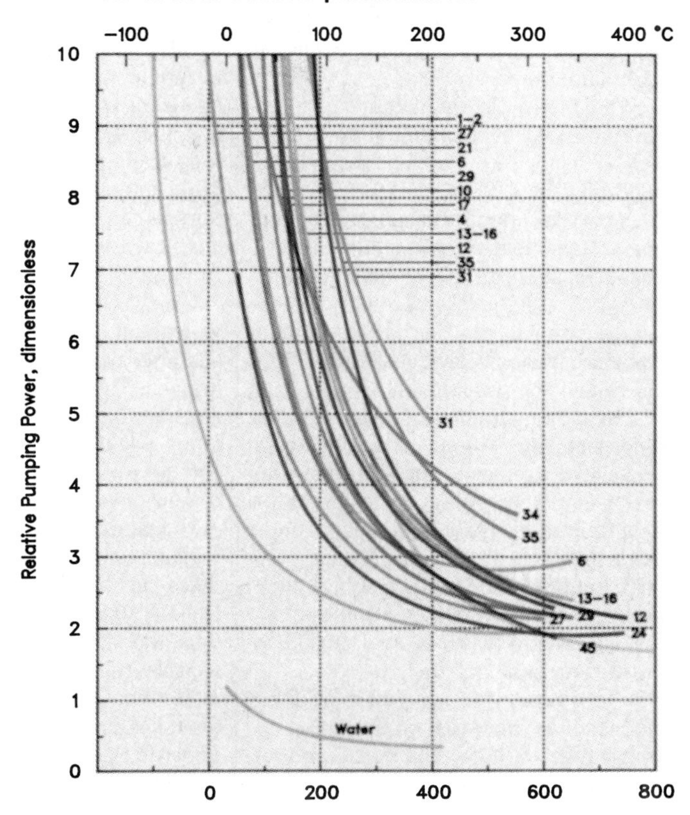

40), polyphenyls (Fluids 12-16), alkylated biphenyl (Fluid 6), mineral oil (Fluid 35), and polymethyl siloxane (Fluid 34).

The relative pumping power needed for various heat-transfer fluids, as a function of temperature, is shown in Fig. 7 for a cost-optimized exchanger with the heat-transfer fluid on the tube side. For turbulent flow inside tubes, the pressure drop is given by:

$$\Delta P = (4fL/D)\,(G^2/2\rho g_c) \qquad (5)$$

where the friction factor, f, is given by:

$$f = 0.046/(DG/\mu)^{0.2} \qquad (6)$$

and where L is the length. Eliminating f and replacing L/D and G with their cost-optimized values yields:

$$\Delta P = [(T_1\text{-}T_2)/g_c\Delta T_m]\times [1.6K_s g_c D^{0.2}/0.023K_p]^{2/2.8}\times [c^{0.6}\mu^{0.46}\rho^{0.43}/k^{0.6}] \qquad (7)$$

where T_1 and T_2 are the inlet and outlet temperatures of the heat-transfer fluid, and ΔT_m is the mean temperature difference between the hot and cold streams [15]. Based on this equation, the relative pumping power shown in Fig. 7 is defined as:

$$F_p = [\Delta P/(\rho c)]/[\Delta P/(\rho c)]_{ref} = (c_{ref}/c)^{0.4}(\mu/\mu_{ref})^{0.46}\times (\rho_{ref}/\rho)^{0.57}(k_{ref}/k)^{0.6} \qquad (8)$$

where the reference properties have the same values as for Eq (4).

Fig. 7 shows that water has the most favorable fluid properties for minimizing pumping power. Comparison with Fig. 6 shows that fluids with desirable pumping properties are generally those with favorable heat-transfer properties.

Basing the heat-transfer-fluid comparison on cost-optimized exchangers with the heat-transfer fluid on the shell side does not significantly alter the fluid ranking shown in Fig. 6. If the heat-transfer fluid does not provide the dominant resistance to heat transfer, then the relative-heat-transfer-performance parameter of the fluid will be lower than that shown in Fig. 6.

The operating environment

If heat-transfer systems are operated within the temperature range of applicability, the fluids will generally operate for years without significant physical or chemical changes that would adversely affect performance.

However, contamination by moisture, air or process materials can cause fluid-related problems, such as volatiles generation, high-boiler formation, coking, sludging, and accumulation of corrosion-particulate debris. Moisture and air infiltration can be reduced or eliminated by including an inert-gas-blanketed expansion tank in the system. The system should be designed to have some hot fluid flow through the expansion tank, where volatile contaminants, noncondensable gases and low-boiling decomposition products will have a free surface for separation from the fluid, after which they can be vented from the system. In the case of fluids with high boiling points and narrow boiling ranges, only minimal amounts are lost during venting operations.

Heaters can expose the fluids to temperatures above their thermal-stability limits. This can happen when the flow past the heat-exchange surfaces is not turbulent, which may cause formation of a thick thermal-boundary layer and excessive wall temperatures. Fully-developed turbulent flow generally can accommodate the heat fluxes found in fuel-fired or electric heaters. Free-convection and fire-tube heaters often overexpose the heat-transfer media to high wall temperatures.

A heat-transfer fluid that can tolerate low levels of contamination can often be used until the next scheduled maintenance.

Operation above the normal boiling point of the heat-transfer fluid, as with vapor-phase systems, typically results in 25-30% leakage loss per year, whereas operating below the normal boiling point produces leakage losses of 5-10% per year. The amount of leakage and the leakage-makeup cost are related to fluid volatility. Pump-shaft seals are a major leakage area. These mechanical seals should, of course, be designed for long operating life, and all seal components need to be compatible with the heat-transfer fluid and the high-temperature environment. A number of operation-design features are helpful in preventing seal leakage:

- The pump packing gland (stuffing box) should have jacketed cooling to keep the fluid in the shaft seal area at 300°F or lower.
- To prevent particulate debris that has accumulated within the system from wearing the shaft seal faces, a seal flush can be employed to supply a low-flow slip stream of filtered fluid to the area. The best filters for this application are fiberglass string-wound filter cartridges having a nominal particle-removal rating of 1 to 30 micrometers.
- A vent-and-drain gland can be used to bathe the outside of the seal with inert nitrogen or steam, to prevent fluid leakage from oxidizing and thereby generating solids, which can change the flat face-sealing geometry by deposition.
- Canned or sealless magnetic-drive pumps can eliminate the need for mechanical seals, especially in low-capacity pumping.

Other critical seals in the system are valve-stem packing, and gaskets for sealing parting surfaces, such as flanges. The best current sealing materials employ flexible graphite valve-stem packing and gaskets. Gaskets should be reinforced with metal mesh; or spiral-wound gaskets conforming to API Standard 601 should be used.

Flash-point-related fires are rare in fluids having flash points above 300°F. Good leakage-control practices can reduce the likelihood of fires involving insulation soaked with heat-transfer fluid. And, of course, safety controls on fuel-fired heaters reduce the risk of fire in general. Such controls should include a heater-outlet high-temperature alarm, low-heater-flow alarm to shut off the heater burner, and a low-level expansion tank switch (alarm) to shut off the system pumps and the burner. Safety relief valves should be employed in the system heater and expansion tank to protect against sudden system-volume expansions. The safety-relief-valve vents should be ducted to a safe area.

If the heat-transfer system has problems during operation, the system-component manufacturers and heat-transfer-fluid suppliers should be consulted. An extensive analysis of the used heat-transfer fluid can identify causes of system inefficiencies, especially if they involve contaminants or fluid-decomposition products. Also, systems with decomposed or contaminated heat-transfer fluid can be brought back to acceptable performance levels by a range of treatments: For instance, simple *in situ* filtration can remove particulate debris, chemical cleaning can disperse sludges, and mechanical methods can remove hard cokes [16]. If the fluids have excessive amounts of debris and high boilers, they can often be brought back into useful condition by vacuum-distillation reclamation.

Fluid-replacement cost, reclamation expense, and disposal cost should not be overlooked during fluid selection.

When two or more heat-transfer fluids appear equally likely to meet the needs of the process and can boast similarly acceptable operational experience, the selection often depends on the services provided by the fluid suppliers. Services such as design guidance, operational guidance, problem resolution, and comprehensive fluid analysis can pay dividends in maximum system efficiency, minimum operating cost, extended fluid and component life, and dependable system performance. ■

References

1. Seifert, W. F., others, Organic Fluids for High-Temperature Heat-Transfer Systems, *Chem. Eng.*, Oct. 30 , 1972, pp. 96-104.
2. Fried, J. R., Heat Transfer Agents for High-Temperature Systems, *Chem. Eng.*, May 28, 1973, pp. 89-98.
3. Singh, J., Selecting Heat-Transfer Fluids for High-Temperature Service, *Chem. Eng.*, June 1, 1981, pp. 53-58.
4. Minton, P. E., and Plants, C. A., Heat Exchange Technology: Heat-Transfer Media Other Than Water, "Kirk-Othmer Encyclopedia of Chemical Technology", 3rd ed., Vol. 12, pp. 171-191, Wiley, New York, 1980.
5. Geiringer, P. L., "Handbook of Heat Transfer Media", Reinhold, New York, 1962.
6. ASTM D 93 Standard: Pensky Martens Closed Cup Flash Point.
7. ASTM D 92 Standard: Cleveland Open Cup Flash and Fire Point.
8. ASTM E 659 Standard: Autoignition Temperature of Liquid Chemicals.
9. Krawetz, A. A., Determination of Flammability Characteristics of Aerospace Hydraulic Fluids, *Lubrication Engineering*, 37, pp. 705-714, 1981.
10. Vincent, G. C., and Howard, W. B., Hydrogen Mist Explosions. Part I. Prevention by Explosion Suppression, in "Loss Prevention", CEP Technical Manual, Vol 10, pp. 43-47, American Institute of Chemical Engineers, New York, 1976.
11. Vincent, G. C., others, Hydrogen Mist Explosions. Part II. Prevention by Water Fog, *ibid*, pp. 55-71.
12. Wagner, W., "Heat Transfer Technology with Organic Media", Technischer Verlag Resch KG, Gräfeling-Munich, 1975 (*Konus Handbook*, Vol. 1, Konus Systems, Marietta, Ga.).
13. "Engineering Manual for Dowtherm Heat Transfer Fluids," Form No. 176-1334-85, Dow Chemical U.S.A., Midland, Mich.
14. "Therminol Fluid Heat Systems," Pub. No. IC/FP-211, Monsanto Co., St. Louis, Mo., 1982.
15. Null, H. R., others, "Process Design for Energy Conservation," AIChE Today Series course notes, American Institute of Chemical Engineers, New York, 1976.
16. "Organic Thermal Fluid Heat Transfer System Cleanout," Therminol Information Bulletin No. 1, Monsanto Co., St. Louis, 1987.

The authors

Richard L. Green, a Senior Research Engineering Specialist for Monsanto Chemical Co., 800 N. Lindbergh Blvd., St. Louis, MO 63167, Tel. (314) 694-1000, currently provides worldwide technical services for Monsanto's Therminol heat-transfer fluids. He has developed functional fluids for lubrication and heat-transfer applications, including recent ones for the solar power industry. He is also inventor of a new generation of phosphate-ester-based, high-performance hydraulic fluids for aircraft.

Alvin H. Larsen is Principal Engineering Specialist in Monsanto's, Engineering Technology Dept., where he provides corporate support for physical-properties and process simulation. He was instrumental in organizing AIChE's Design Institute for Physical Properties Data and leading its Data Compilation Project. He holds B.S.Ch.E. and B.A. degrees from the U. of Utah, and a Ph.D. in chemical engineering from California Institute of Technology, and is a member of AIChE, ACS and Sigma Xi.

Allen C. Pauls is a Fellow in Monsanto's Engineering Technology Dept., responsible for development of design methods, equipment specifications and consultation in heat transfer. A member of AIChE and Technical Chairman of Heat Transfer Research, Inc., he holds B.S. and Ph.D. degrees in chemical engineering from Purdue U.

MATCHING THE FLUID WITH THE PROCESS

W. F. Seifert,
Dow Chemical U.S.A.

To select a high-temperature heat-transfer fluid for a given application, first define the general properties needed in the fluid to be selected, then gather information on various manufacturers' available fluids.

Unfortunately, the number of ways to present fluid data is nearly equal to the number of fluid manufacturers. Units of measurement and even the methods for measuring fluid characteristics may not be directly comparable. Before an intelligent selection can be made, there must be some assurance that important specifications can be accurately compared and evaluated.

The fluid information presented in this article is based on gathering recent available data from various manufacturers. The data were converted as necessary to assure like units of measurement. Test methods, on the other hand, may still not be comparable because some of the data were developed before industry testing-standards were adopted. These exceptions are noted in the data presented here, if measurement methods were detailed in the specific manufacturer's literature.

First, consider the temperature

The first thing to look at in fluid selection is, quite naturally, the process temperature range.

If your application involves chilling or freezing, the most common heat-transfer fluids are halogenated hydrocarbons, ammonia, air, brines, and glycol/water solutions. Glycols and brines are also acceptable for a limited high-temperature range above the boiling point of water.

If the application involves higher temperatures, an organic fluid is frequently the choice. But before you delve further, be sure that your application conditions do in fact warrant the cost of an organic fluid. For temperatures between 32° and 350°F, water is the most economical and efficient heat-transfer fluid. In some situations, even freezing temperatures can be tolerated. For example, if system shutdown would be required should ambient conditions fall below the freezing point, freeze protection could be provided by heat-tracing the piping for equipment exposed to these low temperatures.

At temperatures above 350°F, the vapor pressure of water increases significantly, and so do the associated process-equipment costs. By way of comparison, most organic heat-transfer fluids, even at 600°F, have vapor pressures below 180 psig, which makes these special fluids more attractive for high-temperature applications. In fact, low vapor pressure at high temperatures is the major reason for choosing an organic fluid over steam.

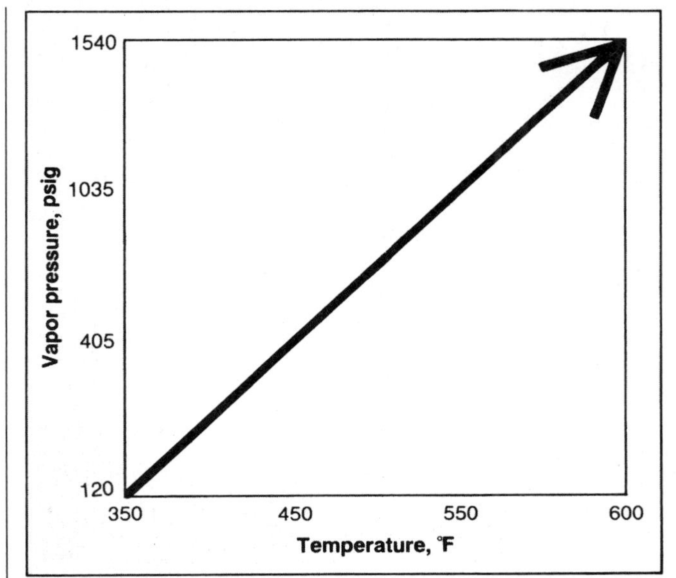

FIGURE 1. The vapor pressure of water rises rapidly with temperature (figure is not drawn to scale).

An array of factors must be kept in mind when choosing the right fluid

Organic fluids also offer advantages at temperatures below the freezing point of water. Some high-temperature fluids can maintain pumpability all the way down to −100°F. In cases where both low- and high-temperature capabilities are necessary, the high-temperature organic heat-transfer fluids can often be the optimum choice.

The low-temperature capabilities of organic fluids may even be important when the process normally requires only high temperature capabilities. When fluids are used in locations where ambient temperatures can drop well below freezing, ability to withstand low temperatures can prevent a system shutdown. Using an organic fluid that remains pumpable at low temperatures can eliminate the need and cost of steam tracing.

The fluid selected should be able to operate at temperatures that extend beyond those of the process. Choose a fluid with at least 50°F higher temperature capabilities at the high end of the

Originally published February 1989

TABLE 1. Properties of major types of heat-transfer fluids[1]

Composition	Temperature use range, °F		Minimum pumping temp.[2],°F	Flash point[3] °F	Fire point[3] °F	Auto-ignition temp.[4],°F	Vapor pressure, psig.[5]	Physical properties of liquid at 600°F			
	Liquid	Vapor						Thermal conductivity, Btu/(h)(ft²)(°F/ft)	Specific heat, Btu/(lb)(°F)	Density, lb/ft³	Viscosity, cP
Mixtures[6] of diphenyl and diphenyl oxide	60/750	495/750	53.6	255	275	(1139)	30.6	0.061	0.579	49.29	0.19
	60/750	495/750	53.6	240	260	1150	31.0	0.056	0.564	49.78	0.183
Mixture of di- and tri-arylethers	20/700	-	−4	285	295	(1083)	9.5	0.059	0.540	53.91	0.30
Aromatic blend	−40/650	-	−50	275	295	(932)	17.3	0.055	0.590	48.67	0.24
Alklylated aromatic	−100/600	358/600	< −100	145	155	(788)	159.8	0.0625	0.720	35.46	0.11
Hydrogenated terphenyls	25/650	-	31	355	375	(662)	(434)	0.062	0.630	49.55	0.38
	15/650	-	30	350	380	705	(434)	0.0545	0.628	50.30	0.40
Dimethyl siloxane	−40/750	-	< −50	320	350	725	50.0	0.052	0.49	42.00	0.43
Polyaromatic	−60/600	-	−50	310	320	835	6.9	0.0540	0.622	49.05	0.30
Synthetic hydrocarbon	0/600	-	4	350	410	675	(341)	0.0535	0.699	42.71	0.36
Paraffinic oils	15/600	-	< 15	390	430	670	(110)	0.046	0.70	44.42	0.40
	40/600	-	< 40	430	ND[1]	675	(150)	0.064	0.70	42.4	0.54
Mineral oils	5/600	-	< 35	350	ND	ND	(210)	0.060	0.70	47.30	0.45
	25/600	-	< 25	380	ND	ND	(160)	0.065	0.67	42.43	0.43
	5/600	-	20	400	ND	ND	(120)	0.065	0.670	42.43	0.47
Isomeric dibenzyl benzenes	−4/662	-	−18	374	ND	932	(134)	0.0644	0.608	50.85	0.38
Isomeric dimethyl diphenyl oxide	−4/626	-	−46	275	ND	1013	15.3	0.0534	0.552	49.79	0.273
Alkyl diphenyl	−39/707	-	−22	266	ND	ND	11.3	0.0569	0.625	47.96	0.322
Alkyl benzene	32/590	-	−49	358	367	ND	(310)	0.0609	0.645	43.70	0.35

Notes:
1. Data for a given fluid are based on input from its manufacturer. An ND entry means no data available.
2. For mixtures of diphenyl and diphenyl oxide, this is the freezing point; for the other products, this is the temperature at which the fluid viscosity is 1000 centipoise.
3. Cleveland Open Cup (COC).
4. Values in parentheses determined by ASTM Method E-669-78; other values determined by ASTM D-2155.
5. Values in parentheses are mm Hg rather than psig.
6. Each of the two products listed is a eutectic mixture of 26.5% diphenyl and 73.5% diphenyl oxide.

range than required, to safeguard the system against unexpected overheating and degradation of the fluid. A similarly large margin is not necessary at the low end of the range, as long as the lowest anticipated process temperature is provided for.

The minimum temperature required for the system's fluid may be dictated by either the process or ambient conditions. When low process temperatures are not a concern, the fluid should remain pumpable at the lowest anticipated temperature of your geographic location in the event of an extended shutdown. It is possible to trace heat-transfer lines with steam or electric heaters to keep fluid pumpable for cold-weather startups, but a fluid that remains pumpable without tracing will prove far more economical in the long run.

It is important to make the distinction between pour point and minimum pumping temperature when comparing fluids. The pour point is determined by an ASTM procedure, the fluid being cooled to the temperature where it will not flow when the container is tilted on its side. For determining pumpability, on the other hand, the viscosity is a major factor to evaluate. The minimum pumping temperature is determined by the temperature at which the fluid viscosity reaches 1,000 cP. A standard centrifugal pump, designed for normal operating conditions, will experience a significant reduction in flowrate when the viscosity increases above this limit. Viscosity is also very relevant as regards heat-transfer efficiency, as discussed in more detail further in this article. Pump and heater bypass loops can be designed into the system; but they, too, are ineffective at high viscosity.

Establishing the fluid temperature range is merely the starting point in the process of selecting a fluid. There are a number of other factors, including vapor pressure, flammability, corrosivity, toxicity, thermal stability, engineering properties and cost that also need to be considered to complete the selection process. These factors are discussed in the sections following. An example of how they relate to a given situation appears in the box on p. 104.

A variety of properties for several key kinds of heat-transfer fluids appear in Table 1. (It was necessary to convert some manufacturers' data to standard English and metric units to facilitate direct comparisons of products.)

Importance of vapor pressure

The pressure requirements and capabilities of the system must be taken into account when selecting a heat-transfer fluid. In new system designs, the fluid itself can dictate the pressure that must be withstood. For systems already in operation, however, the pressure capabilities of such components as the expansion tank or process vessels may strongly influence which fluids can be substituted, depending on the application.

The expansion tank should have pressure capabilities sufficient to take into account the fluid vapor pressure at planned operating temperatures, plus whatever pressure is created when the fluid thermally degrades or if water or other contaminants get into the system. In sizing an expansion tank for a new system, good design practice requires that the vessel be able to withstand pressure of at least 15 to 20 psi over the highest anticipated fluid vapor pressure.

Most expansion tanks available today are designed as pressure vessels and will have their ratings clearly specified by the manufacturer. On the other hand, an older tank enlisted for a retrofitted system may or may not be designed for pressure, or may not be able to withstand the pressures created within the heat-transfer system.

Most high-temperature heat-transfer fluids used at or below 500°F do not, under normal operating conditions, create unusually high vapor pressure. In the 600°F to 750°F operating range, however, the situation changes drastically, and vapor pressures of 30 to 140 psig are not at all unusual, depending on the fluid selected.

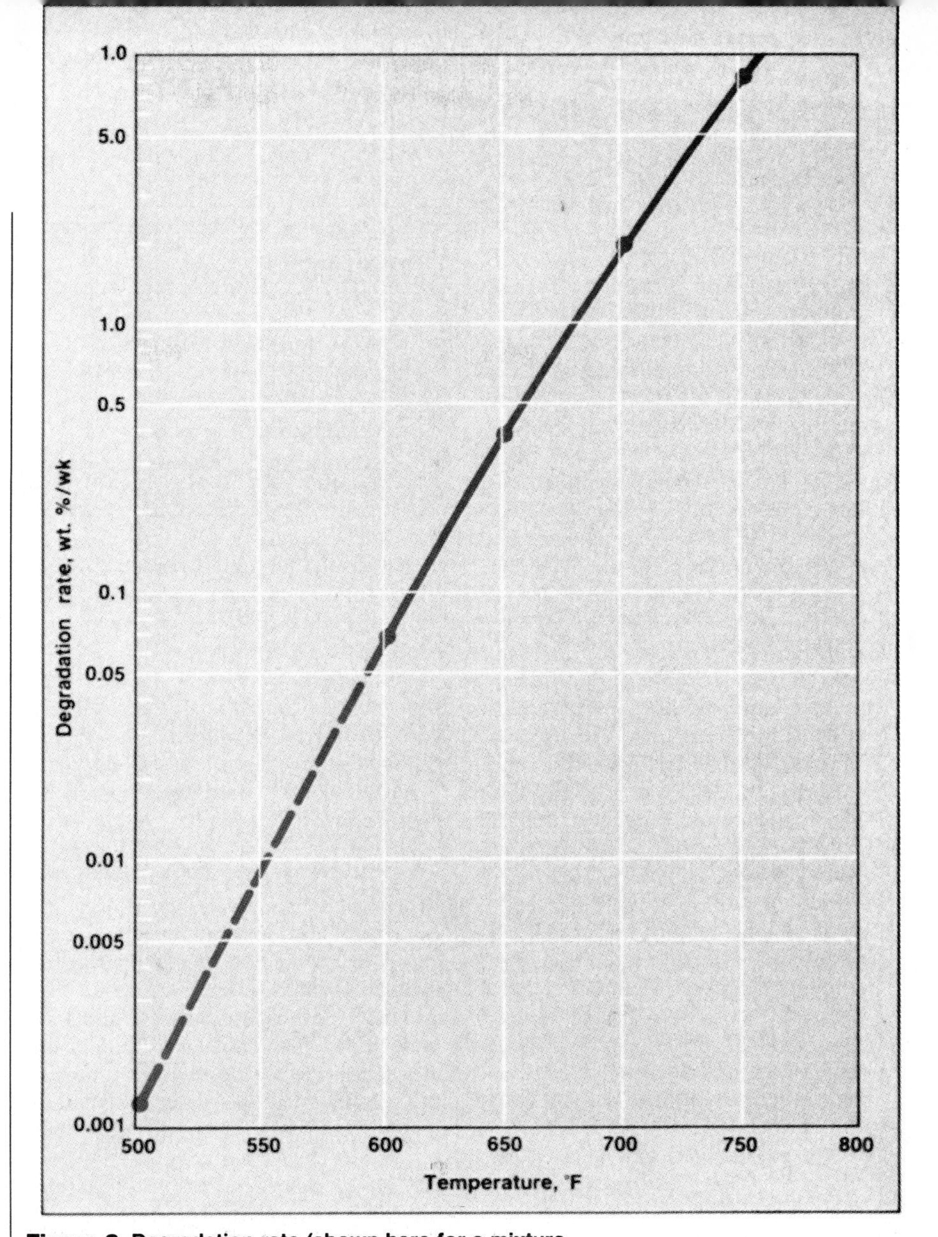

Figure 2. Degradation rate (shown here for a mixture of di- and triaryl ethers) increases with temperature.

Flammability and corrosivity

Most organic heat-transfer fluids are flammable when the temperature is sufficiently high and a source of ignition is present. At its flash point, the fluid ignites on application of a flame or spark. At its fire point, vapor is generated at a rate sufficient to sustain combustion. At the autoignition temperature, no ignition source is required to initiate combustion. The autoignition temperature of most organic fluids is far above their maximum recommended use temperatures. In all cases, air must be present at sufficient concentrations to support combustion.

The heat-transfer fluids listed in Table 1 are all noncorrosive to mild steel, so material selection can generally be based on operating temperatures and pressures of the process medium. For example, at low temperatures stainless steel may be required for strength even though neither the process fluid nor the heat-transfer fluid is corrosive. Specific material-compatibility data are available from most fluid and heat-transfer-equipment manufacturers.

Toxicology and environment

Most high-temperature heat-transfer fluids commercially available are moderately toxic, but they should present no unreasonable risk to operating personnel if normal operating precautions are taken. The various fluid manufacturers provide detailed information on specific health and environmental concerns that pertain to the fluids they market.

Since the fluids are normally used in closed systems, they do not ordinarily enter the atmosphere. Should the material accidentally contaminate the air, water or soil, it is well to know what effects it might have. In general, the greatest threat is to fish if the fluid is discharged into a body of water. Often, however, when the fish are later exposed to uncontaminated water, the compounds from heat-transfer fluids disappear from their tissue in a relatively short time.

A rupture in a system carrying the high-temperature fluids poses the threat of severe burns. If the fluid is contaminated by the material being processed (or by other materials), those contaminants may introduce their own health and safety hazards.

In other respects, most heat-transfer fluids present few health problems. For instance, unwitting inhalation of noxious fluid vapors is not likely, because most fluids have distinct aromatic odors even at low concentrations in air. Levels much higher than the odor-threshold level may irritate the eyes and breathing passages, yet still be below concentrations unsafe to breathe. Whenever unusual concentrations of fumes might occur, workers should wear respiratory protection suitable for organic mists and vapor. And whenever there may be a chance for eye contamination, suitable eye protection should be employed so as to avoid the discomfort that may result from direct contact. If the eyes become accidentally contaminated, they should be thoroughly washed with water, and medical attention should be promptly sought.

Single short exposures of heat-transfer fluids are generally not irritating to the skin. Prolonged or repeated skin contact, however, may cause slight to moderate irritation, or even a burn, and should therefore be avoided. Some fluids that have been used at high temperatures for extended periods of time may cause dermatitis or other skin irritation. Contaminated skin should be thoroughly washed with plenty of water, and contaminated clothing should be removed and decontaminated before reuse.

TABLE 2. Stabilities of several heat-transfer fluids

Fluid composition	Relative stability
Mixture of diphenyl and diphenyl oxide	0.0024
Mixture of di- and tri arylethers	1.0
Akylated aromatic	1.5
Hydrogenated terphenyl	1.6
Aromatic blend	2.2

(Relative-stability data generated in forced-circulation unit and/or ampoule test at 600°F at Dow Chemical Company.)

Ingestion of small quantities of high-temperature heat-transfer fluid will probably not be injurious, although ill effects may result if large amounts are swallowed. Again, the fluid manufacturer is the best source of information on precautions against ingestion and steps to be taken in the event that the material has been swallowed.

Degradation and its causes

When high process temperatures are required, fluid degradation can significantly affect system performance and operating costs. Degradation occurs when the temperature is so high that molecular bonds within the fluid break down. Some of the breakdown products have significantly lower molecular weights and volatize at process temperatures; they are known as "low boilers" or "lights". In normal operation, low boilers are periodically vented, and the lost fluid replaced by makeup. Fluids that degrade to produce a large percentage of lights incur a relatively high makeup volume.

Not all degradation products are lights, however. As degradation occurs and bonds break, some of the molecules can recombine to form high-molecular-weight components. These materials are good heat transfer agents until they exceed the manufacturer's recommended level of concentration.

The rate at which degradation occurs increases with the temperature. A general rule with organic fluids is that degradation doubles for every 18°F to 21°F increase in fluid temperature. Figure 2 shows the degradation rate versus temperature for a particular heat-transfer fluid. Under normal operating conditions, the fluid film temperature at the wall of a heater tube is designed to range between 50°F and 75°F above the bulk temperature of the fluid. Flame impingement on heater tubes can raise fluid temperature by hundreds of degrees and cause accelerated degradation. Low fluid velocity or complete loss of fluid circulation can cause a considerable volume of fluid in the heater tubes to become overheated.

By choosing a fluid with high thermal stability, however, the engineer can forestall most problems associated with localized or temporary temperature excursions beyond the upper process-temperature limit. Apart from minimizing problems associated with fluid overheating, a high-stability fluid will result in better thermal efficiency and less-frequent fluid replacement due to degradation. In some cases, fluids can be reprocessed to remove degradation products, allowing a large percentage of the fluid to be recovered, and thus improving the process economics.

TABLE 3. Comparing two typical fluids for degradation in heat-transfer loop

Fluid location (and temperature) within system	Volume percent of total system (Col. A)	Degradation of Fluid A* Pure fluid at stated temp. (Col. B)	Degradation of Fluid A* Within system (A×B)/100	Degradation of Fluid B* Pure fluid at stated temp. (Col. D)	Degradation of Fluid B* Within system (A×D)/100
Film in heater (650)	1.0	0.396	0.0040	7.92	0.079
Heater discharge (600)	34.0	0.0696	0.0237	1.39	0.473
Return loop (550)	34.0	0.0103	0.0035	0.206	0.070
Expansion tank (450)	31.0	0.0001	0.0000	0.002	0.001
Total degradation, weight % per week			0.0312		0.623

*Degradation is in weight % per week. *Fluid A is a mixture of di- and triaryl ethers (see also Fig. 2). Fluid B is a petroleum oil.

The cost of degradation

Within a given system, it is possible to determine the degradation rate of a given fluid, and thence the fluid make-up cost, with a reasonable degree of accuracy. This makes it possible to compare different fluids.

First, establish the temperature profile for the system. This profile specifies the temperature of the heat-transfer fluid in various segements of the system, and the percentage of the total volume contained in each of those segments. In existing heat-transfer systems, there may already be records of the temperature at various points; in new-system designs, one can estimate temperatures from the heat source and through each user in the process.

Second, assemble the thermal stability data for the fluids to be compared. For at least one of the fluids, the data should include degradation rates at several temperatures within the temperature range of the system, as shown for instance in the aforementioned Figure 2.

For calculation purposes, the relative degradation rates of the various fluids under consideration are assumed not to vary with temperature. For instance, if Fluid A is five times more stable than Fluid B at 550°F, the same relationship is assumed to hold true at 600°F. This makes it possible to determine the degradation rates of the various fluids at various temperatures from one thermal-stability plot (such as Figure 2) and a set of relative-stability multiplier factors, such as those in Table 2. The final piece of information required is the estimated system operating time per year. A system run intermittently throughout the year will naturally have a lower annual degradation percentage than a system that runs continuously.

The system's degradation rate is calculated by adding the degradation for each volumetric component. A sample calculation comparing two fluids appears in Table 3.

Total fluid makeup per year can be estimated by adding the degradation losses to the mechanical losses. The latter, of course, depend on the process requirements and operating practices. Annual mechanical losses for liquid systems are generally in the range of 5% to 10% of the total fluid in the system. Obviously, a system properly monitored and maintained, and utilizing a fluid of suitable stability, will minimize operating cost and provide designed heat transfer efficiency.

TABLE 4. Engineering properties of major heat-transfer fluids

Fluid composition	Maximum usage temp. °F*	Minimum pumping temp. °F	Film coeff. at 600°F Btu/(h)(ft²)(°F) At 5 ft/s	Film coeff. at 600°F At 7 ft/s	Pressure drop at 600°F psi per 100 ft At 3 ft/s	Pressure drop at 600°F At 7 ft/s
Mixture of diphenyl and diphenyl oxide	750	53.6	464	607	1.3	6.9
Mixture of diphenyl and diphenyl oxide	750	53.6	449	588	1.3	7.0
Dimethyl siloxane	750	< −50	239	313	1.2	6.2
Mixture of di- and triaryl ethers	700	−4	387	506	1.5	7.6
Hydrogenated terphenyl	650	31	353	462	1.4	7.2
Hydrogenated terphenyl	650	30	319	417	1.4	7.3
Aromatic blend	650	−50	384	503	1.3	6.9
Alkylated aromatic	600	< −100	522	684	0.9	4.9
Polyaromatic	600	−50	355	465	1.4	7.0
Synthetic hydrocarbon	600	4	302	395	1.2	6.2
Paraffinic oil	600	< 15†	268	351	1.3	6.5
Paraffinic oil	600	< 40†	280	367	1.3	6.4
Mineral oil	600	< 35	319	417	1.4	7.0
Mineral oil	600	< 25†	310	406	1.3	6.3
Mineral oil	600	20	298	390	1.3	6.3
Isomeric dibenzyl benzenes	662	−18	366	479	1.4	7.4
Isomeric dimethyl diphenyl oxide	626	−46	359	469	1.4	7.1
Alkyl diphenyl	707	−22	350	458	1.4	6.9
Alkyl benzene	590	−49	330	433	1.3	6.4

* Maximum use-temperature recommended by fluid manufacturer. † Reported pour-point temperature.

Effects of contamination

Contamination of an organic heat-transfer fluid can greatly accelerate the degradation and fouling in the system. The most frequently experienced contaminant is oxygen from the air. Above 300°F, most organic fluids oxidize at a measurable rate. As is the case (already noted) with thermal degradation, the oxidation products include low-boiling and high-molecular-weight products alike. These can increase the fluid viscosity, foul the system, and contribute to corrosion. Inert gas blanketing is recommended to prevent oxygen contamination in the expansion tank.

Besides oxygen, any strong oxidizing agent will also have a significant effect on a heat transfer system. In practice, the presence of such materials may be due to cross-contamination in the process, or they may be introduced during chemical cleaning procedures. To avoid the latter problem, cleaning-chemicals and inorganic salts must be carefully flushed from the system after cleaning. The effects of such materials are aggravated if trace amounts of water are also present.

Most organic process contaminants are not thermally stable when exposed for an extended time to the temperatures associated with a heat-transfer system, and their fouling of the system can be caused by their own cracking and polymerization. The heat-transfer system should be monitored routinely to detect foreign organic materials.

Process performance

An engineering evaluation is a key step in the selection process. The purpose of this evaluation is to determine the heat-transfer film coefficient and fluid-flow pressure drop for each respective fluid.

Many studies have been made to define correlations that relate the film coefficient (h_f) to the fluid's physical properties. For instance, the Sieder and Tate correlation [1] gives reliable coefficients for organic fluids. It is expressed by the following equation:

$$h_f = (F)(k/d)(dv\rho/\mu)^{0.8} \times (C_p\mu/k)^{1/3}(\mu/\mu_w)^{0.14}.$$

where C_p is specific heat, d is diameter, F is a units factor, k is thermal conductivity, v is velocity, μ is viscosity at the fluid bulk temperature, μ_w is viscosity at the wall temperature, and ρ is density.

Based on physical properties from Table 1, the film coefficient calculated for several fluids is tabulated in Table 4. This relative study is based on a fluid temperature of 600°F and velocities of 5 and 7 ft/s. The optimum velocity in any given situation will be an economic balance between cost of power and cost of heating surface; the two velocities selected here are within the range of typical designed exchangers, as discussed following.

In applications where low-temperature capabilities are directly required by the process — such as for cooling exothermic reactions in batch processing — there is an additional factor to keep in mind: The heat-transfer efficiency will drop off sharply when the fluid dynamics of the system switch from turbulent to laminar flow. This is likely to occur as fluids with relatively high viscosity are cooled. Because of the drop in heat-transfer capability, it is impractical to select a fluid whose lowest pumpable temperature is close to the chilled-loop temperature — indeed, in order to ensure turbulent flow and maintain heat-transfer efficiency, the fluid selected should not only be pumpable but also have a low viscosity at the minimum required process temperature. In particular, the design fluid velocity should yield a Reynolds number ($dv\rho/\mu$) greater than 3,000.

PUTTING THE PRINCIPLES TO WORK

In selecting an organic heat-transfer fluid for a high-temperature process, first narrow the selection field by defining the operating- and ambient-temperature requirements. For example, if the operating-temperature range is between 55°F and 450°F, all of the 19 fluids listed in Table 4 will perform adequately, according to published temperature data from their various manufacturers, and the choice will depend on other considerations.

If the high-temperature requirement is raised to 650°F, the oils as well as some of the other formulations are eliminated. If it rises another 50° to around 700°F, the selection is automatically narrowed to just five fluids. The same elimination process takes place when the low-temperature limit drops from 55°F to 0°F. Even if the high-temperature limit remains at a relatively low 450°F, all the oils and some of the other formulations can be eliminated. When the high-temperature limit is raised to 650°F with a low temperature limit remaining at 0°F, there are only four fluids left that fit this profile.

The next consideration is the film coefficient, which is a measure of heat transfer efficiency. Ideally, of course, the fluid selected should exhibit maximum efficiency. The application itself, however, will determine the minimum acceptable film coefficient and thus which fluids would meet this criterion. Film-coefficient data, shown in Table 4, has been calculated at velocities of 5 and 7 ft/s.

Assume, for instance, that the maximum required operating temperature is near 700°F, and that the low-temperature requirement is not extreme. Of the five fluids that seem to be appropriate for this situation, the film coefficient at 7 ft/s. varies from 313 to 607. This variation should be kept in mind as the selection process continues.

The third property to examine is the pressure drop, also listed in Table 4. The ideal fluid would, of course, exhibit minimum pressure drop. But notice that in the present example the fluid with the lowest film coefficient also has the lowest pressure drop, while that of the most efficient fluid of the five is somewhat higher. Again, the application and the geometry of the heat transfer system will help resolve which trade-offs are acceptable.

Finally, after temperature range, film coefficient and pressure drop, the thermal stability should be the next major consideration. If the fluid cannot pass this requirement, it need not be evaluated for such properties as flammability, toxicity, or even cost. If a less-stable fluid fouls the system, cost of the fluid is insignificant. Damage to the system, lost productivity, and the cost to remedy the problem become the overriding considerations.

The design pressure drop in an organic heat-transfer system is based on process restrictions or on adoption of standard design practices. Fluid-transfer lines are frequently sized by assuming a reasonable velocity [2]. In process exchangers, the velocity is dictated by process requirements and economics; typical fluid velocities range between 3 and 10 ft/s. Relative pressure-drop data for several fluids are tabulated in Table 4, based on standard calculation procedures [3] and at selected velocities of 3.0 and 7.0 ft/s. ■

The author

Walter F. Seifert, Research Associate at Dow Chemical Co.'s Larkin Laboratory in Midland, Mich., has been involved in research and development for his entire 28 years at Dow, much of it related to high-temperature heat-transfer fluids. Before joining Dow, he spent ten years in the petroleum-refining industry, in engineering design and manufacturing operations. He has four patents on new heat-transfer fluids, plus two on friction-reducing agents. Author of numerous papers on such subjects as high-temperature fluids, oil recovery and fluid mechanics, Mr. Seifert holds a B.S.Ch.E. degree from the U. of Colorado. He is a member of AIChE.

References

1. Sieder, E. N., and Tate, G. E., *Ind. Eng. Chem.*, 28, 1936, p. 1429.
2. Perry, R. H., and Chilton, C. H., "Chemical Engineers' Handbook", 5th Ed., McGraw-Hill, Sec. 5, p. 31.
3. "Flow of Fluids," Technical Paper No. 410, Crane Co., New York, 1969.

STEPHEN WALSH

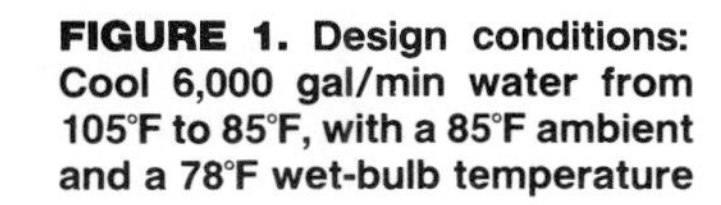

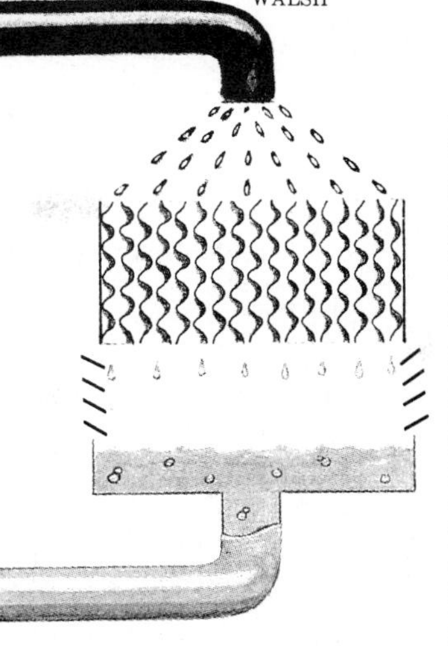

FIGURE 1. Design conditions: Cool 6,000 gal/min water from 105°F to 85°F, with a 85°F ambient and a 78°F wet-bulb temperature

COOLING WATER PAYS OFF

Robert Burger, Burger Associates

To ensure an economic return on the investment in a cooling tower, the engineer must continually review operating procedures and costs and upgrade tower efficiency. To accomplish these objectives, he or she needs to understand construction and operation.

Cooling towers are used in the chemical process industries (CPI) to reject waste heat that is absorbed from process streams by cooling water. The water carrying the waste heat can be discharged into a body of water or recycled and cooled. When recycled to a cooling tower, the water cools through the release of latent heat of vaporization and the exchange of sensible heat with the ambient air.

Recycling conserves water resources. This is important because about 500 billion gallons of water a day is used by industry in the U.S. alone. Of course, the unwanted heat from the water must be returned to the environment in an ecologically acceptable way. The cooling tower does this, and enables the CPI to recycle most of the water it requires with a reasonable capital investment.*

Towers can be upgraded

Not the lowest price, but lowest total cost (capital and operating), should be the criterion in purchasing a cooling tower. This involves comparing fan-motor and pump horsepower, as well as maintenance and life-cycle costs. The savings in the simple purchase price could result in a thermal performance that can boost expenses a hundred-fold.

An upgraded cooling tower can hike the performance of exchangers, turbines and compressors

Purchasing agents sometime buy the least expensive equipment available without taking into account operating costs. A cooling tower acquired this way can be costly because it may diminish the efficiency of overall plant operations. Fortunately, such towers can be upgraded.

One guide to efficiency is, of course, how much energy the tower itself consumes. The engineer should look into how much energy is needed to power a tower, and seek ways to improve its efficiency.

However, a more-important cost aspect that the engineer should investigate is how the cooling water is used, and how colder water could enhance equipment efficiency. For example, the colder the water to a refrigeration unit, the less energy is required to power it. Similarly, condensers function more efficiently with colder water, as do condensing turbines. Therefore, the engineer's primary objective should be to make the cooling water colder because there is value in doing so.

For example, water sent 1°F (0.6°C) colder to the compressors and condensers in air-conditioning and refrigeration equipment will save about 2½% electrical energy. Thus, water a little more than 4°F (2.2°C) colder can save 10% of the electrical energy. A 2,000-ton (1,815-metric ton) refrigeration system circulating 6,000 gal/min (22,700 liters/min) could consume power at the rate of $450,000/yr. A 10% savings in power consumption would be $45,000/yr realized from chilling the water a further 4°F could contribute significantly to a quick payback of money spent to upgrade a tower.

The criteria for gauging cooling-tower performance are noted in Figure 1. These design conditions, which are specified when a cooling tower is purchased or rebuilt, include the flowrate of the water to the tower, the entering and exiting temperatures of the water, the wet-bulb temperature, the *range* and the *approach*.

The range (also known as the ΔT) is the difference between the entering

*See the editorial "Managing Water," *Chem. Eng.*, January 1991, p. 5.

Originally published March 1991

and exiting temperatures of the water. The specified quantity of water must be cooled in accordance with the range at the stated wet-bulb temperature. The wet-bulb temperature, which represents the ambient and the dewpoint temperatures (or ambient temperature and the relative humidity), indicates the evaporative quality of the air. The difference between the temperature of the cold (discharge) water and the wet-bulb temperature is called "the approach to the wet bulb."

As stated, the basis of cooling-tower operation is evaporative cooling and the exchange of sensible heat. Contact with air releases latent heat from the water (i.e., evaporates some water). The heat thus released per pound of evaporated water amounts to approximately 1,000 Btu. Additional heat taken from the water lowers its temperature. Evaporative cooling accounts for about 75% of the heat transfer, with sensible heat accounting for the remainder.

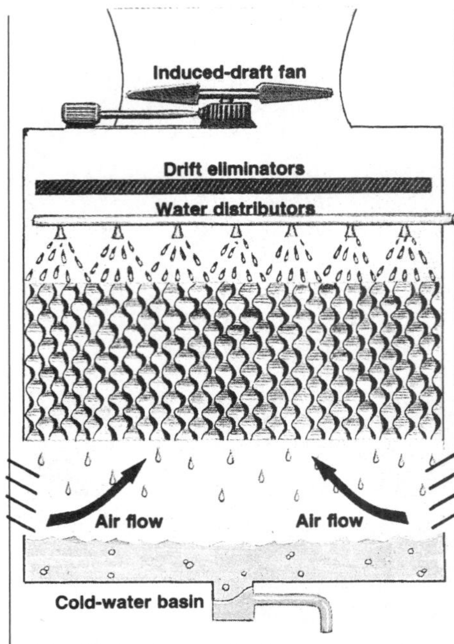

FIGURE 2. Air velocity through crossflow towers must be greater than through counterflow ones

One penalty in this operation is the water lost to the atmosphere as hot vapor. The evaporation loss at typical design conditions is about 1% of the water for each 12.6°F (7°C)[*1*].

Except small specialty models, all cooling towers are custom designed. Their shape, size and configuration vary with the particular sets of thermal parameters. But they are available in two primary operational designs based on the air flow in relation to the water flow: counterflow and crossflow (as diagrammed in Figure 2). The tradeoffs between them are discussed later.

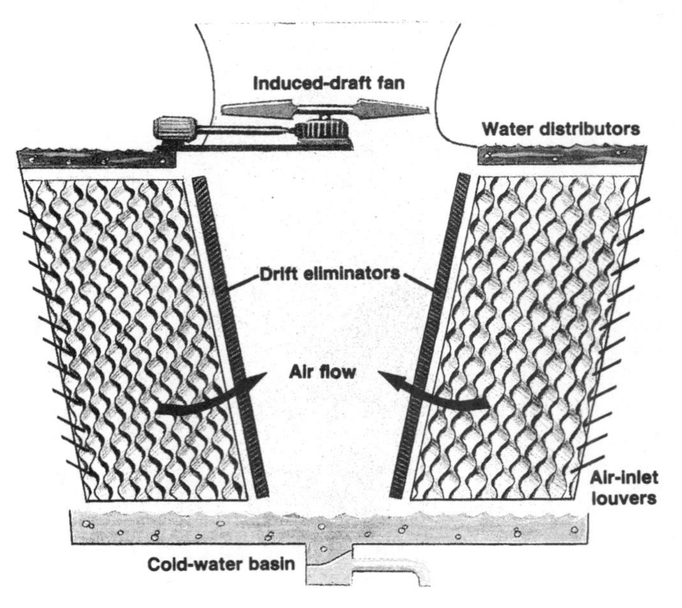

Upgrading a tower

There are three major components of a tower to investigate when seeking to upgrade it:

• Wet-decking fill, which makes up the surface for transferring heat from the water to the air entering the tower (Figure 2)

• Water-distribution system, which spreads the water uniformly over the wet-decking fill (Figure 2)

• Drift eliminators, which prevent most of the water entrained in the exiting air from being carried into the atmosphere (Figure 3)

Frequently, the greatest improvement in tower performance can be made by simply changing the wet-decking fill to a modern cellular fill. It is first necessary to evaluate the potential for change in heat transfer in terms of the fill's characteristics, as indicated by the manufacturer's performance curves. Developed painstakingly by trial-and-error experimentation, the curves are expressed as KaV/L — i.e., as a proportion of the vapor to the liquid [*2*].

With a fill of splash bars made of wood slats, the water droplets bounce from one layer to another (Figure 3). The counterflowing air simply cools the exterior of droplets. With a cellular fill, on the other hand, the droplets are spread into a very thin film. Now the air contacts a much larger surface for heat exchange. The air also passes through the cellular fill at smaller pressure drops. Standard polyvinyl chloride (PVC) makes a viable fill up to about 125°F. From 125°F to 140°F, chlorinated PVC is recommended, and polypropylene above 140°F.

Clean water is essential because dirty water can quickly clog cellular fill. If dirty water cannot be avoided and bacteriological buildup is possible, the passages in the fill should be larger. If chemicals, such as solvents, could attack the fill, a resistant type should be chosen. If the water quality is very poor, a cellular fill might not be the choice at all.

The water-distribution system should deliver water uniformly over the fill for maximum contacting between the air and water. The troughs of earlier systems distributed the water evenly to splash plates if both were precisely placed. With time, however, this delicate balance became undone by the deterioration of the the troughs and the plates. This resulted in many columns of water falling as much as 4 feet before being broken up. Obviously, this condition seriously undermined tower efficiency.

Water distribution must be uniform in towers fitted with cellular fill, because the fill constitutes nearly all of the tower surface for heat and mass transfer. Water distribution via spray nozzle works better, except when a nozzle becomes clogged. This leaves a dry spot in the fill through which the air is detoured (following the path of least resistance), wasting energy and cooling capacity.

Preferred for crossflow towers are spray nozzles that can be exchanged, if necessary, for others having different opening sizes and patterns.* The nozzles should be located about about three to four inches above the fill, and come off both sides of the distribution headers via 1-ft long pipes. The spray patterns should overlap to ensure thorough coverage.

Water-distribution arrangements can be more flexible for counterflow towers than for crossflow ones. In many systems, low-pressure-drop (3- to 15-psi) nozzles provide solid-cone patterns having little or no overlap. One spray nozzle developed specifically for counterflow towers produces a square pattern.

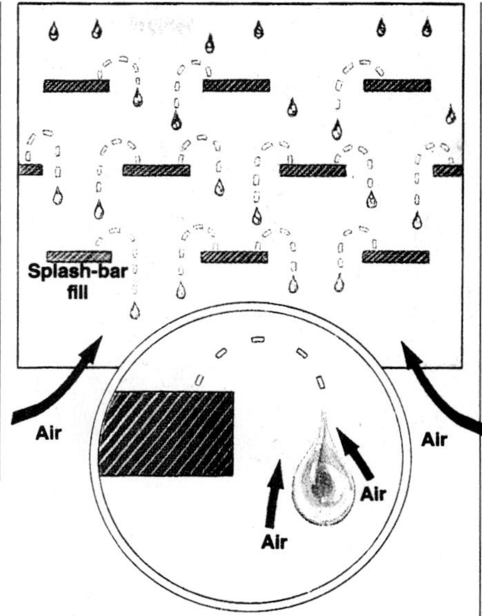

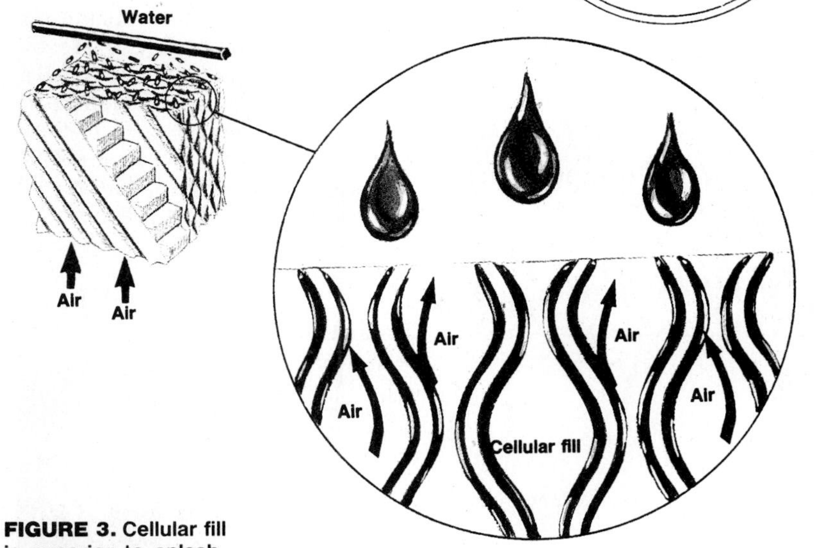

FIGURE 3. Cellular fill is superior to splash-bar fill because it enhances heat transfer

For some towers, nozzles of less than 4 psi pressure drop that produce a hollow-cone pattern are used. With these nozzles, the overlap must be considerable to eliminate dry areas. Such overlap is generally appropriate only for large towers having little internal bracing and few support columns. This arrangement is usually limited to concrete structures, because redwood towers have too many internal structures that could block off sprays and create dry areas. Whether the tower is crossflow or counterflow, the key to performance is uniformity: in water distribution, static pressure drop, fill configuration, and air velocity.

Drift eliminators prevent water droplets from being drawn or blown out of the tower. The mist or fog produced by the evaporation process cannot be eliminated from the tower, but a good drift-elimination system can prevent water droplets from leaving with the discharged air.

The elimination system may consist of baffles or cellular fill. It is placed between the water-distribution system and the point of air discharge to limit dispersal of entrained water droplets into the atmosphere. Figure 4 shows a three-pass drift eliminator made of baffles, and a new, more-efficient, six-pass type fabricated from cellular PVC. In the same space, the cellular type improves contact between water and air dramatically.

The performance of a cooling tower (i.e., the rate of heat removal) depends on the balance between the water and air volumes. Consequently, drift eliminators are normally designed to be efficient through a range of air velocities. An air flow that is too high could cause water to drift excessively from the tower. An air flow that is too low will, of course, limit the heat removal from the water.

A poorly designed drift-eliminator system increases the static pressure through the tower, slowing the air and lowering the cooling. A well-designed system can improve tower performance by up to 0.5°F (0.3°C). The lower pressure drop through the cellular drift eliminator allows the air to pass through at a higher velocity without additional cost.

Attempts to improve cooling-tower performance are frequently made by speeding up the fan rotation and increasing the pitch angle of the blades so as to raise air velocity. With the slat-type eliminator, this not only consumes more power but also increases drift loss, because the eliminators usually were not designed for the higher velocity. This is not the case with the cellular drift eliminator. Power consumption does not rise, because the pressure drop through cellular material will not increase.

Drift elimination could theoretically be eliminated but not practically. An acceptable level generally is stated to be "not in excess of 0.002% of the circulating water." Crossflow towers are more susceptible to drift loss than counterflow ones because the air velocity through them tends to be higher.

Counterflow vs. crossflow

Counterflow towers with cellular-film fill are more efficient than crossflow towers with splash-bar fill in all respects. Even with the same fill, however, the counterflow tower's average driving force for heat and mass transfer is greater because the coldest water contacts the coldest air and the warmest water the warmest air.

For the same cooling duty, a crossflow tower always requires more air

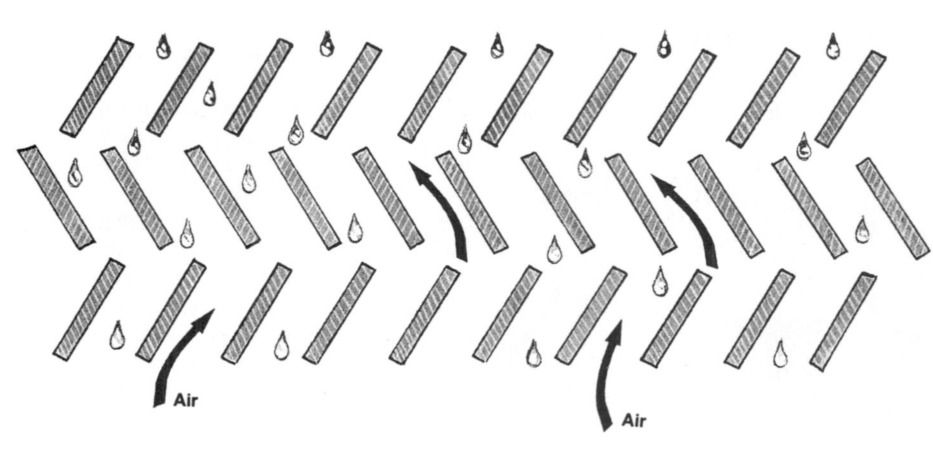

FIGURE 4. Cellular drift eliminator curbs the loss of entrained water better than baffles

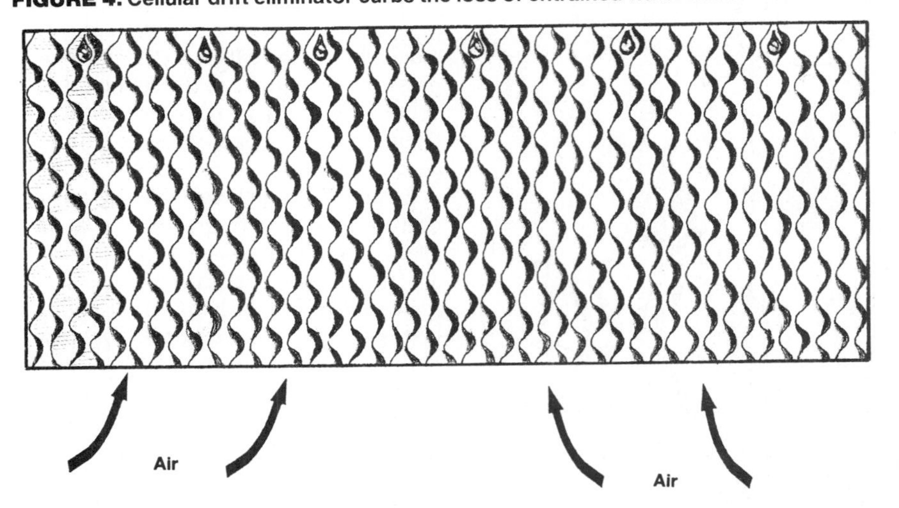

flow or fill, or both. It tends to recirculate hot air at its top because the higher air velocity there creates a lower pressure at the top of the air-intake louvers.

Additionally, wind may return hot discharge air into the upper air-intake louvers, lowering tower performance. The windage loss through crossflow towers is greater because the air-inlet velocity is higher and the air distribution is less uniform.

Winter icing in counterflow towers is more readily controlled because of such options as an auxiliary spray system beneath the wet-decking fill, better heat retention via reverse fan operation, and more tightly controllable operation. And when the towers do become iced, the counterflow tower typically sustains less damage. Moreover, inspections and maintenance for large counterflow towers are easier than for crossflow ones because the fill and drift eliminators are more accessible.

The pumping head required to raise the water to the hot-water distribution basins of crossflow towers can be twice as high. This can boost operating costs over a small tower's service life by hundreds of thousands of dollars, and a large one's by millions. Although many crossflow towers as high as 60 feet are still being built with wood or plastic splash bars, the cellular fill has gained prevalence because of its greater heat-transfer efficiency and lower pumping-head requirement.

The author

Robert Burger is president of Burger Associates, Inc. (2815 Valley View Lane, Suite 220, Dallas, TX 75234; tel.: 1-800-44-TOWER; FAX: 214-243-5869), a consulting, engineering and contracting firm, which is a leader in retrofitting towers to upgrade performance. The author has had more than 35 years of experience with cooling towers, and has had more than 40 articles published, as well a book entitled "Cooling Tower Technology." He was a founding editor of the Cooling Tower Institute's journal. ■

QUALITY STEAM, NOT STEAM QUALITY, DELIVERS MAXIMUM HEAT

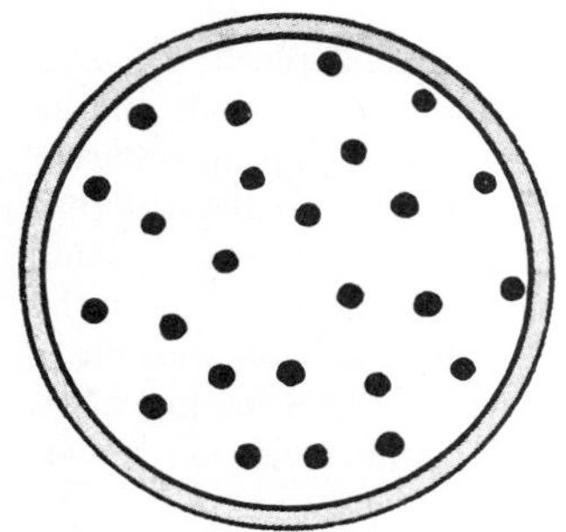

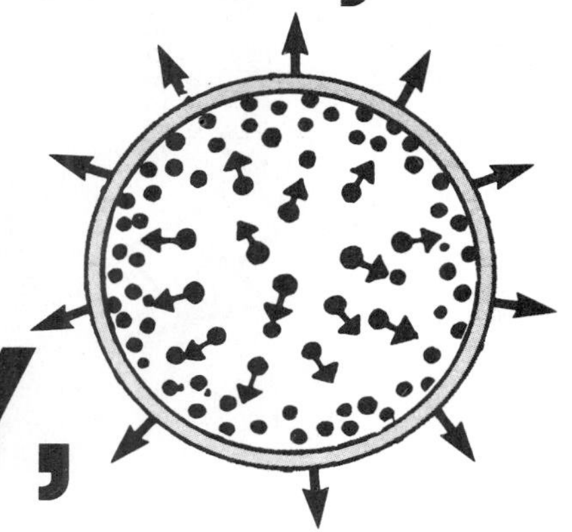

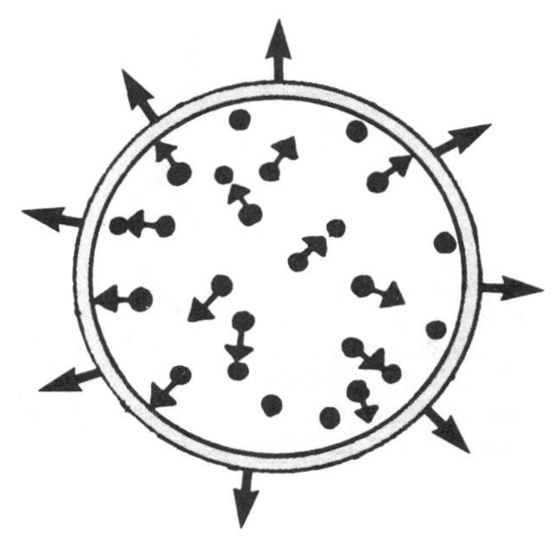

FIGURE 1. Condensing steam moves air to the surface of a heat-transfer pipe, where it collects to form an insulating film

There is more to quality steam than keeping it free of water

Walter T. Deacon,
Armstrong Machine Works, Inc.

Traditionally, the term "steam quality" defines the percentage of gaseous water flowing through a pipe. For example, if 95 lbs of gaseous water and 5 lbs of liquid water pass through a pipe, steam quality is said to be 95%. Quality steam, on the other hand, is not only dry but also clean, free of noncondensable gases, and at the right pressure. The distinction is not academic. Dirty steam causes valves to leak, and entrapped gases and poor pressure control undermine heat-transfer efficiency.

Flow and pressure controllers (valves and orifices) benefit most from cleanliness, although heat transfer is also better with clean steam. Entrained particles clog ports, hinder pressure and temperature control, and erode and corrode valve body parts, resulting in leaks.

Accumulations of noncondensable gases (including air) in steam piping limit steam flow, temperature and heat transfer. Air enters a steam system that is shut down and through the vacuum breakers of temperature-controlled processes. Noncondensable gases are liberated in the boiler, and carbon dioxide and oxygen come into a system dissolved in the boiler feedwater as carbonates and bicarbonates.

Noncondensable gases reduce the steam's temperature by contributing to the mixture's total pressure (as per Dalton's Law of Partial Pressure, which states that the pressure of a mixture of gases is equal to the sum of their partial pressures). Of course, less heat is transferred at a lower temperature.

The other gases are forced to the heat-exchange surface by the condensing flow of the steam. They form a stagnant film on the surface (Figure 1). Air is, of course, a good insulator. A 0.01-inch film of air has the same resistance to heat transfer as a 0.2-inch film of water and a copper panel 11 feet thick. If allowed to accumulate, the noncondensable gases will take up enough volume to effectively block steam flow and stop all energy transfer. Blocked condensate flow can cause dangerous water hammer.

When cooled in the presence of condensate, carbon dioxide can combine with water to form carbonic acid. This formation is highly probable because accumulated gas causes a temperature drop. The corrosion of iron forms a soluble bicarbonate, which leaves no protective coating on the metal. If oxygen is also present, rust forms and CO_2 is released, which can cause more corrosion. After the gas-

Originally published June 1990

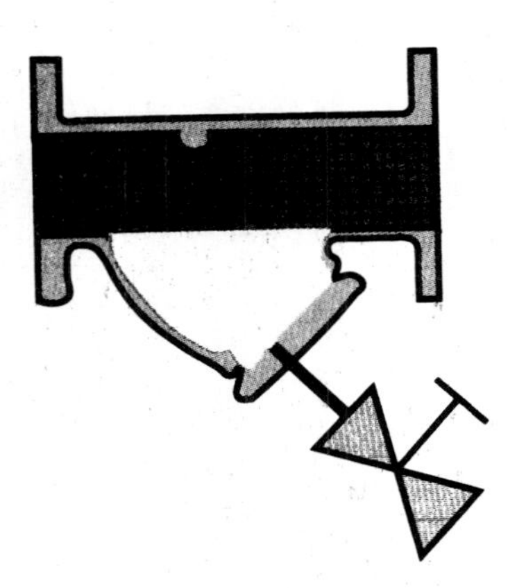

es become dissolved, they can be drained and removed. Until removed, they will continue to corrode metal surfaces.

Pressure is an important factor in maximizing heat transfer via steam. The higher the steam pressure, the higher are its temperature and energy content. However, the quantity of latent heat in steam increases as steam pressure decreases. To optimize latent heat transfer, steam should be used at as low a pressure as possible.

Pressure also controls the temperature of saturated steam. Because temperature difference governs heat transfer, higher temperatures make it easier to transfer heat, allowing heat exchangers to be smaller. The trade-off between a smaller exchanger and higher steam pressure could depend on steam consumption, which rises with declining latent heat.

Steam headers and branch lines are sized to distribute steam without excessive pressure drop. As the pressure declines, the total energy content of the steam decreases. A lower temperature slows heat transfer, creating a demand for more steam. Faster flow increases the pressure drop even more, wasting more total energy. Higher demand creates faster velocity, which contributes to more erosion and noise in the piping.

Reasons for measuring quality

Steam quality must be measured to qualify flowmeter readings for billing, to check boilerhouse practices, and to troubleshoot process operations. The trend in boilerhouse design has been to place larger heat-transfer surfaces in more-compact equipment packages. Because more steam is coming from smaller units, there is a greater likelihood of water being carried over with the steam. As the steam flows through the distribution system, heat radiates from the piping causing some condensation. The treatment of the water with chemicals can upset boiler operation and contribute to water overflow.

Field procedures for gauging steam quality are described in ASME Performance Test Code 19.11—Steam and Water Purity in the Power Cycle. Via four procedures, the solids content of steam, including carried-over water droplets, can be determined. Another procedure, based on the throttling calorimeter, is the only one by which steam quality can be found directly. It is most accurate with steam that is below 600 psi and whose moisture content is above 0.5%.

The calorimeter is a simple, easy to use instrument. When properly installed and insulated, it is also very accurate. Its functioning is based on the fact that when steam expands without doing work, its heat content (enthalpy) does not change. If dry steam is discharged to atmosphere, the pressure drop will cause it to become superheated. If moisture is in the steam, the atmospheric discharge will reduce its temperature. By comparing the two temperatures, one can determine the steam quality.

In the other methods (ion exchange, electrical conductivity, sodium-tracer flame photometry and specific-ion electrodes), the quantity of solids in the steam is measured to gain an indirect indication of quality. The solids content of a steam sample is determined and compared with the solids content in the boiler. This procedure is also accurate but not practical for field testing when condensed steam, not boiler carryover, is a factor.

Steam quality objectives

The detection of low steam quality often indicates a problem with water level control, feedwater treatment, even improper boiler size, as well as the failure of such equipment as separators and mist eliminators. The following objectives should be a part of a program to achieve steam quality.

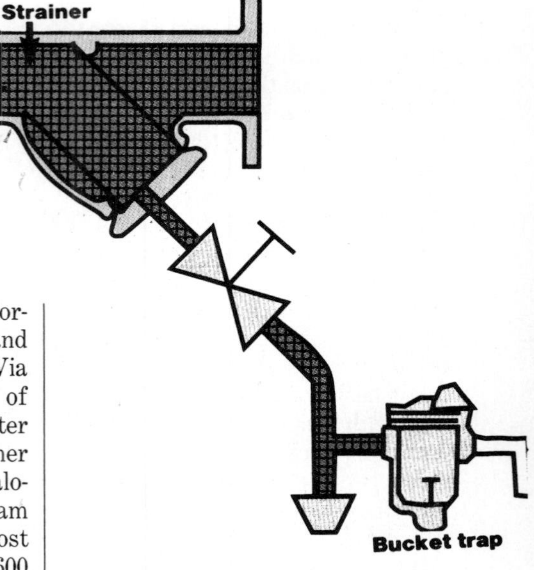

A strainer protects a steam trap by catching particles; for a strainer to do this most effectively, a downstream bucket trap should keep it free of condensate

Steam

Condensate

Allowing for pipe pressure drop, deliver the steam at a pressure suitable for the highest-pressure consumer. Even look into generating steam at a still higher pressure (up to 600 psig) to take advantage of specific volume, which decreases with higher pressure. Thus, at lower specific volume, more steam can flow through the piping or, for the same flowrate, the pressure drop can be less.

The steam velocity in distribution

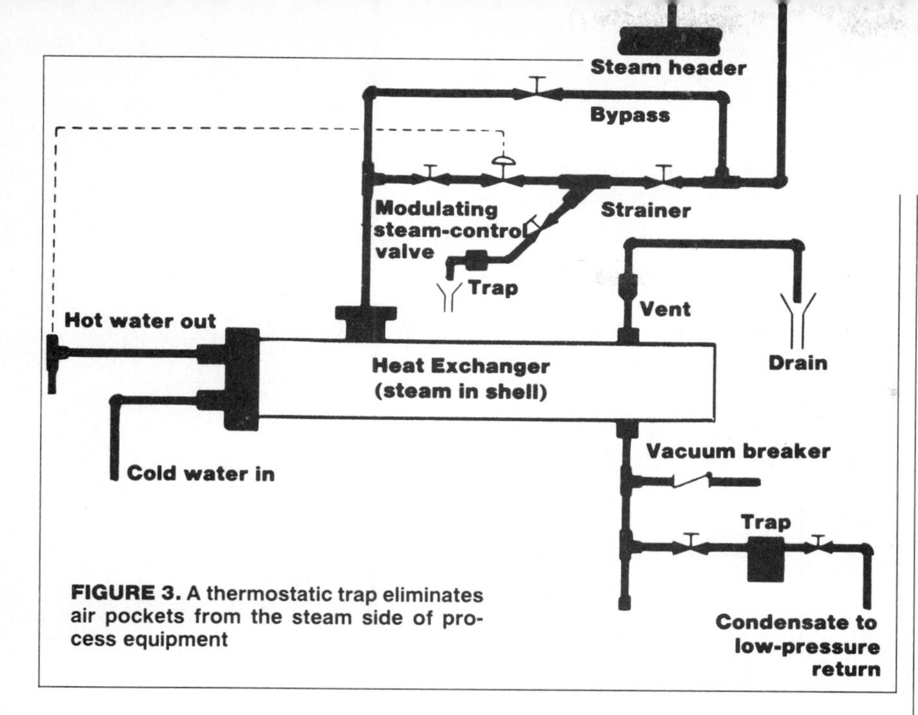

FIGURE 3. A thermostatic trap eliminates air pockets from the steam side of process equipment

piping should not exceed 12,000 ft/min, and the total pressure drop in the piping should not exceed 20% of boiler pressure. Of course, pressure drops should be avoided to preserve energy content and retain steam temperature. In one plant, engineers found that a boilerhouse expansion could be avoided by adding and repiping several lines to reduce pressure drops.

On the other hand, steam pressures should be reduced at points of usage to pressures as low as is practical to take advantage of steam's latent heat (which, as noted, increases with lower pressure). Local pressure-reducing or temperature-control valves can improve process control as well as the exploitation of latent heat. Pressure-reducing stations for entire areas might be considered, but a pressure or temperature regulator at the equipment can often achieve the same results with some savings (for example, the valve and the distribution line can be smaller).

Trap dirt with strainers

To avoid the accumulation of liquid and dirt, drip legs with dirt pockets should be installed at 500-feet intervals of the distribution system, ahead of risers and elevated expansion loops and at the end of headers. Strainers should be placed upstream of control valves and steam traps to protect them from dirt, even though a strainer is usually located a low point in the distribution piping. A strainer at a low point collects condensate, which reduces the effective area of the strainer screen (Figure 2). Some condensate will be picked up by the steam flow when the valve is opened, striking the valve seat with fluidized dirt. Installing an inverted-bucket trap on the strainer blowdown will drain the condensate, freeing the strainer screen from blockage and keeping it clean.

Bellows-type thermostatic steam traps can serve as automatic air vents on heat exchangers. Air in a system usually gets pushed into quiet zones by the flow of condensing steam. At such places, a thermostatic trap will sense the temperature reduction caused by air. To eliminate air accumulations, such equipment as batch process cookers, large shell and tube heat exchangers and large process air heaters are furnished with automatic air vents (Figure 3).

Steam traps should discharge condensate at or near its saturation temperature. Traps that back up and allow condensate to cool accelerate carbonic-acid corrosion. Properly sized, traps that do not back up condensate, such as inverted-bucket traps and float-and-thermostatic traps, also help maximize heat transfer.

Air vents and steam traps must be properly located to function efficiently. A trap should be installed so that it will be accessible for inspection and repair, and below and close to a drip point. Locate the proper places for air vents by means of a pyrometer or by visualizing the flow pattern in, for instance, a heat exchanger.

With a pyrometer, gauge the skin temperature of the heat exchanger to find the gas accumulations as cold spots. Lacking a pyrometer, visualize the flow in an exchanger by imagining how the steam must rush in and condense (Figure 4), and how the condensate must flow down and out the drain. Locate the air vent away from these active flow areas. Usually, vents are best located at high points away from inlets. A properly located steam trap that does not allow condensate to cool usually takes care of lower flow-quiet zones because it will also vent air. ■

FIGURE 4. One way of locating vents is to visualize the steam flow in equipment

The author

Walter T. Deacon is a senior applications engineer and assistant sales manager for Armstrong Machine Works (Three Rivers, MI 49093; tel.: 616-273-1415). His main responsibility is steam system troubleshooting and training. He works closely with customers, providing technical advice and counsel on complex steam-energy management applications. He holds a B.S. degree in mechanical engineering from Purdue University.

GET THE MOST OUT OF STEAM

Glenn E. Hahn
Spirax Sarco, Inc.

Remove particles, condensate and gases to increase system life, and tap condensate energy

Steam is the most popular heat-transfer medium in the CPI. It is easy to generate, safe to use and very efficient for heating processes or facilities.

Steam gives up heat when it condenses. Nearly 1,000 Btu per pound of steam is released by condensation at heat-transfer coefficients 10–100 times higher than those of convection.

Most engineers are familiar with steam generation, but often overlook precautions when transporting steam from the boiler to heat-transfer equipment such as valves and heat exchangers. For instance, solid particles, noncondensable gases (air or carbon dioxide) and condensate should be removed from distribution lines and heat-transfer equipment. If they are not removed, erosion and corrosion can occur, reducing the equipment's service life and the system's efficiency.

Condensate can also be a valuable resource. The liquid can be reused, either as a low-pressure steam source or as boiler feed water. If condensate is not collected, recoverable energy is down the drain.

Many factors must be considered if steam-distribution systems are to have long and efficient lives. These include: pressure between the boiler and heat-transfer equipment; solid particles that can cause erosion; and condensate removal and reuse.

Maximize generation pressures

Steam can be produced and distributed in either saturated or superheated forms. Saturated steam is in equilibrium with water at a given temperature and pressure. Superheated steam is saturated steam that is heated above its saturation temperature. It must cool to the saturation temperature before condensate can form.

Whether saturated or superheated, steam is usually generated at the boiler's maximum design pressure, since high-pressure steam occupies less volume than does low-pressure steam. For example, one pound of steam has a volume of 28 ft^3 at 14.7 psia (1 atm). At 150 psia, the volume of that same pound of steam is 3 ft^3.

As a result, high-pressure steam systems cost less to build than low-pressure systems, because high-pressure distribution pipes can have smaller diameters.

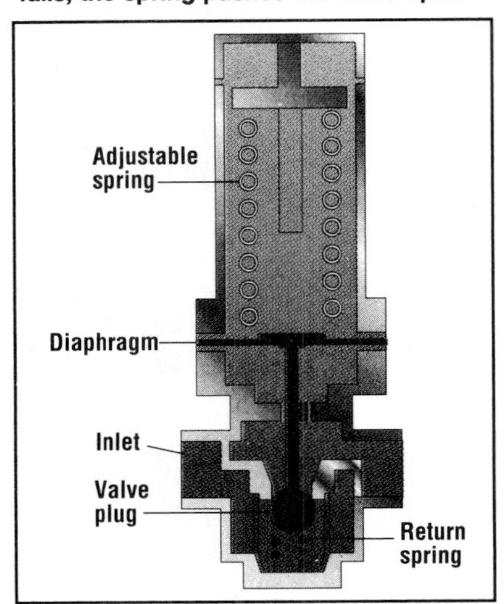

FIGURE 1. In the direct-acting reducing valve, the downstream pressure on the diaphragm balances the spring, closing the valve. When the downstream pressure falls, the spring pushes the valve open

STEVEN STANKIEWICZ

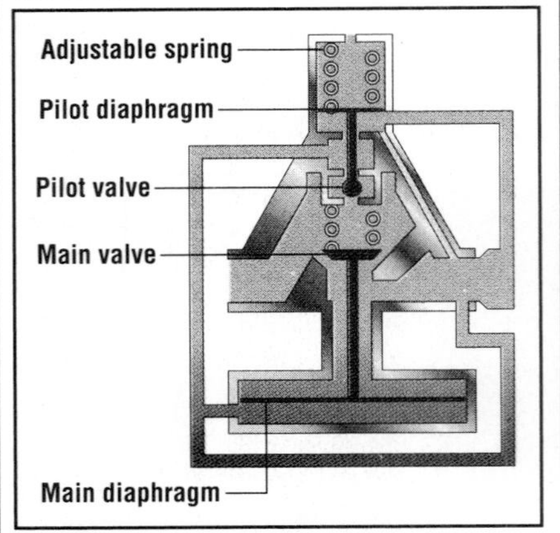

FIGURE 2. The pilot-operated PRV employs two diaphragms to provide improved control over pressures and flows

Originally published January 1994

However, high-pressure steam systems have some drawbacks. Although smaller pipes can be used, their walls must be thicker to handle the pressures. Similarly, valves, fittings and other heat-transfer equipment must also have thicker walls.

The potential for heat loss is another drawback to using high-pressure steam. As the steam pressure increases, its temperature also increases, so that high-pressure steam systems lose more heat to the atmosphere than do low-pressure systems.

Pressure regulation devices

Although steam should be generated and distributed at the highest possible pressure, it is actually used at many different pressures in a facility. A 150-psia distribution line may feed processes requiring five or six different pressures, depending upon the application. Steam tracing, for example, generally uses steam at about 20 psia.

There also are practical reasons to lower the steam pressure before it reaches heat-transfer equipment. The higher the steam pressure, the higher its temperature, so that if the steam pressure is too high, you run the risk of overheating your processes.

Paradoxically, low-pressure steam gives off more heat upon condensing than does high-pressure steam. Thus, less low-pressure steam is necessary for a given heating load.

Pressure-reducing valves (PRVs) often stand between the steam-distribution line and heat-transfer equipment. They are "gates" that reduce the pressure of the steam entering heat-transfer equipment. There are two basic types of PRVs: direct-acting and pilot-operated ones.

The direct-acting PRVs are the simplest available, and are used where accurate pressure control is not critical (Figure 1, p. 80). Inside the valve is an adjustable spring that presses down on a diaphragm. The pressure downstream of the valve presses up on the diaphragm.

As steam use increases downstream of the PRV, the pressure below the diaphragm falls and the spring pushes the valve open. This allows more steam to pass through the valve and balances the downstream pressure.

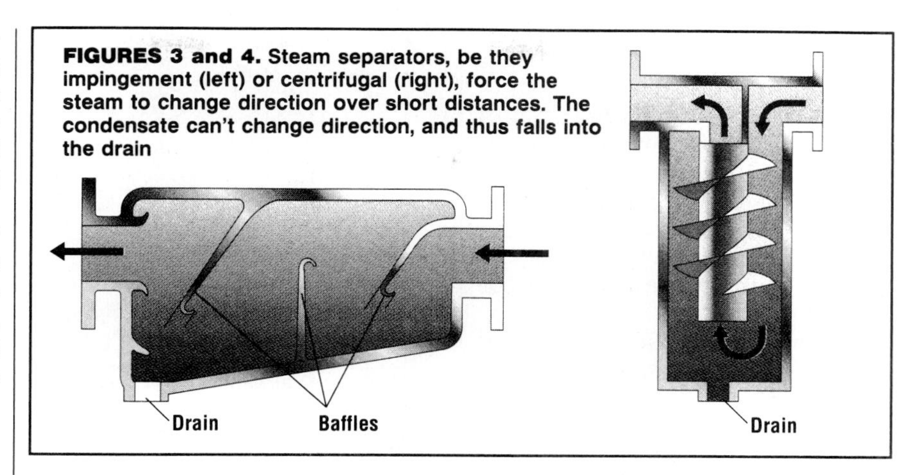

FIGURES 3 and 4. Steam separators, be they impingement (left) or centrifugal (right), force the steam to change direction over short distances. The condensate can't change direction, and thus falls into the drain

Pilot-operated PRVs (Figure 2, p. 182) contain two diaphragms. As in direct-acting units, downstream pressure balances the force of a spring on a diaphragm. When the downstream pressure falls, the spring pushes the valve open.

However, instead of going downstream to equalize the pressure, the steam passes through a small pipe inside the valve to another, larger diaphragm. The steam pressure against this diaphragm opens the main valve and allows steam to pass downstream. Any pressure changes are felt first by the pilot diaphragm, which adjusts the main diaphragm and valve.

Pilot-operated PRVs are more accurate than direct-acting ones, because small changes in downstream pressure act on the pilot diaphragm to produce large changes in steam flow. For this reason, pilot-operated PRVs should be used when very accurate pressure control is required.

Remove solids and liquids

Steam moves through distribution pipes at velocities of around 100 ft/s. Any solid particles or droplets of liquid carried by the steam strike fittings, valves and other equipment, causing erosion. The higher the steam velocity, the more severe the erosion. The system should contain removal devices for solids and liquids.

Newly installed pipes, or those that have just been repaired, frequently contain dirt, weld slag or mill scale. Rust on pipe walls is another source of solids that can be carried through steam pipes.

Solids are removed by inline strainers. The devices, which are installed directly upstream of control or pressure-reducing valves, contain mesh screens that capture particles. The screens should be removed and cleaned periodically.

Condensate can be just as damaging to steam systems as solids can. It

RULES OF THUMB TO USE WHEN SELECTING STEAM TRAPS

APPLICATION OR CONDITION	RECOMMENDED TRAPS
Steam distribution lines	Thermodynamic, balanced-pressure
Heat exchangers that cannot tolerate back-up	Mechanical float or inverted-bucket
Water hammer	Bimetallic, inverted-bucket or thermodynamic
Batch processes with frequent startups	Balanced-pressure or bimetallic thermostatic
Heat exchangers with varying loads and pressures	Mechanical float or inverted-bucket
Systems that shut off abruptly	All except inverted-bucket traps, which can lose their seals
Freezing	Thermostatic, balanced-pressure, bimetallic or thermodynamic

forms almost continuously in the system and can cause erosion and water hammer.

Water hammer occurs when a slug of water has too much momentum to turn a corner and crashes into a valve or fitting. The effect is equivalent to hitting the equipment with a hammer. Water hammer severely weakens welds and can cause catastrophic failure in steam systems.

Condensate can also be corrosive. Trace amounts of carbon dioxide can be introduced to the boiler feed water when it is treated. The CO_2 remains in the steam and lowers the pH of condensate, making it corrosive to metals.

To prevent erosion and corrosion problems, it is critical to keep condensate formation as low as is practical. Insulating the steam pipes does well to limit condensate formation, but won't eliminate it altogether.

Removing condensate that does form is equally critical. There are several steps that can be taken in the design stage to provide for condensate removal from distribution pipes and heat-transfer equipment.

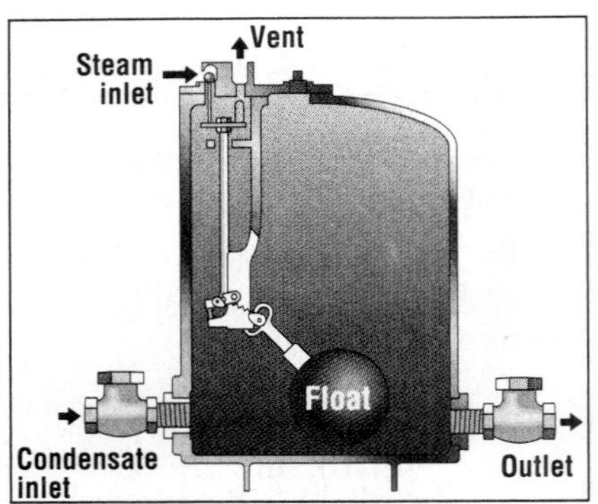

FIGURE 5. Operation of the pressure-powered condensate pump is triggered by a mechanical float. Condensate buildup raises the float, opening a steam valve that pushes out the liquid

Six steps taken in the design stage will prolong equipment life by reducing solids and condensate buildup

• ***Install drains and collection legs.*** Collection legs are vertical lengths of pipe that are attached to the bottoms of steam pipes. These legs are fitted with drains from which the collected condensate passes into a return pipe or out of the system altogether. Collection legs should be at least 28 in. long and located 150–300 ft apart.

• ***Slope distribution lines.*** Steam pipes should be angled downward in the direction of the flow, so that condensate falls into the collection legs. The lines should drop at least ½ in. every 10 ft. For distribution lines up to 4 in. dia., legs should be the same diameter as the main line. Lines larger than 4-in. dia. can be fitted with collection legs that are one or two standard diameters smaller than the main line.

• ***Fit every collection leg with a steam trap.*** Steam traps are self-actuating valves that are closed in the presence of steam and open to drain condensate. Non-condensable gases such as air and CO_2 can also be removed through steam traps. Several types of traps are available, each with different methods for detecting of steam and condensate (box, p. 84). Traps should be selected carefully, because they can operate incorrectly or be damaged when superheated steam is used.

• ***Install separators.*** Some condensate gets past collection legs. To remove it from the steam, inline separators should be installed. Two types of separators are popular: impingement and centrifugal (Figures 3 and 4, p. 81).

Both types change the steam's direction. In the impingement separator, droplets strike the baffles and fall to the bottom, from where they are drained.

The centrifugal separator operates like a cyclone. A spin is imparted to the steam, causing the condensate droplets to strike the side walls and fall to the bottom of the unit.

• ***Distribute superheated steam.*** Superheated steam, which is heated above the saturation temperature, loses heat without producing condensate. Therefore, distributing it can reduce erosion and corrosion in piping.

However, the heat-transfer coefficient for superheated steam is less than 10% that of saturated steam. Superheated steam should be cooled to its saturation temperature, or "desuperheated," before it is used.

Desuperheating is performed by spraying water directly into the steam. The evaporation of the water cools the steam to its saturation temperature. Any remaining water is removed by a separator.

• ***Start up systems slowly.*** Most condensate forms in saturated and superheated steam systems at startup, when steam comes in contact with cold pipes. The potential for erosion is also greatest at startup because steam velocities are higher than normal as the system is filling up.

Thus, systems should be filled slowly. If filled too quickly, boiler pressure can drop, causing water entrainment.

Gases cause problems, too

In addition to solids and liquids, non-condensable gases — air and carbon dioxide — must be removed from steam because they reduce system efficiency in two ways. First, they dilute the steam and lower its condensation temperature. Thus, more steam is needed to maintain process temperatures and the system's efficiency is reduced. Second, non-condensable gases transfer heat by convection, which is less efficient than condensation (conduction).

Air enters steam systems from the outside. When the system is opened for repair, it fills with air. Similarly, when the steam supply to an exchanger is shut off, the steam in the exchanger condenses and creates a vacuum that draws air in through valve packings, flanges or other joints.

Carbon dioxide is introduced to a steam system when boiler feedwater is treated. Hydrochloric acid is commonly used to remove carbonates from the feedwater, and CO_2 is formed when carbonates react with HCl.

Both air and CO_2 dissolve in condensate, lowering its pH and increasing the risk of pipe-wall corrosion. Air, which contains oxygen, speeds up the rate of corrosion.

Steam systems must be equipped to remove non-condensable gases. This is commonly done in two ways: by installing vents on top of the heat-transfer equipment, or by using steam traps with thermostatic sensors.

Thermostatic steam traps are most-commonly used. These devices continually measure the steam temperature and determine if it is below the saturation temperature for a given pressure. If the temperature is too low for the pressure — indicating the presence of a non-condensable gas — the trap opens until the steam temperature rises to its proper temperature.

Draining heat exchangers

Condensate removal from heat-transfer equipment should be handled the same way as from distribution lines. Condensate must readily drain to collection legs, and the steam traps should match the application.

Unlike distribution lines, heat-transfer equipment doesn't always have sufficient steam pressure to push out condensate. If condensate is not drained from heat-transfer equipment, the available heat-transfer area is reduced.

To compensate for the reduced area, the control valve, which regulates the pressure, opens the steam valve, admitting steam to increase the pressure in the exchanger. When the pressure is high enough, condensate is pushed out of the steam trap.

When the exchanger fills with steam, the process heats up and the control valve closes the steam valve. The pressure in the exchanger falls and condensate accumulates again.

This repeating cycle is called "water logging" or "stalling," and its symptom is an uncontrollable rise and fall in the heat exchanger's temperature. If stalling is ignored, water hammer and corrosion can occur in the heat exchanger.

The lower the pressure in the heat-transfer equipment, the greater the risk of stalling. The phenomenon is most common in low-pressure (below 25 psia) equipment such as process-water and space heaters and reboilers. If the process temperature is below 212°F, the steam pressure may have to be throttled below atmospheric pressure, which assures stalling.

Use pumps to drain exchangers

When the condensate cannot drain from heat exchangers as quickly as it forms, pumps are required to remove

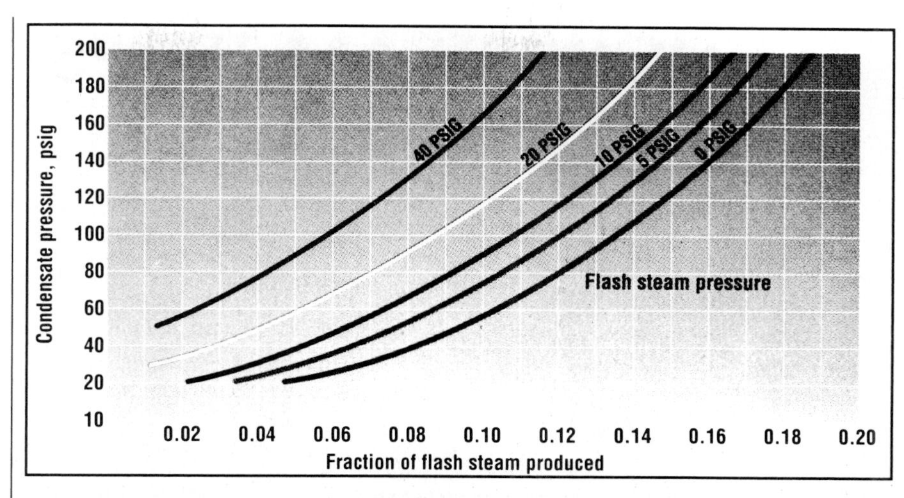

FIGURE 6. The graph shows the yield of flash steam at various pressures as a function of the condensate pressure.

it. Selection of the proper condensate pump is critical.

Centrifugal pumps and vapor-powered pumps are the most common because they are cost-effective and can handle condensate temperatures. Centrifugal pumps provide continuous flow, but require a minimum inlet pressure, called the net positive suction head (NPSH). The NPSH is calculated by the following equation:

$$H = H_p + H_z - H_f - H_{vp}$$

where H = net positive suction head
H_p = pressure at fluid surface
H_z = static head of liquid at pump inlet
H_f = friction loss in piping
H_{vp} = vapor pressure of liquid

Insufficient pressure at the inlet of a centrifugal pump causes cavitation. This occurs because the fluid boils as it enters the impeller, creating bubbles. When these bubbles pass the impeller, the surrounding pressure increases and the bubbles condense or implode violently. These implosions can reduce the service life of the seals, bearings and impeller.

In most condensate pumping applications, the NPSH is equal to the static head less the friction loss. The vapor pressure of condensate at the pump inlet is equal to that at the fluid surface because the liquid is at or near its boiling point.

Thus, the pump is usually placed directly under the steam trap and a minimum head of condensate is provided in the piping. This maintains the NPSH available to the pump.

Another way to provide sufficient NPSH is to cool the condensate before it reaches the pump. This can be done by evaporation, convection losses from the pipe, or even by installing a cooling coil. However, this removes the heat that gives condensate its value.

To eliminate the need for complex suction piping schemes, vapor-powered pumps can be used to remove condensate from heat exchangers (Figure 5, p. 82). Vapor-powered pumps cost about as much as centrifugal pumps, but have no rotating parts, and require no electrical connections or minimum head. When vapor-powered pumps are used, there should always be a condensate reservoir just upstream of the pump to prevent condensate from backing up into the exchanger while the pump is being discharged.

The pump, installed under the heat exchanger, contains a collection chamber into which condensate falls. A float in the chamber gradually rises, eventually opening a steam valve. The steam pressurizes the chamber, forcing condensate into the return line. As the chamber empties, the float drops, the steam valve closes and the chamber is vented. When the pressure of the chamber and the heat exchanger are equalized, condensate collection resumes.

When a heat exchanger always employs steam at pressures below conden-

HOW STEAM TRAPS DETECT AND

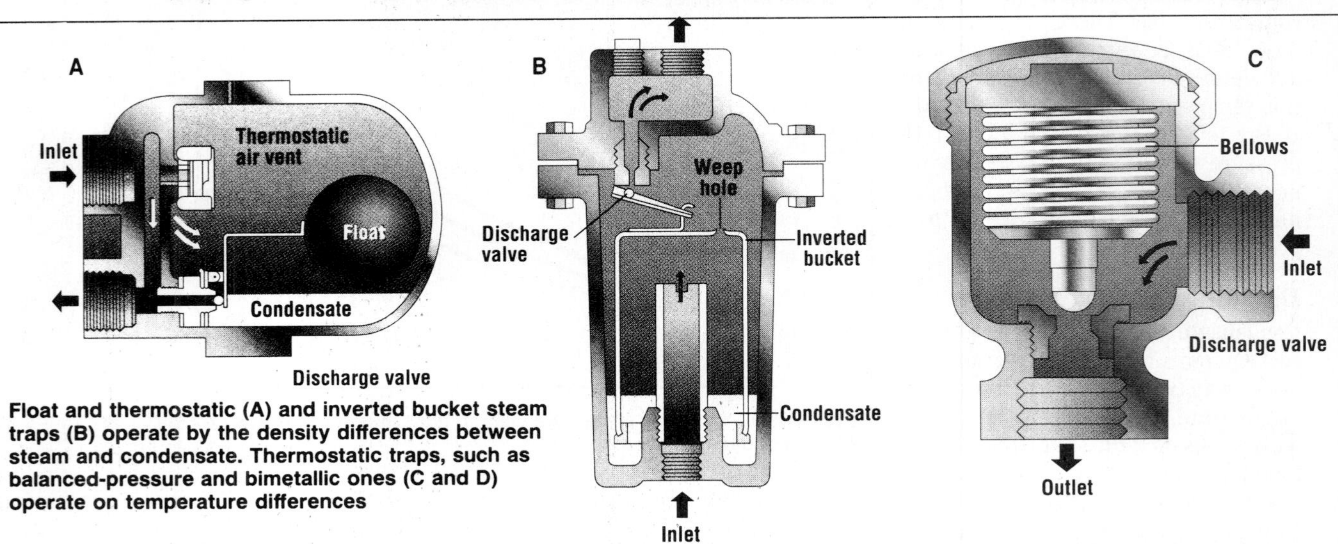

Float and thermostatic (A) and inverted bucket steam traps (B) operate by the density differences between steam and condensate. Thermostatic traps, such as balanced-pressure and bimetallic ones (C and D) operate on temperature differences

The steam trap is the gate through which condensate leaves the process. It is an automatic valve that closes when steam is present, and opens when condensate is present. Discharge occurs as long as the condensate pressure is higher than the downstream pressure.

There are three basic types of steam traps. Each employs a different method of detecting condensate and steam and a different means of draining condensate. For each application, there are usually one or two types of traps that will operate most effectively (Table).

Mechanical traps take advantage of the different densities of steam and condensate. In these traps, the discharge valve is connected to a float or a "bucket" that responds to condensate levels.

Thermostatic traps contain temperature sensors that trigger operation of the discharge valve. The valve opens when the sensor comes in contact with condensate.

In *thermodynamic traps*, flash steam in the condensate operates the discharge valve. The principle of operation in these traps is that gaseous steam moves faster than liquid condensate.

Mechanical traps

The simplest type of mechanical trap is the float trap. In this trap, the discharge valve is connected directly to a float that rises and falls with the condensate level. As condensate collects in the trap, the float rises, opening the valve. The size of the opening is proportional to the level of condensate, so that changes in flow adjust the discharge rate. Float traps operate independently of temperatures and pressures.

While open, the discharge valve is "sealed" with condensate. As a result, neither steam nor air can escape the trap. As a rule, float traps also have thermostatic air vents that permit non-condensables to escape. When the steam temperature drops — indicating the presence of air or another non-condensable gas — the vent opens until the temperature rises. Such traps are called float and thermostatic traps (A, above).

Float traps handle a wide variety of condensate loads and steam pressures but are susceptible to corrosion and water hammer. Condensate can also back up into this type of trap unless check valves are installed at the outlet. Float traps are used in systems with frequent or sudden changes in steam pressure.

Inverted-bucket steam traps (B) withstand water hammer better than float traps. These traps are operated by steam entering an inverted bucket that floats on top of the condensate in the trap.

When the steam is turned on, the bucket is at the bottom of the trap and the discharge valve is open. As condensate enters the trap, the water level rises inside and outside of the bucket and condensate passes through the open valve. Steam and air are displaced through a small hole in the top of the bucket, called the weep hole.

As steam enters the trap and fills the bucket, the bucket floats, closing the discharge valve. The steam slowly escapes the bucket through the weep hole. As more condensate enters the trap, the bucket sinks, the valve opens and the steam pressure forces condensate out of the trap.

Inverted-bucket traps are sealed with condensate. However, sudden pressure changes can break the seal, in which case the bucket remains at the bottom of the trap and the discharge valve stays open. Steam and condensate are discharged until condensate flow increases sufficiently to reseal the bucket.

Thermostatic traps

Condensate, once out of direct contact with steam, will cool to a temperature slightly less than that of steam. Thermostatic traps employ this reduced temperature to discharge condensate.

The simplest thermostatic trap, the liquid-expansion trap, opens and closes by the expansion and contraction of a thermostat filled with oil. The thermostat is connected to a piston that opens and closes the valve. The piston can be adjusted by a nut to regulate the temperature at which the valve opens and closes.

A disadvantage of the liquid-expansion trap is that it has a narrow capacity range and does not respond quickly to changes in flow. For example, if the opening is set to allow a certain flowrate of condensate, a quick increase in flow will invariably cause condensate backup at the valve. Therefore, this type of trap should not be used with heat exchangers or distribution lines because of the risk of backup.

Balanced-pressure thermostatic traps

RELEASE CONDENSATE

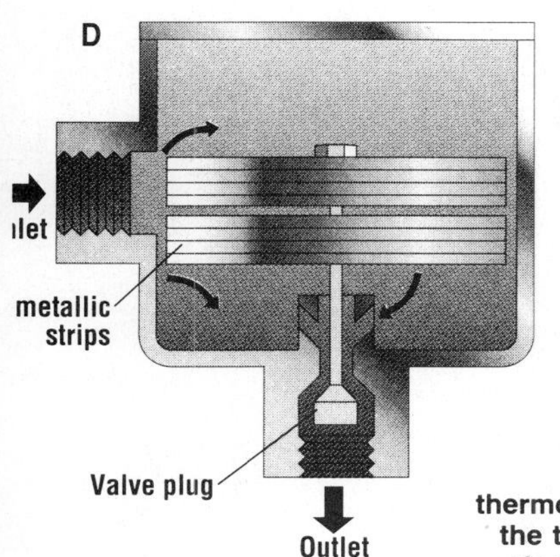

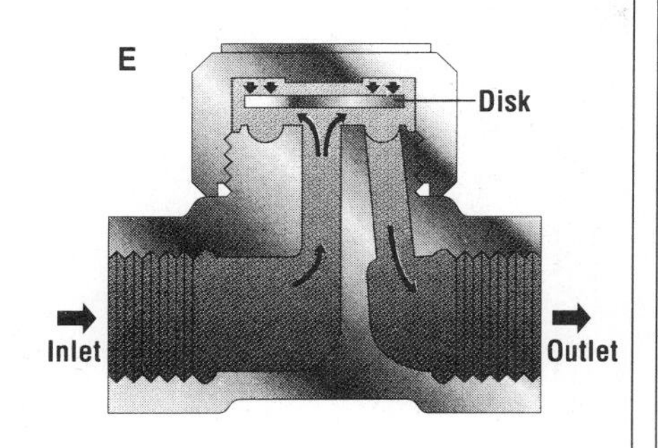

Flash steam opens and closes the disc in a thermodynamic trap (E). Steam and condensate exiting the trap raise the disc. As the temperature increases, steam flashes, increasing the pressure and snapping the disc shut

(C) are also activated by the cooler temperature of condensate. These traps are also operated by fluid expansion, but respond to steam pressure as well as temperature.

The valve in a balanced-pressure trap is operated by a flexible bellows containing an alcohol mixture whose boiling point is below that of water. When steam is present in the trap, the mixture in the bellows boils, expanding the trap and closing the discharge valve. As condensate fills the trap, the liquid condenses and the bellows contracts to open the valve.

Balanced-pressure traps are excellent air vents. In fact, they are used as the air vents in float and thermostatic traps. However, balanced-pressure traps are susceptible to some condensate backup. These types of traps can be damaged by water hammer because the wall of the bellows is thin. But a wafer-type balanced-pressure thermostatic trap is now available that is less susceptible to damage.

A more-rugged type of thermostatic trap is the bimetallic trap (D). This trap contains strips made from two metals with different thermal-expansion coefficients.

As steam fills the trap, one metal expands more than the other metal, causing the strip to bend. This movement lifts the valve plug and closes the valve. When condensate enters, the strips cool, revert to their original shapes, and open the valve to discharge liquid.

Bimetallic traps respond slowly to temperature and pressure changes. As a result, they can create condensate backup in drain legs.

Thermodynamic Traps

In a thermodynamic trap (E) the discharge valve is a free-floating disc that lies on a flat seat. When the system is started up, air and cool condensate raise the disc and exit the trap.

As the temperature of the condensate increases, some of the liquid flashes into steam. This increases the flow velocity because steam has a higher volume than an equivalent weight of condensate. The increase in velocity reduces the pressure on the underside of the disc, causing the disc to snap shut against the seat.

Once the disc has shut, it is held closed by the pressure of the steam in the chamber above it. When that steam condenses, the disc rises and discharge resumes. This cycle is repeated every 5–10 seconds.

Thermodynamic traps are extremely rugged because they have only one moving part: the disc. Their operation is also easy to verify, because the snapping of the disc is audible.

Thermodynamic traps are also effective at removing air from steam systems, as long as the systems are filled slowly. High-velocity air passing through the trap closes the disc just as flash steam does. Since air doesn't condense, the disc won't reopen. For this reason, thermodynamic traps should be supplemented with thermostatic air vents if used for rapid startups. □

sate-line pressure, the pump can be used without a steam trap. When the exchanger operates only occasionally at low pressures, steam traps should be placed downstream of the pump.

Turn condensate into steam

The condensate drained from distribution lines and heat exchangers need not be thrown away. The liquid contains heat, and is relatively pure. Thus, it can be reused either as a heat source or as boiler feed water without additional treatment.

Condensate is an effective heat source when it is used to produce flash steam. Since the temperature of condensate is just below the boiling point of water, dropping the pressure causes it to boil, or "flash," creating a mixture of hot water and low-pressure steam. The heat generated from flash steam can replace or supplement a low-pressure steam supply.

Figure 6 (p. 185) shows how much flash steam can be produced from condensate. The curves on the graph represent flash-steam pressures. Condensate pressures are along the y-axis, and flash-steam yield is along the x-axis. To determine the yield of flash steam from a given amount and pressure of condensate, find the intersection of the desired flash-steam pressure line and the condensate pressure. The yield of flash steam will be found on the x-axis.

So, for example, flashing 100 lb of 150-psig condensate to 40 psig produces 9 lb of steam, leaving 91 lb of water at 285°F. The steam provides about 900 Btu/lb of heat, or 8,100 Btu. The hot water delivers 185 Btu/lb if cooled from 285 to 100°F, for 16,835 Btu. Thus 100 lb of condensate yields 24,935 Btu.■

Edited by Wayne Grinthal

The author

Glenn E. Hahn is a technical training manager for Spirax Sarco, Inc., P. O. Box 119, Allentown, PA 18105, phone:(215) 797-5830, fax:(215)433-1346. After receiving a mechanical engineering degree from Penn State Univ. (State College) in 1968, Hahn worked as a project and sales engineer for Ingersoll-Rand Co. and Ecolaire Heat Transfer Co. (Easton, Pa.). He joined Spirax Sarco in 1984 and has moved from application engineer to training manager. He has been published twice previously on the subject of steam utilization and currently conducts training seminars for Spirax Sarco at the Allentown facility.

Section V

The Proven Techniques for Energy Accounting

REBECCA PERRY

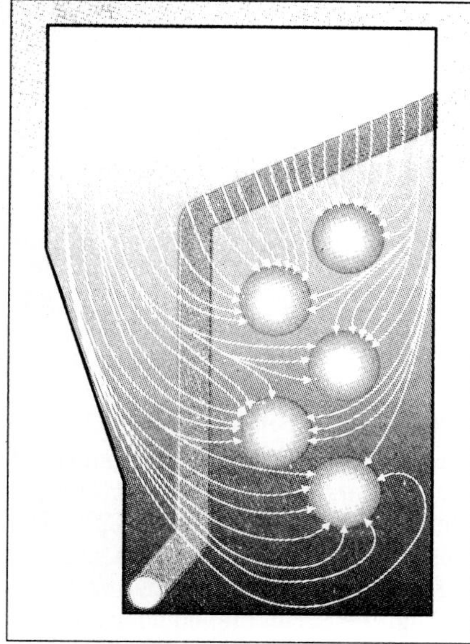
Air flow in uncorrected windbox is uneven

Fired heaters and boilers have been made more efficient in the past decade, but further gains can be expected

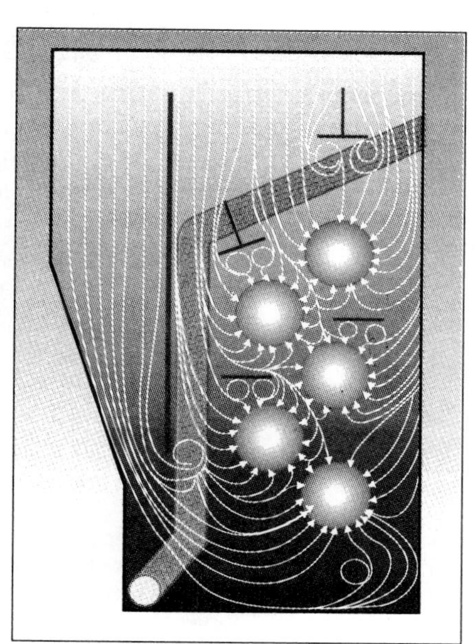
Splitters distribute windbox flow evenly

MAKE EVERY BTU COUNT

Ashutosh Garg
Kinetics Technology International Corp., and
H. Ghosh
Engineers India Limited

Although the world has been awash with oil, recent events in the Middle East show how tenuous that condition can be. Forecasters had predicted that energy prices would be climbing by the middle of the decade, if not sooner, even without the spur of Iraq's invasion of Kuwait. Energy conservation is expected to again become popular and fuels for fired heaters to become heavier, dirtier and more difficult to burn. Complicating matters, pollution control laws will further restrict both gaseous and solids emissions.

Heat-recovery equipment is, of course, vital to energy conservation. In addition to seeing larger numbers of air preheaters and waste-heat boilers, operations and maintenance supervisors and engineers are likely to encounter improved burners and soot blowers in fired heaters, as well as many more exchangers for recovering low-level heat at faster paybacks. Potential improvements in energy efficiency for heat-recovery equipment and how they can be realized are listed in Table 1.

At the heart of fired heaters

When fuel was cheap, most fired heaters were fitted with natural-draft burners, which were operated with high excess air (30%-50%) and long flames. They have always been bottom fired, with steam atomizing guns. Boilers, on the other hand, were mostly fitted with forced-draft mechanically atomized oil burners that worked under positive pressure. They have usually burned low-viscosity fuels.

Many differences between boilers and fired-heater burners have vanished. Burners are now designed for both maximum combustion efficiency and minimum pollution. Fired heaters are being fitted with boiler burners modified to suit heater requirements. Burners are also being fitted with steam atomizers so they can burn high-viscosity fuels.

For efficient combustion, burners must be not be fed excessive fuel or air. Excess fuel results in unburned fuel being exhausted from the stack. This boosts corrosion because it creates localized reducing areas and contributes to the formation of CO and particles, which are released to the atmosphere. Excess air adds to the loss of sensible heat from the stack, lowers heat-transfer efficiency, contributes to fan power loss, hikes corrosion by forming SO_3, which also increases air pollution.

Combustion losses are minimized by keeping excess air at 3%-5%. This can be achieved through the optimum mixing of air and fuel (liquid or gaseous). Modern heaters can hold nitrogen oxide emissions to 100 ppm with gaseous fuel and to 200 ppm with oil feed, and particle emissions to less than 50 mg/m^3.

For maximum combustion efficiency, it is important to establish equal air

Originally published October 1990

TABLE 1. Potential efficiency improvements in heat-recovery equipment

Equipment	Operation	Improvement
Oil and gas burners	Promote flame conditions that result in complete combustion at lower levels of excess air	0.25% for each 1% decrease in excess oxygen
Air preheaters	Transfer energy from stack gases to combustion air	2.5% per 55°C drop in stack gas temperature
Economizers and waste-heat boilers	Transfer energy from stack gases to feed water	1% for each 55°C increase in in feedwater temperature
Combustion control systems	Precisely regulate flow of fuel and air flow	0.25% for each 1% decrease in excess oxygen
Sootblowers	Remove deposits that retard heat transfer from the external surfaces of boiler tubes	2.5% for each 55°C decline in stack gas temperature

distribution to all the burners, especially when the excess air is limited to less than 5%. One way of accomplishing this is via directional splitters(p. 191).Where to locate the splitters can be determined by means of flow modeling in an accurately scaled transparent model of the ducts and windbox. One study based on such a model reduced flow variations to the burners of a fired heater to ±2% of the mean air flow. Such flow modeling of flue ducts can also help achieve uniform fluegas distribution in air preheaters, avoiding the dead pockets that cause localized dewpoint corrosion.

Firing heavier fuels

With improved methods, petroleum refiners are extracting more product from each barrel of crude oil, leaving a much poorer quality residue to burn in heaters. Fuel specifications in one refinery before and after a recent expansion are noted in Table 2. The differences are typical of changes in fuel specifications.

In conventional atomizers, fuels of inferior quality tend to convert oil droplets into small coke particles. The droplets burn slowly and are not totally consumed by the flames. Entrapped in the fluegas, the particles are carried out through the stack. The oil fuel should be filtered if it contains solids.

Three things can be done to enable fired heaters to burn heavier fuels more efficiently: raise the pressure and quality of the atomizing steam to improve atomization; mix the fuel oil and atomizing fluid thoroughly; and reduce the size of the spray droplets so that they will burn quickly without producing unburned carbon. These changes are limiting particle emissions to less than 50 mg/m^3, even when heavy fuel oils and vacuum residues are burned.

A recent trend is to use fuel gas, if available at a high enough pressure, as the atomizing fluid. Not only is gas inexpensive, it also enriches the flame.

TABLE 2. Refiners now burn a crude residue of poorer quality

Residue properties	Residue burned in 1980	1990
Specific gravity	0.935	0.974-0.984
Pour point, °C	50-52	48-69
Kinematic viscosity, cSt 100°C	20-25	1,100-2,900
Sulfur, wt%	0.30	0.2-0.6
Ash, wt%	0.08	0.236-0.25
Trace metals, ppm		
Vanadium	5	35
Nickel	130	220
Iron	5.5	212
Copper	0.07	13
Sodium	90	94
Heating value, kcal/kg	10,000	9,600

Preheat combustion air

Heat is recovered from the stack of fired heaters and used to warm up the combustion air (bottom, p. 220). The most common air preheaters are the regenerative and recuperative exchangers and the circulating-liquid system.

Frequent in steam-generation service, the regenerative exchanger has been successfully adopted for fired heaters. Seals have been improved in design and in material to reduce the leakage of air to fluegas, a hazard. A newer approach to leakage control in large preheaters involves automatically activated deflecting sector plates at the hot end. Different heat-transfer shapes have also been adopted to suit alternative fuels. Other improvements pertain to modular motors, support bearings and soot blowers.

A static exchanger consisting of circular tubes or a rectangular cross-section of cast-iron tubes, the recuperative exchanger has been proven in service as a combustion-air preheater. The cast tubes may have fins inside and outside and be located at the fluegas inlet, or have fins only on the outside and be placed at the outlet to avoid dewpoint condensation.

The heat-transfer fluid in the circulating-liquid preheater is warmed in the fired heater's convection section and passes through a liquid-to-air exchanger, which heats the combustion air. Such a unit is the preheater of choice when space is limited (because it does not need an induced-draft fan and fluegas ducts) or when a heat-transfer fluid is available in the plant. This preheater can also be used to preheat process and utility streams.

Exchangers having a transfer surface of Pyrex are gaining acceptance. This borosilicate glass makes an ideal surface for exchangers that are designed for cooling a fluegas below its dewpoint. Chemically inert and resistant to scaling, the glass is also inexpensive. Glass tubes will be increasingly installed in recuperative exchangers for recovering energy now vented out of stacks.

Plate heat exchangers, which can operate below the fluegas dewpoint, can lower fluegas temperatures to about 65°C. Because condensation occurs below 65°C, this exchanger recovers both sensible and latent heat. Protected by an acid-resistant coating, the heat-transfer metal surface is not contacted by condensate. Even when the fluegas temperature drops to 55°C, the fluegas leaves the air preheater superheated by about 20°C. This prevents downstream condensation and corrosion.

Waste-heat boilers represent a good choice for raising fired-heater efficiency if steam is needed, because heat recovered as steam is more valuable than heat saved as fuel. A boiler can be installed in the convection section of a fired heater (above). The stack temperature depends on the boiler steam pressure and feed-water temperature. The inlet temperature of the boiler feedwater must be controlled to avoid dewpoint corrosion.

Heat recovered from fluegas can be used to warm up the combustion air (below); convection section of fired heater can be adapted to generate steam (right)

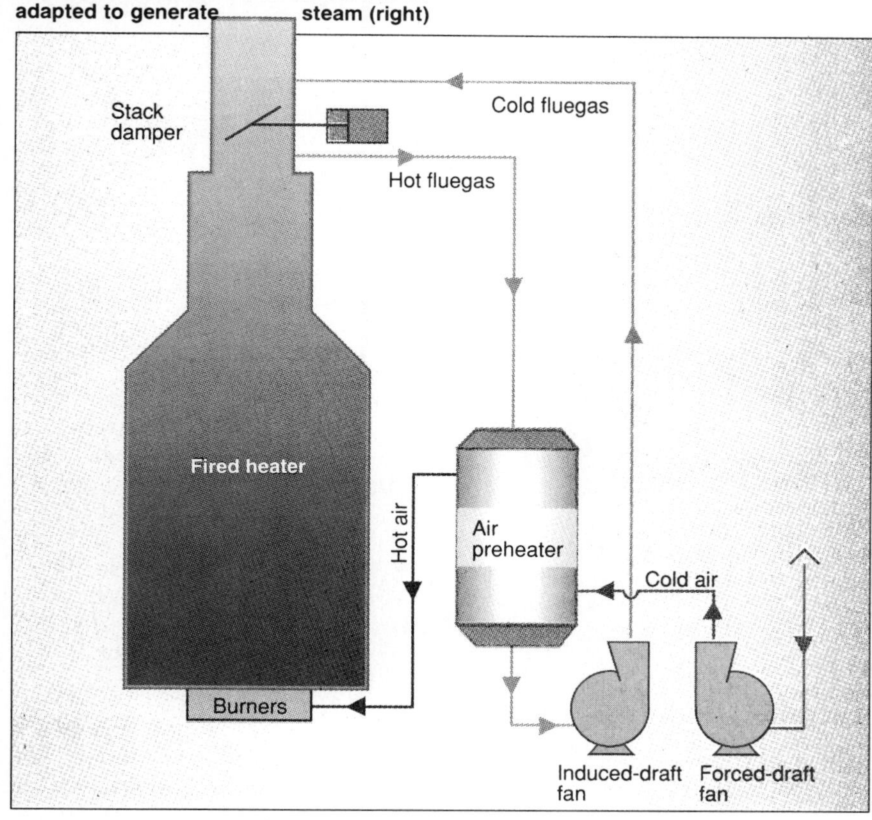

Direct-contact recovery of condensate heat holds promise for recovering low-level heat. In this new system, hot fluegas is passed through a cold water stream (p. 194). This cools the fluegas to between 32°-43°C, and heats the water stream to about 55°C, recovering both latent and sensible heat from the fluegas, and boosting the efficiency of the fired heater from 5% to 8%. One side benefit of the scrubbing is a reduction of SO_2 and SO_3. The limitation of this equipment is that low-level heat is useful mostly for warming water.

Sootblower advances

Soot deposits on tube surfaces can reduce heat-transfer efficiency by as much as 10%. Conventionally, high-pressure steam blown across tube surfaces (usually, about once per shift) removes ash deposits. Steam blowers require continuous maintenance. Microprocessor-based control has boosted blower flexibility, making it possible to blow areas where heat transfer has dropped below a critical level and to operate the blowers in varying sequences.

A new method of keeping a heat-transfer surface clean is to suspend the ash and soot in a gas stream by means of nearly continuous sound pressure waves. These sonic sootblowers generate sound by means of a flexible diaphragm or a motor connected to a frequency converter (see below).

The major advantages of sonic sootblowers are that they can be mounted anywhere, cover a large area, provide continuous cleaning and be fully automated. They also need little maintenance, because they have only one moving part. Their chief limitation is that they can remove only light, friable deposits of slag or soot. If the fuel oil being burned contains sodium and vanadium, the deposits in the convection section will be sticky because of their low melting point. Such deposits are difficult to remove by steam or sound. Fuel treatment can make the deposits friable.

Closer combustion control

Little attention was paid to efficient combustion control when fuel was cheap. With the development of forced-draft firing and air preheaters, controlling excess air became important. Microprocessors have made efficient control easy.

Onstream oxygen measurement provides the dynamic feedback needed for closely controlling combustion in fired heaters. The oxygen content in exhaust gases represents a direct measure of combustion efficiency. When the oxygen content is low, the measurement of carbon monoxide provides a sensitive,

Heat can be recovered from fluegas via direct-contact spraying (left); sonic sootblowers prevent tube deposits by means of pressure waves (below)

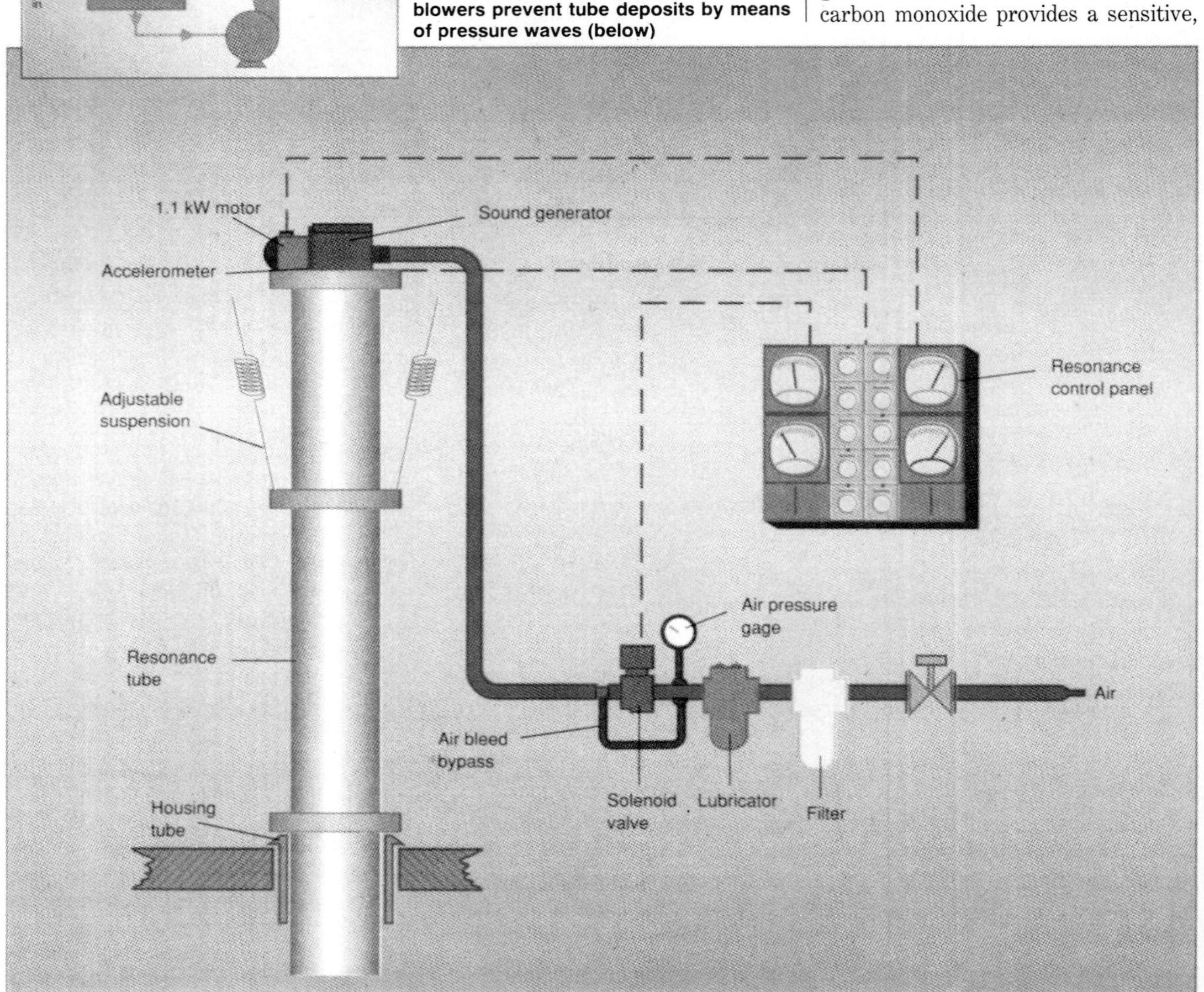

Clyde Blowers PLC, Clyde Bank, Scotland

reliable guide to combustion efficiency. With oil-fired furnaces, CO contents range from 150 to 200 ppm. Fluegas analyzers now measure both oxygen, carbon monoxide and other combustible gases. With the cost of microprocessor-based controllers continuing to decline, even small heaters will be equipped with advanced control systems.

Nitrogen oxides, major air polluters, range from 100 to 500 ppm in fluegas, predominantly as nitric dioxide. Factors that influence the quantity of NO_x in fluegas are: quantity of nitrogen in the fuel, furnace temperature, excess air and the temperature of the air entering the heater. The formation of NO_x can be reduced by recirculating the fluegas, conducting staged combustion, decreasing the quantity of excess air, and lowering the temperature of the combustion air.

Reducing the excess air to a heater fouls the convection section, boosting the emission of particles. The fouling of the convection section lowers the draft, and this progressively decreases the air flow, increasing the quantity of particles. At turndown operation, heaters emit greater quantities of solids because of the poor mixing of air and fuel. Also influential is fuel quality, particularly the content of asphaltenes in liquid fuels. Particle emissions from modern forced-draft burners being fed preheated air can be controlled to 50 mg/m^3 even when heavy fuels are fired with 5% excess air.

The authors

Ashutosh Garg is a senior thermal engineer with Kinetics Technology International Corp. (1333 So. Mayflower Ave., Monrovia, CA 91016-4099; tel.: 818-303-4711). Formerly, he was a deputy manager with the Heat & Mass Transfer Div. of Engineers India Ltd. His career responsibilities have included the design, sales and commissioning of fired heaters and combustion systems He holds a degree in chemical engineering from the Indian Institute of Technology.

H. Ghosh is a chief consultant with Engineers India Ltd. (P.T.I. Bldg., 4, Sansad Marg., New Delhi-110 001, India). He is responsible for the design of all heat-transfer equipment. He received a degree in chemical engineering from Jadavpur University, and worked for Fertilizer Corp. of India and A.P.V. Calcutta, before joining Engineers India Ltd. ■

A PRACTICAL GUIDE TO ENERGY

Kenneth E. Nelson
KENTEC Inc.

Over the past two decades, the chemical process industries (CPI) have made dramatic progress in cutting energy consumption. However, documentation of the results, especially on an individual-plant basis, has often been invalid and even misleading. That is because the techniques used for calculating energy savings fail to distinguish between the different forms of energy.

For example, saving a million Btu in the form of warm water (low-level energy) is not as worthwhile as saving a million Btu of electricity (high-level energy). Similarly, saving a million Btu of low-pressure steam is not equivalent to saving a million Btu of fuel gas.

In this article (Part 1), the author defines each type of Btu and explains the differences between them, clarifying their relationships. Also presented here are calculation methods for converting power, steam (at different temperatures and pressures), and various types of fuels to a single common basis: methane. Btu subscripts are used to assist the reader in following the logic. Effective techniques for tabulating energy use and reporting performance will be covered in Part 2, Btu Accounting on pp. 203–208.

When a company purchases all its electricity from a utility and makes all its steam in boilers using only pure methane as the fuel, energy accounting is relatively straightforward. The calculations become more complex, however, when companies cogenerate steam and electricity, and use a mixture of industrial fuels. Understanding the relative values of fuel, electricity and different levels of steam is especially important when justifying and evaluating cogeneration plants.

Accurate documentation and reporting of energy performance focuses attention on trends and provides valuable insights into plant operations. The techniques are applicable to an individual plant, a site that includes many plants, a product group, or an entire company. A technically sound methodology allows valid comparisons among plants that use different levels of steam, different forms of energy, and various types of energy-conversion devices. The energy consumption of steam-turbine drives, for example, can be compared with that of electric motors. The recommendations put forth in this article should have wide applicability throughout the CPI.

The procedures developed are not intended to serve as a model for setting steam and electricity prices, although they may be applied to allocating fuel costs. Many factors not directly related to energy consumption enter into setting the price for a pound of steam or a kilowatt-hour of electricity. Operating and maintenance costs, for example, together with distribution costs and capital depreciation are normally added to the fuel cost associated with generating electricity.

For consistency and clarity, the author has chosen units, such as Btu and lb, that are commonly used in the U.S. In countries where different systems of untis are employed, the procedures presented here will need only simple unit conversions.

Originally published September 1994

Monitoring energy use

The most useful way to calculate energy use is on an individual-plant basis. There are two general guidelines that should be observed when setting up such a tracking system:

1. *Relate all energy consumption to equivalent methane*

In calculating the energy consumed by a plant, one usually needs to add the energy content (Btu) of fuel of various compositions, the energy content (Btu) of steam at various temperatures and pressures, and the energy content (Btu) of electricity that is purchased or generated at the site. These are different types of Btu. To avoid the problem of adding "apples and oranges," one must convert these individual Btu to an equivalent basis.

For such a common basis, the higher heating value (HHV) of methane is recommended because methane is a widely used fuel and its properties are well documented. The reason for choosing HHV, instead of lower heating value (LHV), is that methane is normally purchased on HHV basis in the U.S.*

In later examples, it is shown that the methane values attributable to specific quantities of steam or electricity (that is, their methane equivalents) are not fixed conversion factors. They depend on how that steam or electricity is produced. In other words, when calculating methane equivalent, one should ask: How much methane has been used to make this quantity of steam (or electricity)?

2. *Calculate the energy consumed*

The most effective way way of reporting energy performance is to calculate the amount of energy *consumed*. The following examples are intended to clarify how energy consumption is determined in typical plant situations (Figure 1):

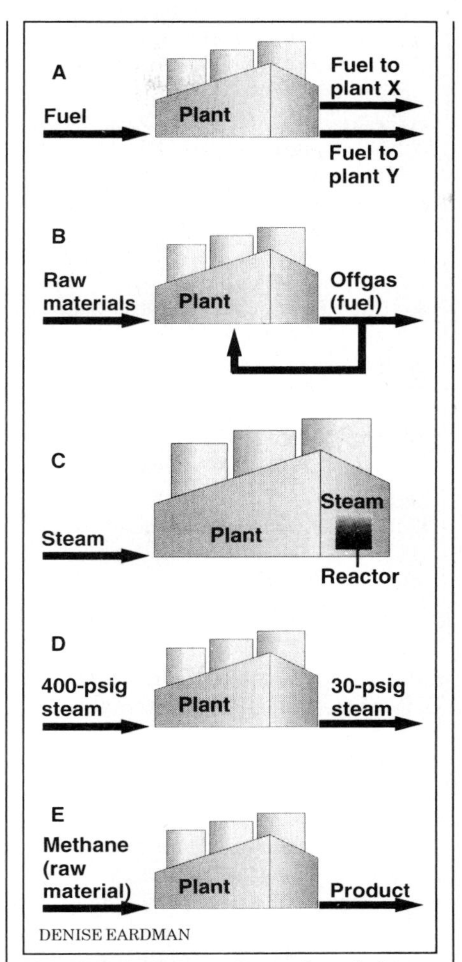

FIGURE 1. There are a number of situations where it becomes important to carefully define what is meant by energy consumed

A. If a plant imports a fuel, but sends part of the fuel to Plants X and Y, Plant A is only charged for the amount of fuel it consumes. Similarly, Plants X and Y are charged for the amount of fuel each consumes. The fuel can be natural gas, offgas, tars, oil, hydrogen and so on.

B. For a hydrocarbon-processing facility, such as an ethylene plant, the offgas is essentially a byproduct. When the plant burns some or all of its offgas, it is charged for the amount that is burned within the plant.

If a plant exports some or all of its offgas, however, the plant receives no fuel credit for the export. Instead, the plant that burns that offgas is charged. This is consistent with the notion of charging a plant only for the energy consumed.

Similarly, if a chlorine plant is sending H_2 to a power plant (where the H_2 is burned), the chlorine plant receives no fuel credit. As is the case with offgas, the H_2 is a byproduct of the process. The efficiency of the chlorine plant does not change because H_2 is sold rather than burned.

Note that if fuel credit is given for exported offgas, it will be canceled by the charge for its use by another plant when energy consumption for the entire site is totaled. This will result in understating the total energy used by the site.

C. Charge only for the net steam imported. If a plant generates steam internally (from an exothermic reaction, for example), less steam must be imported. If the reaction is endothermic (requiring heat), we charge for the steam or fuel used to supply that heat. Therefore, it makes sense to give credit for exothermic heat, and to charge only for the net steam imported. Similarly, if the plant exports part or all of the

ACCOUNTING

Apply a common basis in assessing the relative value of various energy sources used at a plant

*In much of the rest of the world, however, LHV is the standard. Therefore, the LHV of methane would be an appropriate substitution. For purposes of energy accounting, multinational companies should standardize their basis on one or the other.

steam generated internally, it should get credit for the exported steam.

D. If a plant imports steam at a high pressure and exports at a lower pressure, it is charged for the amount of steam imported and receives credit for exported steam. Thus, one is in effect charging the plant for the value it extracted from the steam.

E. If a plant uses methane (or any fuel) as a raw material, it should not be included as part of the plant's energy use.

It is also possible to devise a system that involves the inherent energy in a product. The energy attributable to making a pound of vinyl chloride, for example, would include the energy used to make ethylene and chlorine. This type of tracking is very complex, difficult to administer, and is not recommended for Btu-accounting purposes. The added effort required for such analysis adds little to the ultimate goal of Btu accounting, which is to improve efficiency.

Making sense of heating values

The two types of heats of combustion for fuels sometimes cause confusion. One is called the gross or higher heating value (HHV), and the other is known as the net or lower heating value (LHV). The difference can be traced back to whether H_2O produced during combustion is referenced to the liquid or vapor state. For methane:

$$CH_4 + 2O_2 \rightarrow CO_2 + 2H_2O$$

HHV = 1,013 Btu/std. ft^3 of methane (with respect to H_2O in liquid state at 60°F and 1 atm)

LHV = 913 Btu/std. ft^3 of methane (with respect to H_2O in vapor state at 60°F and 1 atm)

The difference = 100 Btu/std. ft^3 of methane (needed to vaporize water at 60°F and 1 atm)

When carbon monoxide is used as a fuel, no water is produced. Therefore, the HHV and LHV of CO are the same.

$$CO + 1/2\ O_2 \rightarrow CO_2$$

HHV = 322 Btu/std. ft^3 of CO
LHV = 322 Btu/std. ft^3 of CO

The methods described below show how to convert non-methane fuel, steam and electricity use to their methane equivalents. For consistency, HHV of methane is used throughout.

NOMENCLATURE

Acronyms

HHV	Higher or gross heating value of a fuel, Btu_H/std. ft^3
HHV_{meth}	Methane-equivalent of a fuel, Btu_{meth}/std. ft^3
LHV	Lower or net heating value of a fuel, Btu_H/std. ft^3

Symbols

H	Enthalpy at given stream conditions, Btu_H/lb
H_0	Enthalpy at heat-sink conditions, Btu_H/lb
S	Entropy at given stream conditions, Btu_H/lb·°R
S_0	Entropy at heat-sink conditions, Btu_H/lb·°R
T_0	Heat-sink temperature, °R
W_{act}	Actual work available from a given stream, Btu_W/lb
W_{max}	Maximum work available from a given stream, Btu_W/lb

Btu varieties

Btu_{meth}	Btu of a fuel, expressed as its HHV methane-equivalent
Btu_H	Btu of heat
Btu_W	Btu of work

A. From Btu of fuel to Btu_{meth}

In practice, one is interested in the energy value corresponding to the LHVs from fuels because H_2O formed during combustion goes up the stack in the vapor state (along with the heat of H_2O vaporization). Therefore, one could argue that it makes more sense to determine methane-equivalents of fuels on the basis of the LHV of methane. For the reasons given earlier, however, HHV of methane has been chosen here. Therefore, it is necessary to convert the LHV of various fuels to their CH_4-equivalents relative to the HHV of methane.

This is easily accomplished by multiplying the LHV of a fuel by the ratio of the HHV of CH_4 to the LHV of CH_4. The resulting CH_4-equivalent relative to methane's HHV is abbreviated as HHV_{meth}.

For H_2:

$$HHV_{meth} = \frac{(\text{HHV of } CH_4)}{(\text{LHV of } CH_4)} \times (\text{LHV of } H_2) = (1{,}013/913) \times 275 = 1.11 \times 275 = 305\ Btu_H/\text{std. } ft^3$$

For CO:

$$HHV_{meth} = \frac{(\text{HHV of } CH_4)}{(\text{LHV of } CH_4)} \times (\text{LHV of CO}) = 1.11 \times 322 = 357\ Btu_H/\text{std. } ft^3$$

The HHV, LHV and HHV_{meth} for common gaseous fuels are given in Table (p. 125). We have used Btu_H to indicate that these are enthalpy (heat) Btus. Note that the HHV_{meth} of CO is higher than either its LHV or HHV (357 vs. 322 Btu_H/std ft^3). Most fuels are more valuable (in terms of HHVmeth) than their HHVs would suggest. This is because most fuels, when burned, produce less H_2O per mole than methane does. Hydrogen, however, is less valuable because it is totally converted to H_2O during combustion.

B. From kWh of electricity to Btu_{meth}

The appropriate factor for converting electricity to equivalent methane depends on the source of electricity. If electricity is purchased from a utility, the logical thing to do is to use the efficiency of the power plant supplying the electricity.

Because of the national power grid in the U.S., however, the exact source of electricity may be difficult to determine. The average power-plant efficiency in the U.S. is about 34% (HHV basis), and is normally expressed as a "heat rate" of 10,000 Btu_{meth}/kWh. Unless there is a specific reason for using a different value, standardizing on 10,000 Btu_{meth}/kWh for purchased power in the U.S. is recommended.

When power is generated at a plant site, it is normally cogenerated with steam. In this case, the efficiency of the cogeneration plant should be used for the portion of electricity coming from the cogeneration plant. Calculating the heat rate of a cogeneration plant is discussed later in this article. Any additional power that is purchased should be valued at 10,000 Btu_{meth}/kWh.

If cogenerated power is sold to a local utility or to another operating company, the fuel attributable to power sales should be deducted from the site's energy consumption. The quantity of fuel is calculated by multiplying the power sales (kWh) by the heat rate of the cogeneration plant (Btu_{meth}/kWh).

If a site imports and exports power at various times during the year, imported power is valued at 10,000

TYPES OF HEATING VALUES				
FUEL	LHV Btu_H/std. ft^3	HHV Btu_H/std. ft^3	HHV_{meth} Btu_{meth}/std. ft^3	HHV DIFFERENCE %
Methane	913	1,013	1,013	0 %
CO	322	322	357	+ 10.9 %
Hydrogen	275	325	305	- 6.2 %
Ethane	1,641	1,792	1,821	+ 1.6 %
Propane	2,385	2,590	2,646	+ 2.2 %

Table. Comparison of heating values for various fuels shows the difference between lower and higher heating values and the methane-equivalent HHVs

Btu_{meth}/kWh. Exported power is valued at the actual efficiency of the cogeneration plant. Over the course of a year, the fuel attributable to power distributed within a plant site should be calculated using the weighted-average heat rates of purchased and cogenerated electricity.

As an example, suppose a site purchases 20 million kWh annually from the electrical grid (at 10,000 Btu_{meth}/kWh) and generates 100 million kWh in its own cogeneration plant where the heat rate is 9,000 Btu_{meth}/kWh. During the year, 10 million kWh is sold to the local utility. The weighted-average heat rate for power distributed within the site is calculated as follows:

Fuel for purchased power = 20 million kWh × (10,000 Btu_{meth}/kWh) = 200,000 million Btu_{meth}.

Fuel for cogenerated power used = 90 million kWh × (9,000 Btu_{meth}/kWh) = 810,000 million Btu_{meth}

Total fuel attributable to site power use = 200,000 million Btu_{meth} + 810,000 million Btu_{meth} = 1,010,000 million Btu_{meth}

Total power distributed to site = 20 million kWh + 90 million kWh = 110 million kWh

Weighted-average heat rate of power distributed to site = 1,010,000 million Btu_{meth}/110 million kWh = 9,182 Btu_{meth}/kWh

The heat rate applicable to steam distributed within the site (all from the cogeneration plant) is 9,000 Btu_{meth}/kWh. Electricity sold to the local utility has been generated at 9,000 Btu_{meth}/kWh.

Note that we are not using the standard textbook conversion factor of 3,413 Btu_H/kWh. This is a theoretical conversion factor. While it is possible to convert electricity into heat at close to 100% efficiency, we cannot convert heat (or, more specifically, the heat of combustion of a fuel) into electricity at 100% efficiency. Thus, no power plant operates at a heat rate of 3,413 Btu_{meth}/kWh. The average U.S. power-plant efficiency of 34% is calculated by dividing the theoretical conversion factor of 3,413 Btu_H/kWh by the average power-plant heat rate (typically 10,000 Btu_{meth}/kWh). Thus,

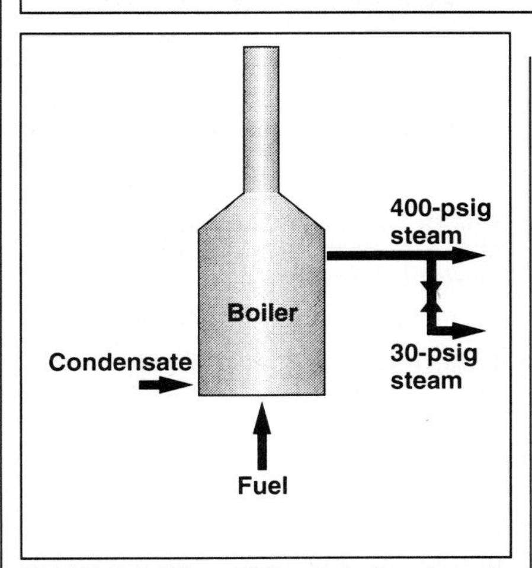

FIGURE 2. When 400-psig boiler steam is reduced to 30 psig, the fuel used per pound of 400-psig steam is the same as that per pound of 30-psig steam

DOING THE MATH WITH Btu_{meth}

We use the notation Btu_{meth} to refer to the unit quanity of energy expressed in HHV of pure methane. To obtain the total methane-equivalent of a fuel gas, multiply the given quantity (std. ft^3) or flowrate (std. ft^3/yr) of the gas by its HHV_{meth} (in Btu_{meth}/std. ft^3). For example, if a plant burns 1 million std. ft^3/yr of hydrogen, the amount of pure methane (purchased on a HHV basis) needed to replace the hydrogen = [1 million std. ft^3/yr hydrogen] × [305 Btu_{meth}/std. ft^3 hydrogen] = 305 million Btu_{meth}/yr.

[(3,413Btu_H/kWh)/(10,000 Btu_{meth}/kWh)] × 100 = 34%

C. From lb of steam to Btu_{meth}
If steam is generated in a boiler, the fuel used in the boiler — expressed as equivalent methane (in Btu_{meth}) — is charged directly to the steam produced. If the steam pressure is reduced through a valve from, say, 400 to 30 psig (Figure 2), the same fuel equivalent (Btu_{meth}/lb of steam) is applicable to both levels of steam. No further calculations are necessary.

In a cogeneration unit, however, the situation is more complex. Here converting cogenerated steam to methane-equivalents is a two-step process. First, the given quantity (in lb) of steam is converted to its equivalent kWh by using a thermodynamic function known as available work and applying an efficiency factor. Then the steam kWh is converted to Btu_{meth}, using the efficiency factor of the cogeneration facility. This two-step procedure is outlined below:

1. *From lb of steam to equivalent kWh*
The formula for calculating the maximum work available from a stream is given by:

$W_{max} = \Delta H - T_0 \Delta S$

or: $W_{max} = (H - H_0) - T_0(S - S_0)$

The enthalpy and entropy values for steam are readily available from steam tables. Selecting a heat-sink temperature is somewhat arbitrary. In principle, it is the lowest temperature available for cooling. In practice, however, this value changes daily. For consistency, a heat-sink temperature of 101°F (560.67°R) is recommended. (In cold climates, a lower heat-sink temperature may be selected.) This is analogous to the 2-in. Hg standard used for rating steam turbines.

The maximum available work from 1 lb of 1,250-psig steam at 950°F is calculated as follows:

$W_{max} = (H - H_0) - T_0(S - S_0)$

For 1,250-psig steam at 950°F:
H = 1,468.0 Btu_H/lb
S = 1.60210 Btu_H/lb·°R

For saturated steam at 101°F and 2-in. Hg:
H_0 = 1,105.5 Btu_H/lb
S_0 = 1.97970 Btu_H/lb·°R
T_0 = 560.67°R
W_{max} = (1,468.0 – 1105.0) – 560.67 (1.6021 – 1.9797)
W_{max} = 574.7 Btu_W/lb

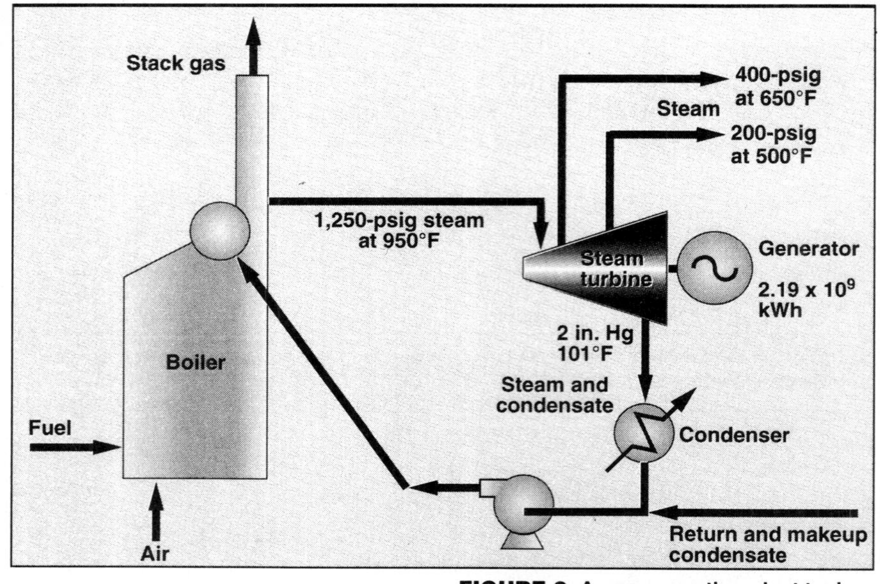

FIGURE 3. A congeneration plant typically exports electricity and one or more levels of steam

To convert *maximum* available work to *actual* available work, multiply by 0.80. This gives a realistic amount of work that could be accomplished by the steam. (Modern steam turbines are capable of achieving 80% efficiency. If a facility has less-efficient turbines, a lower percentage value may be more appropriate.)

$W_{act} = 0.80 \times W_{max} = 459.8$ Btu_W/l

To convert to kWh/lb, divide W_{act} (in Btu_W/lb) by 3,413 Btu_W/kWh:

W_{act} = (459.8 Btu_W/lb) / (3413 Btu_W/kWh)} = 0.1347 kWh/lb

Note that the use of 3,413-Btu_W/kWh is legitimate here because this conversion factor represents a theoretical maximum of 100% efficiency. It can also be expressed as 3,413 Btu_H/kWh.

2. *Convert steam kWh to* Btu_{meth}

To convert steam kWh to Btu_{meth}, multiply by the efficiency of the cogeneration facility. In the next section, we describe the technique for calculating the efficiency (heat rate) of a cogeneration plant. In the example calculation below, a heat rate of 10,300 Btu_{meth}/kWh is assumed.

Btu_{meth}/lb = 0.1347 kWh/lb (10,300 Btu_{meth}/kWh) = 1,387.5 Btu_{meth}/lb

Note that 1,250-psig steam at 950°F has three different Btu/lb values and that each has its own reference temperature:

1,468.0 Btu_H/lb (above an arbitrary 32°F base)

574.7 Btu_W/lb (using a 101°F heat-sink temperature)

1,387.5 Btu_{meth}/lb (above a 60°F base)

Using subscript notation for these three different types of Btu helps avoid confusion.

Efficiency of a cogeneration plant

In order to assign appropriate Btu_{meth} values to steam and power produced by cogeneration, we must know the efficiency of the cogeneration plant. Using the available-work approach described above, this is a straight forward calculation. The idea is to convert various levels of steam production into equivalent kWh so that they can be added to the kWh of electrical output. The procedure is outlined below.

Example calculations

Suppose we have a cogeneration plant (Figure 3) with the following annual performance:

Fuel used = 26,280,000 million Btu_{meth}

Net electrical energy exported = 2,190 million kWh

Steam exported = 1,752 million lb of 400-psig steam at 650°F, and 2,628 million lb of 200-psig steam at 500°F

Step 1. Convert steam exported to equivalent kWh

$$W_{max} = (H - H_0) - T_0(S - S_0)$$

For 400-psig steam at 650°F:
H = 1,333.9 Btu_H/lb
S = 1.61066 Btu_H/lb·°R

For 101°F saturated steam at 2 in. Hg:
H_0 = 1,105.5 Btu_H/lb
S_0 = 1.97970 Btu_H/lb·°R
T_0 = 560.67°R

W_{max} = (1,333.9 – 1,105.0) – 560.67 (1.61066 – 1.97970) = 435.8 Btu_W/lb

W_{act} = [(435.8 Btu_W/lb)/(3413 Btu_W/kWh)] (0.80) = 0.1022 kWh/lb

Total actual work = 0.1022 kWh/lb (1,752 million lb) = 179 million kWh

For 200-psig steam at 500°F:
H = 1267.2 Btu_H/lb
S = 1.61476 Btu_H/lb·°R

For 101°F saturated steam at 2 in. Hg:
H_0 = 1,105.5 Btu_H/lb
S_0 = 1.97970 Btu_H/lb·°R
T_0 = 560.67°R

W_{max} = (1,267.2 – 1105.0) – 560.67 (1.61476 – 1.97970)

W_{max} = 366.8 Btu_W/lb

W_{act} = [(366.8 Btu_W/lb)/(3,413 Btu_W/kWh)] (0.80) = 0.0860 kWh/lb

Total *actual work* = 0.0860 kWh/lb (2,628 million lb) = 226 million kWh

Step 2. Divide fuel use by sum of power kWh and steam kWh.

The total equivalent power output = Exported power + 400-psig steam + 200-psig steam = 2,190 million kWh + 179 million kWh + 226 million kWh = 2,595 million kWh

The heat rate for the cogeneration plant = (26,280,000 million Btu_{meth})/(2,595 million kWh) = 10,127 Btu_{meth}/kWh

This heat rate can now be applied to steam and power used by an individual plant, and to electricity sales. If, for example, the cogeneration plant sells 5 million kWh to a local utility company, the fuel allocation would be:

5 million kWh × 10,127 Btu_{meth}/kWh = 50,635 million Btu_{meth}

Individual plant's energy use

In the following example, we apply the procedures described thus far to calculate the total energy used by an individual plant, expressed as Btu_{meth}. We also divide that energy by total production to obtain Btu_{meth}/lb of product.

Example calculations

A typical plant's utility streams are shown in Figure 4. One can usually ignore the energy associated with water and condensate because it is small, compared with other energy uses. If we wanted to be rigorous, however, we could relate water use to the energy used for pumping and treating it. Similarly, we could determine how much energy would be needed to supply condensate. If either amounts to a significant portion of a plant's total energy consumption or export, of course, it should be included.

Problem: Calculate the energy used (in Btu_{meth}) and energy consumption for each unit of product (in Btu_{meth}/lb) for a plant that has the following annual production and utility requirements:

Annual plant production = 470 million lb of product

Fuel gas used = 200 million std. ft^3

Fuel gas composition = 75 vol. % methane, and 25 vol. % hydrogen

Electricity used = 12 million kWh (cogenerated)

400-psig steam at 650°F used = 80 million lb

200-psig steam at 500°F used = 70 million lb

30-psig steam at 350°F exported = 60 million lb

The electricity and all three levels of steam are produced in an onsite cogeneration plant having a heat rate of 10,127 Btu_{meth}/kWh. Because 30-psig steam is exported by the production plant, the cogeneration system needs to generate less 30-psig steam for other plants at the site.

Step 1. Convert fuel use to Btu_{meth}

Methane = 200 million std. ft^3 × (0.75) (1,013 Btu_{meth}/std. ft^3) = 151,950 million Btu_{meth}

Hydrogen = 200 million std. ft^3 × (0.25) (305 Btu_{meth}/std. ft^3) = 15,250 million Btu_{meth}

Step 2. Convert electricity use to Btu_{meth}

Electricity = 12 million kWh × (10,127 Btu_{meth}/kWh) = 121,524 million Btu_{meth}

Step 3. Convert steam uses to Btu_{meth}

$$W_{max} = (H - H_0) - T_0(S - S_0)$$

For 400-psig steam at 650°F:

H = 1,333.9 Btu_H/lb

S = 1.61066 Btu_H/lb·°R

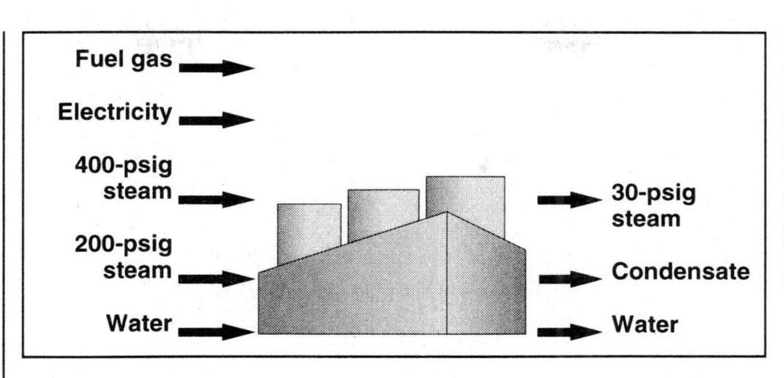

FIGURE 4. A production plant at a large site uses fuel, electricity, and different levels of steam. In this example, 30-psig steam is exported to the site's steam header

For saturated steam at 101°F and 2 in. Hg:

H_0 = 1105.5 Btu_H/lb

S_0 = 1.97970 Btu_H/lb·°R

T_0 = 560.67°R

W_{max} = (1,333.9 – 1,105.0) – 560.67 (1.61066 – 1.97970) = 435.8 Btu_W/lb

Fuel consumed for each lb of steam used = [(435.8 Btu_W/lb)/(3,413 Btu_W/kWh)] 3 (10,127 Btu_{meth}/kWh) (0.80) = 1,034.5 Btu_{meth}/lb

Annual fuel consumption = 1,034.5 Btu_{meth}/lb (80 million lb) = 82,760 million Btu_{meth}

For 200-psig steam at 500°F:

H = 1267.2 Btu_H/lb

S = 1.61476 Btu_H/lb·°R

For saturated steam at 101°F and 2 in. Hg:

H_0 = 1105.5 Btu_H/lb

S_0 = 1.97970 Btu_H/lb·°R

T_0 = 560.67°R

W_{max} = (1267.2 – 1105.0) – 560.67 (1.61476 – 1.97970) = 366.8 Btu_W/lb

Fuel consumed for each lb of 200-psig steam used = [(366.8 Btu_W/lb)/3,413 Btu_W/kWh)] 3 (10,127 Btu_{meth}/kWh) (0.80) = 870.7 Btu_{meth}/lb

Annual fuel consumption = 870.7 Btu_{meth}/lb (70 million lb) = 60,949 million Btu_{meth}

For 30-psig steam at 350°F:

H = 1211.0 Btu_H/lb

S = 1.71815 Btu_H/lb·°R

For saturated steam at 101°F and 2 in. Hg:

H_0 = 1105.5 Btu_H/lb

S_0 = 1.97970 Btu_H/lb·°R

T_0 = 560.67°R

W_{max} = (1,211.0 – 1,105.0) – 560.67 (1.71815 – 1.97970) = 252.6 Btu_W/lb

Fuel consumed for each lb of 30-psig steam used = [(252.6 Btu_W/lb)/(3,413 Btu_W/kWh)] × (10,127 Btu_{meth}/kWh) (0.80) = 599.6 Btu_{meth}/lb

Annual fuel consumption = 599.6 Btu_{meth}/lb (60 million lb) = 35,976 million Btu_{meth}

Step 4. Calculate total fuel consumed (in Btu_{meth})

Methane = 151,950 million Btu_{meth}

Hydrogen= 15,250 million Btu_{meth}

Electricity = 121,524 million Btu_{meth}

400-psig steam = 82,760 million Btu_{meth}

200-psig steam = 60,949 million Btu_{meth}

30-psig steam = – 35,976 million Btu_{meth} (exported)

Total = 396,457 million Btu_{meth}

This value represents the fuel consumed at the cogeneration plant for making 470 million lb of product in the production plant. It is expressed as equivalent methane.

Step 5. Calculate the energy consumed (in Btu_{meth}/lb product)

Energy used per pound of product = (396,457 million Btu_{meth})/(470 million lb)} = 843.5 Btu_{meth}/lb product

When tracking individual-plant performance over a period of years, there is a drawback to using actual fuel usage because the efficiency of the cogeneration plant may change. As energy consumed per pound of product (Btu_{meth}/lb) changes from year to year, it may not be clear how much of the variation was caused by changes in the cogeneration plant and how much by changes in the production plant.

It is, therefore, desirable to know how these plants would perform with a fixed power plant — one whose efficiency remains the same year after year. There are two main cases that need to be considered: when steam and power are cogenerated; and when power is purchased from an outside utility and steam is made in boilers.

A. *Cogenerated steam and power*

For cogeneration plants, an efficiency of 10,000 Btu_{meth}/kWh can be used and applied to both power and steam.† This calculation is similar to Steps 2 and 3 above, substituting 10,000 for 10,127 Btu_{meth}/kWh.

The recalculated values are as follows:

Methane = 151,950 million Btu_{meth}

Hydrogen = 15,250 million Btu_{meth}

Electricity = 120,000 million Btu_{meth}

400-psig steam = 81,720 million Btu_{meth}

200-psig steam = 60,186 million Btu_{meth}

30-psig steam = –35,526 million Btu_{meth}

Total = 393,580 million Btu_{meth}

Energy consumed per pound of product = (393,580 million Btu_{meth})/(470 million lb)} = 837.4 Btu_{meth}/lb

The application of a fixed heat rate is explored further in Part 2.

B. *Purchased electricity and steam made in boilers*

If electricity is purchased from an outside utility and steam is produced in boilers, one can use a heat rate of 10,000 Btu_{meth}/kWh (the U.S. average) for electricity, and a typical boiler efficiency. In the following example, the methane-equivalent of 400-psig steam is calculated using a boiler efficiency (HHV basis) of 70%. Condensate enters the boiler at 150°F.

Step 1. Calculate the heat added to the steam in Btu_H

For 400-psig steam at 650°F:

H = 1,333.9 Btu_H/lb

For condensate at 150°F:

H = 118.0 Btu_H/lb

Heat added = 1333.9 Btu_H/lb – 118.0 Btu_H/lb = 1,215.9 Btu_H/lb steam

Step 2. Calculate the fuel required using a 70% thermally efficient boiler

Fuel required = (1215.9 Btu_H/lb) / [(70 Btu_H)/(100 Btu_{meth})] = 1737.0 Btu_{meth}/lb of steam

Note that we divide the amount of heat added to the steam (1,215.9 Btu_H/lb of steam) by the furnace efficiency, which is is expressed as Btu_H/Btu_{meth}.

If 200- and 30-psig steam are made in separate fired boilers, similar calculations can be made. If 200- and 30-psig steam are obtained by reducing 400-psig steam, the 1,737 Btu_{meth}/lb steam factor calculated for 400-psig steam also applies to the 200- and 30-psig steam.

The 400-psig steam produced by cogeneration (at a heat rate of 10,127 Btu_{meth}/kWh) has a methane equivalent of 1,034.5 Btu_{meth}/lb, while the steam produced in a 70%-efficient boiler has a methane equivalent of 1,737.0 Btu_{meth}/lb. That is, 68% more fuel is required when 400-psig steam is made in a 70%-efficient boiler. This comparison illustrates the fuel economy of cogeneration in producing steam efficiently.

There are some who believe that cogeneration makes electricity cheaper. But as the above calculations demonstrate, the real value of cogeneration is that it makes steam, not electricity, cheaper. That is because the alternative of making the same in a fired boiler is less efficient. ■

Edited by Gulam Samdani

†If the site's cogeneration plant has a heat rate significantly different from 10,000 Btu_{meth}/kWh, choose a value that is close to a recent average.

The author

Kenneth E. Nelson is president of KENTEC Inc. (P.O. Box 45910, Baton Rouge, LA 70895; tel: (504) 273-1524; fax: (504) 275-7207), a consulting firm that specializes in working with companies to help reduce energy use, improve yields, and cut waste. His industrial experience includes 31 years with the Dow Chemical Co., where he designed, operated and managed production plants, analyzed and evaluated projects and processes, instituted and led energy-savings and waste-reduction programs, and managed offsite contract engineering efforts. Prior to retiring from Dow last year, he was energy conservation manager for Dow's U.S. plants. He was named Energy Manager of the Year in 1990 by the Industrial Energy Technology Conference, and subsequently joined its Advisory Board, serving as program chairman for the 1992 and 1993 conferences and organizing Energy Managers' Workshops in 1993 and 1994. He received a B.S. in chemical engineering from Illinois Institute of Technology in 1962.

Btu ACCOUNTING: SHOWING RESULTS

Kenneth E. Nelson, KENTEC Inc.

In the preceding article on pp. 196–202 we showed how to calculate the energy consumed to make a pound of product. To realize a payoff, however, the results must be presented in graphs or tables that clearly display what has happened. They must call attention to plant performance and ultimately lead to more efficient use of energy.

Energy-consumption reporting is particularly valuable when viewed over a period of time. We recommend compiling data annually and maintaining a ten-year performance history.

PRODUCTION AND ENERGY-CONSUMPTION DATA FOR PLANT PRODUCING PRODUCT ABCD

	A	B	C	D	E	F	G	H	I	J
1	Acct. 115									
2	ABCD Plant	1983	1984	1985	1986	1987	1988	1989	1990	1991
3										
4	ABCD production, Mlb/yr	375,000	400,400	415,000	425,000	430,000	500,000	540,000	525,000	490,000
5										
6	Basic input:									
7	Electricity used, MkWh/yr	14,100	13,100	13,700	11,800	11,900	12,200	13,500	13,000	11,500
8	400-psig steam used, Mlb/yr	101,600	100,600	102,500	84,200	83,000	87,000	107,400	90,000	78,000
9	200-psig steam used, Mlb/yr	85,390	83,300	81,200	71,500	71,000	80,000	92,300	80,300	68,000
10	30-psig steam used, Mlb/yr	-55,200	-57,200	-56,000	-54,000	-53,000	-60,000	-85,000	-65,000	-58,000
11	Fuel gas used, MMBTUmeth/yr	155,300	158,300	155,000	160,300	170,000	197,000	240,000	190,000	163,000
12										
13	Converting to equivalent fuel gas at actual cogeneration-plant fuel equivalents:									
14	Electricity equiv., MMBTUmeth/yr	152,139	161,130	166,455	134,520	126,854	127,124	142,020	135,980	119,060
15	400-psig steam equiv., MMBTUmeth/yr	112,915	126,213	127,775	97,716	90,867	92,467	116,148	96,588	82,368
16	200-psig steam equiv., MMBTUmeth/yr	80,619	89,139	84,846	70,506	65,166	71,273	83,894	73,663	60,967
17	30-psig steam equiv., MMBTUmeth/yr	-36,451	-42,354	-41,164	-36,936	-33,673	-36,449	-52,400	-40,386	-35,248
18	Fuel gas used, MMBTUmeth/yr	155,300	158,300	155,000	160,300	170,000	197,000	240,000	190,000	163,000
19	Total equiv. fuel gas, MMBTUmeth/yr	464,522	492,428	492,912	426,106	419,214	451,415	529,662	455,844	390,147
20	Btu/lb ABCD @ actual pwr. plt. eff.	1,239	1,230	1,188	1,003	975	903	981	868	796
21										
22	Converting to equivalent fuel gas using 10,000 Btu/kWh as basis:									
23	Electricity equiv., MMBTUmeth/yr	141,000	131,000	137,000	118,000	119,000	122,000	135,000	130,000	115,000
24	400-psig steam equiv., MMBTUmeth/yr	104,648	102,612	105,165	85,716	85,241	88,740	110,407	92,340	79,560
25	200-psig steam equiv., MMBTUmeth/yr	74,716	72,471	69,832	61,848	61,131	68,400	79,747	70,423	58,888
26	30-psig steam equiv., MMBTUmeth/yr	-33,782	-34,434	-33,880	-32,400	-31,588	-34,980	-49,810	-38,610	-34,046
27	Fuel gas used, MMBTUmeth/yr	155,300	158,300	155,000	160,300	170,000	197,000	240,000	190,000	163,000
28	Total equiv. fuel gas, MMBTUmeth/yr	441,882	429,949	433,117	393,463	403,784	441,160	515,344	444,153	382,402
29	Btu/lb ABCD @ 10,000 Btu/kWh	1,178	1,074	1,044	926	939	882	954	846	780
30										
31	Annual Production, MMlb/yr	375	400	415	425	430	500	540	525	490

Originally published October 1994

*See “A Practical guide to energy accounting,” pp. 196–202.

Four cases may be considered:

- Individual plant performance
- Site performance, for sites having more than one plant
- Company performance, for companies having more than one site
- Performance based on product, for identical or similar products made at different plants or sites

Of these, individual plant performance is inherently the most useful. It also serves as the best basis for site, company and product performance reports.

A key element in energy accounting is the relating of all energy consumption to a common basis. As developed last month in Part 1 in this series, we choose Btu_{meth} (i.e., Btu of methane equivalent, expressed as its higher heating value) for this purpose. It represents the amount of methane that would be needed to replace (in the case of fuels) or generate (in the case of steam and power) the energy being used.

Plant performance

A typical plantsite includes either a steam-and-power cogeneration plant or a steam plant, plus a number of production plants. We consider these utilities and production plants separately.

Steam-power plants: If the site has a cogeneration plant, calculate the total equivalent power (kWh/yr) and actual heat rate (Btu_{meth}/kWh) as explained in Part 1. Make a ten-year summary and show the results graphically. If the site instead has steam boilers (and purchases its electricity), report the annual efficiency (Btu_{meth}/lb steam), as well as the total steam produced.

Production plants: For most production plants, we calculate the Btu_{meth} required to make a pound of product, as described in Part 1. Occasionally it

ENERGY DATA FOR SITE COGENERATION POWER PLANT

	A	B	C	D	E	F	G	H	I
33	Acct. 203								
34	Site Cogeneration Power Plant	1983	1984	1985	1986	1987	1988	1989	1990
35									
36	Fuel gas used, MMBTUmeth/yr	14,329,532	16,505,886	16,558,893	15,922,179	15,109,266	14,969,213	15,459,588	15,354,880
37	Offgas used, MMBTUmeth/yr	22,700	34,100	26,800	33,100	22,600	32,500	31,000	27,900
38	Vent gas used, MMBTUmeth/yr	62,600	58,600	59,300	65,300	47,200	52,800	68,500	67,300
39	Total fuel used, MMBTUmeth/yr	14,414,832	16,598,586	16,644,993	16,020,579	15,179,066	15,054,513	15,559,088	15,450,080
40									
41	Power produced, MkWh/yr	1,170,950	1,180,320	1,205,760	1,222,720	1,231,820	1,243,780	1,258,330	1,262,450
42	Power used, MkWh/yr	130	120	140	130	140	130	140	140
43	Net power exported, MkWh/yr	1,170,820	1,180,200	1,205,620	1,222,590	1,231,680	1,243,650	1,258,190	1,262,310
44									
45	400-psig steam generated, Mlb/yr	355,790	378,200	322,070	381,800	393,740	412,500	419,300	427,200
46	400-psig steam power equiv., kWh/lb*	0.1030	0.1020	0.1026	0.1018	0.1027	0.1020	0.1028	0.1026
47	400-psig steam power equiv., MkWh/yr	36,646	38,576	33,044	38,867	40,437	42,075	43,104	43,831
48	200-psig steam generated, Mlb/yr	1,237,500	1,246,300	1,269,900	1,399,500	1,477,300	1,573,800	1,744,800	1,642,800
49	200-psig steam power equiv., kWh/lb*	0.0875	0.0870	0.0860	0.0865	0.0861	0.0855	0.0864	0.0877
50	200-psig steam power equiv., MkWh/yr	108,281	108,428	109,211	121,057	127,196	134,560	150,751	144,074
51	30-psig steam generated, Mlb/yr	330,000	370,000	365,000	380,000	413,000	420,000	460,000	452,000
52	30-psig steam power equiv., kWh/lb*	0.0612	0.0602	0.0605	0.0600	0.0596	0.0583	0.0586	0.0594
53	30-psig steam power equiv., MkWh/yr	20,196	22,274	22,083	22,800	24,615	24,486	26,956	26,849
54									
55	Total equivalent power, MkWh/yr	1,335,944	1,349,479	1,369,958	1,405,314	1,423,927	1,444,771	1,479,001	1,477,063
56									
57	Actual heat rate, Btu/kWh	10,790	12,300	12,150	11,400	10,660	10,420	10,520	10,460
58	Standard heat rate, Btu/kWh	10,000	10,000	10,000	10,000	10,000	10,000	10,000	10,000
59									
60	*These values vary from year to year because of differences in average steam temperatures and pressures.								

TABLES 1 (opposite page) **and 2:** In practice, these two tables are part of the same spreadsheet. Each table ordinarily shows the latest ten years of performance; in this case, however, the final columns have been omitted because of space and legibility limitations

	BASES FOR CALCULATING TABLES 1 AND 2	
	A	B
1	Acct. 115	
2	ABCD Plant	1983
3		
4	ABCD production, Mlb/yr	375,000
5		
6	Basic input:	
7	Electricity used, MKWH/yr	14,100
8	400 psig steam used, Mlb/yr	101,600
9	200 psig steam used, Mlb/yr	85,390
10	30 psig steam used, Mlb/yr	-55,200
11	Fuel gas used, MMBTUmeth/yr	155,300
12		
13	Using actual cogeneration plant fuel equivalents:	
14	Electricity equiv., MMBTUmeth/yr	=B7*B57/1000
15	400 psig steam equiv. MMBTUmeth/yr	=B8*B46*B57/1000
16	200 psig steam equiv., MMBTUmeth/yr	=B9*B49*B57/1000
17	30 psig steam equiv., MMBTUmeth/yr	=B10*B52*B57/1000
18	Fuel gas used, MMBTUmeth/yr	=B11
19	Total equiv. fuel gas, MMBTUmeth/yr	=SUM(B14:B18)
20	Btu/lb ABCD @ actual pwr. plt. eff.	=B19/B4*1000
21		
22	Using 10,000 Btu/KWH as basis:	
23	Electricity equiv., MMBTUmeth/yr	=B7*B58/1000
24	400 psig steam equiv. MMBTUmeth/yr	=B8*B46*B58/1000
25	200 psig steam equiv., MMBTUmeth/yr	=B9*B49*B58/1000
26	30 psig steam equiv., MMBTUmeth/yr	=B10*B52*B58/1000
27	Fuel gas used, MMBTUmeth/yr	=B11
28	Total equiv. fuel gas, MMBTUmeth/yr	=SUM(B23:B27)
29	Btu/lb ABCD @ 10,000 Btu/KWH	=B28/B4*1000
30		
31	Annual Production, MMlb/yr	=B4/1000
32		
33	Acct. 203	
34	Site Cogeneration Power Plant	83
35		
36	Fuel gas used, MMBTUmeth/yr	14,329,532
37	Off-gas used, MMBTUmeth/yr	22,700
38	Vent-gas used, MMBTUmeth/yr	62,600
39	Total fuel used, MMBTUmeth/yr	=SUM(B36:B38)
40		
41	Power produced, MKWH/yr	1,170,950
42	Power used, MKWH/yr	130
43	Net power exported, MKWH/yr	=B41-B42
44		
45	400 psig steam generated, Mlb/yr	355,790
46	400 psig steam power equiv., KWH/lb*	0.103
47	400 psig steam power equiv., MKWH/yr	=B45*B46
48	200 psig steam generated, Mlb/yr	1,237,500
49	200 psig steam power equiv., KWH/lb*	0.0875
50	200 psig steam power equiv., MKWH/yr	=B48*B49
51	30 psig steam generated, Mlb/yr	330,000
52	30 psig steam power equiv., KWH/lb*	0.0612
53	30 psig steam power equiv., MKWH/yr	=B51*B52
54		
55	Total equivalent power, MKWH/yr	=B43+B47+B50+B53
56		
57	Actual heat rate, Btu/KWH	=B39/B55*1000
58	Standard heat rate, Btu/KWH	10,000

TABLE 3: The numbers in Tables 1 and 2 (pp. 203, 204) are based on the input data and formulas shown above. The row numbers above correspond with those of Tables 1 and 2

makes more sense to track Btu_{meth} per pound of feed.

This can, for instance, be the case when a plant makes several products having similar chemical functionality (e.g., mono-, di- and tri-substituted molecules), and the energy consumption varies more with the feedstock that provides the substitution than with total pounds of product. When choosing this option, make sure that the feedstock is adequately metered.

Calculating the Btu_{meth}/lb of product is done slightly differently for the two power-and-steam options discussed on page 131.

• If steam and power are cogenerated onsite, use the actual heat rate of the power plant to calculate Btu_{meth}. Also, calculate it using a reference standard heat rate, such as 10,000 Btu_{meth}/kWh. This separates the individual plant efficiencies from fluctuations in steam-and-power-generation efficiency.

Divide both Btu_{meth} values by the annual production. Report actual energy consumption per pound of product, as well as energy consumption per pound based on the reference standard rate for the cogeneration plant, both for a ten-year period

• For plants that purchase electricity and employ steam generated in central boilers, use a heat rate of 10,000 Btu_{meth}/kWh for electricity, in both the actual and standard cases. Calculate the Btu_{meth} equivalents as explained in Part 1, using both the actual (based on the actual efficiency of the site's central boiler plant) and standard (based on a constant boiler power plant efficiency) values. Add electrical Btu_{meth} and steam Btu_{meth} together and divide by the total production, to give Btu_{meth} per pound of product for both the actual and standard cases

Useful formats for organizing and displaying the data are illustrated in Tables 1 and 2. As the first column of each implies, these two tables would in practice be parts of the same spreadsheet. Table 1 pertains to the production plant; Table 2 pertains to the associated cogeneration plant. Table 3 (in part recapitulating information from Part 1) includes equations showing how various lines in Tables 1 and 2 are calculated.

In addition to this tabular detail,

CALCULATING CARBON DIOXIDE EMISSIONS

Closely related to Btu accounting is the calculation of carbon dioxide emissions, generated whenever fuels are burned. The amount of CO_2 emitted can be calculated if the quantity and composition of the fuel are known. As explained below, calculation is preferable to stack-gas monitoring. The calculated emissions can be expressed in terms of either the mass of a fuel or its Btu content.

Mass of fuel: Carbon dioxide is formed when carbon atoms in fuel react with oxygen in the air. Except for very small amounts of carbon monoxide (CO), every carbon atom reacts with two oxygen atoms to form CO_2. Therefore, a direct stoichiometric relationship exists between the pounds of carbon burned and the pounds of carbon dioxide emitted.

For methane, the most common fuel, this relationship is:

$$CH_4 + 2O_2 \rightarrow CO_2 + 2H_2O$$

The ratio of carbon dioxide's molecular weight to that of methane is (44.010/16.043), i.e., 2.7433. For every 1,000 lb of methane burned, 2743.3 pounds of CO_2 are emitted.

Similar ratios can be calculated for ethane (2.9272) and propane (2.9941), the other two major components of most natural-gas supplies. Accordingly, once a natural gas composition and usage is documented, the amount of CO_2 emitted is easily calculated.

When coal is burned, the calculation is somewhat different. Although coal varies in composition, any given shipment is typically accompanied by an "ultimate analysis" that gives the amount of each component, reported on a dry basis.

A typical ultimate analysis for coal is:

Carbon	71.6%
Oxygen	8.2%
Hydrogen	4.9%
Nitrogen	1.3%
Chlorine	0.1%
Sulfur	3.1%
Ash	10.8%

The percent moisture is also reported. In this example, the coal contains 12.8% water. For every 1,000 pounds of coal received (wet basis), then, the pounds of carbon is (1,000)(0.716)(1 − 0.128), or 624 lb.

As with methane, virtually all carbon is converted to carbon dioxide during combustion. The ratio of carbon dioxide's molecular weight to the formula weight of carbon is (44.010/12.011), i.e., 3.664, so 624 lb of carbon yields (624)(3.664), i.e., 2,286 lb of CO_2.

Finally, since 624 lb of carbon corresponds to 1,000 pounds of wet coal, we can relate the amount of carbon dioxide produced to the pounds of this particular wet coal burned: 2,286 lb/1,000 lb = 2.286 lb CO_2 per pound of coal.

The same procedure can be followed with other fuels. An important example is No. 2 fuel oil, whose typical ultimate analysis is:

Carbon	87.3%
Oxygen	0.04%
Hydrogen	12.6%
Nitrogen	0.006%
Sulfur	0.22%
Ash	$< 0.01\%$

Although the main focus in this article is on fuels, the engineer involved with tracking carbon dioxide may well wonder about the CO_2 content of hydrocarbons used as feedstock. Most major feeds, including naphtha, butane, and propane-butane mixtures, do not contain significant amounts of carbon dioxide.

However, LPG (liquefied petroleum gas) feedstocks, consisting mainly of ethane-propane mixtures, typically contain 3-4% carbon dioxide. This is removed and vented to the air before the feedstock is sent to cracking furnaces. As is the case with fuels, the pounds of carbon dioxide vented to the atmosphere can be calculated if the total quantity of feed and the percent carbon (including carbon in the form of dissolved CO_2) are known .

Btu Content of Fuel: Keeping track of carbon dioxide emissions is in many cases simpler if the emissions are expressed in terms of the Btu content of the fuel instead of its weight. Following is a summary of CO_2 emissions per million Btus (MMBtu), higher heating value (HHV), of various major types of fuels, with the calculations shown for reference. For fuels such as coal or oil, the CO_2 emissions will depend on fuel composition, and we have assumed typical compositions.

Methane:

(1 MMBtu CH_4)(1 lb CH_4/23,879 Btu CH_4)(44 lb CO_2/16 lb CH_4)(1 ton CO_2/2,000 lb CO_2) = 0.0576 ton CO_2

Thus, burning 1 MMBtu CH_4 emits 0.0576 tons CO_2, (or 0.0157 tons carbon).

Ethane:

(1 MMBtu C_2H_6)(1 lb C_2H_6/22,320 Btu C_2H_6)(88 lb CO_2/30 lb C_2H_6)(1 ton CO_2/2,000 lb CO_2) = 0.0657 ton CO_2

Thus, burning 1 MMBtu C_2H_6 emits 0.0657 tons CO_2 (or 0.0179 tons carbon).

Coal:

Assume that the coal is 75% carbon and that one ton is equivalent to 27 MMBtu (reasonable assumptions for U.S. western subbituminous grades). First, determine the CO_2 emissions per ton:

(0.75 ton C/ton coal)(44 tons CO_2/12 tons C) = 2.75 tons CO_2

Then, (1 MMBtu coal)(1 ton coal/27 MMBtu)(2.75 tons CO_2/ton coal) = 0.1019 ton CO_2

Thus, burning 1 MMBtu coal emits 0.1019 tons CO_2 (or 0.0278 tons carbon).

Offgas:

Numerous processes, especially ones involving hydrocarbons, generate offgases suitable for use as fuel. These gases may vary widely in composition.

For illustrative purposes, consider a typical ethylene-plant offgas whose composition on a volume basis is 60% CH_4 and 40% H_2. Then, [1 std ft^3 offgas][(0.60 std ft^3 CH_4/std ft^3 offgas)(1,013 Btu/std ft^3 CH_4) + (0.40 std ft^3 H_2/std ft^3offgas)(325 Btu/std ft^3 H_2)] = 737.8 Btu.

And (1 MMBtu offgas)(1 std ft^3 offgas/737.8 Btu)(0.60 std ft^3 CH_4/std ft^3 offgas)(1 std ft^3 CO_2/std ft^3 CH_4)(44 lb CO_2/359 std ft^3 CO_2)(1 ton CO_2 /2,000 lb CO_2) = 0.0498 tons CO_2

Thus, burning 1 MMBtu offgas emits 0.0498 tons CO_2 (or 0.0136 tons carbon).

Fuel oil:

Assume that the oil contains 87.3% carbon as indicated earlier, and that 1 lb is equivalent to 18,000 Btu. Then,

(1 MMBtu oil)(l lb oil/18,000 Btu)(0.873 lb C/lb oil)(44 lb CO_2/12 lb C)(1 ton CO_2 /2,000 lb CO_2) = 0.0889 ton CO_2

Thus, burning 1 MMBtu oil emits 0.0889 tons CO_2 (or 0.0243 tons carbon).

Syngas:

Coal-derived synthesis gas (syngas) is normally used as a fuel. Calculating its CO_2 emissions requires a somewhat different approach, because its heating value is expressed on a volumetric basis: approximately 250 Btu/std ft^3 (HHV).

Its CO_2-producing components are CO and CH_4, and it itself contains some CO_2. The volumetric percentages are typically:

CO	28%
CO_2	23%
CH_4	1.5%

Together, these account for 52.5% of the syngas volume.

Now, 1 mole of CO yields 1 mole of CO_2 upon combustion, and the same as true for the CH_4. Accordingly, recalling that a mole of any gas occupies 359 st ft^3, we can say with regard to the 52.5% of the gas consisting of CO, CO_2 and CH_4, that :

(*Continued, next page*)

CALCULATING CARBON DIOXIDE EMISSIONS

(Continues from previous page)

FUELS	Tons of CO_2 per million Btu burned	Tons of carbon per million Btu burned
Hydrogen	0	0
Offgas	0.0498	0.0136
Methane	0.0576	0.0157
Ethane	0.0657	0.0179
Fuel oil	0.0889	0.0243
Coal	0.1019	0.0278
Syngas	0.1287	0.0351
Carbon	0.1301	0.0355

(1 std ft^3 CO_2/std ft^3 gas)(44 lb CO_2/359 std ft^3 CO_2) = 0.1226 lb CO_2/std ft^3 gas (or 0.0613 tons CO_2/1,000 std ft^3 gas)

Therefore, with regard to the entire syngas stream:

(1 MMBtu syngas)(1 std ft^3 syngas/250 Btu syngas)(0.525)(0.1226 lb CO_2/ std ft^3 syngas) = 257.4 lb CO_2, or 0.1287 tons CO_2

Thus, burning 1 MMBtu syngas emits 0.1287 tons CO_2 (or 0.0351 tons carbon).

Carbon:

We include carbon here as a baseline reference. One pound is equivalent to 14,093 Btu. Accordingly:

(1 MMBtu carbon)(1 lb carbon/14,093 Btu)(44 lb CO_2/12 lb carbon)(1 ton CO_2/2,000 lb CO_2) = 0.1301 tons CO_2

Thus, burning 1 MMBtu carbon emits 0.1301 tons CO_2 (or 0.0355 tons carbon). Note, thus, that carbon is the highest CO_2 emitter.

These CO_2 emissions and carbon contents are summarized in the table. In comparing them, be sure to keep the overall plant picture in mind. Syngas, for example, is generated by coal gasification, a net exporter of steam. In the table, no CO_2 or carbon credit has been taken for fuel that would need to be burned to make the equivalent amount of steam if the syngas process were eliminated.

The better way

The alternative to calculating carbon dioxide emissions from fuel usage is direct monitoring. There are at least three reasons why we consider the calculation approach more accurate and otherwise better.

In the first place, fuel users and vendors alike tend to be very careful that (purchased) fuel is measured accurately. Plant pay meters (meters upon which fuel-gas purchases are based) are well maintained and are checked regularly. In many cases, there is double metering. The same kind of care is found with many other sources of carbon dioxide emission.

Secondly, monitoring CO_2 requires knowing not only the percent CO_2 in a vent but also the mass flow through it. While analyzers are available to monitor vent composition (and often optimize furnace operation), accurately measuring the hot gas flow is difficult and expensive.

Finally, calculating carbon dioxide emissions based on fuel use usually requires far fewer measurements than does monitoring. For example, a given site may have ten furnaces but only one gas meter.

With waste burners, some estimating may admittedly need to be done. And with fuels that vary in composition, some composition averaging over time may be necessary. But overall, calculating the carbon dioxide emissions based on fuel or even waste composition will be far more accurate than virtually any attempts at monitoring.

In specific cases, such as vent streams containing carbon dioxide that did not result from combustion, monitoring may be the only solution. But it should be a last resort. ❐

graphical summaries of key data are highly recommended. For examples of useful graphs, see Figures 1 and 2.* Once a spreadsheet format has been established, graphs can be generated almost entirely by subroutines (macros) within the spreadsheet program itself.

Site performance

Accounting for Btus throughout a multi-plant site in a meaningful way is more difficult. The choice of what to include depends on the purpose of the report.

The simplest technique is merely to add together all the fuels purchased in a given year and divide by the total pounds of output produced during the year. For some purposes and some sites, this may be appropriate. However, this method masks a number of factors that may be relevant:

- Some plants may have consumed fuels generated internally, in addition to — or instead of — purchased fuels
- The relative operating rates of plants having high Btu_{meth}/lb and those having low Btu_{meth}/lb may change from year to year
- Some plants at the site may have been shut down and not replaced
- Obsolete plants may have been shut down and replaced with more efficient ones making the same types of products
- New plants, making new products, may have started up

Therefore, in most cases there is no substitute for a more-detailed approach: tracking site performance by totaling individual plant performance. This is useful and valid even if some small energy consumers on the site must be omitted for one reason or another. The data that are thus developed can be presented in several forms.

When reporting Btu_{meth}/lb performance for the past ten years, at least three options are available. The first consists of reporting the performance of all plants, making year-by-year notations as to specific changes (new plants, plant shutdowns, and so forth) that are likely to have affected the corresponding annual figure.

A second option is to report site Btu_{meth}/lb for the past ten years but exclude from the year-to-year numbers the energy consumption (and product output) of all plants that were shut down and not replaced during the ten years. Again, note the causes of any aberrations. Both actual and standard Btu_{meth}/lb plant conversions may be shown.

Finally, a variant of the second approach is to report site Btu_{meth} usage including only plants that have been running during the entire ten years and assuming that both production and steam-and-power-generating efficiencies have remained constant. Only individual-plant efficiencies change.

To eliminate variability caused by power plant efficiency, use the standard Btu_{meth}/lb values. Then, multiply each plant's annual standard Btu_{meth}/lb by that plant's production in the *most recent* year to give annual Btu_{meth}. Finally, add the various plant Btu_{meth} values for each year. This approach does not reflect actual Btu_{meth} used for the

*Note that these graphs refer to Btu, rather than Btu_{meth}. This is for simplicity.

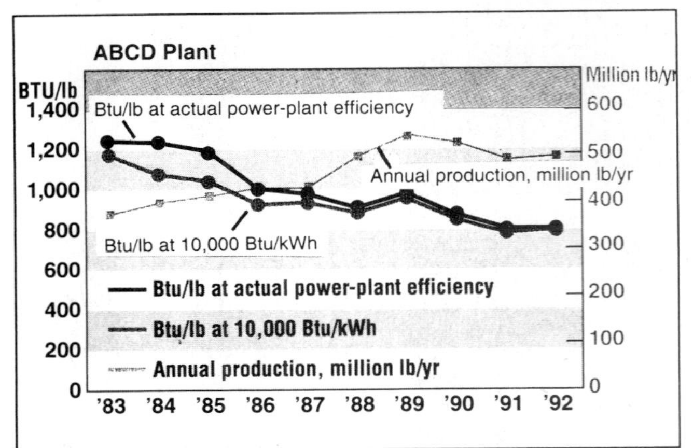

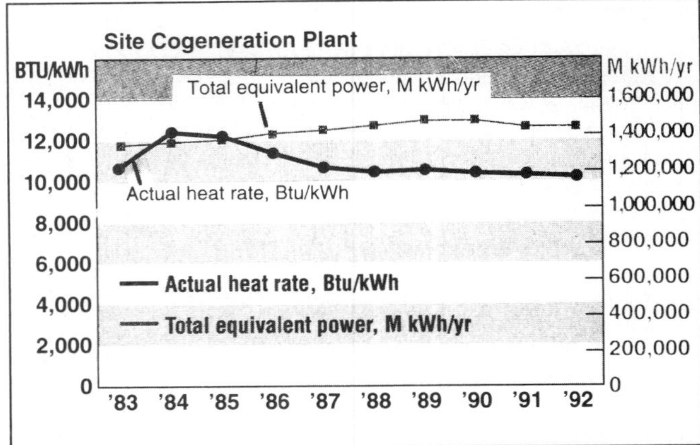

MARLENE JAEGER

FIGURES 1 (left) and 2: The energy-related performance of a production plant and of a cogeneration plant can conveniently be captured by graphs. Such graphs can in most cases be readily generated via the spreadsheet program associated with Tables 1 and 2

past ten years, but it does clearly show variations in site efficiency. Some of the variation may be caused by operating rates.

Company performance

Tracking overall company energy performance is far less useful than individual-plant or even site tracking. The problems summarized above for tracking sites are compounded. For instance, besides the addition or shutdown of individual plants, entire sites may have been added or eliminated.

If the company-tracking effort must nevertheless be made, a reasonable estimate of companywide fuel use can be obtained by adding total fuel purchased, total off-gas and vent gas burned, total waste oils and tars burned, and any other plant byproducts that are used as fuel. Except perhaps for fuel purchased, all of these data will probably need to be obtained on a plant-by-plant basis. Their sum can then be divided by total production expressed as pounds, as dollars of sales, or as dollars of value added.

Using dollars of sales or dollars of value added introduces additional variables, because changes in raw-material and product prices influence the results. (Using this measure is analogous to rating automobile fuel efficiency as "miles per dollar of dealer profit.")

Another alternative is to simply summarize the top 90% of company sites in terms of their energy consumption. If the plants at these sites are already being tracked on an individual basis, this approach eliminates a lot of extra effort.

RANKING OF A FIRM'S PLANTS BY ENERGY USAGE				
	PLANT	Million Btu/yr	% Total	Cum. %
1	NO	632,170	30.9	30.9
2	TU	429,600	21.0	51.9
3	ABCD	393,640	19.3	71.2
4	KLM	137,690	6.7	77.9
5	XYZ	97,280	4.8	82.7
6	PQ	88,620	4.3	87.0
7	RS	75,220	3.7	90.7
8	HIJ	70,830	3.4	94.1
9	VW	66,470	3.3	97.4
10	EFG	53,330	2.6	100.0
	TOTAL	2,044,850		

TABLE 4: Listing plants by energy consumption can pinpoint targets for improvement

As in the case of site reporting, we recommend developing a company-wide-performance chart that includes only plants that have been running for the previous ten years. Again, use both the actual and the 10,000-Btu_{meth}/kWh standard for both power and steam, and recalculate the individual plant energy usage based on the production in the most recent year (using actual and standard Btu_{meth}/lb for the past ten years).

If a company has several sites, it is useful to show a pie chart illustrating which sites use the most energy. Such a pie chart helps target sites having large energy usage. The same information can be spelled out in tabular form, as in Table 4 above.

Product performance

A valuable analysis of energy efficiency may consist of comparing plants making similar products. Typically, these are located at different sites within a company.

To compare plant performance, fix the power- and steam-generation efficiencies at 10,000 Btu_{meth}/kWh. In this way, variations in power-plant efficiencies at the different sites will not distort the production-plant efficiencies.

A graph showing the ten-year Btu_{meth}/lb history of each plant together with the weighted average will highlight opportunities for improvement. The weighted average is obtained by summing the annual fuel consumed by the plants (at standard heat rate) and dividing by total production. ■

Edited by Nicholas P. Chopey

For information about the author, see page 202.

PUMP UP YOUR ENERGY SAVINGS

Cut steam-heating costs with an ejector thermocompressor

Robert B. Power
Consultant

Rising fuel costs and the necessity of operating more efficiently are forcing engineers to find innovative ways to conserve energy. A device called an ejector thermocompressor can help. This component recycles waste steam into steam for heating.

The simple device, which can be used in many CPI applications, uses high-pressure steam to compress low-pressure, waste steam to a usable level of pressure. When attached to steam headers, for example, an ejector compresses waste steam that can then heat an evaporator, still, dryer roll or heater.

Potential applications occur in any situation where a flow of vapor or gas is supplied at a pressure higher than the acceptable minimum. An ejector thermocompressor can accomplish a useful pumping effect at such a location.

The new sizing and cost-estimating methods in this article* make it easy for engineers to select appropriate ejector thermocompressors for their own applications. For companies, these methods translate directly to saved energy and money in the bank.

What is it and how does it work?

The name "thermocompressor" applies to any compressor that raises vapor-stream pressure, thus condensing the

*Adapted from *Steam Jet Ejectors for the Process Industries* by R. B. Power. New York: McGraw-Hill, Inc., 1994

Originally published February 1994

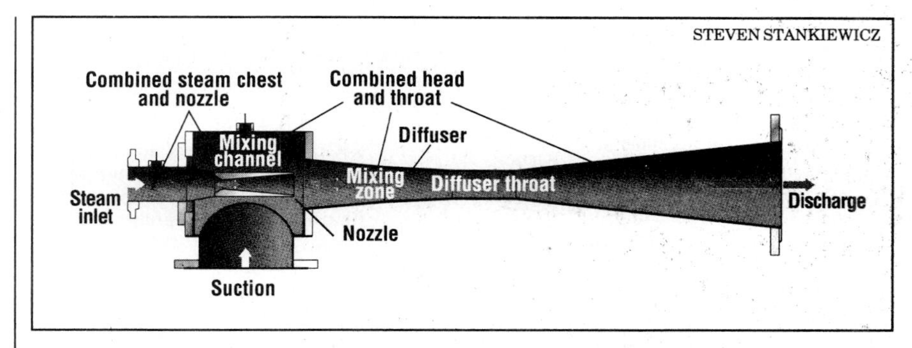

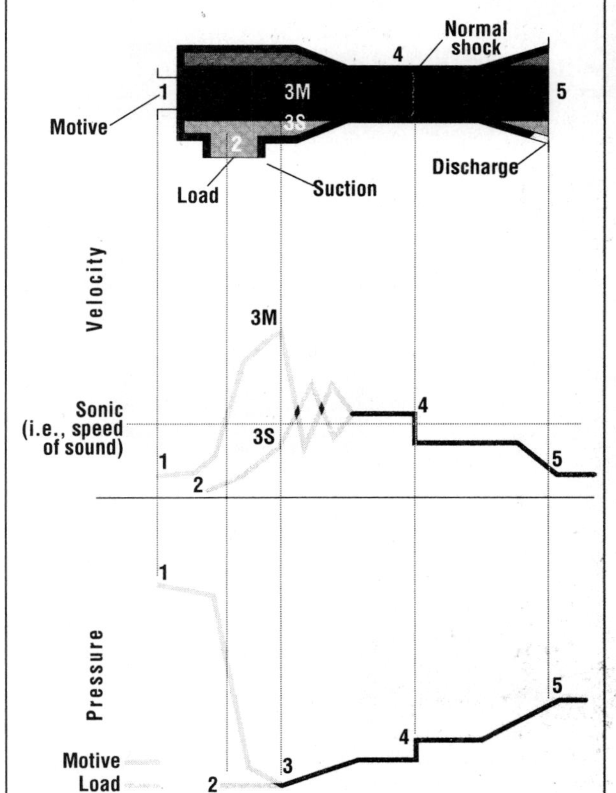

FIGURE 1. A basic, single-stage unit, above, illustrates an ejector thermocompressor's general working principles. Load and motive fluids combine inside the unit's mixing chamber and emerge at the discharge. Other types of units are shown on p. 121

FIGURE 2. An ejector's major strategic points, left, are shown vis-à-vis velocity and pressure for the load and motive fluids. The fluids have equal pressure in the mixing chamber, and equal velocity in the diffuser throat

R. B. Power

vapor at a higher temperature for heating. Thermocompressors, often called "heat pumps," include ejector and mechanical models. Mechanical models, which include reciprocating, rotary displacement, centrifugal, and axial-flow types, are typically more energy efficient than ejectors. However, they are generally more expensive, more complex, and less reliable.

When evaluating thermocompressors, consider mechanical models as well as ejectors. This article, however, discusses only ejector thermocompressors, referring to them simply as "thermocompressors." These are members of the ejector family.

An ejector is a fluid-pumping device in which a high-pressure motive fluid (e.g., steam) performs a pumping effect (Figure 1, p. 209). It has no moving parts. The motive fluid passes through a nozzle and emerges as a low-pressure, high-velocity jet. The nozzle directs this jet along the axis of a venturi-shaped diffuser. A second fluid, called the "load," enters at a suction point.

As the jet approaches the diffuser, it mixes with the second stream in a low-pressure mixing zone. The velocity after mixing is lower than that at the nozzle. The mixture flows through the diffuser, emerging from the discharge at a pressure between the suction and motive-fluid pressures.

Important variations in velocity and pressure for the motive (primary) fluid, the load (secondary) fluid and the mixture of the two are shown in Figure 2 (p. 209). Precisely describing these and other parameters at various points through the ejector involves tedious calculations [*1*]. A simplified description of the process, however, is adequate for preliminary estimates and has only three steps:

1. The motive fluid expands isentropically through a nozzle, which converts pressure energy to velocity energy at 100% efficiency. Steam velocity here is typically about 3,000 ft/s. The motive fluid's velocity upon entering the nozzle is negligible
2. The load fluid enters the suction chamber, where it mixes with the motive fluid at the suction pressure. Momentum is conserved, as the friction at the walls is negligible. The load-fluid velocity before mixing is negligible
3. The mixture of motive and load fluids passes through the diffuser, which converts the velocity energy to pressure energy with an efficiency of 65 to 95% (lower velocities and larger-size thermocompressors generally yield higher efficiencies). The velocity at the discharge is negligible

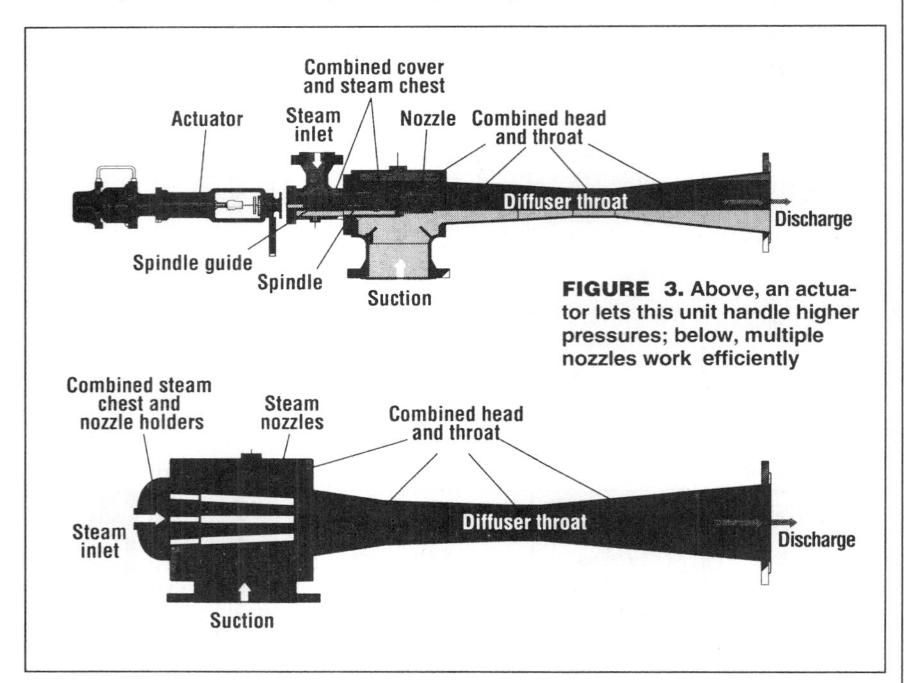

FIGURE 3. Above, an actuator lets this unit handle higher pressures; below, multiple nozzles work efficiently

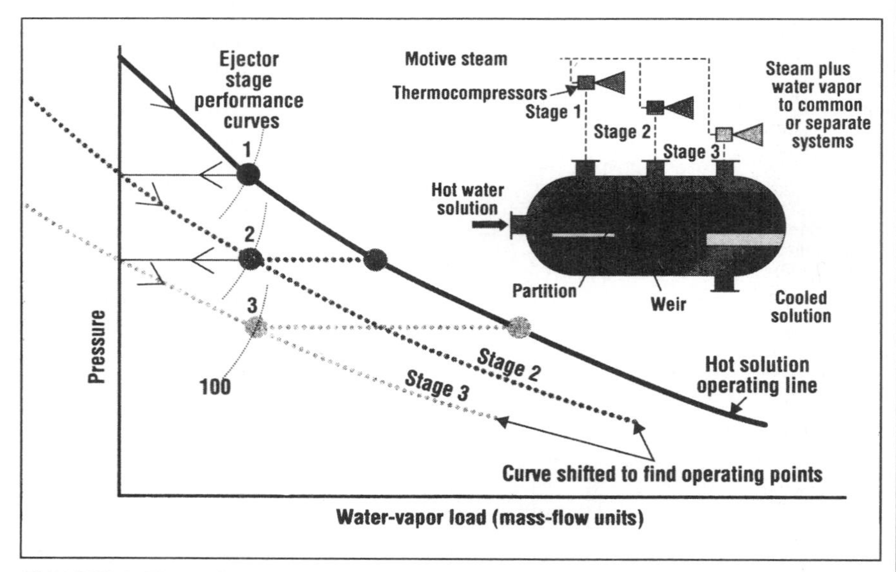

FIGURE 4. Three ejector stages are shown in this flash-cooling system. For stages 2 and 3, operating curves can be shifted to the left to simplify graphing

Thermocompressor construction

Thermocompressors are single-stage units (Figures 1 and 3). The simplest of the three is designed for moderate pressures (Figure 1). A more-complex model has an actuator that positions a spindle in the nozzle for capacity control (Figure 3, top); this construction, which features heavy walls, is suitable for higher pressures.

A multiple-nozzle ejector of a larger size and low pressure (Figure 3, bottom) is generally the most efficient of the three. Vendors of this type of model claim that it requires less high-pressure, motive steam. In addition, nozzles and diffusers for gas-jet applications

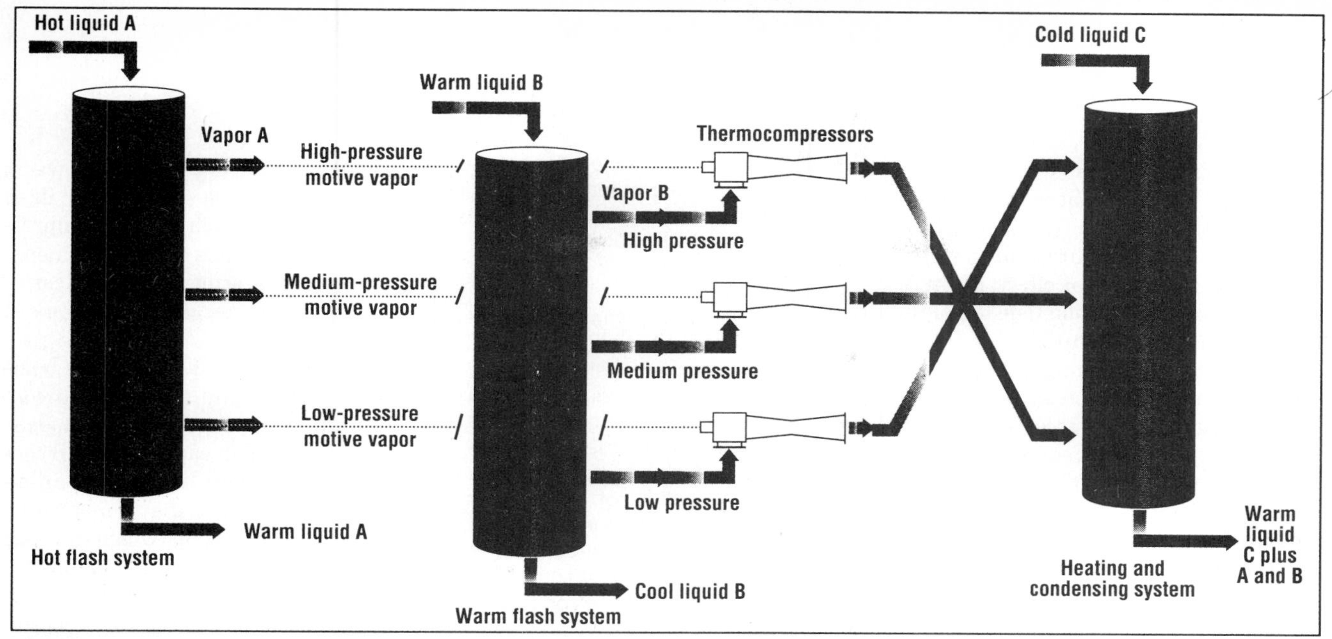

FIGURE 5. Multiple streams are cooled, flashed and heated in this integrated configuration

sometimes include steam jacketing to prevent ice or hydrate formation.

Singly, thermocompressors have a limited control range for adjustments [2–6]. Therefore, the most efficient way to maintain efficiency over a wide range of flows is to use multiple elements (i.e., two or more stage assemblies in parallel).

Pressure range and optimization

The following application example illustrates how a thermocompressor may take advantage of excess steam. Your plant may have sources of waste steam that can be similarly employed:

A process-heater application is to use 40,000 lb/h of 150-psig steam, condensing in a heat exchanger. However, an engineer observes that 8-psig (22.7-psia) steam is hot enough to deliver heat to the process, if a larger heat exchanger is used. Low-pressure, waste steam is available nearby at a pressure of –1 psig (13.7 psia).

The engineer decides to use the 150-psig steam to compress the waste steam at –1 psig. A thermocompressor using only 20,000 lb/h of 150-psig (164.7 psia) steam will compress 20,000 lb/h of steam from 13.7 psia to 22.7 psia. The high and low pressure steam mix within the thermocompressor, and the 40,000-lb/h steam mixture emerges at 22.7 psia.

The thermocompressor that the engineer selects costs $11,000 and has no moving parts. It has 12-in. connections, is 10 ft long and weighs 1,500 pounds.

Here, the application has a subatmospheric (vacuum) suction pressure and a positive pressure discharge. This example emphasizes that atmospheric pressure has no special significance for thermocompressor applications. A thermocompressor can operate below atmospheric pressure, above it, or both below and above it.

If the discharge vapors are condensed in a surface condenser, the use of a larger condenser permits a lower condensing temperature and pressure. This thus reduces the amount of motive steam required by the ejector.

NOMENCLATURE

d	diameter, in inches
DCF	density-correction factor, to convert load gas to equivalent motive gas
E_d	diffuser efficiency
h	enthalpy, Btu/lb
KE	kinetic energy, Btu/lb
lb	pound mass
P	pressure, psia
Re	mass-entrainment ratio, load/motive, dimensionless
Re_s	mass-entrainment ratio for steam, = $1/R_s$
R_s	mass ratio of motive steam to load steam, dimensionless
T	temperature, °Rankine
W	mass flow, lb/h
MW	molecular weight
Subscripts	
l	load
m	motive
s	steam
$suct$	suction

Keep in mind that the rules of thumb presented in this article are not precise. The object is to approximate the design that yields the lowest total cost over the life of the project.

Flash cooling of process liquids

If there is no nearby source of low-pressure steam, a hot aqueous stream can be used instead. This stream is cooled by flashing some vapor at low pressure. This vapor can be compressed by the high-pressure steam.

A hot-water solution containing a nonvolatile solute can be flash cooled in this way. Flashing the water vapor in a series of chambers minimizes the compression work and maximizes the steam economy. The configuration is shown schematically in Figure 4 (p. 210). Three flash stages are shown (for simplicity, only one ejector element is shown in each flash stage; multiple elements are often used for capacity control over a wide range of capacities).

In this application, the pressure differential between stages is large enough to pump the liquid through each partition and weir between stages. The stages thus may be arranged side by side in a horizontal tank. The size and configuration of the weirs and vapor outlet minimize the entrainment of liquid droplets.

The performance curves at the bottom of Figure 4 describe the operation quantitatively. The liquid-operating line is curved, reflecting the effect of the changing concentration of the solute. Here, the performance curves show pressure versus water-vapor mass flow — a preference of the process engineer and operator. Or, the vertical scale can be temperature, and the horizontal scale can be heat effect, in Btu/h or in tons of refrigeration (one ton of refrigeration = 12,000 Btu/h).

To simplify graphing, this liquid curve is shifted horizontally so the liquid exiting one stage becomes the feed to the next stage. Transparent overlays can be useful for graphs, especially if multiple liquid rates and varying feed conditions are to be evaluated. The overlays display the liquid curves for combinations of flow rates and feed conditions.

The overlay is positioned to represent the feed conditions to the first flash stage, and the intersection of the liquid curve with the ejector curve defines the vapor flashed in the first stage and the condition of the exit liquid. The overlay is then shifted to the left so that the condition becomes the feed to the next stage, and so forth.

Performance curve shapes

Figure 4 shows only a portion of each ejector performance curve, in the vicinity of the design points. On a linear plot of suction pressure versus load, each ejector-performance curve is tangent to a line from the origin to the design point for that curve. For critical ejectors, with a discharge pressure greater than twice the suction pressure, the suction pressure at no load is typically 20 to 30% of the design pressure.

Noncritical thermocompressors , i.e., those having a discharge pressure less than twice the suction pressure, have more-complicated performance curves. Their suction pressure varies with the load, the discharge pressure, and the motive pressure. By holding one pressure constant, a family of curves can be drawn for combinations of the other two pressures.

Integrating thermocompressors

A thermocompressor can sometimes be effectively integrated into a multiple-stream process system. Useful cooling, flashing and heating functions can be performed simultaneously. One such example is shown in Figure 5 (p. 211) in a highly schematic manner.

High-pressure motive vapor is obtained by flashing a hot, high-pressure stream of liquid A. The designation "high-pressure" simply means that the motive pressure is usefully higher than that of the vapor it will pump. It can be higher or lower than, or equal to, atmospheric pressure. This example has three flash stages, yielding motive vapor at three pressure levels.

One important byproduct of flashing is that the remaining stream of Liquid A is cooled and has less of the vapor A. Note that, if the stream is not a pure material, the flashed vapor will be rich in the more-volatile components.

The motive vapor A is used in three thermocompressors here, removing flashed vapor B from a stream of warm liquid B in a three-stage warm flash system. Within each thermocompressor, the two streams mix, and emerge at a pressure intermediate to the corresponding pressures in flash systems A and B.

The mixtures are delivered to a heating and condensing system. The condensers may be direct-contact models, as shown, in which vapors A and B condense and mix with liquid C. In an alternative figuration, they can also be surface condensers, in which the AB mixture does not contact liquid C.

Liquid C warms as it flows downward. This requires the thermocompressor discharge-stream positions to be reversed, as shown, to achieve maximum pumping effectiveness. The downward-flow arrangement in the flash and condensing equipment is required only if the fluid flow must be pumped by gravity from stage to stage.

Performance calculations

When evaluating a potential application, begin your performance estimates with the questions, "How much high pressure steam is required?" or "What suction pressure or discharge pressure can be created?" If preliminary estimates confirm the feasibility of using a compressor, then design attention can be focused on maximizing the application's effectiveness.

For quick estimates, you must know the pressures of the motive gas at the suction and discharge; the desired mass-flow rate of motive, suction or discharge gas; and the temperatures and molecular weights of both streams. The manufacturer will need to know the specific heats and specific-heat ratios of the gases, plus the compressibility factors or critical temperatures and pressures if the gases are uncommon.

Figure 6 (p. 213) is useful for estimating the design-point performance of steam-jet compressors. This graph shows expansion work versus compression work, relating motive and discharge pressure to suction pressure. It is based on published data [2-6] that have been considerably smoothed.

These data are generally accurate to within 20% over the range of compression ratios to 5, expansion ratios to 1,000, and motive/load ratios from 0.25

NAME THAT EJECTOR

A brief explanation of ejector nomenclature will help us avoid confusion in our discussion of ejector thermocompressors. The basic ejector principle is employed in a variety of devices. These range in size from tiny pocket units to 10-ft-dia-by-200-ft-long monsters, which test rocket engines for space vehicles.

Ejectors perform a variety of functions. They remove air from process and power-vacuum systems; prime pumps; vacuum-chill water, wetted vegetables or sand and gravel; pump cool water into steam boilers; and convey ashes and other granular solids. Placed at the bottom of a well, they boost up water to be pumped by an above-ground mechanical pump. The motive fluid can be gas or liquid. The load can be any of those, or can be granular solids.

Generic names and trade names are commonly used interchangeably in the industry, and different manufacturers use different names to describe some ejector applications. Accordingly, it is often useful to describe a specific ejector in an unambiguous and generic manner by naming the motive and load fluids: (motive) jet (load) ejector. Because the word "jet" identifies the device as an ejector, the word "ejector" often may be replaced by a descriptive word such as "pump," "compressor" or "conveyor."

Thus, a thermocompressor is a "steam jet steam ejector," or "steam jet steam compressor." The process-industry vacuum producer is a "steam jet air ejector," and a vacuum-refrigeration system that uses water as the working fluid is a "steam-jet refrigeration" system. Other example combinations: water-jet water ejector, water-jet air ejector, gas-jet gas compressor, and gas-jet solids conveyor. ❒

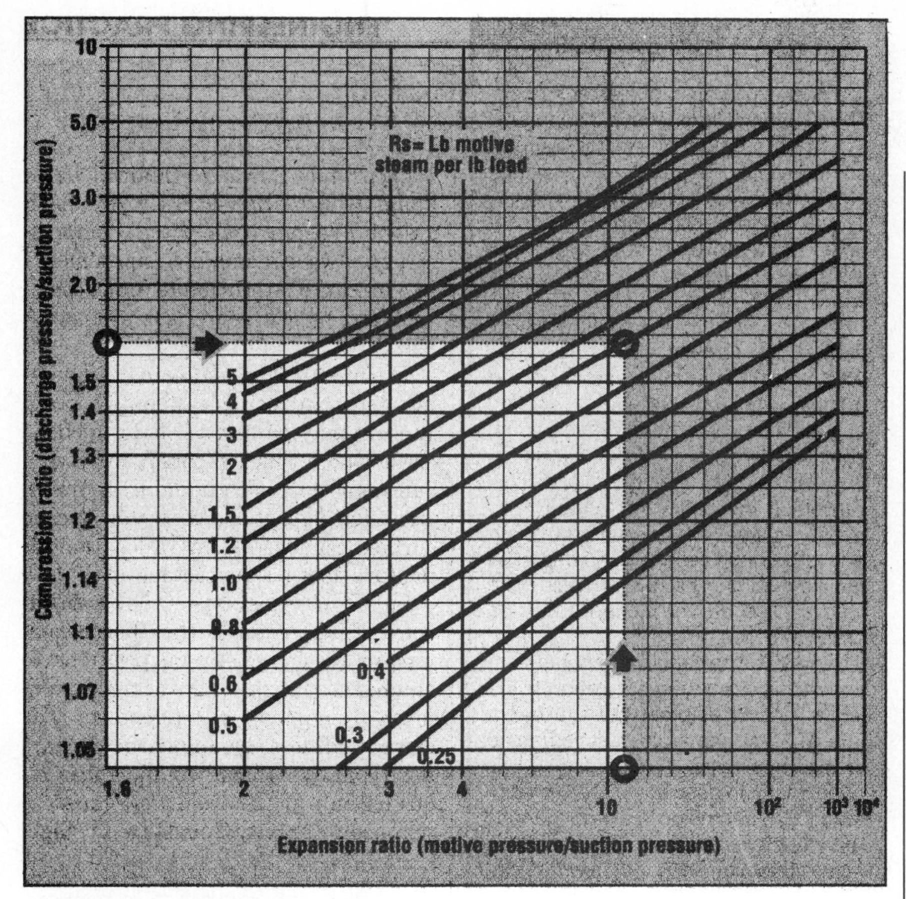

FIGURE 6. A plot of compression vs. expansion ratios helps in design work

to 5. Figure 6 is useful for proposed applications for feasibility and for rough economic estimates. A more-detailed calculation can then fine-tune the estimate and to size the equipment.

Absolute pressures and temperatures are used in the following ejector calculations. For the equivalent calculations, ideal-gas laws are used.

Stage size, weight and cost

The design of suction and discharge connections is dependent on the velocities of the gases being used. A typical design velocity at the inlet and outlet is 200 ft/s (however, this is lower for stages with a compression ratio less than 1.2).

Thermocompressor stage sizes are usually based on discharge conditions. They commonly have suction and discharge connections of equal size. A connection's diameter, d, is determined by the mass-flow rate and absolute pressure at that connection. The overall length of the stage assembly is about 10 diameters (7 diameters for multiple nozzles), and the diameter is found as follows:

$$d = 0.26\ (W_s / P)^{0.5} \qquad (1)$$

The cost of a thermocompressor is similar to that of a vacuum-ejector stage. The price will increase for the following features: high-pressure design, variable-area nozzle, spindle actuator, and special testing (if the application is outside the range for which good prototype test data are available). The following equations, which are valid for d of 2 to 48 in, estimate the weight and the 1992 price of a thermocompressor in \$U.S.:

$$\text{approximate cost} = 600\ (d)^{1.2} \qquad (2)$$

$$\text{weight in lb} = 40\ (d)^{1.4} \qquad (3)$$

The following application example demonstrates how to calculate the expansion ratio (motive pressure/suction pressure) and compression ratio (discharge pressure/suction pressure). Then, Figure 6 is used to locate the mass-flow ratio of motive-to-load gases. Finally, Eqs. (1)–(3) estimate the size and cost of a thermocompressor unit.

Example:

Use 150-psig motive steam to compress load steam from 20 psia to 40 psia, delivering 20,000 lb/h. Both steam sources are dry and saturated at the supply conditions. First determine the steam flow rates; and then find the approximate size and price of such a unit.

The expansion ratio is (150 + 14.7)/20 = 8.2. The compression ratio is 40/20 = 2.0. Checking Figure 6 for these values, obtain the mass ratio, R_s = 1.7 lb of motive steam per lb of load steam. The motive steam requirement is [1.7/(1 + 1.7)] (20,000) = 12,600 lb/h. The load steam is 20,000 – 12,600 = 7,400 lb/h.

The suction size is $0.26\ (7400/20)^{0.5}$ = 5 in. The discharge size is $0.26(20{,}000/40)^{0.5}$ = 5.8 in. Assume that the suction and discharge diameters will be the same; approximate both as 6 in. The length is 10 d = 60 in, or 5 ft. The weight is 40(6)1.4 = 500 lb, and the 1992 price is $600(6)^{1.2}$ = \$5,000 U.S.

Figure 6 can be used for the first application example given on p. 122: 150-psig steam compresses load steam from –1 psig to 8 psig. The expansion ratio is 164.7/13.7 = 12; the compression ratio is 22.7/13.7 = 1.66. From the chart, R_s is 1.0, or 1 lb motive per lb load.

Optimization and confirmation

If the process and cost numbers make the target application look attractive, try different combinations of flow and pressures to find the optimum conditions. Then, do a detailed calculation to confirm the performance. An adequate check involves using the three-step model described on p. 210.

To perform such a check, you can calculate energy values with ideal-gas laws, steam tables or charts. The following example, illustrated in Figure 7 (p. 214), uses a Mollier chart. We will use the scenario described in the previous application example on this page.

The enthalpy of saturated motive steam, h_{motive}, is 1,196 Btu/lb, and the suction-steam enthalpy, h_{suct}, is 1,156. The motive steam expands isentropically (along a vertical line) from the starting point down to the 20-psia pressure line, where it has an enthalpy of 1,040. The kinetic energy available at the nozzle outlet, KE_{noz}, is 1,196 – 1,040 = 156 Btu per lb of motive steam. After mixing, the kinetic energy of the mixture, KE_{mix}, is:

$$KE_{mix} = KE_{noz}/(1 + 1/R_s)^2 \qquad (4)$$

R_s = 1.7 and KE_{noz} = 156 Btu/lb

$KE_{mix} = 156/(1 + 1/1.7)^2 = 62$ Btu/lb

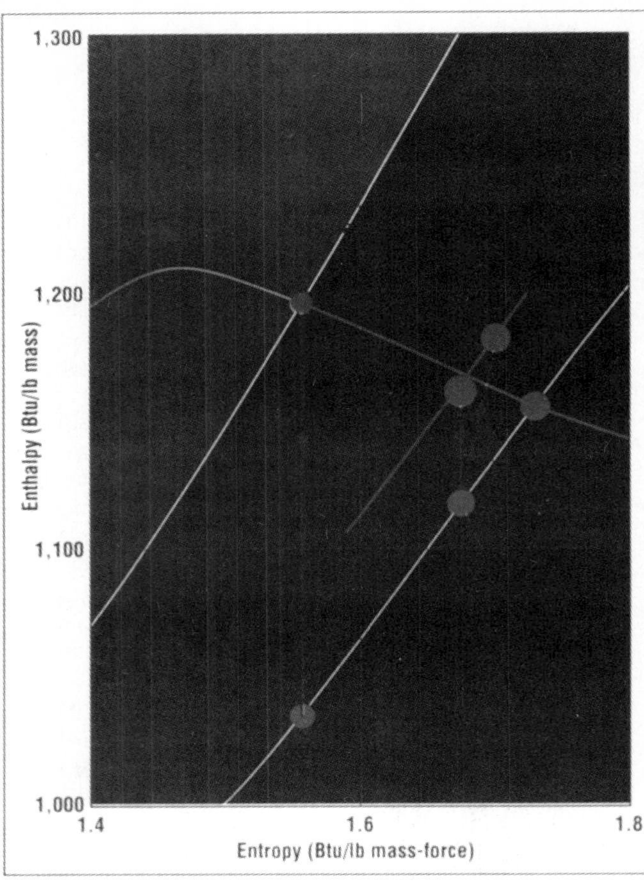

FIGURE 7. Optimum conditions for the example on p. 124 confirm a proposed design's performance

Next, calculate the enthalpy of the mixture, $h_{discharge}$, emerging from the diffuser. Kinetic energy is again zero, and the overall process is assumed to be adiabatic, so the enthalpy at the discharge is:

$$h_{discharge} = [R_s(h_m) + h_{suct}]/(R_s + 1) \quad (5)$$

$= [1.7\,(1{,}196) + 1{,}156]/(1.7 + 1)$
$= 1{,}181$ Btu/lb

Diffuser efficiency

The diffuser efficiency, E_d, is defined as a ratio. This is the ideal (isentropic) enthalpy increase during compression divided by kinetic energy of the mixture entering the diffuser:

$$E_d = \text{Ideal enthalpy increase}/KE_{mix} \quad (6)$$

The enthalpy of the mixture entering the diffuser, $h_{diffuser}$, equals the discharge enthalpy, $h_{discharge}$, minus the kinetic energy of the mixture, KE_{mix}, so 1,181 – 62 = 1,119 Btu/lb. To determine the ideal (isentropic) enthalpy increase during compression, follow these steps: Locate the point on the Mollier diagram where the 1,119-enthalpy line crosses the 20-psia suction pressure line. Then proceed up (isentropically) to the 40-psia line, where the enthalpy is 1,170. Thus, the ideal enthalpy increase, or compression work, is 1,170 – 1,119 = 51 Btu/lb.

The diffuser efficiency, E_d = 51/62 = 82%. Is this reasonable, or are you using an unrealistically high or low efficiency? A check of the actual efficiency rates offered by thermocompressor manufacturers is in order.

Figure 8 (p.215) shows a range of diffuser efficiencies (E_d) plotted versus mixture kinetic energy (KE_{mix}) entering the diffuser. The "high" and "low" curves bound most of the single-nozzle data sets examined by the author. Multiple-nozzle efficiencies may be higher. The values in this figure should be regarded as approximate.

Now look at Figure 8 and decide whether a diffuser efficiency, E_d, of 82% is reasonable with a mixture kinetic energy, KE_{mix}, of 62 Btu/lb. At a kinetic energy of 62 Btu/lb, note that the efficiency range is 76-84%. Note that, although the E_d of 82% is within that range, it is uncomfortably high.

To gain confidence in the feasibility of this design, increase the amount of steam. Since a small change in the steam rate makes a big difference in efficiency, try increasing R_s from 1.7 to 1.8 and repeat the calculations.

The mixture's kinetic energy, KE_{mix}, is $156/(1 + 1/1.8)^2$ = 64 Btu/lb. So the discharge enthalpy, $h_{discharge}$, is [1.8(1,196) + 1,156]/(1.8 + 1) = 1,182 Btu/lb. The enthalpy of the mixture entering the diffuser, $h_{diffuser}$, is 1,182 – 64 = 1,118 Btu/lb. The ideal enthalpy increase or compression work is 1,168 – 1,118 = 50 Btu/lb. E_d = 50/64 = 78%. This falls well within Figure 8's high-to-low band and is more likely to be confirmed by ejector manufacturers.

If you want a more-accurate estimate, check the example calculations using a larger Mollier chart. Note the contributions to error made by round-off and chart-reading uncertainties.

Calculations for other gases

The best estimates are those prepared by ejector manufacturers. However, your design schedule may not include enough time to wait for manufacturer response. The detailed method described in the previous section is useful, but requires some calculation time. It is limited further because Figure 8 gives diffuser efficiencies based on steam, and will be less accurate for other gases.

The steam curves in Figure 6 will often be sufficiently accurate for preliminary estimates. The effects of changes in physical properties will tend to cancel out if the load gas is the same as the motive gas. This method is generally conservative — actual performance is usually better.

If the load gas is different from the motive gas, convert the load to equivalent motive gas using a density-correction factor, *DCF*. The equation for *DCF* is based on the generalization that the design-load-handling capability varies roughly with the square root of the load density. The ejector will handle more pounds of more-dense gas.

$$DCF = [(MW_l/MW_m)(T_m/T_l)]^{0.5} \quad (7)$$

You'll find that your detailed calculations using Equations 4 through 7 provide a useful reality check in preliminary estimates. Large Mollier charts or ideal-gas laws are useful for finding values for the motive and load gases.

Contact an ejector manufacturer to confirm your estimates for size, mass ratios and diffuser efficiencies before you invest much time in estimating and

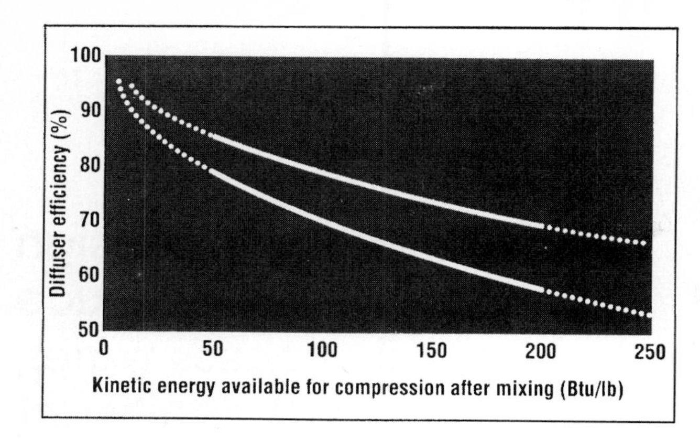

FIGURE 8. Compiled data from manufacturers' catalogs show low and high limits for actual ejector efficiencies

designing. Your discussions with manufacturers and your specifications will be improved by your understanding of the subject.

Diffuser efficiency is an important variable to establish early, and one good example that lies within your operating conditions will be a useful benchmark. Practice will help you improve your judgment in adjusting the results from Figure 6 for your gases.

Estimate for dissimilar gases

The following example illustrates an application in which the motive and load gases are not the same. Use Equation 7 first to find *DCF*. Then, calculate the heating value of the resulting mixed gases:

Use pressurized propane gas to compress atmospheric air to 5 psig, and find the heating value of the mixture. The propane-supply conditions are 60 psig and 105°F. The air conditions are 14.27 psia and 60°F. Propane has a molecular weight of 44 and a heating value of 21,000 Btu/lb. Air has a molecular weight of 29.

The expansion ratio is (60 + 14.27)/14.27 = 5.2. The compression ratio is (5 + 14.27)/14.27 = 1.35. For these values, you find that Figure 6 gives R_s of 0.9 (lb motive/lb load). This then translates to an entrainment ratio, Re_s, of 1.11. The density correction factor, *DCF,* is:

$$DCF = \{[29(460 + 105)]/[44(460+60)]\}^{0.5} = 0.85$$

and the corrected entrainment ratio, $Re = DCF\ (Re_s) = 0.85(1.11) = 0.94$; so there will be 0.94 lb air per lb propane.

In addition, the mixture of 1.0 lb propane and 0.94 lb air will have a volume of approximately 21 ft^3 at standard temperature and pressure. Its heating value will be 21,000/21 = 1,000 Btu/ft^3. This falls within 10% of the value that one finds in a thermocompressor catalog [*5*].

Capacity control

Noncritical thermocompressors (i.e., those having a discharge-to-suction-pressure ratio less than 2:1) permit some capacity control by allowing the motive gas to be throttled or by using a variable-area motive nozzle [*2–6*].

For critical and noncritical thermocompressors, capacity control over a broad range of capacities is best achieved by using multiple elements and turning individual elements on and off to match the changing load.

Efficiencies, ethics and testing

Because thermocompressors are often justified on a cost-reduction basis, the choice among competitive bids can often be influenced by small differences among vendor quotes. Each manufacturer simply presents the best design available. Diffuser efficiency is crucial.

The situation is complicated by industry's perception of users. Some industry members believe that users know very little about ejectors and are generally unwilling to spend more money for better operating economy. In addition, a common perception is that most users do little or no performance testing for ejectors.

Sometimes, customers complain that manufacturers claim unattainable efficiencies for their designs. In a situation where the only ethical constraint is self-monitoring, some questionable situations are likely to develop.

The best strategy is to state pragmatically what you are willing to pay for performance, select the bid that appears realistic and that offers a low evaluated cost, and add a price-performance clause to your purchase contract if appropriate.

So, first witness the test to assure yourself that the performance is appropriate. Then, if the actual performance differs significantly from the contract performance, adjust the price.

Witnessing the test can be costly. But by being present at testing, you ensure that you get what you pay for. For example, while watching the test of a batch of thermocompressors, I discovered that one was undersized. The manufacturer had neglected to make a conversion from equivalent air to equivalent steam for that unit.

The flipside of this discussion of ethics is to encourage you to treat the ejector manufacturers evenhandedly. Listen impartially to each of them, and respect the confidentiality of the information they share with you.

Don't try this at home

Designing your own ejector thermocompressor sounds like a simple proposition. However, there are many good reasons why you shouldn't undertake this deceptively easy-looking task [*1*]. Basically, it is simply not cost effective.

You are looking for high performance, not merely low cost. The price you pay for your unit reflects knowhow and expertise, which deliver that performance. ■

Edited by Irene Kim

References

1. Power, R.B., "Steam Jet Ejectors for the Process Industries," McGraw-Hill, New York, 1994
2. "Thermocompressors," Bulletin TC-V80-A, Croll-Reynolds Co., Inc.
3. Freneau, P., "Steam-Jet Thermocompressors Supply Intermediate-Pressure Processes," Jan-Feb 1945 *Power*
4. "Ejector Manual," Graham Mfg. Co., Inc.
5. "Jet Compressors," Performance data supplement to Bulletin 4F, Schutte and Koerting Div., Ketema
6. "Jet Pumps and Gas Scrubbers" catalogue, GEA Wiegand GmbH

The author

Robert B. Power is a consulting mechanical engineer (915 Alynwood Cir., Charleston WV 25314; Tel. 304-344-2887). He previously worked for 33 years in Union Carbide's central engineering department, where he specialized in the design and troubleshooting of ejector and fluid-machinery equipment and systems. He has a B.S. from Carnegie Mellon University (Pittsburgh, Pa.) and an M.S. from West Virginia University (Morgantown, W.V.). He is a member of the American Soc. of Mechanical Engineers (ASME) and the National Soc. of Professional Engineers (NSPE), a registered engineer in West Virginia, and an adjunct faculty member at WVU College of Graduate Studies.

COOLING TOWERS

Seemingly minor design upgrades can produce colder process water, and significant savings

Robert Burger
Burger and Associates, Inc.

THE OFTEN OVER-LOOKED

Because of their apparent simplicity, cooling towers are all too often taken for granted in the chemical process plant, electric-power generating station, and refrigeration or air conditioning system. Engineers assume that they will provide the desired cold water return.

In reality, cooling towers are complex devices that require careful design and maintenance. Slight deviations from design specifications can have a real impact on overall plant economics.

Design conditions are specified before a cooling tower is purchased. To be decided upon are the volume of circulating water (gpm), hot water temperature (HWT) at the inlet, cold water temperature (CWT) at the discharge, and wet-bulb temperature (WBT).

The WBT measurement is a reflection on the actual temperature at a given relative humidity or dewpoint. It is an indication of the evaporative capabilities of the atmosphere. This is significant, since evaporation is an important mechanism in the function of a cooling tower (Figure 1).

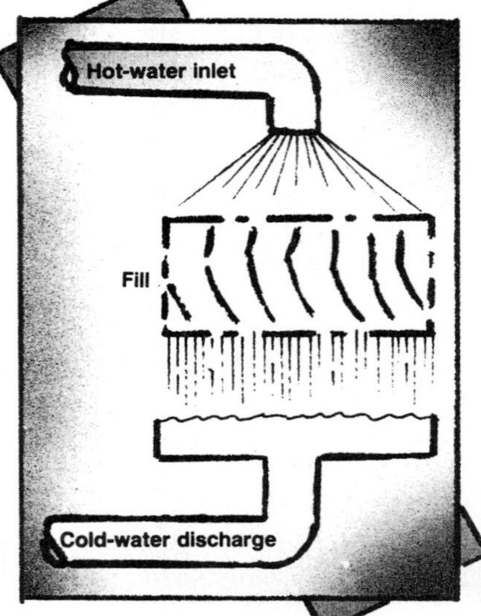

Figure 1. All cooling towers rely on some type of fill to increase the turbulence and surface area of the hot water, in order to maximize air contact and promote evaporative cooling

While all cooling towers are purchased to function at 100% of capacity in accordance with design conditions, actual onstream utilization often translates into lower levels of operation — sometimes as much as 30% lower than design. This happens for a variety of reasons:

- New plant expansion needs additional water volume and colder temperatures from the tower
- Deficient maintenance has reduced the performance of the tower
- The cooling tower was underdesigned when purchased, or was originally undersized, due to so-called "low bidder syndrome [1]"

Seek lower outlet temperatures

The cooling tower plays the major role in waste-heat removal in manufacturing, air conditioning, chemical and petrochemical processes. During the condensation process, the colder the condensing water, the higher the unit production and the lower the unit cost. In many process operations, a reduction in operating temperatures — often desirable for economic reasons — may be obtained by increasing the capacity of the cooling tower [2].

The importance of colder water for gas compression, for example, is evident in that all compressors share one common characteristic: A major portion (nearly 80%) of energy is converted to heat. This rejected heat must be contin-

Originally published May 1993

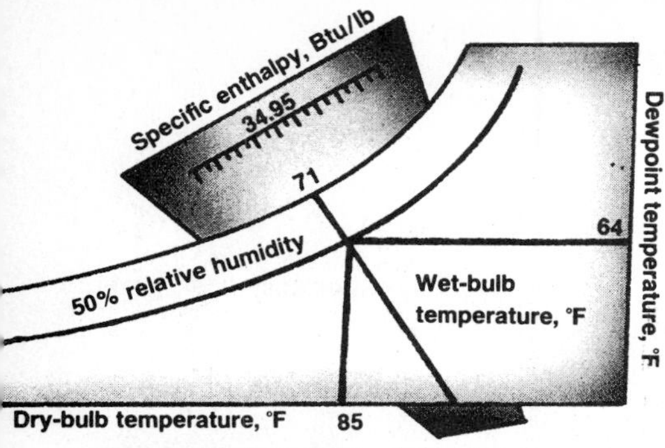

Figure 2 (above). To determine the wet-bulb temperature using a psychrometric chart (found in engineering reference books), the WBT is located at the intersection of any two of the ambient dry-bulb temperature, the relative humidity or the dewpoint temperature

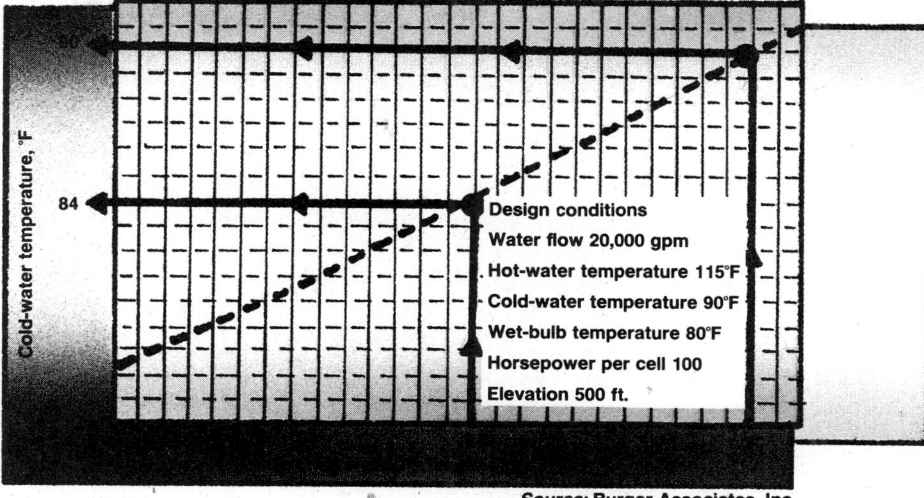

Figure 3 (right). The performance curve (dashed line) is a function of design specifications. Using it, the design WBT (in this case 80°F) can be used to determine the cold-water temperature the tower is capable of producing (90°F). This relationship lets the user predict cold-water availability as a function of any WBT (for instance, 84°F, at a WBT of 69°F as shown)

Source: Burger Associates, Inc.

PROFIT CENTER

uously removed at the same rate it is generated, or the compressor will overheat and shut down. Reducing the operating temperature of a compressor will proportionately reduce energy input requirements. In other words, the colder the water delivered to the equipment, the less energy is required to produce the same degree of work, at lower costs [3].

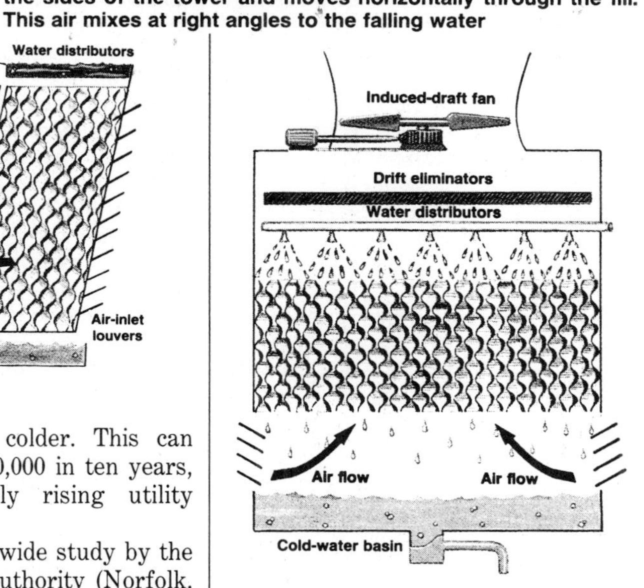

Figure 4 (left). In a crossflow tower, cooling air is drawn in from the sides of the tower and moves horizontally through the fill. This air mixes at right angles to the falling water

In refrigeration systems for process or comfort cooling, the use of colder water reduces the energy required to operate the system. Enthalpy charts [4] indicate that within the operating range of the tower, for every 1°F reduction in the temperature of the cold water produced by the cooling tower, a compressor enjoys a 2.25% energy savings, to produce the same cooling results.

Thus, a cooling system consuming $450,000 of electric energy to power its turbines can reduce utility costs by roughly $45,000/year, just by producing a water return from the cooling tower that is 4°F colder. This can amount to over $500,000 in ten years, assuming constantly rising utility costs.

In a recent nationwide study by the Tennessee Valley Authority (Norfolk, Tenn.) and the Electric Power Research Institute (Palo Alto, Calif.), 50% of power-plant cooling towers checked were determined to be suffering from inadequate thermal performance — in the area of 20% deficiency. This substandard operation, according to the survey, is costing the electric utilities an estimated $25 million per year in lost revenue and higher fuel costs [5].

Induced-draft fan
Drift eliminators
Water distributors
Air flow
Air flow
Cold-water basin

Figure 5 (above). In a counterflow tower, the air is drawn up from the bottom of the tower, and mixes with falling water over the entire height of the tower

Of the plants surveyed, 65% had towers that had either failed to meet design specifications, or were cited as the cause of decreased plant efficiency or capacity. When the back pressure of the turbine in the air-intake system in-

Use an Upgrade to Uncover Hidden Savings

Case history I

Recently, a U.S. Gulf Coast CPI plant wanted to double its product output. Analysis of the operation indicated that the quality and quantity of cold water from the cooling tower was the limiting factor.

The existing, three-cell crossflow tower was well constructed, but was equipped with outdated thermal and aerodynamic systems. Analysis indicated that the existing 21,000-gpm complex could be upgraded to cool another 10,000 gpm, in order to meet the new production requirements.

To accomplish this, each of the three 75-horsepower (hp) motors were replaced by 125-hp motors, and the 18-ft-dia., cast-aluminum, bladed fans were replaced with 22-ft-dia., eight-blade fans made of fiberglass-reinforced-epoxy resin. Fiberglass velocity-regain (VR) cylinders, each 12 ft high, were also added to assist the upgrade, and the old-fashioned wood splash bars were replaced with extruded V-shaped splash bars.

Thinner water film promotes cooling

Such bars are oriented parallel to the air stream, with their V-shaped spines facing upwards. Falling water breaks over the bars, forming an easily cooled, thin film. Compared with step-like splash bars, the V-shaped ones provide lower pressure drop and improved film cooling, thanks to their angularity.

Following the retrofit, the tower was tested in accordance with the Cooling Tower Institute Acceptance Test Code ATC-105, and the new capacity goal was met. Capital expenditures for the retrofit were $650,000. This compares favorably with the estimated $1.1 million required to build the two new cells, and the concrete basin, piping and electric service required to support new tower construction. Also, the total horsepower requirement of the upgraded tower is 30% less than that associated with the addition of two new cells, so the upgrade helped contain energy costs, as well.

Case history II

At a petrochemical plant, the design specifications for a two-cell counterflow tower were to cool 6,500 gpm from an inlet temperature of 108°F to a discharge temperature of 90°F (ΔT = 18°F), at an ambient WBT of 80°F. According to the manufacturer's data, the existing 18-ft-dia. aluminum fans, operating at 40 hp provide air throughput of 300,000 actual cubic feet per minute (acfm).

During a field inspection, operating personnel indicated that at best, at the full heat load and design WBT, the tower was operating at 70–75% capability; the outlet temperatures were typically deficient by at least 3°F.

Retrofits double tower capacity

During plans to double the capacity of the plant, the cooling tower again was identified as the limiting factor. To meet the needs of the proposed expansion, tower production had to be increased, to handle 10,500 gpm of water entering at 108°F at 80°F WBT — roughly 60% more than the existing 6,500-gpm capacity.

Construction of a new two-cell tower was ruled out, due to scheduling problems, space constraints and budgetary requirements. Engineering calculations [7] indicated that the existing tower could be upgraded to accommodate this increased throughput.

To do this, the water trough and wood splash bars were removed. High-thermal-transfer, cellular fill, made of polyvinyl chloride (PVC), and a PVC water-distribution system, including large-orifice nozzles were installed. Such nozzles are inherently nonclogging, thanks to their large-diameter orifices.

The new nozzle system produced extremely uniform water distribution, and new cellular, drift eliminators helped the unit to function at lower static-pressure levels. These changes produced a 50% boost in tower capability, and let the facility proceed with its expansion. □

creases, a plant experiences decreased cycle efficiency.

Wet-bulb determination

Cooling towers cool by evaporation, so the amount of humidity in the air affects a tower's efficiency. In cooling tower design, the wet-bulb temperature (WBT) is a critical factor. It reflects the capability of the atmosphere to evaporate moisture. The wet-bulb temperature (WBT) is measured with a special wet-bulb thermometer, or can be determined from a psychrometric chart (Figure 2), which converts ambient temperature into WBT, as a function of relative humidity or dewpoint.

The regional average WBT is different for each geographical area. In the U.S., 30-year averages are compiled by the U.S. government, using weather data gathered from military and civilian facilities. This information is published by the Government Printing Office (Washington, D.C.), in a document entitled "Army, Navy, and Air Force Engineering Weather Data."

Use WBT to judge performance

At an industrial meeting, two operators complain about the performance of their cooling towers. Each operator complains that the ΔT, the difference in water temperature between the hot-water inlet and the cold-water discharge, has failed to meet design specifications. One operator complains that the tower is rated for a ΔT of 20°F at 78°F, but only produces a differential of 18°F. The other tower, also rated for a 20°F ΔT, produces a 22°F differential.

Each of these engineers has failed to consider the WBT, relying instead on ambient temperature measurements. The first tower was rated for top performance when the WBT was at 78°F. It was then cited for poor performance at a measured temperature of 78°F (when the WBT was actually 80°F). In this case, when the WBT falls to the design temperature of 78°F, the ΔT will meet specifications of 20°F.

Conversely, the second tower, with the reported 22°F T, might have been designed for a 78°F WBT. If the wet bulb was actually at 71°F, but the WBT was at the design specifications of 78°F, the tower would actually produce a deficiency of about 6°F.

Table 1 (right). In this excerpt of a government publication listing average WBT data, region-to-region fluctuations are evident

TABLE 1: WET-BULB TEMPERATURES: GEOGRAPHIC FLUCTUATIONS

STATE	LOCATION Latitude		Longitude		Elev. ft	WET-BULB TEMP., °F 1%	2.5%	5%
Texas (cont.)	N		W					
Beaumont Army Hospital	31	49	106	28	4,185	69	68	68
Beeville, Chase Field NAS	28	22	97	40	190	82	81	79
Bergstrom AFT, Austin	30	12	97	40	541	79	78	77
Brooke Army Medical Center	29	28	98	27	785	77	76	76
Brooks AFB	29	21	98	27	598	78	77	77
Kelly AFB, San Antonio	29	23	98	35	690	78	77	76
Kingsville NAS	27	29	97	49	50	81	80	80
Lackland AFB	29	23	98	37	670	78	77	76
LaPorte ANG Station	29	40	95	04	24	81	80	80

Because the average WBT fluctuates from region to region, and throughout the year, WBT charts provide design engineers with a range of options. WBT figures are classified into three categories: 1%, 2.5% and 5% (Table 1). The percentages indicate the number of hours "during the air-conditioning (AC) season" that WBT equals or exceeds the listed averages. Industry defines the AC season as June through September, or a total of 2,930 hours.

All industries have standardized the use of the 2.5% figure, which assumes that the actual WBT equals or exceeds the listed figure by only 73.25 hours (0.025 × 2,930 hours) for the entire year. This is a rather small window in which the tower is not likely to produce the CWT specified in the design.

Table 2 illustrates how size varies with cost for different WBT requirements. Figure 3 illustrates how the ever-fluctuating WBT affects the projected CWT discharge.

Nearly all cooling towers can be upgraded to perform at higher levels of efficiency (box, left), which can provide a rapid, cost-effective payback. Most towers built 10–15 years ago lend themselves very well to thermal or water-volume upgrading. Before installing any new cooling tower, all existing systems should be evaluated to determine the upgrade potential.

State-of-the-art upgrading

Cooling towers are fabricated in two major configurations: crossflow and counterflow towers. Each design shares the same elements — an air-handling system, a water-distribution system, and multiple heat-transfer surfaces — but each arranges these elements differently.

In the crossflow tower (Figure 4), cooling air moves horizontally through the fill and mixes at right angles to the water that falls from the top of the tower. In counterflow towers (Figure 5), countercurrent mixing takes place, as air rises from the base to meet the water that is falling.

Table 2 (below). Comparing towers with the same throughput and cooling capacity, the difference in WBT affects the size and the cost of a cooling tower installation

TABLE 2. SIZE AND COST VARIATIONS, AS A FUNCTION OF WET-BULB TEMPERATURE

	Area 1	Area 2
Flowrate, gpm	20,000	20,000
Wet-bulb temperature, °F	74	80
Hot-water temperature, °F	115	115
Cold-water temperature, °F	90	90
Motor horsepower, hp	100	100
Fan dia., ft	26	30
Cooling range, °F	25	25
Approach to wet bulb, °F	16	10
Airflow, ft³/min	825,350	979,400
Box size, ft	36×72	36×72
Erected cost, $	350,000	425,000

The aim of any cooling tower upgrade is to produce larger volumes of colder water, so as to better conserve energy during the operation of associated equipment. Following is a brief review of each of the three major elements of any counterflow or crossflow tower. Each is subject to upgrading; the upgrade options are discussed in order of ascending cost.

- ***Air-handling system*** Since airflow promotes evaporation in the tower, maximizing air throughput is a goal. This airflow is created by fans. By closely matching the pitch of the fan blades to the motor-plate amperage, the blades will draw more air through the tower. In general, the least expensive way to improve air handling is to increase volume by pitching the blades to the maximum angle that is consistent with the motor-plate amperage.

Another option is to install velocity-regain (VR) fan cylinders, which relieve the exit pressure that the fan must work against. This simple addition can create roughly 7% more air throughput. Similarly, a larger motor or fan can increase the air flow, creating higher levels of performance [6].

- ***Water-distribution system*** Crossflow water-distribution patterns are set by the flow requirements and the orifices located in the hot-water distribution basin, which are fixed in position. The conventional orifice drops a solid column of water onto a lattice of wood (Figure 6a), which causes the water to splash over the top of the fill.

A higher level of performance can be obtained by replacing the wooden redistribution or splash decks with efficient, cellular decks (Figure 6b). These provide more-uniform water distribution at the top of the tower, thereby utilizing the entire tower height — rather than just a portion of it — for breakup, splash and, ultimately, cooling. By optimizing heat transfer this way, the discharge water has a lower net outlet temperature.

In counterflow towers, water troughs or enclosed flumes should be replaced by low-pressure spray piping, to distribute the water into finer droplets. Existing spray systems can also be greatly improved by installing noncorroding polyvinyl chloride piping in conjunction with non-clogging, non-corroding, large-orifice spray nozzles, or gravity splashers. Similarly, variable-orifice nozzles can be installed to break the water into finer droplets, to increase surface area and promote better heat transfer.

- ***Heat-transfer surfaces*** The most dramatic improvement in the performance can be obtained by installing cellular fill. In both the counterflow or crossflow configurations, conversion to high-efficiency film fill can result in improvements as much as 4–6°F colder water, which could increase the capacity of an existing tower by up to 50%.

Cellular fill consists of a series of corrugated plates whose channels are oriented vertically. The channels cross each other at 60-deg angularity.

As water falls through the fill, each droplet is stretched into a thin film. This creates more surface area and allows the available crossflow or counterflow air to cool the entire droplet more rapidly.

Clearly, replacing wooden splash bars with cellular film fill is a major undertaking, often the most expensive fixed-cost retrofit procedure. However, among the three options discussed here, this procedure produces the highest level of improvement in performance, which will provide for a more rapid return on investment and greater energy-conservation profits through the use of colder water.

No stone should be left unturned in the search to reduce energy consumption. Cooling towers are hidden bonanzas of energy conservation and dollar savings when properly engineered and maintained. In many cases, the factor most limiting production is the quality and quantity of cold water produced in the tower.

Nearly all cooling towers can be upgraded to perform at higher levels of efficiency. The savings accrued in energy conservation and additional product manufactured can be an important factor on an operator's profit and loss sheet. A periodic, professional, hands-on inspection is strongly recommended for every tower. ■

Edited by Suzanne Shelley

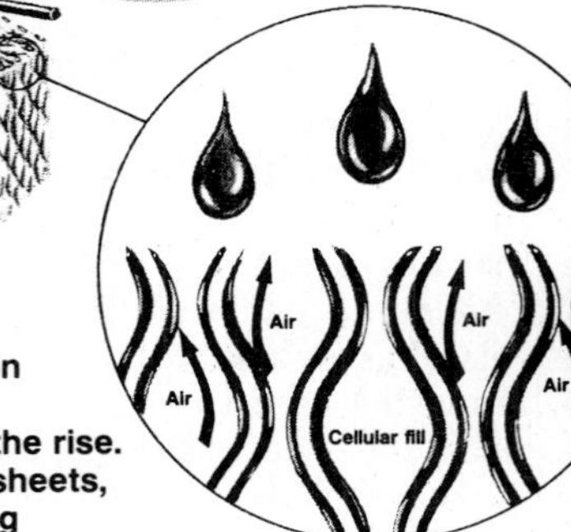

Figure 6a (top). Used for more than 70 years, wood splash bars break the falling water into smaller droplets, to promote cooling in the tower
Figure 6b (right). Today, use of cellular-film fill is on the rise. As water falls through the closely spaced, corrugated sheets, the fill stretches the droplets into a thin film, maximizing surface area and optimizing heat transfer

References

1. The Marley Cooling Tower Co., Kansas City, Mo., "Energy Considerations in Cooling Tower Applications," Apr. 22, 1976.
2. LeFevre, Marcel, "Evaluation of Cooling Tower Test Accuracy," Cooling Tower Institute publication, Feb. 4, 1981.
3. Stoven, K., "Energy Evaluation in Compressed-Air Systems," *Plant Engineering*, June 1979.
4. Allied Chemical Co., "The Pressure Enthalpy Diagram: It's Construction, Use and Value," 1982.
5. Tennessee Valley Administration and Electric Power Research Institute report, "Performance Monitoring Power Plant Operations," *Power Engineering*, August 1988.
6. Monroe, Robert, "Consider Variable-Pitch Fans," *Hydrocarbon Processing*, December 1980.
7. Burger, Robert, "Cooling Tower Technology, Maintenance, Upgrading, and Rebuilding," Prentice-Hall, Chapter 6, Thermal Analysis, Revised 1989.

The author

Robert Burger is president of an international consulting and construction firm, Burger Associates, Inc. (2815 Valley View Lane, Suite 220, Dallas, TX, 75234; Tel. 1-800-44-TOWER, or 1-800-448-6937). He received his mechanical engineering education at Cornell University, and has been active in cooling towers for nearly four decades. Burger has authored more than 50 articles and a textbook, entitled "Cooling Tower Technology" (Prentice-Hall, revised 1989; available from the author) and was named "Engineer of the Year" in 1989 by the National Assn. of Power Engineers.

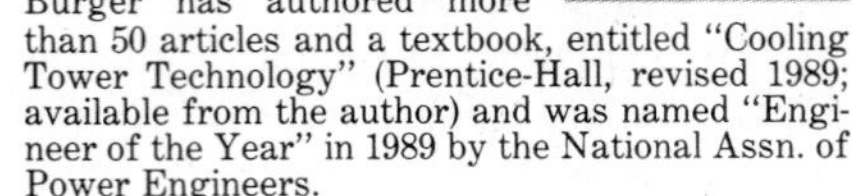

Section VI

Energy Management Practices

RECOVERING HEAT WHEN GENERATING POWER

V. Ganapathy
Abco Industries

This steam generator for use in Cheng-cycle units (see p. 226) typifies the efficient recovery of heat in power plants. As the table shows, generators can be unfired, supplementary-fired or furnace-fired

TYPE OF SYSTEM	GAS INLET TEMPERATURE, °F	GAS-TO-STEAM RATIO
Unfired	800 - 1,000	5.5 - 7
Supplementary fired	1,000 - 1,700	2.5 - 5.5
Furnace fired	1,700 - 3,200	1.2 - 2.5

Originally published February 1993

Intelligent use of heat-recovery steam generators (HRSGs) is vital for the efficient operation of cogeneration plants, which furnish both thermal energy (usually in the form of steam) and electric energy. HRSGs are similarly important in combined-cycle power plants, in which the thermal energy rejected from the primary electric-power-generation step is harnessed (as discussed below) to produce additional electrical energy.*

In these facilities, the HRSG is typically heated by gas-turbine exhaust, because both cogeneration and combined-cycle plants are likely to employ these turbines as their prime movers. Gas turbines are simple and efficient, incur low installed cost per kilowatt, require relatively little space, can start up quickly, and require relatively little cooling water. Their electrical output ranges from 5 to 150 MW, so both large and small plants find applications for them.

Natural gas is the fuel most widely used for gas turbines in the U.S., whereas fuel oil is the main fuel in other countries. In either case, the turbine exhaust is clean and thus usually does not pose corrosion or related problems for the HRSG.

*For an explanation of how HRSGs work, how they are employed during manufacture of ammonia, sulfuric acid and other chemicals, and how they can be used in incineration systems, see "Effective use of heat-recovery steam generators," pp. 234–238. That article also provides guidelines on specifying HRSGs.

In cogeneration plants, HRSGs generally convert this turbine-exhaust energy into low-pressure saturated steam, around 10 to 300 psig, suitable for applications such as drying, process heating or cooling. In combined-cycle plants, by contrast, the HRSG generates high-pressure, high-temperature steam, usually exceeding 750 psi and 705°F, which drives steam turbines to produce additional electricity. If the cogeneration or combined cycle plant is large, the gas-turbine exhaust might be used not only for steam generation at multiple pressure (including low-pressure steam for deaeration) but also to heat condensate or heat-transfer fluids.

Due to the large mass-flows associated with gas turbines, water-tube rather than fire-tube HRSGs are generally the choice (for more discussion about this tradeoff, see the article cited in the footnote preceding). However, fire-tube boilers may be more economical if the installation is of low capacity. With water-tube designs, extended surfaces can be used to make the units compact.

Depending on the amount of steam to be produced, HRSGs for gas-turbine-exhaust applications may be unfired, supplementary-fired or furnace fired. In the last-named two options, additional fuel is sprayed into the turbine-exhaust stream. Note that this is in fact feasible without need for additional oxygen, as the exhaust itself typically contains about 16% oxygen by volume.

The table gives the steam-generation capabilities of each of the three options. In a typical situation, simulation and temperature-profile analyses are carried out to determine which one is most appropriate.

Unfired HRSGs

Typical temperature of gas entering an unfired HRSG ranges from 800 to 1,050°F. In the case of natural-circulation units (discussed below), two of the widely used configurations are the single- and two-gas-pass designs.

In the two-pass version, a horizontal baffle plate divides the evaporator into two portions. The exhaust gases enter the lower portion of the shell side of the evaporator, make a 180-deg turn and then flow across the top half of the evaporator.

This design requires relatively little floor space. It is ordinarily the choice when exhaust-gas flow is less than 200,000 lb/h and when steam generation at a single pressure is adequate.

Various gas inlet and exit configurations are possible. It is possible also to incorporate a steam superheater and a water-heating economizer.

When gas flow exceeds 200,000 lb/hr or when steam must be generated at multiple pressures, it is preferable to use a single gas pass; otherwise the HRSG would have to be inconveniently high. A side-benefit with single-pass HRSGs is that they are easier to outfit with catalytic reduction systems (discussed later) to remove NOx or carbon monoxide.

The type of circulation, whether natural or forced, also affects the boiler configuration. With natural circulation, the difference in thermal head between the comparatively cool water in the downcomer pipes and the hotter steam-water mixture in the evaporator is responsible for the circulation through the system. Vertical tubes are used for the evaporator in such designs.

The circulation ratio (mass ratio of circulating steam-water mixture to generated steam) in such a unit is arrived at by balancing the thermal head available between the cooler and hotter columns of steam-water mixture against the losses in the system. It varies from about 10 to 40 depending on the system losses and configuration [*2*].

In the forced-circulation design, pumps circulate the steam-water mixture through horizontal evaporator tubes to and from the steam drum. Circulation ratio is selected to be in the range of 3 to 10.

The forced-circulation approach, widely used in Europe, is more compact. However, savings in floor space must be weighed against the capital costs, operating costs and reliability risks associated with the pumps. The startup time for the two options do not differ significantly — in both cases, the overall heat transfer rate is governed mainly by the (lower) gas-side coefficient, which is not significantly affected by the tube orientation.

Vertical tubes, as in natural circulation designs, provide a natural path for the steam bubbles. With horizontal tubes, on the other hand, separation of steam from the steam-water mixture poses a risk of overheating. This is particularly true when the heat flux is very high (as in the supplementary-fired systems, discussed below). Operating personnel should periodically check for signs of departure from nucleate boiling (boiling that involves distinct vaporization nuclei on the heating surface), especially with the horizontal-tube option.

Proper use of fins

As mentioned earlier, extended surfaces are used to make the unfired HRSG compact, whether it is of natural- or forced-circulation design. Fin densities range from 2 to 5 fins per inch, fin height from 0.5 to 1.0 in., and fin thickness from 0.05 to 0.12 in. Both solid and serrated fins are widely used. Gas pressure drop across the tubes increases with increase in fin density or height.

Steam generators, properly applied, can produce big savings

A significant consideration is the maximum temperature the fin will incur, as this limits the materials of fin construction that can be used. The maximum temperature will be at the fin tip. The thinner the fin, the higher this temperature will be, other things being equal. A large ratio of external to internal surface (i.e., a high fin density) also increases the fin-tip temperature, which likewise raises that of the tube wall.

This effect is particularly pronounced when the tubeside coefficient is low, as it is in steam superheaters. Accordingly, a low fin density is recommended for superheaters. On the other hand, economizers and evaporators can accommodate a higher density because the tubeside coefficient is very large, on the order of 1,000 to 3,000 Btu/(ft^2)(h)(°F) [*2, 3*].

Choosing a large amount of fin sur-

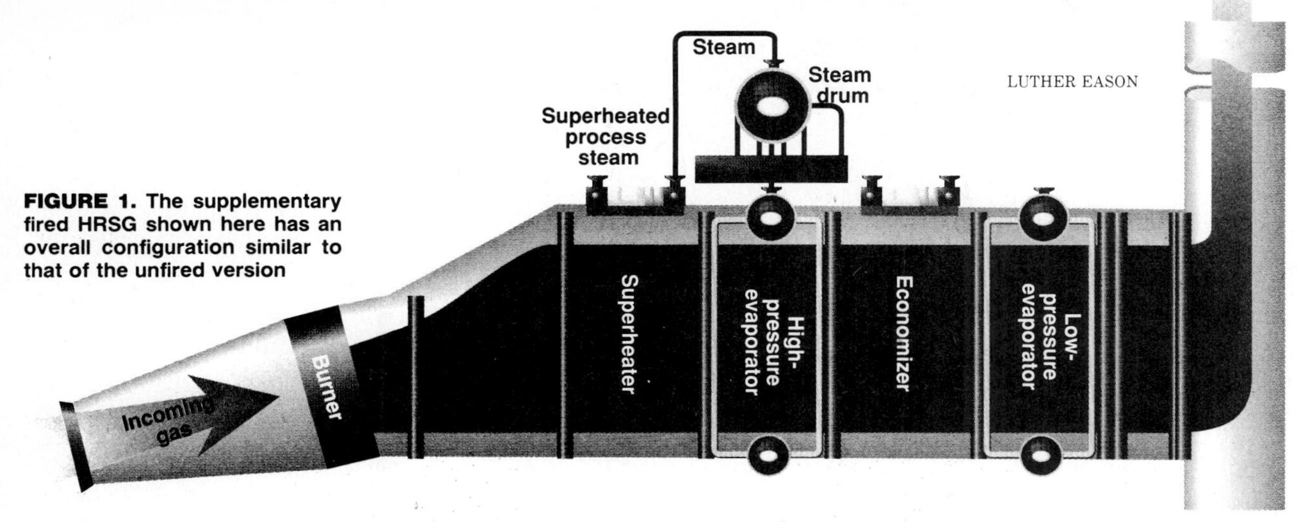

FIGURE 1. The supplementary fired HRSG shown here has an overall configuration similar to that of the unfired version

face does not automatically mean that more energy will be transferred — as the engineer compares alternative designs having successively more fin area, it becomes important to look not just at the area but instead at the product of area and overall heat-transfer coefficient. The gas-side heat transfer coefficient decreases as the ratio of external to tube internal surface increases [*2*, *3*].

The metal casing for unfired HRSGs is internally insulated with 4 to 5 in. of ceramic fiber insulation and is protected from the hot gases by a liner made of stainless steel, alloy steel or carbon steel depending on the gas temperature. The design of this liner must allow for expansion.

The same approach can be used for the duct that leads from the turbine to the HRSG. However, some manufacturers prefer to use alloy steel for the duct material and place the insulation on the outside. Though this reduces maintenance problems with the insulation liners, the expansion problems due to higher casing temperature has to be dealt with.

Supplementary fired HRSGs

The term "supplementary fired" is used somewhat loosely, but it generally implies an HRSG that is outfitted with a duct burner to raise the temperature of the entering gas from the turbine duct burner and thus increase steam production. A supplementary-fired HRSG using such a burner can be seen in Figure 1.

The HRSG that is shown in Figure 1 has a single-pass design and generates steam at both high and low pressure. Apart from having the duct burner, it closely resembles the single-pass unfired designs discussed on the previous page.

Indeed, the HRSG design for supplementary firing does not differ much from that for the unfired HRSG. The main exceptions are the sizing of drums and piping to accommodate the larger steam flows, and the selection of tube and fin materials to accommodate the higher gas temperatures. Casing-design considerations limit the maximum gas temperature after firing to 1,700°F — beyond this temperature, the liner material starts warping and thus exposes the bare insulation to the hot gases.

In line with the precautions discussed earlier for unfired HRSGs, the superheater and evaporator portions of the supplementary-fired versions are designed with varying combinations of fin densities to minimize tube wall and fin tip temperature. For instance, one common arrangement employs a few rows of bare (unfinned) tubes closest to the flame, then has a few rows of tubes with low fin density, followed finally by ones with higher density.

When specifying an HRSG, the engineer should make sure that the distribution of the gas flow across the duct does not distort the flame pattern and thereby overheat the tubes or ductwork. Discussions with the burner supplier regarding gas uniformity, distance from burner to HRSG surfaces and duct configuration are important in this vein.

Furnace firing

Depending on steam demand, the aforementioned simulations and temperature profile analyses may dictate a firing temperature over 1,700°F. In that case, furnace-fired HRSGs may be used.

For temperatures up to 2,300°F, a suitable arrangement consists of an HRSG equipped with a duct burner and employing a water-cooled, integral membrane wall (photo, p. 97). The evaporator portion includes a radiant section that is followed by a convection section; after they leave the evaporator, the gases pass through an economizer.

As discussed under supplementary firing, the convection section is designed with a few bare tubes followed by finned tubes, with a varying fin configuration that takes into account the tube wall and fin tip temperatures and heat flux inside the tubes. If de-

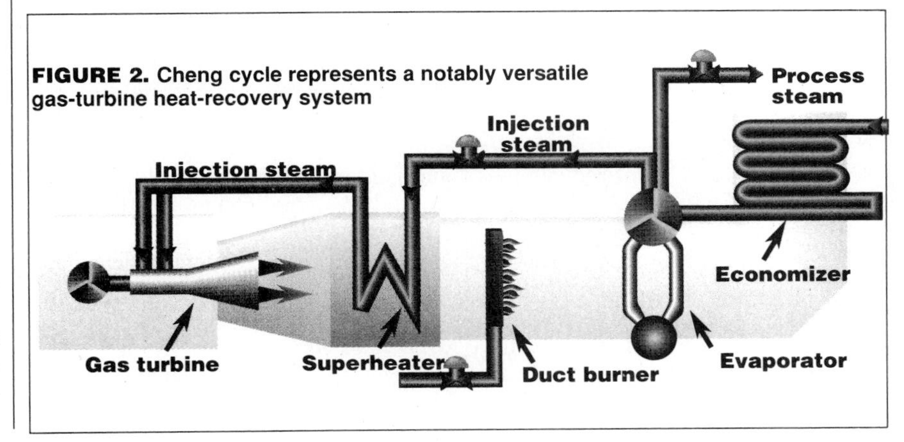

FIGURE 2. Cheng cycle represents a notably versatile gas-turbine heat-recovery system

sired, superheater tubes can be placed in the same area as the convection tubes.

When the required firing temperature exceeds 2,300°F, a register burner is required. This consists of a burner that is outfitted with its own air chamber (wind box), of the type that is used in packaged steam generators. Such a burner system can fire up to the adiabatic combustion temperature, leaving a residual oxygen content of less than 3% (dry basis) in the exhaust. As seen from the table, one can maximize the amount of steam generation with such a design.

Firing and system efficiency

As indicated earlier, the decision between unfired, supplementary-fired or furnace-fired HRSGs is based mainly on the quantity of steam required. It should be noted, however, that supplementary or furnace firing in any case improves the energy efficiency of the operation; for instance, the efficiency as defined by the American Soc. of Mechanical Engineers Power Test Code PTC 4.4 [*7*]. The explanation is straightforward: By consuming oxygen in the turbine exhaust, one is reducing the amount of excess air. [*2, 3*]

In some installations, an HRSG must be able to operate even when the gas turbine is off. Such a system requires a forced-draft fan to supply the air for combustion to the burner. With such fresh-air operation, the burner duty is much higher, because the air temperature has to be raised from ambient to the high firing temperature.

Isolating dampers are employed to shield the hot gases from the fan whenever the gas turbine is running. This mechanism should be designed to provide for quick switchover to the fan mode of operation.

Pressure drop

Whether the HRSG is unfired, supplementary fired or furnace fired, it is important to keep pressure drop in mind when preparing the specifications. As a rule of thumb, each additional 4 in. w.c. (water column) of pressure drop through the HRSG reduces the power output of the upstream gas turbine by nearly 1%.

HRSGs associated with relatively small gas turbines (up to about 10 MW) typically incur a pressure drop of 5 to 6 in. Toward the other end of the scale, larger multiple-pressure units equipped with catalytic reduction systems involve a drop as high as 12 to 14 in. At any event, every effort should be made to minimize the figure.

Duct burners typically offer a low resistance to gas flow, on the order of 0.3 in. w.c. For register burners, however, the figure is around 4 in.

Cheng cycle

An interesting application of HRSGs with gas turbine exhaust is the Cheng cycle, Figure 2. In conventional gas-turbine-HRSG systems, whenever the steam demand falls off, one has to either bypass the exhaust gases to the stack using a diverter, or else vent (and thereby waste) the steam, or operate the turbine at a lower electrical output (which may not be desired). The Cheng cycle avoids those drawbacks: The steam not required for the process is superheated and injected into the gas turbine, thus increasing the electrical power output of the latter. The ratio of injection steam to process steam can be varied to suit the process requirements.

Since steam is being injected into the gas turbine, its purity is of utmost importance. Cheng-cycle units accordingly employ a combination of internal and external steam separators to remove noncondensibles and other impurities and thus to achieve steam purity in the parts-per-billion range.

In a Cheng cycle installation, the steam generation can be increased by including a duct burner between the superheater and evaporator. Since the amount of water vapor present in the turbine exhaust gases can be as high as 26% when operating in the steam-injection mode, with oxygen only in the range of 11 to 12% by comparison, an augmenting air fan is usually required to stabilize the flame in the duct burner.

Since the increased electrical output is obtained by steam injection, there is no need for a steam turbine-condenser system. The Cheng cycle achieves the electrical power output of a comparable combined cycle system or slightly more power, with less equipment and complexity. A two-gas-pass HRSG that is intended for Cheng cycle usage appears in the photo on p. 94.

Catalytic reduction

Due to air-pollution regulations, catalytic reduction units are required on turbine systems in several U.S. states, to limit nitrogen oxides and carbon monoxide in the exhaust gases. Since these units require a narrowly defined temperature range for efficient operation (e.g., 600 to 750°F for certain NOx catalysts), their placement within the HRSG is particularly important.

A typical HRSG exhibits variations in gas flow and exhaust-gas temperature over time, so numerous performance runs must be made to determine the best location for the reduction unit. Often an evaporator may have to be split into two or more sections to achieve this objective.

Boiler fabricated with water-cooled membrane wall is suitable for furnace firing

Performance testing

In the same vein, bear in mind that the performance data generated during proposal stages of a HRSG project may not cover all of the operating regimes that can be anticipated during the life of the equipment. A gas turbine may operate at different ambient conditions or load, so the exhaust-gas flowrate, composition and temperature may change, which in turn affects the HRSG performance [*9*]. Also, the steam parameters may be different during plant commissioning.

Another uncertainty to be aware of is

the error that is due to the instruments and measuring tolerances. It is very difficult, for instance, to measure the gas flow; in fact, the aforementioned ASME PTC 4.4 code cites a 3 to 5% error in gas-flow measurement with different methods.

Also, depending on the duct size, a variation of 30 to 50°F can be experienced in gas temperature across the duct cross section. Too, the exhaust-gas temperature could vary by 10°F due to instrument tolerances.

Given these facts, one could argue that the HRSG can receive up 5% less mass flow at a 10°F lower exhaust gas temperature than the design values and still show that the design flow and temperature conditions are being met. As a result, the HRSG could be generating less steam even though correctly sized for the design conditions.

Hence it is prudent for both the supplier and end user to arrive at a consensus on HRSG performance, operating regimes and testing procedures before installing the unit. Performance data of the HRSG at different gas and steam parameters may be generated and discussed before testing. ■

Edited by Nicholas P. Chopey

For information on suppliers of waste-heat-recovery boilers, see the Chemical Engineering Buyers Guide for 1993.

The author

For biography of author, see January issue, p. 238.

References

1. Ganapathy, V., Simplify heat recovery steam generator evaluation, *Hydrocarbon Processing*, March 1990.
2. Ganapathy, V., "Applied Heat Transfer," Pennwell Books, Tulsa, 1982.
3. Ganapathy, V., Evaluate extended surface exchangers carefully, *Hydrocarbon Processing*, October 1990.
4. Ganapathy, V., "Waste Heat Deskbook," Fairmont Press, Atlanta, 1991.
5. Ganapathy, V., Chart estimates supplementary fuel parameters, *Oil and Gas Journal*, June 25, 1984.
6. Ganapathy, V., Program computes fuel input, combustion temperature, *Power Engineering*, July 1986.
7. ASME Power Test Code PTC 4.4, "Gas turbine heat recovery steam generators," American Soc. of Mechanical Engineers, New York, 1981.
8. "HRSGs — Software for simulation of design and off-design performance of HRSGs," Ganapathy, V., available from author.
9. Ganapathy, V., How to evaluate HRSG performance, *Power*, June 1991.
10. Ganapathy, V., Heat recovery boilers — the options, *Chemical Engineering Progress*, February 1992.

CAPTURE HEAT FROM AIR-POLLUTION CONTROL

The widespread consumption of fossil fuels has been blamed for a global warming trend, which is attributed to heat trapped in the atmosphere by carbon dioxide. Acid rain and volatile organic compounds (VOCs) that result from industrial operations have also been implicated in the destruction of our fragile environment.

Emissions problems are common in petrochemical, chemical, pharmaceutical and other industries where toxic gases and liquids are handled. Rigorous pollution-control laws have become a primary concern to the operators of such process plants. Several technologies can be considered to solve the problem. One of the best known, most widely accepted and proven techniques is thermal oxidation or incineration.

While stringent environmental regulations are being legislated worldwide to reduce pollution, their implementation is potentially hazardous because it mandates the collection of toxic and often flammable vapors.

To protect plant personnel, equipment and surrounding communities, ultra-reliable deflagration and detonation arrestors must be incorporated into the heat-related control technologies used to capture and control VOCs.

Traditionally, VOC capture rates of 70%, combined with destruction efficiencies of 90%, have produced typical overall control rates of 63%. In the U.S., Phase 1 and Phase 2 of the 1990 Clean Air Act Amendments mandate a minimum 80% control rate for VOCs. And, many states now require a 100% capture rate.

Many technologies have been adapted to solve industrial emission problems. Thermal oxidation offers a simple, cost-effective way to eliminate combustible waste gases, and it provides the operator with an energy-conservation opportunity, since valuable thermal energy can be recovered during operation. The fuel required for efficient incineration in a thermal oxidizer may be the waste product itself, or, if the heat value of the waste gas is too low, too dilute, or inert, supplemental fuel can be mixed with the waste gas.

Calculate energy and material balances to maximize efficiency

Two types of oxidizers are used for air-pollution control—thermal (or direct fired) and catalytic. In each case, supplementary fuel may be required, since the volatiles in the gas stream are almost always below concentrations required to sustain combustion. Due to potential operating problems, catalytic incinerators are severely limited in those applications that have a high concentration of particulates.

The supplementary fuel can be either natural gas or fuel oil. A natural-gas-fired unit is relatively simple to operate, and will produce a cleaner exhaust stream than one using fuel oil.

For destruction or removal efficiency (DRE) of hydrocarbon vapors and liquids in excess of 95%, a thermal oxidizer typically operates at roughly 1,400°F. At higher operating temperatures (on the order of 1,800°F), with adequate mixing and residence time, the incineration efficiency can be increased to 99.99%.

The operating cost is a serious consideration when selecting a thermal oxidizer. One way to substantially offset this cost is to reuse the heat generated by the unit. While this concept is not new, it is often overlooked by users because of the relatively high capital investment required.

Two major types of heat-recovery equipment can be combined with a thermal oxidizer: A heat exchanger, which preheats waste fumes, combustion air or fluids used in the plant; or a heat-recovery boiler, which creates steam using captured heat. When it is economically feasible, both can be used for maximum recovery.

Add a heat exchanger

A typical configuration of a thermal oxidizer with a heat exchanger is depicted in Figure 1. To ensure safe performance, the operator must carry out the following sequence of operations:

1. Purge the unit with a volume of fresh air that is at least four times the unit's internal volume
2. Light the pilot after all start-up interlocks are satisfied
3. Start the supplementary fuel burner after the pilot flame has been approved by an ultraviolet (UV) scanner. The pilot will be turned off automatically
4. Modulate the fuel control valve with a temperature controller, until a preset temperature has been reached
5. Once the minimum temperature setpoint has been reached, replace the fresh air with the waste gas

The system must also have shutdown interlocks to respond to the following pre-set parameters:

1. Fuel high pressure
2. Fuel low pressure
3. Air low pressure (or waste gas low pressure, if the gas serves as the source of combustion air)

Originally published October 1993

The addition of heat exchangers or heat-recovery boilers offsets a plant's fuel usage, and cuts the operating cost of pollution control

John F. Straitz III, P.E.
NAO, Inc.

4. Flame failure
5. High temperature

If the waste gas is mostly air with a very low VOC concentration (typically in the 1,000- to 3,000-ppm range), no additional air has to be supplied to ensure combustion, and no dilution is needed to prevent flashback or explosion. Such a heat exchanger will operate without additional instrumentation and controls, which will keep the total cost down.

During operation, the cold and hot sides of the exchanger must be in balance. If a waste stream varies in flow and vapor concentration, system design can be complicated. If the hot side is carrying substantially more heat than the design specifications, or the cold side is too low in flow, the hot gas may overwhelm the capacity of the exchanger. In this case, it must be forced to partially bypass the exchanger and go directly to the stack. The temperature control and instrumentation systems would then be more complex.

When gas pressures are low (2- to 10-in. water column, for example), as is often the case for a thermal oxidizer, a plate-type heat exchanger should be used. Such a unit provides the most practical design for the cost-efficient recovery of sensible energy.

The heat-transfer surfaces are composed of parallel metal plates separated by tabs. The pattern on the plates creates thin channels, through which the hot and cold fluids flow. In spite of low pressures, a plate-type heat exchanger provides an efficient heat-transfer surface on a volumetric basis.

The transfer surfaces for plate-and-frame exchangers can be readily formed from a wide variety of alloys. A heat-transfer matrix is welded to the internal expansion joints and to the frame. A 3-in. ceramic blanket, placed between the inner plates of the heat exchanger and its outer casing, will serve as insulation to reduce heat loss.

A heat exchanger used with a thermal oxidizer typically creates a 2- to 5-in. pressure drop on each side of the inlet and outlet. The thermal transfer effectiveness (TTE; %) can be calculated as:

$$TTE = (t_2 - t_1)/(T_1 - t_1) \times 100\% \quad (1)$$

where:

t_1= Inlet waste gas temperature (hot side)

t_2= Waste gas temperature after the heat exchanger (cold side)

T_1= Oxidation temperature

A simplified payback calculation for a heat exchanger investment is shown in the box below. That calculation is based on the following design parameters: The thermal oxidizer is to be furnished with a gas-to-gas heat exchanger with an efficiency of 60% and a gas flow of 10,000 scfm. Natural gas is used as a supplementary fuel at cost of $5/million Btu. The electricity cost is $0.08/kWh. The system runs 24 h/d, 350 d/yr. Compared with a simple thermal oxidizer, the additional cost for one outfitted with a heat exchanger is roughly $200,000. The additional maintenance and equipment costs are roughly $15,000/yr.

Add a heat-recovery boiler

When steam is called for onsite, a waste-heat-recovery boiler will provide an energy efficient source. The most practical configuration is a horizontal oxidizer chamber with an outlet transition piece, that connects the oxidizer to the boiler. (Figure 2). This transition is equipped with an expansion joint to allow anticipated thermal expansion

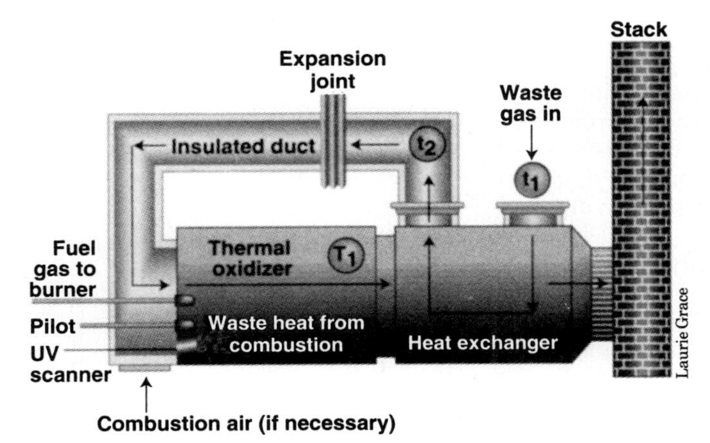

Figure 1. By adding a heat exchanger to a thermal oxidizer, operators can capture waste heat to preheat waste fumes or combustion air entering the oxidizer, or to heat other process fluids used in the plant

	Natural gas, MM kcal/h	Electricity, kW
Thermal oxidizer (TO)	1.25	3.75
TO plus heat exchanger (HX)	0.5	15

Payback (yr) = Equipment cost/[Energy savings - Parts cost]
= Equipment cost/{[(Gas usage without HX - Gas usage with HX) x (h/d operated) x (d/yr operated)]+[(Electric usage without HX - Electric usage with HX) x (h/d operated) x (d/yr operated)] - Parts cost}
= 200,000/{[(1.25-0.5) x 20 x 24 x 350]+[(3.75-15) x 0.08 x 24 x 350] - 15,000}= 0.415 yr

DESIGN SPECS

Waste gas and liquid entering the Northern California pulp mill thermal oxidizer

Non-condensible gas

Maximum gas flowrate = 2,000 acfm

Gas temperature = 130°F

Gas pressure = 30 in. water column

Maximum gas concentration: H_2S=0.2 lb/min.

CH_3SH =2.3 lb/min.

CH_3SCH_3 =3.6 lb/min.

$CH_3S_2CH_3$ =0.9 lb/min.

Air =Balance

Waste liquid (pinene)

$C_{10}H_{16}$= 17.8 lb/min.

Natural gas pressure

15 psig at the battery

Oil #2 (backup fuel)

Pressure specified by vendor

from the chamber. The expansion joint material (stainless steel or exotic alloy, for instance) is chosen as a function of temperature and the thermal expansion of the chamber. The chamber's support is fixed at the burner side, while the boiler is fixed at the stack end, so that both units expand toward the expansion joint.

The rate of heat transfer from the fluegas to the boiler depends on the temperature and the specific heat of the fluegas, its velocity, and the direction of flow over absorbing surfaces. Generally, the thermal efficiency of a waste-heat-recovery-boiler is between 60 and 70%.

Boiler size depends on fluegas composition, temperature, and feedwater conditions. Final instrumentation and all features of the boiler must be specified by the company that provides the complete system, in compliance with user specifications.

Boiler controls must be interlocked to the thermal oxidizer's instrumentation. Fluegas from the boiler is released to ambient air at about 500°F, if there is no additional gas use or treatment.

Consider a catalytic oxidizer

A typical catalytic oxidizer configuration is shown in Figure 3. The unit is a thermal oxidizer that uses a catalyst to promote oxidation of VOCs at lower tempeatures. The primary benefit of a lower process temperature is the ability to operate using less supplemental fuel than is required by a thermal oxidizer.

Major factors affecting the performance of a catalytic oxidizer include:

- Operating temperature
- Combustion chamber velocity (the reciprocal of residence time)
- VOC type and concentration (high concentration can overheat the catalyst)
- Catalyst characteristics (material, structure, resistance to poisoning)
- Presence of waste contaminants (heavy metals, sulfur, chlorides)
- Type of heat recovery used

Catalytic oxidizers are used most widely in the control of VOC vapors emitted by dryers and ovens used in surface-coating processes. In general, the use of such a device is favorable when the process generates large volumes of waste gas with low VOC concentrations—typically less than 25% of the lower explosive limit.

While the emission-control performance of a thermal oxidizer is similar to that of a catalytic oxidizer, there are three situations in which a thermal oxidizer is favored over a catalytic one:

1. For exhaust streams that contain significant amounts of catalyst poisons or fouling agents, thermal oxidizers may be the only technically feasible control equipment

2. Where extremely high VOC destruction efficiencies (99.9+%) of difficult-to-control VOC types are required, thermal oxidizers provide higher destruction efficiency

3. For relatively rich VOC waste gas streams (20-25% LEL or higher), the potential for catalyst overheating may call for the waste gas stream to be diluted with additional air. To handle the added volume, a larger catalyst would be required, which would increase both operating and capital equipment cost

One plant's experience

At a major pulp mill in Northern California, routine process operations generate a number of hazardous gases and liquids. Recently, the plant opted for a thermal oxidizer with a heat-recovery boiler. Reactions by the local community and EPA regulations forced plant management to improve the treatment technology for both non-condensible gases and turpentine liquid (pinene) generated by the plant process.

Historically, the non-condensible gases were burned in an enclosed flare, which had no controls to ensure complete destruction and no device to remove SO_2 from the fluegas. When the enclosed flare was in service, foul odors were produced. Turpentine liquid was burned in a rotary kiln-type incinerator without any problems.

Recently, the mill operators opted for a thermal oxidizer with a waste-heat-recovery boiler, for two reasons. First, the fluegas from the thermal oxidizer has to be quenched from 1,600°F to 200°F before it enters the facility's packed-tower scrubber. This can be done with an expensive quench section, which uses high-flowrate water

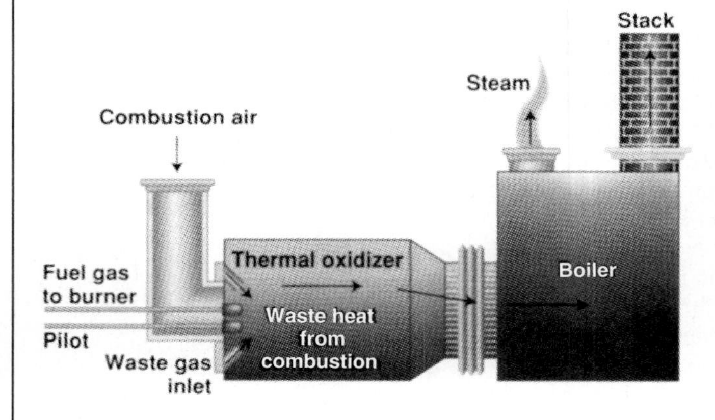

Figure 2. With the addition of a waste-heat-recovery boiler, incinerator offgases can be used to create an economical source of steam onsite

INPUT TO THE OXIDIZER AT THE CALIFORNIA PULP MILL

Component	Flowrate, scfh	Flowrate, lb/h	HHV*(comb.), Btu/lb	Total heat, Btu/h
CH_4	2,000	84.9	21,535	1,828,321
H_2S	135	12.1	6,548	79,230
CH_3SH	1,092	138.5	10,111	1,400,374
CH_3SCH_3	1,321	216.5	10,003	2,165,650
$CH_3S_2CH_3$	219	54.0	9,358	505,332
Air (in gas)	110,870	8,500	NA	NA
$C_{10}H_{16}$	3,254	1,071	19,217	20,581,407
Air (blowers)	604,809	46,370	NA	NA
TOTAL	**723,700**	**56,447**	**NA**	**26,560,314**

*HHV = High heat value = chemical heat energy of each constituent

sprayers made of Hastelloy C-276. Second, the plant uses a large volume of steam in its routine operations. The design specifications are shown in the box above.

Heat recovered from the thermal oxidizer (TO) reduces the demand for natural gas to fire the main plant boiler. The temperature of the fluegas leaving the heat-recovery device must be lower than 350°F, according to specification. By adding an economizer after the boiler, fluegas temperature can be decreased, while boiler feedwater is preheated. A boiler economizer, part of the waste-heat-recovery boiler, uses a gas-to-liquid heat exchanger to recover heat from combustion exhaust gases.

To meet permitted emissions standards, the minimum destruction efficiency required by the state of California was calculated to be 99.95%. To achieve target destruction efficiencies, the design must optimize the following factors, each of which has a direct effect on thermal destruction:

- Oxidation temperature
- Residence time
- Mixing or turbulence

Testing and experience show that a residence time of one second and process temperature of 1,600°F with turbulent flow will accomplish DRE = 99.95%, as required.

During operation, the waste liquid (turpentine) will not be supplied to the TO at the stated flowrate all the time. Therefore, the design specifications must consider the variability of the non-condensible-gas phase, which ranges from pure air to the maximum concentration shown in the specifications.

The TO must also be designed for operation from minimum to maximum heat-release conditions. The number one concern for the system, besides destruction efficiency, is safety, which is discussed later. The first evaluation is straightforward application of the law of conservation of matter and the law of conservation of energy.

Calculate the material and energy balances

The California pulp mill uses a natural-gas burner, to provide additional heat energy to ensure flame stability and complete destruction of the waste streams. The burner was designed with a minimum natural-gas firing rate of 2,000 scfh, which will guarantee total destruction at all flowrates, even during periodic surges.

Excess fuel is not wasted, because heat from the oxidizer is recovered for steam production. Sample material and energy balance calculations are shown in the box above, and at the top of p. 13.

The size of the TO is determined directly from the heat and energy balance output. The following equation is used to convert fluegas flowrate from standard conditions to actual flow at process temperature and pressure:

$$Vao\ (acfs) = V(P/Pao)(Tao/T) \qquad (2)$$

where:

V = flowrate at standard conditions (scfs)
Vao = flowrate at actual operating conditions (acfs)
P = absolute pressure at standard conditions (psi)
T = temperature at standard conditions (°R)
Pao = Absolute pressure at actual operating conditions (Pa)
Tao = Actual temperature (°R)

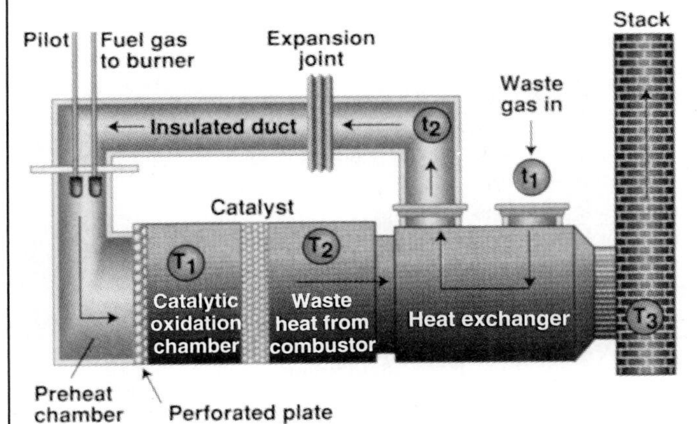

Figure 3. While catalytic incinerators boost fuel efficiency by promoting oxidation at lower temperatures, certain waste streams can poison or overheat the catalyst

The active volume of the TO chamber (Vch; ft^3) is calculated as:

$$Vch = Vao\ (acfs)/Residence\ time\ (s) \qquad (3)$$

Calculation of stoichiometric air requirements is very important. It allows the operator to distribute and control combustion and to quench air to the burner in the most efficient way, ensuring high mixing and flame stability. Stoichiometric air calculations for the maximum flow are shown in the box at the top of p. 14 (remember that the oxygen content of air is 21%). For the pulp

OUTPUT FROM THE OXIDIZER

Component	Flowrate, scfh	Flowrate, lb/h	Cp*, Btu/lb-°F	Total heat, Btu/h
SO_2	2,985.9	506.3	0.147	116,671
CO_2	38,717.3	4,528.4	0.304	2,158,159
H_2O	36,975.3	1,758.7	0.574	1,582,295
O_2	90,141	7,608.4	0.265	3,160,425
N_2	565,385	42,044	0.288	18,980,833
TOTAL	**20,777.5**	**25,599.4**	**NA**	**25,998,383**

Heat loss is taken as approximately 2% of the input

* Cp = Specific heat at constant pressure for each constituent

mill, total stoichiometric air required is 286,622.7 scfh.

The total air required for optimum conditions, as calculated from the heat and energy balance, is 11,928 scfm, of which 10,081 scfm is delivered by the blowers. Since the total stoichiometric air requirements represent less than half of the amount delivered by a single blower, the total air flowrate must be supplied by two blowers.

During operation, the primary and secondary blowers split the air flow coming to the TO, with the primary air blower supplying air to the burner. The secondary blower delivers air directly to the chamber section, bypassing the burner.

Figure 4 shows the customized installation at the California pulp mill. The main TO chamber is a horizontal vessel whose transition section is mated to the boiler with a flanged connection.

The carbon-steel combustion chamber is sandblasted and coated with one coat of primer. The outside shell is rated for 195°F. The chamber is furnished with three sight ports for flame observation. There is an access door and an outside personnel shield.

The inside shell is also coated with coal tar mastic to enhance corrosion protection before it is insulated with a 3-in. ceramic blanket. Mastic protects the wall from condensation and corrosion, particularly during startup and shutdown. The ceramic blanket at the conical section and the cylindrical transition to the boiler are protected with a double layer of Incolloy mesh screen, which protects the inside surface from abrasive particles in the high-velocity fluegas stream. The cylindrical transition zone includes an expansion joint.

Fluegas makes three passes through the boiler to the economizer. The tubing is made of carbon steel. Boiler feedwater enters the economizer at a temperature of 290°F. This temperature is suficient to prevent fluegas condensation, which may cause corrosion on the tubes.

The burner is the heart of the thermal-oxidation system, and is custom-designed for each project. Proper burner design assures thorough mixing for efficient destruction of VOCs. Supplementary fuel, either natural gas or #2 fuel oil, is distributed to the center of the burner.

The waste gas passes through a knockout drum where liquid droplets are extracted, and then it enters the waste-gas-distribution plenum. To create reliable waste-gas mixing and burning, the stream is routed through a circular manifold and is distributed to the burner through 12 pipes connected to that manifold. This design guarantees that the waste gas is directed evenly into the flame.

The end of each pipe is furnished with a flame arrestor. The specific type is based on the explosive potential of the gases or vapors. A detailed classification of such characteristics can be found in "Classification of Gases, Liquids and Volatile Solids, Relative to Explosion-Proof Electrical Equipment," published by the National Material Advisory Board of the National Electric Manufacturers' Assn. (Boston).

Flame arrestors are a crucial part of the system's safety precautions. At these high temperatures, many explosive gases create the enormous potential of flashback. If the flame is allowed to pass back through the pipe, the velocity could accelerate very rapidly and cause an explosion.

The turpentine burner consists of two manifolds with three straight pipes, oriented horizontally. Each manifold has a flexible hose conection for turpentine.

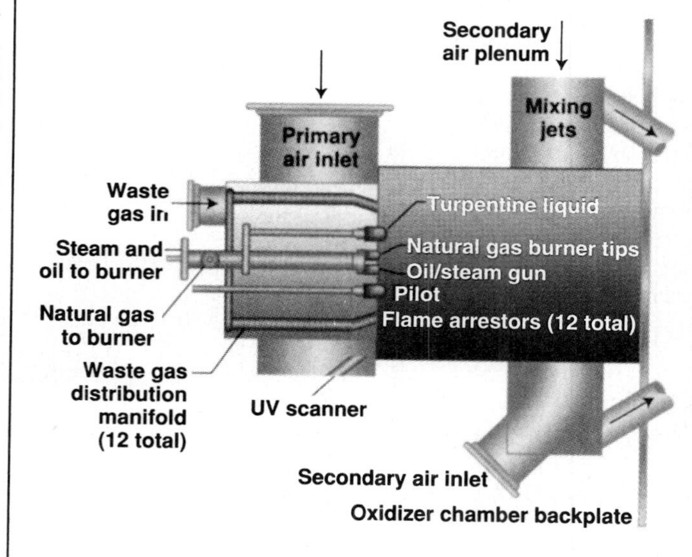

Figure 4. In this thermal oxidizer, custom designed for the Northern California pulp mill discussed here, the waste turpentine is injected through a distribution manifold with 12 nozzles. Incinerator waste gases power a heat-recovery boiler

Turpentine's viscosity is very close to that of water. Each pipe has an atomizing hollow-cone nozzle, which uses direct pressure to produce a foam-like spray of turpentine. No air or steam is required as an atomizing medium.

Nozzle turndown is 3:1, which is adequate for this application. The nozzles are made of 316 stainless steel, and are

STOICHIOMETRIC AIR CALCULATIONS FOR THE MAXIMUM FLOW (input from material and energy balance)

$CH_4 + 2(O_2) = CO_2 + 2(H_2O)$ Air = 2/0.21 x CH_4 = 2/0.21 x 2,000 = 19,047.6 scfh	$CH_3SCH_3 + 4.5(O_2) = 2(CO_2)+3(H_2O)+SO_2$ Air = 4.5/0.21 x CH_3SCH_3 = 4.5/0.21 x 1,321 = 28,319.2 scfh
$H_2S + 1.5(O_2) = H_2O + SO_2$ Air = 1.5/0.21 x H_2S = 1.5/0.21 x 135 = 959 scfh	$CH_3S_2CH_3 + 5.5(O_2) = 2(CO_2) + 3(H_2O) + 2(SO_2)$ Air = 5.5/0.21 x $CH_3S_2CH_3$ = 5.5/0.21 x 219 = 5,737.8 scfh
$CH_3SH + 3(O_2) = CO_2 + 2(H_2O) + SO_2$ Air = 3/0.21 x CH_3SH = 3/0.21 x 1,092 = 15,598.2 scfh	$C_{10}H_{16} + 14(O_2) = 10(CO_2) + 8(H_2O)$ Air = 14/0.21 x $C_{10}H_{16}$ = 14/0.21 x 3,254 = 286,622.7 scfh

oriented with a spray angle of 60 degrees.

The complete burner assembly was tested by firing both natural gas and #2 oil, one at the time. A test simulation for the waste-gas burner was provided by air flow only to monitor flame distribution and mixing.

If the burner is the heart of the system, then the instrumentation and controls are certainly the brains. The control system for this unit consists of a burner monitor, programmable logic controller (PLC) and a series of temperature controllers.

The flame monitor is a microprocessor-based control system wth self-diagnostics. For increased operator flexibility, the system can be configured to include:

- A non-volatile memory that allows the unit to remember its historical and its present position, even when power is interrupted
- A constant flame signal readout, which eliminates the need for a d.c. voltmeter
- An alarm circuit contact for all lockout modes
- A readout of main operational hours and complete cycles
- A run-check switch, which allows the operator to stop the program sequence in any of four different positions
- Remote display capability

The PLC ignition-control unit receives a signal from a UV scanner, which has a built-in, self-checking amplifier. This system includes programmed sequence timing for purge, trial for pilot ignition and trial for main burner ignition. All shutdown interlocks from the oxidizer, boiler and scrubber are wired to the interlock loop of the flame monitor. Flame-failure response time is four seconds.

In the California pulp mill, the thermal oxidizer's PLC was designed to communicate with a main PLC in the plant control room. The thermal oxidizer's PLC is used for monitoring, timing and sequencing of operations, controls and instrumentation.

Since National Fire Protection Assn. (Quincy, Mass.) codes do not allow computer software to be used as a flame-safety shutdown device, an approved flame-safeguard, relay-control device was used to control the shutdown of the TO and the PLC.

The PLC gives the system flexibility for ease of changes in the field. It also eliminates a large number of relays, timers and associated wiring.

The flexibility of the thermal oxidizer's PLC is especially important to initiate firing of the turpentine waste. Turpentine's heat of combustion at full rate is about two-thirds that of the specified heat release for the system. In order to keep the sudden heat release from overwhelming the temperature controller, fuel control valve and air blowers, the waste turpentine must be introduced to the system slowly.

Rather than a standard solenoid valve, the PLC controller responds to a signal to open the turpentine flow by partially opening the liquid waste control valve, using a gradual analog output. This gradually introduces the waste stream from about 30% to full flow over a 3-minute period, to smooth out the startup process.

PLC input and output lights indicate process status. If the TO system is shutdown through interlocks, the PLC will indicate the location of a specific problem by flashing an appropriate alarm light.

Two temperature controllers modulate the fuel control valve (the heating loop) and the primary and secondary blowers' inlet volume control (the cooling loop). The TO system has a pair of thermocouples to control both temperature controllers and a pair of limit controllers.

The fuel control valve receives a 4- to 20-mA signal, and the blowers accept a split signal (4- to 12 mA for the primary blower, 12- to 20-mA for the secondary blower). The limit controllers provide temperature setpoint signals to control the waste-gas input and the high-temperature shutdown.

In the case of this pulp mill, the thermal system, installed primarily to satisfy a pollution-control problem, has allowed the operators to recover 17.2 million Btu/h. The waste-heat recovery boiler and economizer produce 18,615 lb/h of steam at 160 psig during full-capacity performance. Ultimately, the payback depends on various operating parameters, such as steam production, the cost of supplementary fuel, and the cost of electricity and other associated utilities.

Edited by Suzanne Shelley

The author

John F. Straitz III, P.E., is president of NAO, Inc. (1284 E. Sedgeley Ave., Philadelphia, Pa. 19134; Tel.: 215-743-5300), where he directs combustion, pollution and energy-control operations. He holds a B.S. degree in mechanical engineering from Case Institute, and an M.S. degree in mechanical engineering from Massachusetts Institute of Technology. Straitz holds patents on various innovations in combustion, pollution, safety and energy-control systems, including subsea ignition systems for offshore flares and enclosed, smokeless incinerators. He has written numerous papers on these subjects, and has taught seminars worldwide. Straitz is an active member of the American Soc. of Mechanical Engineers, AIChE, the Combustion Institute, the Acoustical Soc. of America, the National Fire Protection Assn., the Physics Soc. of America and the Air Pollution Control Assn.

Suggested readings

1. Straitz, John F., III, Burning vapors, *Mechanical Engineering*, June 1987, pp. 40-44.
2. Brunner, Calvin R., "Handbook of Incineration Systems," publisher and city TK, 1991.
3. Straitz, John F., III, P.E., Flame Arrestors: Selection, Design and Testing, Petrosafe Paper, Pennwell Publishing, Tulsa, Okla., April 1993.
4. Jennings, M.S., et al, "Catalytic Incineration for Control of Volatile Organic Compound Emissions, Noyes, Park Ridge, N.J., 1985.
5. Monsanto Research Corp., Hazardous Waste Incineration Engineering, Technical Bulletin, St. Louis, Mo., 1981.

EFFECTIVE USE OF HEAT-RECOVERY STEAM GENERATORS

V. Ganapathy
Abco Industries

Photo, right, shows water-tube boilers (background) that recover heat from waste incinerators. Such heat-recovery steam generators can be applied to a broad range of gas streams, whose compositions (below) have major implications for the metallurgy of the equipment

COMPOSITION OF TYPICAL WASTE GASES

GAS	TEMP., °C	PRESSURE, ATM	COMPOSITION, %													
			N_2	NO	H_2O	O_2	A	SO_2	SO_3	CO_2	CO	CH_4	H_2S	H_2	NH_3	HCl
1	300-1,000	1	78-82			8-10		8-11								
2	250-500	1	80-82			10-12		0.5-1.0	6-8							
3	250-850	3-10	65-67	8-10	18-20	5-7										
4	200-1,100	1	70-72		16-18	2-3				9-10						
5	300-1,100	30-50	12-13		40-41					6-8	7-9	0.3		30-32		
6	500-1,000	25-50	13-15		34-36					13-15	0.2-1			38-40		
7	200-500	200-450	18-20				1-5							56-60	18-20	
8	300-1,200	40-80	0.2-0.5				0.3-0.5			4-6	46-48	0.2-0.5	0-0.8	45-49		
9	100-600	1	70-80		6-10	13-16				3-4						
10	175-1,000	1	70-75		8-12	5-8				10-13						traces
11	250-1,350	1	75-80		6-10	3-5				6-8						5-7
12	150-1,000	1	65-72		16-25	1-3				4-6						
13	300-1,450	1.5	50-55		20-25			3-5		5-7	2-3		2-3	3-4		

1. Raw sulfur gases
2. SO_3 from converter
3. Nitrous gases
4. Primary-reformer fluegases
5. Secondary-reformer gases
6. Converted gases
7. Synthesis gas
8. Shell gasifier effluent
9. Gas turbine
10. Modular municipal-solid-waste incinerator
11. Chlorinated-plastics incinerator
12. Fume incinerator
13. Sulfur condenser

All photos courtesy of Abco

Originally published January 1993

Heat-recovery steam generators (HRSGs), often called waste-heat boilers, recover energy from gas streams in a wide range of chemical-process plants. They play the same role in cogeneration and combined-cycle plants that generate steam and electric power, and in facilities that incinerate solid, liquid or gaseous waste.

The HRSG is basically a heat exchanger that serves as a boiler. The steam-generation rate and the amount of space available help determine the particular type used in a given situation. So do the quantity, temperature, pressure, chemical composition and purity of the gas. Such properties can vary widely (table). Accordingly, HRSGs are in general custom-designed for each situation, and the purchasing company's engineers must take special care in preparing a well-written specification. Guidelines for doing so appear later.

In cogeneration plants and incineration systems, the overriding purpose of the HRSG is to maximize the amount of energy recovered, consistent with economic and technical limitations (e.g., high- or low-temperature-corrosion problems). In chemical-process plants, the main purpose may instead be to cool the gas stream to a particular temperature level that is needed from a process standpoint, with the energy recovery being a welcome but secondary consideration. If the gasflow in such cases becomes excessive, part of it is bypassed either within or external to the HRSG itself.

HRSGs fall into two broad types, depending on how the gas and steam are contained. In fire-tube boilers, the gas flows inside tubes and its thermal energy is transferred to the steam-water mixture on the outside. In water-tube boilers, conversely, the gas flows outside the tubes and the steam-water mixture flows inside, via either natural or forced circulation.

Fire-tube and water-tube boilers each have their advantages and drawbacks. Key attractions of the fire-tube option:

- Conveniently handles gas at high pressures
- Economical for low gas mass-flowrates (e.g., under 100,000 lb/h)
- Easier to clean, so attractive for dirty-gas service that does not entail slag

Among the major attractions of the water-tube option:

- Handles steam pressures above 1,000 psi; indeed, can exceed 2,800 psi
- Suitable for high gas mass-flowrates, (e.g., millions of pounds per hour)
- The preferred option for situations in which much heat is transferred and the minimum temperature difference between the gas side and the water side is relatively small (i.e., a low pinch point)
- Offers more flexibility as regards placement of a superheater, if one is required
- Faster response to changes in load
- The choice if the gas contains slagging particulates

How to use HRSGs effectively in chemical-process plants can be aptly illustrated by two major examples, both covered below: steam reforming of natural gas to produce hydrogen, as in an ammonia or methanol plant; and manufacture of sulfuric acid by the contact process. Also included below is a look at HRSGs in incineration plants, followed by guidelines for proper specifying of these heat-exchange devices. An article in the next issue will focus on HRSGs in cogeneration and combined-cycle plants.

Heat recovery in reforming

The steam-reforming process offers a few opportunities for HRSG usage, including the cooling of the product streams from the reformer and from the shift converter. A mixture of feed gas and steam is held at high pressure and temperature in a reaction furnace or reformer heated by fluegases. The reaction-product stream, known as reformed gas and usually at 1,600–1,700°F, must be cooled before being sent to a reactor where an exothermic shift-conversion reaction takes place. The shift-conversion product stream, called converted gas, must likewise be cooled.

The HRSG that cools the reformed gas is typically a fire-tube boiler because the gas pressure is high, ranging from 200 to 600 psig. However, water-tube boilers are instead common in large-capacity plants, because big plants tend to generate steam at higher pressures. (The water-tube boilers can employ extended surfaces, because the hot gases are clean.)

Since the stream has to be cooled to a particular temperature (usually about 600 to 700°F) for subsequent reaction, an internal gas-bypass arrangement is provided. This ensures that the temperature of the exiting stream stays within a narrow range even if plant load conditions vary.

The reformed gas consists largely of hydrogen and steam. Because both have relatively high heat capacities and thermal conductivities, this mixture is conducive to a high heat-transfer coefficient (for example, nearly six to eight times that with fluegases from combustion of fossil fuels). As a result, the heat flux at the tube sheets and tubes will tend to be very high. To forestall

mechanical problems, the HRSG is usually designed to keep the flux under, say, 100,000 to 120,000 Btu/(ft^2)(h).

For instance, the tube sheet is protected from the hot effluent by a refractory material, as is the boiler's inlet vestibule. And, ceramic ferrules may be employed to transfer the heat from the gas to the portion of the tubes that are cooled by the steam-water mixture.

The steam-reforming operation generates not only the product mixture but also several waste-gas streams. What's more, the reformer is heated by fluegas, which usually leaves that vessel around 1,800°F. As there are thus numerous hot streams available for heat recovery, it is a general practice in big plants to have one sole steam drum, with a large holdup time (e.g., 5 to 15 min) and located at a high elevation, to feed mixed steam and water to the cold sides of several HRSGs, each heated by one of the gas streams. The circulation ratio (ratio of steam-water mixture circulated to amount of steam generated) might vary from about 10 to 20 depending on system pressure, plant layout, and sizing of downcomers and risers.

Such a system can not only generate steam but also incorporate a provision for preheating the the reformer feed gases. Figure 1, for example, shows an HRSG system consisting of a fluegas boiler, a reformed gas boiler and a feed gas heater all in one package.

In this scheme, fluegas from the reformer preheats the reformer feed gas (upper portion of figure) and then transfers energy to an evaporator-economizer module, while the reformed gas at high pressure transfers energy to a fire-tube boiler with internal gas bypass. Since the steam pressure is the same for both boilers, a common steam-circulation system is used, as shown. The HRSG that recovers heat from the downstream converted gas boiler can also be added to the system if desired. A packaged module such as this is ideal for small-capacity reforming plants.

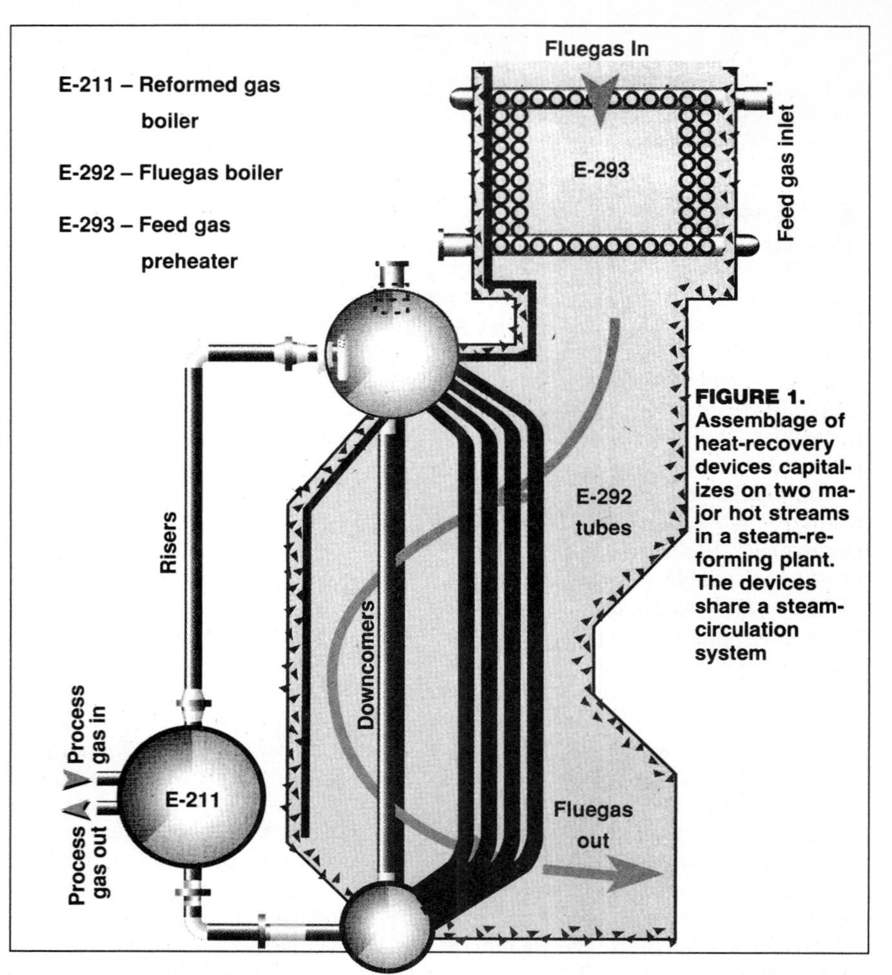

FIGURE 1. Assemblage of heat-recovery devices capitalizes on two major hot streams in a steam-reforming plant. The devices share a steam-circulation system

For troublefree operation...

When selecting materials of construction for HRSGs in this or similar service, bear in mind that the reformed and the converted gas streams each contain large amounts of hydrogen. The Nelson chart discussed in References *2* and *4* offers guidelines for alloy selection. Often, boiler tubes and other materials in contact with reformed gases are made of steels containing at least 1% Cr and 0.5% Mo, which are relatively unaffected by hydrogen. Depending on the temperature, the bypass damper is made of stainless steel or Inconel. So is the feed-gas preheater, which heats the mixture of natural gas and steam to about 1,000°F.

In dealing with high-heat-flux operations, such as reforming, it is especially important to keep water chemistry in mind. If scale or sludge is allowed to build up in the reformed-gas boiler, the resulting increase in fouling factor can cause the tube-wall temperature to rise substantially, possibly leading to mechanical failure. In an HRSG operating at, say, 75,000 Btu/(ft^2)(h), if scaling causes the steam-side fouling factor to rise from 0.001 to 0.003 (ft^2)(h)/Btu, the resulting increase in tube-wall temperature is $75{,}000 \times (0.003 - 0.001)$, i.e., 150°F.

Hence, blowdown and phosphate treatment should be monitored at regular intervals to maintain the dissolved solids and conductivity in the steam drum. American Soc. of Mechanical Engineers (ASME) guidelines on boiler and feed water should be followed.

Since the high-pressure steam from reformer systems is commonly used for driving steam turbines, steam purity is also of great significance. Drum internals should be designed with cyclones and chevron separators (banks of V-shaped corrugated sheets) to remove entrained water and, with the water, impurities, if steam purity in parts per billion range is desired.

HRSGs in sulfuric acid plants

In the manufacture of sulfuric acid by the contact process, products of combustion of raw or spent sulfur can be cooled in HRSGs at various stages of

LUTHER EASON

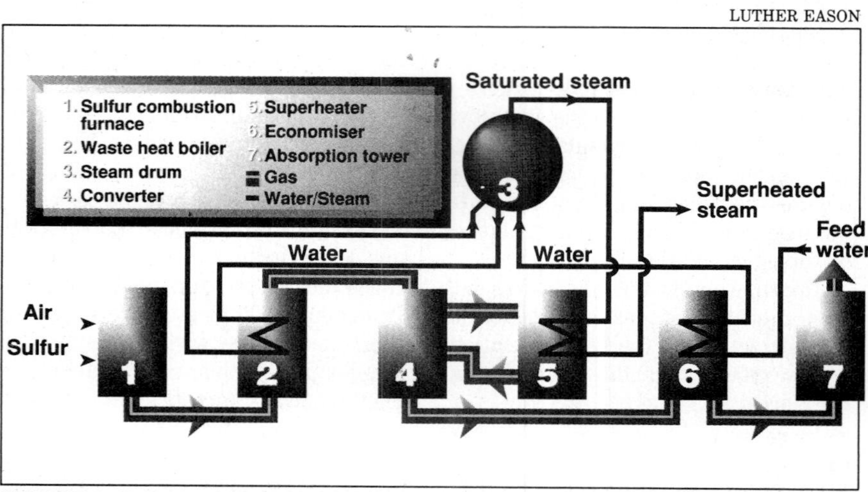

FIGURE 2. Sulfuric acid plant offers at least three major points for heat-recovery

the process. For one typical scheme using three HRSGs, see Figure 2.

In this case, the HRSG just downstream of the sulfur combustor cools the hot sulfur-combustion products (table) from about 2,150 to 750°F. Since the exit-gas temperature has to be controlled over wide load ranges, a gas bypass system is provided; it can be either within or external to the HRSG. In this service, water-tube boilers are preferred for economic reasons when the gas flow exceeds about 400,000 lb/h.

Economizers and superheaters can also be incorporated into the process flowscheme, as Figure 2 illustrates. If the gas pressure is relatively high, on the order of 3 to 10 psig, these heat-recovery devices can be housed inside shells or pressure vessels.

Since the gas stream contains large amount of SO_2 and SO_3, care should be taken to minimize low-temperature corrosion due to possible formation of sulfuric acid. One precaution is to maintain a high feed-water temperature at the economizer inlet, at or close to the dew point of the gas stream. Correlations for computing dew points for various acid vapors may be obtained from the literature [*4*, *10*].

In the U.S., carbon-steel tubes with extended surfaces are usually employed; several have been in operation around the country for decades. Some HRSG vendors outside the U.S. prefer to use cast-iron gills shrunk over carbon-steel tubes to protect the tubes from corrosion attack. However, this does not prevent corrosion altogether, and it increases the weight and cost of the economizers.

Incineration heat recovery

Rotary kilns, fluidized-bed combustors, fixed-bed incinerators and related devices are used today to burn a wide variety of industrial, municipal and other wastes. Incineration generates large volumes of fluegases, at temperatures varying from 1,650 to 2,400°F depending on the process, the type of wastes handled and the level of destruction. Hence, HRSGs are logical adjuncts to them.

Large plants to burn municipal wastes, ranging in capacity from 500 to 2,000 tons/d, resemble conventional coal-fired boilers. Also popular are modular incinerators, suitable for capacities of around 20 to 200 tons/d. Typical gas temperatures from municipal incinerators are around 1,800°F. In specifying an HRSG for this service, the engineer must bear in mind that the gas is dirty, contains particulates that can cause slagging, and is corrosive at high and low temperatures alike. For some typical compositions (for instance, those from incinerating chlorinated-plastics wastes), see the table.

In incineration of chemical wastes, the exit gas temperature is likely to be higher, around 2,000 to 2,400°F. As with municipal wastes, the gas stream poses risks of high- and low-temperature corrosion [*10*].

Gas-side corrosion problems associated with high gas and metal temperatures in incineration systems can be broadly divided into two categories. The first is high-temperature, liquid-phase corrosion, caused by molten alkali-metal salts, such as metal chlorides and their eutectic mixtures, that have low melting points. If fluegases containing these salts are in contact with HRSG tubes above these temperatures, the salts are quite likely to melt and then deposit out. This can not only plug tubes, but also can build up corrosive deposits that may destroy the tubes over time.

The second broad category is corrosion due to hydrogen chloride or chlorine. HCl, formed during combustion of waste plastics, is very corrosive above 800°F. So care should be taken to see that if superheaters are used, their tube wall temperature does not exceed that level. Chlorine, formed when incinerating chlorinated wastes, is more corrosive than HCl on carbon steel at lower temperatures, especially around 400 to 450°F. Accordingly, HRSGs associated with the burning of such wastes are used to generate saturated steam, at pressures that keep the temperature below 400°F (but above the acid-vapor dewpoint). For more information on HRSGs for incineration systems, see References *4* and *10*.

Specifying HRSGs

As noted earlier, HRSGs are custom designed. In order to come up with a reasonable proposal and an effective design, HRSG suppliers need the following information as a minimum.

1. Nature of the gas and how it is generated — for instance, from an incinerator, from a gas turbine, from a chemical-synthesis reactor. This indicates the cleanliness of the stream, whether extended surfaces can be used and whether cleaning devices such as soot blowers are required

2. Gas flow in mass units, inlet gas temperature to the HRSG, gas-composition analysis, and particulate loading and analysis. These data will help the HRSG designer to infer the high- and low-temperature corrosive potential of the gas and and its slagging tendencies. Providing gas flow in mass units helps avoid confusion and errors associated with converting volumetric flow to mass flow. Volumetric units, such as actual or standard cubic feet per minute, should be avoided

3. A description of the system or application, showing how the steam is to be utilized. This information helps with the sizing aspects and steam-purity requirements. If the steam is to be used for drying or heating, for instance, a relatively low purity may be adequate — but if it is to be injected into a turbine, a high purity on the order of parts-per-billion may be required, calling for cyclones and chevron steam separators. If the process considerations require a large holdup for the drum, this should be indicated. Such a system description is particularly helpful in multi-pressure HRSG units

4. Space limitations or other layout considerations, so that the HRSG supplier can consider them in the design. Modifying a design to fit into a given space at a later stage may be expensive

5. A limit on gas pressure drop, or else data on the cost of incurring gas pressure drop, along with energy cost. This enables the HRSG supplier to opti-

In line with their versatility, HRSGs are in general custom-designed for each situation

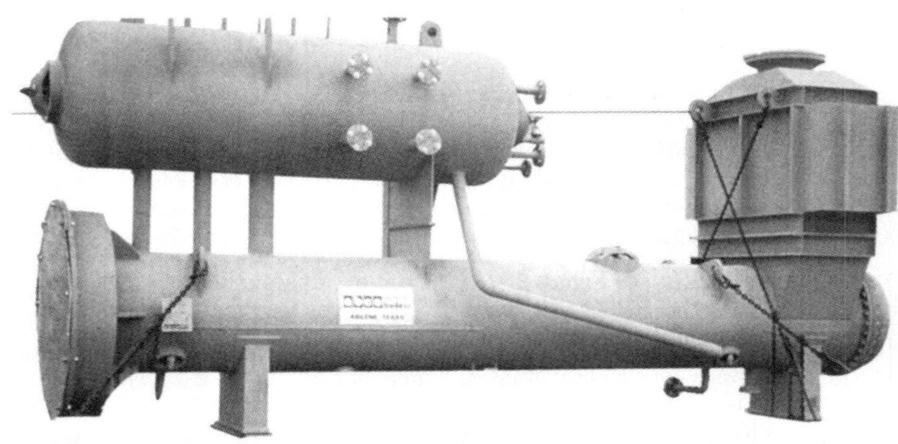

FIGURE 3. Elevated-drum fire-tube boiler that is equipped with a steam superheater is destined for use at an incineration facility. HRSGs fit in well with incinerator operation

mize the design based on initial and operating costs. In this vein, evaluating a bid based simply on surface area or initial cost should be avoided

6. Feed-water analysis. Along with the already-mentioned steam-purity requirements, this enables the HRSG designer to determine if the feed water can be used for desuperheating steam in an attemperator (controlled cooler), and to ascertain the type of drum internals

7. Feed water temperature to the HRSG. Furthermore, in situations where condensate is to be recycled (and accordingly a deaerator included in the system), the temperatures and amounts of both the feed (i.e., makeup) water and the condensate should be shown, as well as the source of the steam to be used in the deaerator, so that the HRSG supplier can make an energy balance around the deaerator

8. Cost of fuel or other energy source. This is especially important in situations where supplemental firing is required. (Such firing crops up most often in cogeneration applications, and will be discussed in the article appearing in the next issue.) If the unit is likely to operate in the unfired versus fired mode for a long period, that should also be stated

Note that surface areas should *not* be spelled out in the specifications. HRSG designers may come up with different surface areas (sometimes differing as much as 200%) for the same duty by varying the fin configuration, gas velocity, tube size or tube spacing.

Similarly, the minimum temperature (pinch point) between gas side and steam side should not be specified. This minimum will vary with gas flow, inlet temperature and extent of auxiliary fuel firing, and it is unlikely that the engineer preparing the specification is in a position to analyze all the potential operating modes. A better alternative is to indicate whether high-pressure or intermediate-pressure steam generation should be maximized, or to present a range of values covering this point.

Finally, if the HRSG will face a varying load, the details should be spelled out. This will help the designer optimize the initial and operating costs over the entire operating regime, rather than basing them on a single operating point. ■

Edited by Nicholas P. Chopey

The author

V. Ganapathy is employed as a heat-transfer specialist by Abco Industries, 2675 I20 Business, P.O. Box 268, Abilene, TX 79604-0268, Tel. (915) 677-2011. A specialist in design of heat-recovery steam generators and of packaged water-tube steam generators, he has written more than 175 articles on those topics, as well as and heat-transfer and steam-plant systems in general. Several of his articles have appeared in this magazine. He is also author of five books in the same area (copies available from him). His work has included development of the software for engineering such systems. His B. Tech. degree in mechanical engineering is from the Indian Institute of Technology, Madras; he also holds an M.Sc.(Eng.) in boiler technology from Madras University.

References

1. Ganapathy, V., Simplify heat recovery steam generator evaluation, *Hydrocarbon Processing*, March 1990.
2. Ganapathy, V., "Applied Heat Transfer," Pennwell Books, Tulsa, 1982.
3. Ganapathy, V., Evaluate extended surface exchangers carefully, *Hydrocarbon Processing*, October 1990.
4. Ganapathy, V., "Waste Heat Deskbook," Fairmont Press, Atlanta, 1991.
5. Ganapathy, V., Chart estimates supplementary fuel parameters, *Oil and Gas Journal*, June 25, 1984.
6. Ganapathy, V., Program computes fuel input, combustion temperature, *Power Engineering*, July 1986.
7. ASME Power Test Code PTC 4.4, "Gas turbine heat recovery steam generators," American Soc. of Mechanical Engineers, New York, 1981.
8. "HRSGs — Software for simulation of design and off-design performance of HRSGs," Ganapathy, V., available from author.
9. Ganapathy, V., How to evaluate HRSG performance, *Power*, June 1991.
10. Ganapathy, V., Heat recovery boilers — the options, *Chemical Engineering Progress*, February 1992.

PUTTING A LID ON EVAPORATION COSTS

Energy-efficient designs can slash utility bills by 50%

Stanley J. Macek
Dedert Corp.

An exceptionally energy-intensive unit operation, evaporation is often used in the chemical process industries for concentrating or purifying a product. Thanks to a number of innovative designs for evaporators, one can now reduce the operating costs by permitting reuse of the otherwise lost or wasted energy of the heating medium, which may be steam or exhaust gases from dryers, scrubbers, ovens, kilns and so on. Although these energy-efficient units often carry higher capital costs than conventional systems, they provide 30%–50% utility savings, ensuring quick payback.

Thermal and mechanical vapor-recompression (TVR and MVR) are two such designs that reuse the water vapor produced during the evaporation process. Another innovative design utilizes the waste heat from a process stream to run the evaporator.

Potential use of each design depends on such characteristics as viscosity, heat sensitivity and the desired level of concentration or purity of the product to be evaporated. Also important are the energy resources, and labor, operating and maintenance capabilities of a particular plant.

In applications where evaporation requirements are small (less than 7,500 lb/h), or steam is relatively inexpensive (less than $3/1,000 lb), conventional steam-heated units may suffice. A single-effect evaporator uses the energy of the steam only once. The vapor from evaporation is separated from the concentrate or product and sent to a condenser, where the energy is rejected from the system by cooling water.

Multiple-effect evaporators reclaim the once-used vapor and use it as the heating medium in subsequent units known as "effects" (Figure 1). As the vapor from one effect drives the following effect's evaporation, the vapor condenses. A vacuum is created over the previous effect's liquid surface, lowering the pressure on the feed undergoing evaporation, thus lowering the boiling point of the liquid.

Each subsequent effect adds to the vacuum. Temperature and pressure are selected to ensure that vapors from one effect are at a temperature sufficient to act as the driving force for the next effect at lower pressure.

Thermal vapor-recompression

TVR evaporators use high-pressure steam to compress the vapor produced

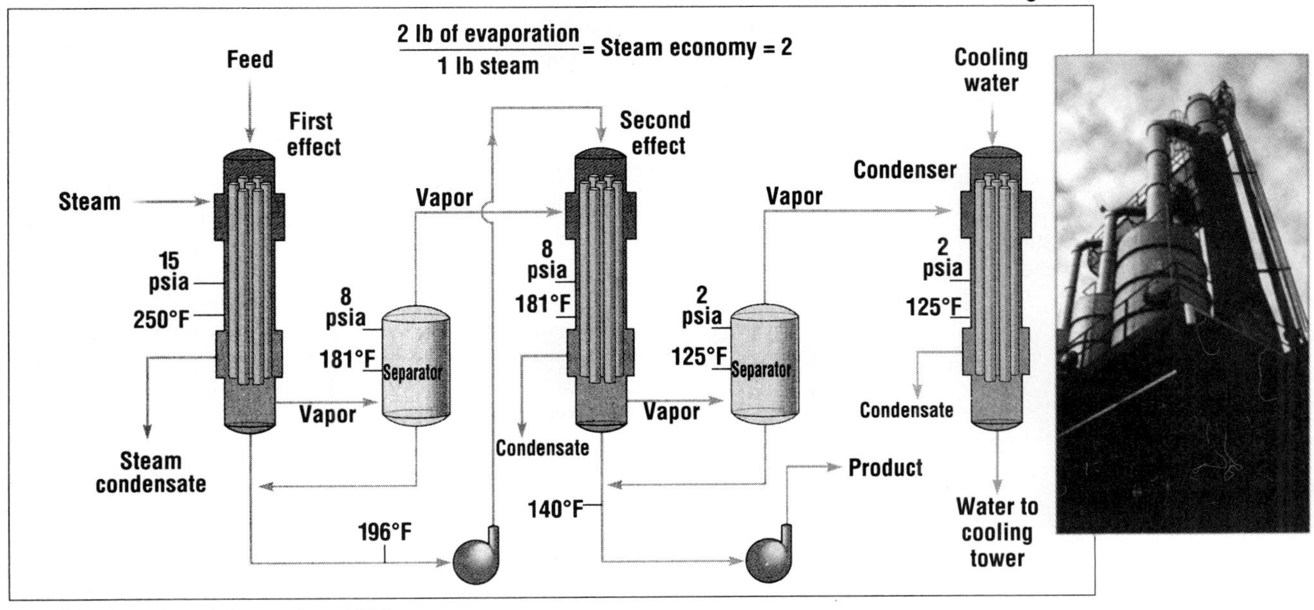

FIGURE 1. For each pound of steam supplied to a double-effect evaporator, two pounds of vapor may be generated by reusing the vapor from the first effect to provide heating for the second. The result is twice as much of steam economy as that for a single-effect unit

Originally published December 1992

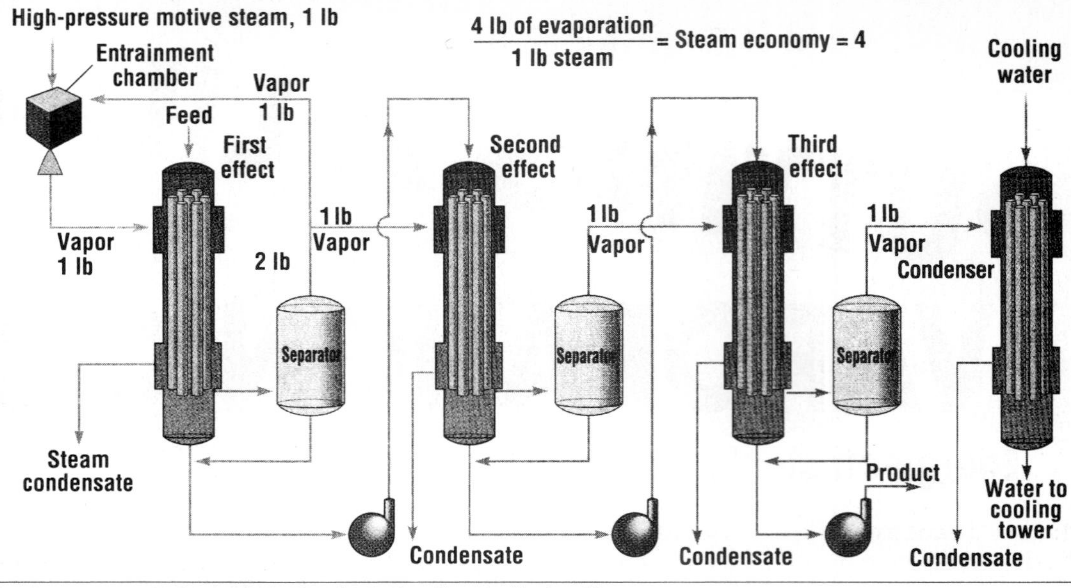

FIGURE 2. A thermal vapor-recompression evaporator featuring three effects offers the equivalent steam economy of a four-effect conventional unit

during evaporation, thus elevating the vapor's temperature and pressure (Figure 2). The high-pressure motive steam expands through one or more straight nozzles into an entrainment chamber. These nozzles are carefully aimed at an exit port on the other side of the chamber. The high velocity of the steam assures virtual straight-line "flight" across the chamber, with little divergence of the stream.

The high-velocity steam itself creates a level of vacuum in the chamber. Its frictional drag entrains any fluid present in the chamber and carries it along. A relatively large quantity of fluid can thus be returned by the motive flow into the first effect.

The exit port is the entrance to a carefully shaped discharge pipe, which is narrow in the beginning to promote intimate mixing of the two streams and thus prevent backflow of the low-pressure component. The pipe then gradually enlarges to permit the combined flow to decelerate and turn its kinetic energy into useful pressure energy.

The mixture of steam and recycled vapor is used as the heating medium in the steam chest for the first effect. This provides more-efficient evaporation because the vapor is now equivalent to steam at a lower operating cost. The vapor used by the thermocompressor is condensed in the first-effect chest and then exits the system.

A triple-effect TVR unit and a conventional steam-heated quadruple-effect evaporator designed for the same capacity use about the same amount of energy. Thus, TVR boosts the steam economy by about one effect.

However, a TVR design requires higher-pressure motive steam. Also, because the first effect of the TVR unit needs to transfer more energy with a somewhat reduced driving force (in terms of temperature difference between "hot" and "cold" streams), the vessels making up the first effect must be increased in size.

Thus, to achieve the same steam economy, the TVR system requires fewer vessels, which may be larger than those employed in the quadruple-effect evaporator. Therefore, there is somewhat of a tradeoff in the initial cost. In other words, TVR calls for fewer, but larger, vessels (with fewer components to maintain), requiring about 10% less floor space than a conventional evaporator.

Retrofitting a conventional steam-heated evaporator with a TVR unit reduces consumption of steam and cooling water. In the case of an existing triple-effect evaporator, the capacity or throughput of the retrofit system may

FIGURE 3. A mechanical vapor-recompression evaporator can reuse all of the vapor produced, sporting a steam economy of 10–20

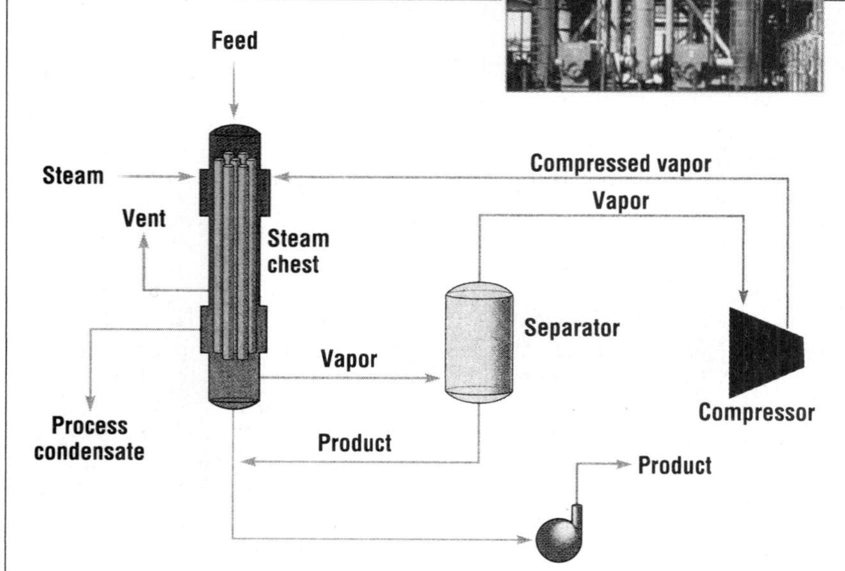

be decreased by 30%, as the temperature-difference driving force is decreased for the existing heat-transfer surface, although steam consumption is decreased by 50%. Therefore, to maintain capacity, additional equipment is required, but the resultant system consumes less energy.

Mechanical recompression

Vapor leaving the separator in an MVR evaporator is compressed mechanically to increase its pressure and condensing temperature so that it may be used as the heating medium in the steam chest. The steam chest generally receives all of the water vapor produced by evaporation (Figure 3). Thus, the MVR approach often obviates a condenser.

While a thermocompressor recycles only a portion of the vapor produced, a mechanical vapor-compressor can reuse all of the vapor in the evaporator system if any steam is used at all. However, in most case no steam supply is necessary after the startup, and the equivalent steam economy on the basis of energy consumption is 10–20. In other words, the system achieves 10–20 lb of evaporation for the consumption of electrical energy that is equivalent to one pound of steam. An MVR unit is more expensive than either TVR or conventional steam-heated equipment, because the compressor typically costs more than the hardware it eliminates.

But MVR evaporators require fewer vessels than steam-powered multiple-efffect systems. As a result, installation costs are lower and less plant space is occupied. In addition, they consume significantly less energy, allowing users to recover the equipment cost over time. They also eliminate, or greatly reduce, water requirements. This is particularly advantageous in water-lean locations. Only a small amount of steam is used to heat the evaporator during startup before sufficient water vapor is generated to begin the recompression cycle.

There are many possible design configurations using MVR technology. Each application must be examined to determine the most cost-effective and energy-efficient system. Compressing over two or more effects can increase efficiency in MVR evaporators. A multiple-effect MVR reduces the required capacity of the compressor, and often its power requirements are low as well.

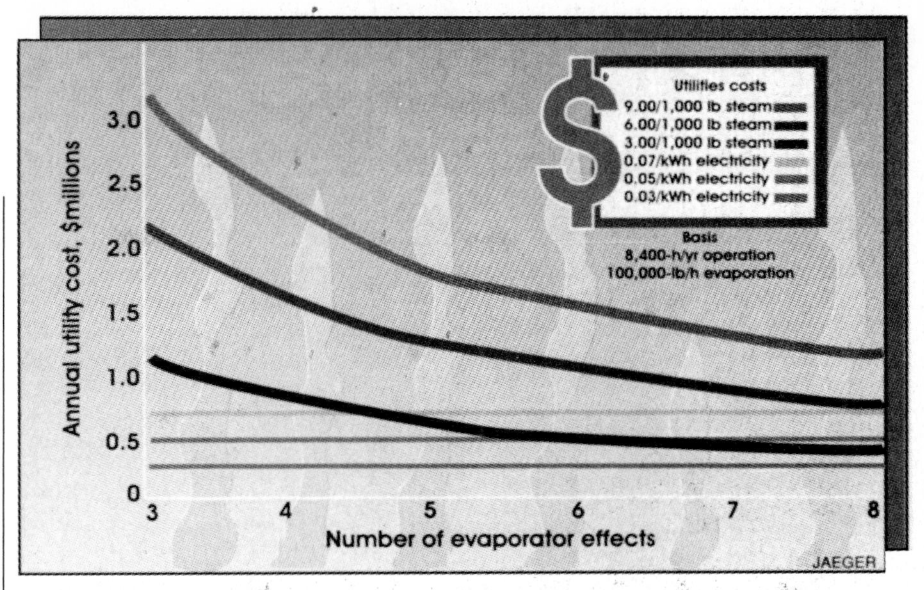

FIGURE 4. The annual utility costs for multiple-effect evaporators drop off with an increase in the number of effects, while the costs of steam and electricity dictate the savings from both steam-heated and mechanical vapor-recompression units

The product's physical properties may call for special considerations. For example, a three-stage, double-effect system accomplishes most of the evaporation with double-effect MVR economy. However, the high-concentration step and the remainder of the evaporation process are carried out at a single-effect MVR economy. This design is useful when the product's boiling-point elevation (BPE; the difference between the boiling temperature of the solution and that of water at the same pressure) is high at the final concentration.

Another commonly used approach for product liquors with a high BPE is to compress much of the vapor with one compressor, and use a second compressor to compress a portion of the discharge vapors from the first compressor to a higher pressure. This provides a greater pressure differential for the high-concentration step.

Recovering the waste heat

Off-gases from existing plant equipment, such as dryers or scrubbers, contain valuable energy that may be recovered for use as the energy source for gas-heated evaporators. Total energy cost is usually much lower than that for either steam-heated or mechanically driven alternatives.

An evaporator operating under the same conditions as the cases used in the illustrative example (box, p. 144) is driven by the exhaust gases from a dryer. The annual utility cost for this system is about $125,000, based on $0.05/kWh for electricity.

In another appliction in a resin plant, waste-heat evaporatoros are able to reduce the volume of a waste stream. Although short on steam and cooling-water supply, the plant does have waste heat in the form of hot methanol vapor. Previously, this hot vapor was reclaimed by a water-cooled final condenser, with the heat being rejected from the system via the condenser's cooling water return.

A double-effect, waste-heat evaporator concentrates the waste stream before it is pumped to a waste-treatment plant by evaporating 55 gal/min of water. As a result, the hydraulic load of the waste-treatment plant is reduced from 130 gal/min to 75 gal/min, and the evaporated water is condensed and sent back to the process for reuse. Using the hot methanol vapor to heat the evaporator also reduces cooling-water consumption because only one water-cooled condenser is required for the two (concentration and treatment) processes.

Consider the major factors

A cost-effective evaporation system must satisfy both operational and economic constraints. Therefore, the factors that must be taken into account for a particular application include:

Energy requirements. Investigate the availability and cost of each energy source for the evaporator system; cost of piping and ancillary equipment to get this energy to the point of use within the plant; and the consequences, if any, of the evaporator's require-

A CONVENTIONAL EVAPORATOR VS. AN MVR UNIT

Before deciding on any particular evaporator system, it is advisable to determine the annual utility costs for each energy source used. This allows comparison between designs on the basis of differences in operating costs. The formulas for steam, electricity and cooling water are:

Cost of steam = steam rate (lb/h) × operating hours (h) × cost of steam ($/1,000 lb)

Cost of electrical power = compressor or pump rating (kW) × operating hours (h) × cost of electricity ($/kWh)

Cost of cooling water = water rate (gal/h) × operating hours (h) × cost of water ($/gal)

Consider a 100,000-lb/h evaporator system running 8,400 h/yr. The installed costs for a quadruple-effect evaporator and a motor-driven MVR unit are $2,400,000 and $3,300,000, respectively. The utility costs are evaluated at $5/1,000 lb for steam, $0.05/kWh for electricity, $0.02/1,000 gal for cooling water.

The utility requirements for the two alternative evaporators are:

	QUADRUPLE EFFECT	MVR
Steam, lb/h	32,000	4,800
Power, kW	43	1,342
Water, gal/min	2,900	None

Based on operating the plant for 8,400 h/yr, the utility costs are:

Quadruple Effect

Steam = ($5/1,000 lb) × (8,400 h/yr) ×(32,000 lb/h) = $1,344,000/yr

Power = (43 kW) × (8,400 h/yr) × ($0.05 kWh) = $18,000/yr

Water = (2,900 gal/min) × ($0.02/1,000 gal) × (8,400 h/yr) × (60 min/h) = $29,000/yr

Total utility cost = $1,387,000/yr

MVR evaporator

Steam = ($5/1,000 lb) × (8,400 h/yr) × (4,800 lb/h) = $200,000/yr

Power = (1,342 kW) × (8,400 h/yr) × ($0.05/kWh) = $560,000/yr

Water = $0

Total = $760,000/yr

The additional first cost = $3,300,000 − 2,400,000 = $900,000 can be offset by the continued utility savings = $1,387,000/yr − 760,000/yr = $623,000/yr

Figure 4 shows the annual operating costs for a number of effects, different steam costs, and the operating costs for a single-effect MVR unit with varying costs of electricity. Note the impact of varying steam costs, and the diminishing operating-cost advantage as the number of steam-heated effects is increased.

In all economic evaluations, the time it takes to pay off the capital expenditures determines whether a particular design is acceptable. When comparing two or more evaporator systems, the added capital costs must be weighed against potential energy savings. The time required to realize energy savings in operating costs has to be balanced against increased capital costs. □

ments on the plant's overall energy balance. For example, a motor-driven MVR unit must be able to start up and operate without putting undue stress on the plant's overall utility load. Besides, MVR compressors require a dependable energy source.

Capital costs. Weigh the added capital cost (for a new unit, a retrofit or expansion of an existing system) against potential energy savings and the time required to realize the savings. System cost is dictated in part by product characteristics, which influence, for example, the size and type of the heat-transfer system, the compressor (if MVR) and amount of power required. One design widely used with high-efficiency evaporators is falling film, which is effective with temperature differences that are smaller than those required by other designs.

Labor and maintenance. Enumerate the ongoing costs for monitoring, cleaning and repairing the unit. ■

Edited by Gulam Samdani

For information on suppliers of evaporators, consult *Chemical Engineering Buyers' Guide for 1993,* pp. 301–303.

The author

Stanley J. Macek is the manager of Evaporator Systems for Dedert Corp. (20000 Governors Drive, Olympia Fields, IL 60461, USA; tel: (708) 747-7000; fax: (708) 755-8815). He has been in the design and sales of chemical process equipment for over 30 years. He received his B.S. degree in chemical engineering from the Illinois Institute of Technology.

INDEX